AF888254

METHODS IN MOLECULAR BIOLOGY

Series Editor
John M. Walker
School of Life and Medical Sciences
University of Hertfordshire
Hatfield, Hertfordshire, AL10 9AB, UK

For further volumes:
http://www.springer.com/series/7651

Gangliosides

Methods and Protocols

Edited by

Sandro Sonnino and Alessandro Prinetti

Department of Medical Biotech & Translational Medicine, University of Milan, Segrate, Milano, Italy

Humana Press

Editors
Sandro Sonnino
Department of Medical Biotech & Translational Medicine
University of Milan
Segrate, Milano, Italy

Alessandro Prinetti
Department of Medical Biotech & Translational Medicine
University of Milan
Segrate, Milano, Italy

ISSN 1064-3745 ISSN 1940-6029 (electronic)
Methods in Molecular Biology
ISBN 978-1-4939-8551-7 ISBN 978-1-4939-8552-4 (eBook)
https://doi.org/10.1007/978-1-4939-8552-4

Library of Congress Control Number: 2018944191

© Springer Science+Business Media, LLC, part of Springer Nature 2018
This work is subject to copyright. All rights are reserved by the Publisher, whether the whole or part of the material is concerned, specifically the rights of translation, reprinting, reuse of illustrations, recitation, broadcasting, reproduction on microfilms or in any other physical way, and transmission or information storage and retrieval, electronic adaptation, computer software, or by similar or dissimilar methodology now known or hereafter developed.
The use of general descriptive names, registered names, trademarks, service marks, etc. in this publication does not imply, even in the absence of a specific statement, that such names are exempt from the relevant protective laws and regulations and therefore free for general use.
The publisher, the authors and the editors are safe to assume that the advice and information in this book are believed to be true and accurate at the date of publication. Neither the publisher nor the authors or the editors give a warranty, express or implied, with respect to the material contained herein or for any errors or omissions that may have been made. The publisher remains neutral with regard to jurisdictional claims in published maps and institutional affiliations.

Printed on acid-free paper

This Humana Press imprint is published by the registered company Springer Science+Business Media, LLC part of Springer Nature.
The registered company address is: 233 Spring Street, New York, NY 10013, U.S.A.

Preface

Gangliosides, sialic acid containing glycosphingolipids, are membrane amphiphilic lipids that are deeply involved in the modulation of the overall properties of cellular membranes, as well as of the functions of membrane-associated proteins.

The term "gangliosides," introduced by Ernst Klenk in 1939 (Medical School, University of Cologne, Germany), referred to a mixture of compounds extracted from the brain of a Niemann-Pick disease patient [1]. Gangliosides attracted very soon the interest of many investigators, and their structures were slowly elucidated using basic

Name	Structure	Symbol	Consortium for Functional Glycomics (CFG)
Glucose	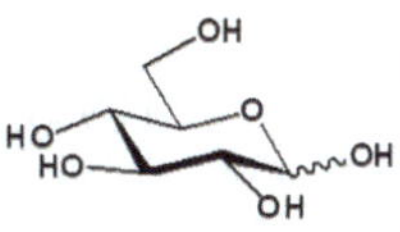	Glc	
Galactose	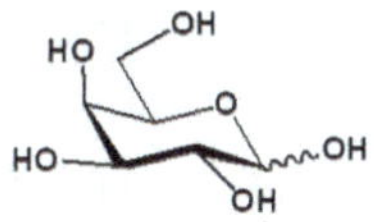	Gal	
N-Acetylglucosamine	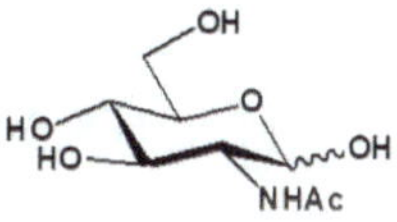	GlcNAc	
N-Acetylgalactosamine	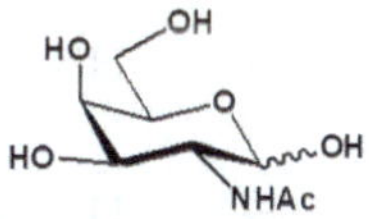	GalNAc	
Fucose	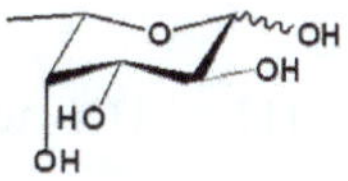	Fuc	

Fig. 1 Structural details of the oligosaccharides structure

Table 1
The sialic acid: the derivatives of neuraminic acid

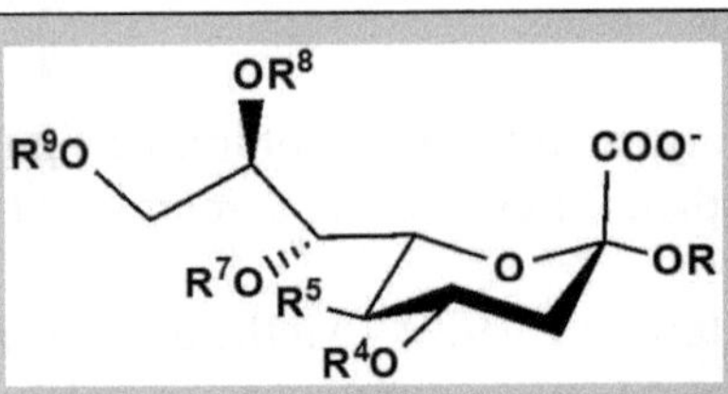

Name	Abbreviation	R4	R5	R7	R8	R9
5-*N*-Acetylneuraminic acid	Neu5Ac	H	HN-Acetyl	H	H	H
5-*N*-Acetyl-4-*O*-acetylneuraminic acid	Neu4,5Ac_2	Acetyl	HN-Acetyl	H	H	H
5-*N*-Acetyl-7-*O*-acetylneuraminic acid	Neu5,7Ac_2	H	HN-Acetyl	Acetyl	H	H
5-*N*-Acetyl-8-*O*-acetylneuraminic acid	Neu5,8Ac_2	H	HN-Acetyl	H	Acetyl	H
5-*N*-Acetyl-9-*O*-acetylneuraminic acid	Neu5,9Ac_2	H	HN-Acetyl	H	H	Acetyl
5-*N*-Acetyl-4,9-di-*O*-acetylneuraminic acid	Neu4,5,9Ac_3	Acetyl	HN-Acetyl	H	H	Acetyl
5-*N*-Acetyl-7,9-di-*O*-acetylneuraminic acid	Neu5,7,9Ac_3	H	HN-Acetyl	Acetyl	H	Acetyl
5-*N*-Acetyl-8,9-di-*O*-acetylneuraminic acid	Neu5,8,9Ac_3	H	HN-Acetyl	H	Acetyl	Acetyl
5-*N*-Acetyl-7,8,9-tri-*O*-acetylneuraminic acid	Neu5,7,8,9Ac_4	H	HN-Acetyl	Acetyl	Acetyl	Acetyl
5-*N*-Acetyl-9-*O*-L-lactylneuraminic acid	Neu5Ac9Lt	H	HN-Acetyl	H	H	L-lactyl
5-*N*-Acetyl-4-*O*-acetyl-9-*O*-lactylneuraminic acid	Neu4,5$Ac_2$9Lt	Acetyl	HN-Acetyl	H	H	lactyl
5-*N*-Acetyl-8-*O*-methylneuraminic acid	Neu5Ac8Me	H	HN-Acetyl	H	Methyl	H
5-*N*-Acetyl-9-*O*-acetyl-8-*O*-methylneuraminic acid	Neu5,9$Ac_2$8Me	H	HN-Acetyl	H	Methyl	Acetyl
5-*N*-Acetyl-8-*O*-sulfoneuraminic acid	Neu5Ac8S	H	HN-Acetyl	H	Sulfate	H
5-*N*-Acetyl-9-*O*-phosphoroneuraminic acid	Neu5Ac9P	H	HN-Acetyl	H	H	Phosphate
5-*N*-Acetyl-2-deoxy-2,3-dehydroneuraminic acid	Neu5Ac2en	H	HN-Acetyl	H	H	H

(continued)

Table 1
(continued)

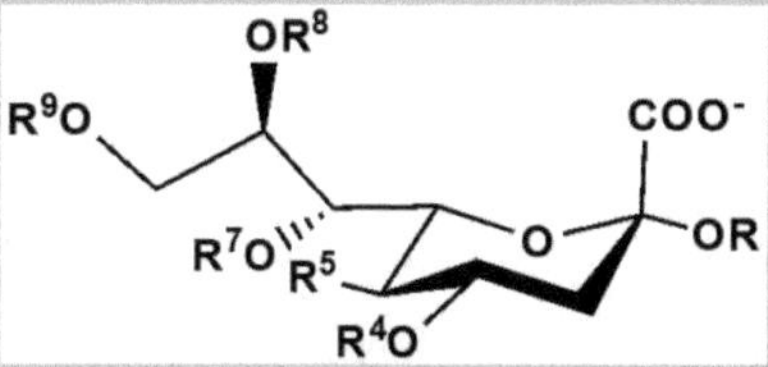

Name	Abbreviation	R4	R5	R7	R8	R9
5-*N*-Glycolylneuraminic acid	Neu5Gc	H	HN-Glycolyl	H	H	H
5-*N*-Glycolyl-4-*O*-acetylneuraminic acid	Neu4Ac5Gc	Acetyl	HN-Glycolyl	H	H	H
5-*N*-Glycolyl-7-*O*-acetylneuraminic acid	Neu7Ac5Gc	H	HN-Glycolyl	Acetyl	H	H
5-*N*-Glycolyl-8-*O*-acetylneuraminic acid	Neu8Ac5Gc	H	HN-Glycolyl	H	Acetyl	H
5-*N*-Glycolyl-9-*O*-acetylneuraminic acid	Neu9Ac5Gc	H	HN-Glycolyl	H	H	Acetyl
5-*N*-Glycolyl-7,9-di-*O*-acetylneuraminic acid	Neu7,9Ac$_2$5Gc	H	HN-Glycolyl	Acetyl	H	Acetyl
5-*N*-Glycolyl-8,9-di-*O*-acetylneuraminic acid	Neu8,9Ac$_2$5Gc	H	HN-Glycolyl	H	Acetyl	Acetyl
5-*N*-Glycolyl-7,8,9-tri-*O*-acetylneuraminic acid	Neu7,8,9Ac$_3$5Gc	H	HN-Glycolyl	Acetyl	Acetyl	Acetyl
5-*N*-Glycolyl-9-*O*-L-lactylneuraminic acid	Neu5Gc9Lt	H	HN-Glycolyl	H	H	L-lactyl
5-*N*-Glycolyl-8-*O*-methylneuraminic acid	Neu5Gc8Me	H	HN-Glycolyl	H	Methyl	H
5-*N*-Glycolyl-9-*O*-acetyl-8-*O*-methylneuraminic acid	Neu9Ac5Gc8Me	H	HN-Glycolyl	H	Methyl	Acetyl
5-*N*-Glycolyl-7,9-di-*O*-acetyl-8-*O*-methylneuraminic acid	Neu7,9Ac$_2$5Gc8Me	H	HN-Glycolyl	Acetyl	Methyl	Acetyl
5-*N*-Glycolyl-8-*O*-sulfoneuraminic acid	Neu5Gc8S	H	HN-Glycolyl	H	Sulfate	H
5-*N*-(*O*-Acetyl) glycolyneuraminic acid	Neu5GcAc	H	HN-Glycolyl-*O*-Acetyl	H	H	H
5-*N*-(*O*-Methyl) glycolyneuraminic acid	Neu5GcMe	H	HN-Glycolyl *O*-Methyl	H	H	H
2-keto-3-deoxynononic acid	Kdn	H	OH	H	H	H
9-*O*-Acetyl-2-keto-3-deoxynononic acid	Kdn9Ac	H	OH	H	H	Acetyl

Table 2
The main oligosaccharide series from mammal gangliosides

Structure	Series		Sugar numbering
ß-Gal-	Galacto	Gal	I
ß-Gal-(1-4)-ß-Glc-	Lacto	Lac	II-I
ß-GalNAc-(1-4)-ß-Gal-(1-4)-ß-Glc-	Ganglio-3	Gg_3	III-II-I
ß-Gal-(1-3)-ß-GalNAc-(1-4)-ß-Gal-(1-4)-ß-Glc-	Ganglio-4	Gg_4	IV-III-II-I
ß-GalNAc-(1-4)-ß-Gal-(1-3)-ß-GalNAc-(1-4)-ß-Gal-(1-4)-ß-Glc-	Ganglio-5	Gg_5	V-IV-III-II-I
ß-Gal-(1-4)-ß-GlcNAc-(1-3)- ß-Gal-(1-4)-ß-Glc-	Neolacto-4	nLc_4	IV-III-II-I

Table 3
The main ganglioside structures from mammals

Accepted short abbreviation	Abbreviations according to IUPAC-IUB nomenclature
GM4	Neu5AcGalCer
GM3	II^3Neu5AcLacCer
GD3	$II^3(Neu5Ac)_2$LacCer
O-acetyl-GD3	II^3[Neu5,9Ac_2-(2-8)-Neu5Ac]LacCer
GM2	II^3Neu5AcGg_3Cer
GD2	$II^3(Neu5Ac)_2Gg_3$Cer
GM1	II^3Neu5AcGg_4Cer
GM1b	IV^3Neu5AcGg_4Cer
Fuc-GM1	$IV^2\alpha$FucII^3Neu5AcGg_4Cer
GalNAc-GM1	II^3Neu5AcGg_5Cer
3'-LM1	IV^3nLc_4Cer
GD1a	IV^3Neu5AcII^3Neu5AcGg_4Cer
GD1α	IV^3Neu5AcIII^6Neu5AcGg_4Cer
GalNAc-GD1a	IV^3Neu5AcII^3Neu5AcGg_5Cer
GD1b	$II^3(Neu5Ac)_2Gg_4$Cer
GD1b-lactone	II^3[Neu5Ac-(2-8,1-9)-Neu5Ac]Gg_4Cer
Fuc-GD1b	$IV^2\alpha$FucII^3Neu5Ac_2Gg_4Cer
GT1a	$IV^3(Neu5Ac)_2II^3$Neu5AcGg_4Cer
GT1b	IV^3Neu5Ac$II^3(Neu5Ac)_2Gg_4$Cer
O-Acetyl-GT1b	IV^3Neu5AcII^3[Neu5,9Ac_2-(2-8)-Neu5Ac]Gg_4Cer

(continued)

Table 3 (continued)

Accepted short abbreviation	Abbreviations according to IUPAC-IUB nomenclature
GT1c	$II^3(Neu5Ac)_3Gg_4Cer$
Chol-1α-a	$IV^3Neu5AcIII^6Neu5AcII^3Neu5AcGg_4Cer$
Chol-1β	$III^6Neu5AcII^3(Neu5Ac)_2Gg_4Cer$
GT1α	$IV^3Neu5AcIII^6(Neu5Ac)_2Gg_4Cer$
GQ1b	$IV^3(Neu5Ac)_2II^3(Neu5Ac)_2Gg_4Cer$
O-Acetyl-GQ1b	$IV^3(Neu5Ac)_2II^3(Neu5Ac)_2Gg_4Cer$
GQ1c	$IV^3Neu5AcII^3(Neu5Ac)_3Gg_4Cer$
GQ1α	$IV^3(Neu5Ac)_2III^6(Neu5Ac)_2Gg_4Cer$
Chol-1α-b	$IV^3Neu5AcIII^6Neu5AcII^3(Neu5Ac)_2Gg_4Cer$
GP1c	$IV^3(Neu5Ac)_2II^3(Neu5Ac)_3Gg_4Cer$

chemical and enzymatic approaches. In 1947, the structure of sphingosine was elucidated [2] and in 1955 that of sialic acid [3]. Finally, in 1963 the first ganglioside structure was described [4].

Gangliosides display bulky oligosaccharide head groups and the double tailed hydrophobic moiety, called ceramide. The oligosaccharide chain is very variable due to the neutral sugar structure, content, sequence, and connections. Sialic acid identifies the derivatives of neuraminic acid (*see* Table 1), the three main forms being the 5-*N*-acetyl-, the 5-*N*-acetyl-9-*O*-acetyl-, and the 5-*N*-glycolyl-derivative and many other minor ones [5]. This, together with the variability of both sphingosine and the fatty acid, makes gangliosides a very large family of compounds. Figure 1 shows some structural details of the oligosaccharides structure, and Tables 2 and 3 show the main ganglioside structures together with the accepted and IUPAC-IUB abbreviations.

Since the time of the structure determination of GM1, mass spectrometry, nuclear resonance spectroscopy, fluorescence spectroscopy, scattering techniques, and, in general, all the analytical procedures have improved tremendously. This, together with the availability of new enzymatic approaches, the knowledge of an incredibly extensive number of glyco-enzymes, and the improvement of multigenetic analyses, makes the following chapters a valuable guidebook for entering in the research world of gangliosides.

Segrate, Italy *Sandro Sonnino*
Segrate, Italy *Alessandro Prinetti*

References

1. Klenk E (1939) Beitrdge zur Chemie der Lipiodosen. Z Phys Chem 262: 128–143
2. Carter HE, Glick FJ, Norris WP, Phillips GE (1947) Biochemistry of sphingolipides. III. Structure of sphingosine. J Biol Chem 170: 285–294
3. Gottschalk A (1955) Structural relationship between sialic acid, neuraminic acid and 2-carboxy-pyrrole. Nature 176:881–882
4. Kuhn R and Wiegandt H (1963) Die Konstitution der Ganglio-N-Tetraose und des Gangliosids GI. Chem Ber 96: 866–880
5. Schauer R (1982) Sialic acids: chemistry, metabolism and function. Springer-Verlag, Wien, New York

Contents

Contributors

DOMENICO ACQUOTTI • *Centro Interdipartimentale Misure, University of Parma, Parma, Italy*

ARUN AGARWAL • *Department of Chemistry and Biochemistry, University of Notre Dame, Notre Dame, IN, USA; Cancer Drug Delivery Research Foundation, Notre Dame, IN, USA*

MASSIMO AURELI • *Department of Medical Biotechnology and Translational Medicine, University of Milano, Milan, Italy*

MANJU BASU • *Department of Chemistry and Biochemistry, University of Notre Dame, Notre Dame, IN, USA; Cancer Drug Delivery Research Foundation, Notre Dame, IN, USA*

SUBHASH BASU • *Department of Chemistry and Biochemistry, University of Notre Dame, Notre Dame, IN, USA; Cancer Drug Delivery Research Foundation, Notre Dame, IN, USA*

BERNADETTE BREIDEN • *Membrane Biology and Lipid Biochemistry Unit, LIMES Institute, Kekulé-Institut für Organische Chemie und Biochemie, Universität Bonn, Bonn, Germany*

PAOLA BROCCA • *Department of Medical Biotechnology and Translational Medicine, University of Milan, Segrate, Italy*

BI-HE CAI • *The Institute of Biomedical Sciences (IBMS), Academia Sinica, Taipei, Taiwan*

LAURA CANTÙ • *Department of Medical Biotechnology and Translational Medicine, University of Milan, Segrate, Italy*

CHIA-CHUN CHAO • *The Institute of Biomedical Sciences (IBMS), Academia Sinica, Taipei, Taiwan*

ELENA CHIRICOZZI • *Department of Medical Biotechnology and Translational Medicine, University of Milano, Milan, Italy*

MARIA GRAZIA CIAMPA • *Department of Medical Biotechnology and Translational Medicine, University of Milan, Milan, Italy*

ELENA DEL FAVERO • *Department of Medical Biotechnology and Translational Medicine, University of Milan, Segrate, Italy*

TOYOSHI FUJIMOTO • *Department of Anatomy and Molecular Cell Biology, Nagoya University Graduate School of Medicine, Nagoya, Japan*

AKIKAZU FUJITA • *Field of Veterinary Pathobiology, Joint Faculty of Veterinary Medicine, Kagoshima University, Kagoshima, Japan*

MIKI HARA-YOKOYAMA • *Department of Biochemistry, Graduate School of Medical and Dental Sciences, Tokyo Medical and Dental University (TMDU), Tokyo, Japan*

HSIANG-CHI HUANG • *The Institute of Biomedical Sciences (IBMS), Academia Sinica, Taipei, Taiwan*

AKIHIRO IMAMURA • *Department of Applied Bio-Organic Chemistry, Faculty of Applied Biological Sciences, Gifu University, Gifu, Japan*

JIN-ICHI INOKUCHI • *Division of Glycopathology, Institute of Molecular Biomembrane and Glycobiology, Tohoku Medical and Pharmaceutical University, Sendai, Miyagi, Japan*
EMI ITO • *Brain Research Center, RIKEN, Wakō, Saitama, Japan*
KAZUHISA IWABUCHI • *Infection Control Nursing, Graduate School of Health Care and Nursing, Juntendo University, Chiba, Japan; Institute for Environmental and Gender Specific Medicine, Graduate school of Medicine, Juntendo University, Chiba, Japan*
REIJI KANNAGI • *The Institute of Biomedical Sciences (IBMS), Academia Sinica, Taipei, Taiwan; Division of Molecular Pathology, Aichi Cancer Center Research Institute, Nagoya, Japan*
KOICHI KATO • *Graduate School of Pharmaceutical Sciences, Nagoya City University, Mizuho-ku, Nagoya, Japan; Exploratory Research Center on Life and Living Systems and Institute for Molecular Science, National Institutes of Natural Sciences, Okazaki, Japan*
IKUO KAWASHIMA • *Department of Advanced Science for Biomolecules, Laboratory of Biomembrane, Tokyo Metropolitan Institute of Medical Science, Tokyo, Japan*
MAKOTO KISO • *Department of Applied Bio-Organic Chemistry, Faculty of Applied Biological Sciences, Gifu University, Gifu, Japan; Institute for Integrated Cell-Material Sciences (WPI-iCeMS), Kyoto University, Kyoto, Japan*
KEN KITAJIMA • *Bioscience and Biotechnology Center, Nagoya University, Nagoya, Japan*
ROBERT LEDEEN • *Division of Neurochemistry, Department of Pharmacology, Physiology and Neuroscience, Rutgers New Jersey Medical School, Newark, NJ, USA*
NICOLETTA LOBERTO • *Department of Medical Biotechnology and Translational Medicine, University of Milano, Milan, Italy*
GIULIA LUNGHI • *Department of Medical Biotechnology and Translational Medicine, University of Milano, Milan, Italy*
RUI MA • *Diagnostic Division, Siemens Corp, Shanghai, People's Republic of China*
JOHN L. MAGNANI • *GlycoMimetics Inc., Rockville, MD, USA*
TERUHIKO MATSUBARA • *Department of Biosciences and Informatics, Keio University, Kouhoku, Yokohama, Japan*
LAURA MAURI • *Department of Medical Biotechnology and Translational Medicine, The Medical School, University of Milano, Milano, Italy*
JOSEPH R. MOSKAL • *The Falk Ctr. for Molecular Therapeutics, Northwestern University, Evanston, IL, USA*
TAKAHIRO NITTA • *Division of Glycopathology, Institute of Molecular Biomembrane and Glycobiology, Tohoku Medical and Pharmaceutical University, Sendai, Miyagi, Japan*
TAKASHI ODE • *Japan Society for the Promotion of Science (JSPS), Tokyo, Japan; Department of Periodontology, Graduate School of Medical and Dental Sciences, Tokyo Medical and Dental University (TMDU), Tokyo, Japan*
ALESSANDRO PRINETTI • *Department of Medical Biotechnology and Translational Medicine, The Medical School, University of Milano, Milano, Italy*
SIMONA PRIONI • *Department of Medical Biotechnology and Translational Medicine, University of Milan, Milan, Italy*
KEIICHIRO SAKUMA • *Division of Molecular Pathology, Aichi Cancer Center Research Institute, Nagoya, Japan*
MAURA SAMARANI • *Department of Medical Biotechnology and Translational Medicine, University of Milano, Milan, Italy*

KONRAD SANDHOFF • *Membrane Biology and Lipid Biochemistry Unit, LIMES Institute, Kekulé-Institut für Organische Chemie und Biochemie, Universität Bonn, Bonn, Germany*
TOSHINORI SATO • *Department of Biosciences and Informatics, Keio University, Kouhoku, Yokohama, Japan*
DOMITILLA SCHIUMARINI • *Department of Medical Biotechnology and Translational Medicine, University of Milano, Milan, Italy*
GÜNTER SCHWARZMANN • *LIMES, c/o Kekulé-Institut f. Organische Chemie und Biochemie, Rheinische Friedrich-Wilhelms-Universität Bonn, Bonn, Germany*
MITSUTOSHI SETOU • *International Mass Imaging Center and Department of Cellular and Molecular Anatomy, Hamamatsu University School of Medicine, Shizuoka, Japan*
SANDRO SONNINO • *Department of Medical Biotechnology and Translational Medicine, The Medical School, University of Milano, Milano, Italy*
EIJI SUGIYAMA • *Department of Biochemistry, Keio University School of Medicine, Tokyo, Japan*
AKEMI SUZUKI • *Division of Glycopathology, Institute of Molecular Biomembrane and Glycobiology, Tohoku Medical and Pharmaceutical University, Sendai, Miyagi, Japan*
MINORU SUZUKI • *School of Integrative and Global Majors, Tsukuba University, Tsukuba, Ibaraki, Japan*
TAKAO TAKI • *AGT & T Co. Ltd., Oujinncho,, Tokushima, Japan*
TAKUMA TSUJI • *Department of Anatomy and Molecular Cell Biology, Nagoya University Graduate School of Medicine, Nagoya, Japan*
HUGH J. WILLISON • *Institute of Infection, Immunity and Inflammation, College of Medical, Veterinary and Life Sciences, Glasgow Biomedical Research Centre, University of Glasgow, Glasgow, UK*
GUSHENG WU • *Division of Neurochemistry, Department of Pharmacology, Physiology and Neuroscience, Rutgers New Jersey Medical School, Newark, NJ, USA*
TAKUMI YAMAGUCHI • *School of Materials Science, Japan Advanced Institute of Science and Technology, Nomi, Japan; Graduate School of Pharmaceutical Sciences, Nagoya City University, Mizuho-ku, Nagoya, Japan*
NORIYUKI YUASA • *Tokyo Chemical Industry Co., Ltd., Tokyo, Japan*

Chapter 1

Chemical and Physicochemical Properties of Gangliosides

Laura Mauri, Sandro Sonnino, and Alessandro Prinetti

Abstract

In this chapter, we briefly describe the structural features of gangliosides, and focus on the peculiar chemicophysical features of gangliosides, an important class of membrane amphipathic lipids that represent an important driving force determining the organization and properties of cellular membranes.

Key words Glycosphingolipids, Gangliosides, Ceramide, Sphingosine, Lipid rafts

1 Chemical Structure of Gangliosides

Gangliosides are amphipathic membrane lipids belonging to the family of glycosphingolipids, which represents the most abundant class of glycolipids in eukaryotic membranes. The collective name of "gangliosides" was given by the German biochemist Ernst Klenk to a mixture of different species isolated from human brain. Klenk was later on also able to demonstrate the presence of an acidic carbohydrate component in gangliosides. He named this residue as "neuraminic acid." The structural complexity of gangliosides was a serious challenge. The structure of sphingosine, the backbone of the hydrophobic moiety of sphingolipids, was elucidated in 1947 [1], and in 1955 that of sialic acid [2]. However, the first complete ganglioside structure was described only in 1963 [3]. Gangliosides are characterized by sialic acid-containing oligosaccharide chains linked via a β-glycosidic linked to the hydrophobic backbone, ceramide, which is responsible for the insertion in cellular membranes. The size of the average glycans in gangliosides in definitely smaller than that of glycans associated with glycoproteins; however, the structural complexity of gangliosides, resulting from the combination of heterogeneous oligosaccharide chains, different types of sialic acids present in the oligosaccharide, and heterogeneous composition of the ceramide backbone (both in its sphingosine and fatty acid components), is amazing, and up to day several hundreds of different gangliosides molecular species have been

Sandro Sonnino and Alessandro Prinetti (eds.), *Gangliosides: Methods and Protocols*, Methods in Molecular Biology, vol. 1804, https://doi.org/10.1007/978-1-4939-8552-4_1, © Springer Science+Business Media, LLC, part of Springer Nature 2018

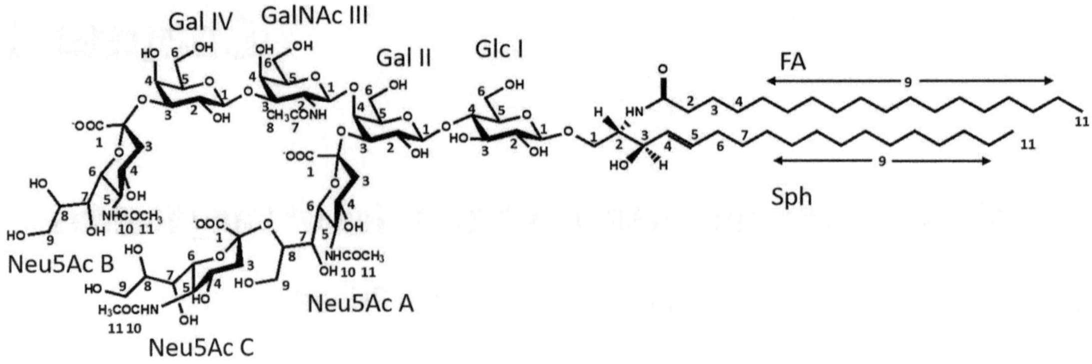

Fig. 1 Structure of the trisialoganglioside GT1b

described. Remarkably, some of the physicochemical properties of gangliosides, responsible for their aggregative properties within biological membranes and relevant for their diverse biological functions, can be deeply affected by even tiny modifications of their chemical structure. Moreover, we are just starting to realize that, within biological structures well known to be highly enriched in gangliosides, the distribution of even closely related molecular species can be dramatically different [4, 5], and that specificity of biological functions for the different molecular species is incredibly elevated [6, 7]. As an example, Fig. 1 shows the structure of the trisialoganglioside.

Gangliosides share with all other sphingolipids the same double tailed hydrophobic moiety, named ceramide. Ceramide is composed of a long chain amino alcohol [8] (2*S*, 3*R*, 4*E)* 2-amino-1,3-dihydroxy-octadec-4-ene, trivially known as sphingosine, linked via an amide bond with a fatty acyl chain (Fig. 2). The C18 molecular species with a trans double bond in the 4,5 position is the most frequent sphingosine found in mammalian sphingolipids. However, less common structures with shorter and longer alkyl chain have been identified, and can be abundant in gangliosides from specific tissues and cell types (e.g., gangliosides containing C20 sphingosine are relatively abundant in mammalian nervous tissue and cultured neuronal cells [9]). In addition, gangliosides containing structures lacking the double bond, which should be more properly named "sphinganine" (while structures containing unsaturations should be properly named "sphingenine"), have been also described. Ceramide structure can be heterogeneous not only for its sphingoid base moiety, but also for the presence of fatty acyl chains with different length and degree of unsaturation. The main players providing complexity to sphingolipids with respect to their acyl chain length seem to be the ceramide synthase enzymes, which catalyze the formation of (dihydro)ceramides. Each ceramide synthase is able to produce specific ceramide species, even if the complexity of the acyl chains present in ceramides is not totally

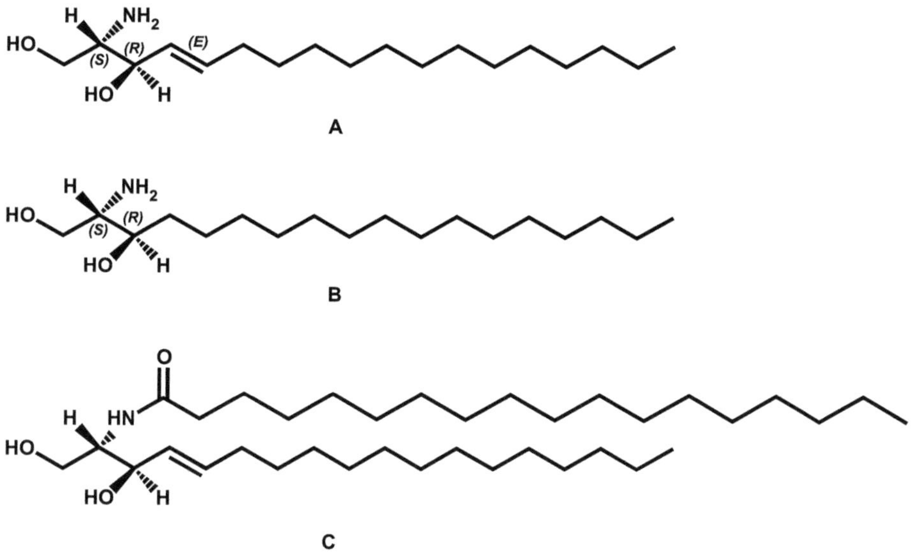

Fig. 2 Structure of sphingosine (**A**), sphinganine (**B**), and ceramide (**C**)

explained by our current knowledge on the acylCoA preference of the different ceramide synthases. In some tissues (for example in the nervous tissue) [10] gangliosides are highly enriched in saturated acyl chains such as palmitic and stearic acid. Hydroxylated fatty acids, highly represented in neutral glycolipids and sulfatides, are less frequently present in gangliosides; however, they are relatively abundant in some tissues. For example, GM3 ganglioside from human liver contains a significant amount of α-hydroxylated C16-C24 chains, which remarkably increase along aging becoming prevalent in mature and elderly individuals [11].

In the wide family of glycosphingolipids, the hydrophilic moiety is represented by an oligosaccharide chain, whose structure can range from a very simple one (e.g., a single galactose residue in galactosylceramide, GalCer, one of the main lipids present in myelin) to structures with a very high degree of complexity (as in polysialylated gangliosides, characteristic of the nervous system and abundant in differentiated neurons). Gangliosides, characterized by the presence of sugar residues containing a carboxyl group, collectively known as sialic acids, represent one of the two families of glycosphingolipids with acidic nature present in eukaryotes (the other being represented by sulfatides [12], containing an *O*-linked sulfate group on a glucose or galactose residue).

Sialic acids (Schauer, 1982), the signature sugar residues present in gangliosides, are derivatives of 5-amino-3,5-dideoxy-D-*glycero*-D-*galacto*-non-2-ulopyranosonic acid, or neuraminic acid (Fig. 3). Three main sialic acids are known to be present in natural gangliosides: the 5-*N*-acetyl-, the 5-*N*-acetyl-9-*O*-acetyl- and the 5-*N*-glycolyl-derivative. In healthy humans, the most abundant

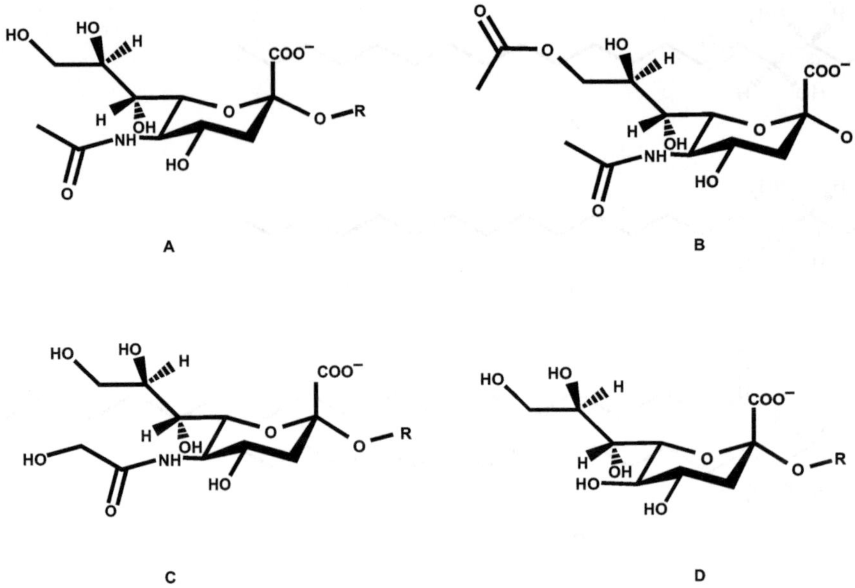

Fig. 3 Structure of 5-*N*-acetyl- (**A**), 5-*N*-acetyl-9-*O*-acetyl- (**B**), and 5-*N*-glycolyl-neuraminic acid (**C**) and of KDN (**D**)

sialic acid is the 5-*N*-acetyl derivative [13–16], but about 10% of the total ganglioside sialic acid is represented by the 5-*N*-acetyl-9-*O*-acetyl-neuraminic acid [17] and polysialogangliosides containing this sugar structure have been characterized in mice brains [18, 19]. Unlike humans, mammals including primates are able to synthesize 5-*N*-glycolyl-neuraminic acid due to the ancestral loss of the *CMAH* gene in the hominid lineage, encoding for the hydroxylase/monooxygenase enzyme responsible for the conversion of CMP-5-*N*-acetyl-neuraminic acid to CMP-5-*N*-glycolyl-neuraminic acid. However, small amounts of gangliosides containing this sialic acid are detectable in human tissue, likely due to the incorporation of 5-*N*-glycolyl-neuraminic acid from dietary sources [20], and 5-*N*-glycolyl-neuraminic acid is usually present at relatively high levels in gangliosides from certain tumors [21]. Ganglioside lactones, gangliosides with polysialyl chains where the sialic acids are linked together with ketosidic and ester linkages, have been detected in fish, whale, and human brain [22–24]. Figure 4 shows the structure of GD1b lactone in comparison with that of GD1b.

Another member of the sialic acid family, characterized by a structure where the *N*-acylamino group at the five position in 5-*N*-acetyl-neuraminic acid is replaced by a hydroxyl group (deaminoneuraminic acid or ketodeoxynonulosonic acid, KDN) is relatively common in gangliosides from different fish tissues [25], but also present in minor amount in gangliosides from various mammalian cells and tissues.

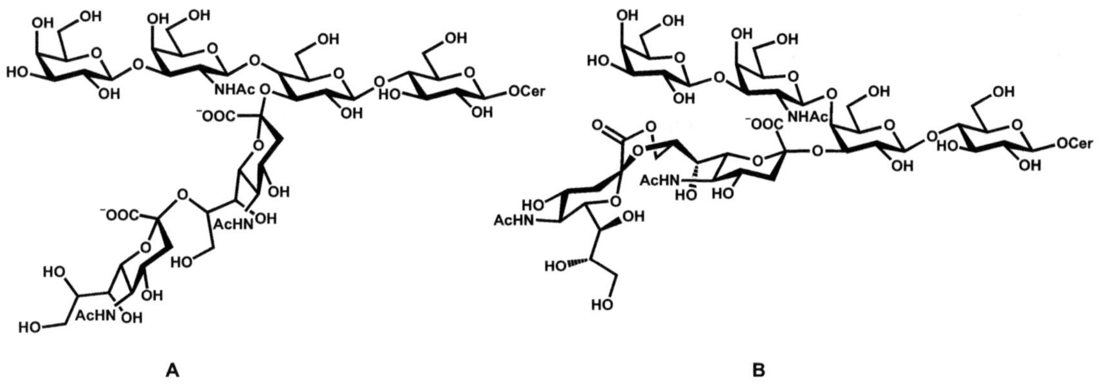

Fig. 4 Structure of GD1b (**A**) and GD1b lactone (**B**)

Table 1
Main ganglioside series

Series	Abbreviation	Structure
Galacto	Gal	ß-Gal-(1-1)-Cer
Lac	Lac	ß-Gal-(1-4)-ß-Glc-(1-1)-Cer
Ganglio-3	Gg_3	ß-GalNAc-(1-4)-ß-Gal-(1-4)-ß-Glc-(1-1)-Cer
Ganglio-4	Gg_4	ß-Gal-(1-3)-ß-GalNAc-(1-4)-ß-Gal-(1-4)-ß-Glc-(1-1)-Cer
Ganglio-5	Gg_5	ß-GalNAc-(1-4)-ß-Gal-(1-3)-ß-GalNAc-(1-4)-ß-Gal-(1-4)-ß-Glc-(1-1)-Cer
Globo-3	Gb_3	α-Gal-(1-4)-ß-Gal-(1-4)-ß-Glc-(1-1)-Cer
Globo-4	Gb_4	ß-GalNAc-(1-3)-α-Gal-(1-4)-ß-Gal-(1-4)-ß-Glc-(1-1)-Cer
Lacto	Lc	ß-Gal-(1-3)-ß-GlcNAc-(1-3)-ß-Gal-(1-4)-ß-Glc-(1-1)-Cer
Neolacto	nLc	ß-Gal-(1-4)-ß-GlcNAc-(1-3)- ß-Gal-(1-4)-ß-Glc-(1-1)-Cer

The nomenclature and structures of the most common gangliosides present in eukaryotes are reported in Tables 1 and 2. Considering the high variability (due to the sugar structure, content, sequence and type of linkages involved) of the oligosaccharide chain of gangliosides together with the abovementioned heterogeneity in their hydrophobic moiety, the resulting number of different ganglioside molecular species so far detected is amazing.

Table 2
Main ganglioside structures

Abbreviations according to Svennerholm [69]	Abbreviations according to IUPAC-IUB [70]
GM4	Neu5AcGalCer
GM3	II^3Neu5AcLacCer
GD3	$II^3(Neu5Ac)_2$LacCer
O-acetyl-GD3	II^3[Neu5,9Ac_2-(2-8)-Neu5Ac]LacCer
GM2	II^3Neu5AcGg_3Cer
GD2	$II^3(Neu5Ac)_2Gg_3$Cer
GM1	II^3Neu5AcGg_4Cer
GM1b	IV^3Neu5AcGg_4Cer
Fuc-GM1	IV^2αFucII^3Neu5AcGg_4Cer
GalNAc-GM1	II^3Neu5AcGg_5Cer
3'-LM1	IV^3nLc_4Cer
GD1a	IV^3Neu5AcII^3Neu5AcGg_4Cer
GD1α	IV^3Neu5AcIII^6Neu5AcGg_4Cer
GalNAc-GD1a	IV^3Neu5AcII^3Neu5AcGg_5Cer
GD1b	$II^3(Neu5Ac)_2Gg_4$Cer
GD1b-lactone	II^3[Neu5Ac-(2-8,1-9)-Neu5Ac]Gg_4Cer
Fuc-GD1b	IV^2αFucII^3Neu5Ac_2Gg_4Cer
GT1a	$IV^3(Neu5Ac)_2II^3$Neu5AcGg_4Cer
GT1b	IV^3Neu5Ac$II^3(Neu5Ac)_2Gg_4$Cer
O-Acetyl-GT1b	IV^3Neu5AcII^3[Neu5,9Ac_2-(2-8)-Neu5Ac]Gg_4Cer
GT1c	$II^3(Neu5Ac)_3Gg_4$Cer
Chol-1α-a	IV^3Neu5AcIII^6Neu5AcII^3Neu5AcGg_4Cer
Chol-1β	III^6Neu5Ac$II^3(Neu5Ac)_2Gg_4$Cer
GT1α	IV^3Neu5Ac$III^6(Neu5Ac)_2Gg_4$Cer
GQ1b	$IV^3(Neu5Ac)_2II^3(Neu5Ac)_2Gg_4$Cer
GQ1c	IV^3Neu5Ac$II^3(Neu5Ac)_3Gg_4$Cer
GQ1α	$IV^3(Neu5Ac)_2III^6(Neu5Ac)_2Gg_4$Cer
Chol-1α-b	IV^3Neu5AcIII^6Neu5Ac$II^3(Neu5Ac)_2Gg_4$Cer
GP1c	$IV^3(Neu5Ac)_2II^3(Neu5Ac)_3Gg_4$Cer

2 Physicochemical and Aggregational Properties OF Gangliosides

Gangliosides are lipids with strong amphiphilic character due to the big oligosaccharide head group, strongly hydrophilic and markedly protruding from the membrane surface, and the double tailed hydrophobic moiety.

One remarkable feature of all gangliosides is the bulkiness of their hydrophilic head groups, as compared to that of phospholipids, representing the matrix of biological membranes. Even one of the simplest ganglioside head groups, such as that of GM3, is much bulkier than the bigger phospholipid head groups. Strong differences in the geometrical features of the head groups of amphipathic lipids organized in the same membrane represent the main driving force for clustering and segregation of different membrane components, allowing to minimize the interfacial free energy. In general, the volume occupied by the sugar head group increases with the complexity of the oligosaccharide chain. With the progressive increase of the volume occupied by the head group, the insertion of the ganglioside molecule in a bilayer membrane the molecules requires a progressively wider interfacial area, and the larger is the interfacial area required by ganglioside oligosaccharide structure, the more pronounced is the segregation of ganglioside molecules within the phospholipid bilayer, and the more positive is the curvature of the resulting ganglioside-rich membrane microenvironment. Aggregates of natural gangliosides or chemically modified gangliosides with homogeneous hydrophobic moiety, mixed aggregates of gangliosides and phospholipids, monomeric gangliosides and the free oligosaccharide chains have been used to extensively explore the geometrical and dynamic properties of gangliosides, in particular GM3 and ganglio-series gangliosides (reviewed in [26]). The results are summarized in Table 3.

Several studies using phospholipid liposomes containing different semisynthetic gangliosides with well-defined composition in terms of ceramide acyl chain length, degree of saturation, and nature of the sphingoid base [27] allowed to elucidate the lateral phase separation properties of gangliosides in phospholipid bilayers. In summary, these studies indicated that (1) gangliosides undergo phase separation in two-components, two-phases as well as in one-component phosphatidylcholine bilayers [28–34]. In addition, GM1 forms segregated domains also in sphingomyelin bilayers [35]; (2) for a given composition of the hydrophobic moiety, the extent of ganglioside phase separation depends upon the number of sugars in the oligosaccharide head group [29, 30]; (3) for a given oligosaccharide composition, the extent of ganglioside lateral phase separation depends upon the length and unsaturation difference between the ganglioside long-chain base and phosphatidylcholine acyl chains [28–30]; on the other hand, a decrease in

Table 3
Hydrodynamic radius R_h (Å), axial ratio R_a/R_b, molecular mass M (kDa) of the aggregates, and surface area a_0 (Å²) of the monomer in aggregate [60–62, 71]

	Aggregate type	R_h (Å)	R_a/R_b	M (kDa)	a_0 (Å²)
Bovine brain GM4 with heterogeneous ceramide	Vesicle	≈300		18,270	≈80
Bovine brain GM3 with heterogeneous ceramide	Vesicle	≈250		16,700	≈80
Bovine brain GM2 with heterogeneous ceramide	Micelle	66.0	3.1	740	92.0
Bovine brain GM1 with heterogeneous ceramide	Micelle	58.7	2.3	470	95.4
Synthetic GM1(*d*18:1,18:0), previously warmed at 60 °C	Micelle	52.8			
Synthetic GM1(*d*18:1,18:0), previously warmed at 40 °C	Micelle	56.2			
Synthetic GM1(*d*18:1,18:0), previously warmed at 25 °C	Micelle	58.7			
Pig brain Fuc-GM1 with heterogeneous ceramide	Micelle	61.0	2.1	394	97.8
Bovine brain GD1a with heterogeneous ceramide	Micelle	58.0	2.0	418	98.1
Bovine brain GalNAc-GD1a with heterogeneous ceramide	Micelle	60.0		509	97.0
Bovine brain GD1b with heterogeneous ceramide	Micelle	52.0	1.8	311	100.8
Synthetic GD1b-lactone with heterogeneous ceramide	Micelle	57.0	2.1	424	97.6
Bovine brain GT1b with heterogeneous ceramide	Micelle	53.2	1.8	378	100.8

Natural compounds have heterogeneous ceramide moiety. Purified species with defined ceramide structure are indicated

the acyl chain length or an increase in its unsaturation induced reduced lateral phase separation of GM1 [32].

In addition to the number of sugar residues, the dynamics of the glycosidic linkages are very relevant in determining the large volume occupied by the oligosaccharide chains of gangliosides. The disaccharide -ß-Gal-(1-4)-ß-Glc-, the lactose, is linked to ceramide in GM3, GD3, and all ganglio series gangliosides. On the basis of molecular calculations and a few available NMR data [36, 37] many conformers in a very reduced range of minimum energy have been determined; however, ^{13}C T_1, T_{1r} and ^{13}C(^{1}H) NOE measurements [38] suggest a reduced mobility for the two linkages. A character of low motional freedom is expected for the glucose residue [39, 40], due the restrictions of motion imposed by the surrounding membrane surface on the oligosaccharide–ceramide linkage [41].

On the other hand, in the trisaccharide of GM3, the sialic acid residue ketosidic linkage to galactose, the α-Neu5Ac-(2-3)-β-Gal, is highly mobile, allowing the existence of at least two main conformations. The trisaccharide -β-GalNAc-(1-4)[α-Neu5Ac-(2-3)-]-β-Gal-, present in gangliosides of the series ganglio-3, ganglio-4 and ganglio-5, behaves as a very rigid unit due to the presence of several interresidual interactions [36, 38, 42–44]. In particular, the strong interaction between the side chain of the sialic acid, blocked in a rigid conformation [36, 45, 46], and the *N*-acetylgalactosamine (stabilized by a hydrogen bond between the GalNAc amide proton and the Neu5Ac carboxyl group [47]) is very important in determining the rigid conformation of the trisaccharide.

The addition of galactose to hexosamine, yielding the disaccharide β-Gal-(1-3)-β-GalNAc- that is present in the structure of the ganglio-4 series gangliosides, leads to a very different situation. In fact, this disaccharide is highly is mobile, allowing the existence of at least two main conformations of the glycosidic linkage. Consequently, GM2, carrying rigid linkages in the outer portion of the oligosaccharide chain, is present only in one preferred conformation, while GM1, featuring the external mobile linkage β-Gal-(1-3)-β-GalNAc, exist in two preferred conformations, and GD1a, with the external α-Neu5Ac-(2-3)-β-Gal-(1-3)-β-GalNAc oligosaccharide, allows four conformations. The volume or solid angle occupied by the different ganglioside monomers at the membrane surface is determined by the number of conformers allowed for a certain ganglioside oligosaccharide moiety. The result of the above-mentioned dynamic properties of the different glycosidic linkages is that the volume required to accommodate the GM1 and GD1a oligosaccharides (characterized by a higher number of preferred conformations) is much higher than the volume that could be predicted based on the features of the GM2 oligosaccharide, simply considering the addition of one or two more sugars.

Among complex polysialylated gangliosides, for the b-series gangliosides, characterized by two sialic acid residues linked to the internal galactose, up to now information is available for GD1b and GT1b only. In the tetrasaccharide -β-GalNAc-(1-4)[α-Neu5Ac-(2-8)-α-Neu5Ac-(2-3)-]-β-Gal-, the interresidual contacts between GalNAc and Neu5Ac linked to Gal are no longer present, however some interactions occur between GalNAc and the external Neu5Ac. As the result of these constraints, the tetrasaccharide chain is arranged in a circle with a diameter of 3 Å, with a central hollow characterized by a highly hydrophobic internal surface, and apparently wide enough to accommodate cations [48]. This conformation confers a bulkier character to the ganglioside portion closer to the hydrophobic–hydrophilic interface. Consequently, GD1b is characterized by a larger surface area than its isomer GD1a. The solid angle occupied by the disialosyl chain is big enough to accommodate host the additional sialic acid unit on the external galactose, present in GT1b and in fact, GT1b surface area is very similar to that of GD1b.

As mentioned above, significant amounts of the inner ester of GD1b, GD1b-lactone have been detected in human brain [22], and it has been proposed that the interconversion of GD1b and GD1b-lactone might play important roles in the regulation of the activity of membrane-associated proteins or of membrane lateral organization. In fact, when the inner ester between the external sialic acid carboxyl group and the inner sialic acid residue is formed, the interresidual interactions between GalNAc and the external Neu5Ac are lost, and consequently the rigid conformation of the trisaccharide -β-GalNAc-(1-4)[α-Neu5Ac-(2-3)-]-β-Gal- observed for example in GM1 and GD1a is restored. This rigid conformation leads to a better lining up of the disialosyl chain with the neutral oligosaccharide chain, reducing the angle between the neutral chain and the inner sialic acid axis [48]. Consequently, the geometry of GD1b-lactone is closer to that of GD1a than to that of GD1b. This implies that the lactonization–delactonization process (occurring for example due to local pH changes in the proximity of the cell surface) might have very deep consequences on the property of membrane microenvironments.

When considering the general features of the head groups of gangliosides inserted in a biological membrane, other factors should be taken into account, in addition to the size of the oligosaccharide chains and the dynamics of the glycosidic linkages. The most relevant is probably the hydrating water associated with the oligosaccharide chain. The amount of water associated with the polar head groups, due to the strong hydrophilic character of the oligosaccharide chains, is remarkably high [49], and it is likely very relevant in avoiding repulsive effects between the negatively charged oligosaccharide chains. A calorimetric study suggests that each oligosaccharide chain is tightly interacting with a number of water molecules ranging from 22 to 60, depending on the oligosaccharide chains [50]. As a term of comparison, 6–7 water molecules are associated with the head group of dipalmitoylphosphatidylcholine. This large amount of interacting water molecules is confirmed by calculations performed on GM2 molecules, indicating a difference of about 5 Å between the dry and hydrated micellar radius [51]. NMR studies indicated the existence of strong interactions between water molecules and the sugar residues sialic acid and inner galactose in the oligosaccharide of GM1 [42]. Taken together, these results would argue against a possible role of direct carbohydrate-carbohydrate interactions in mediating ganglioside-dependent biological events [52].

As described previously in this chapter, gangliosides are highly heterogeneous in their lipid moieties, due to the possible presence of different sphingoid bases, and of different fatty acids linked to the sphingoid base. Bulk membrane glycerophospholipids (e.g., phosphatidylcholine, PC), are also present in a large number of different molecular species. A large proportion of membrane phospholipids is rich in unsaturated fatty acids, ensuring that at physiological temperature the membrane behaves as a bidimensional fluid, allowing a certain degree of lateral motility and conformational freedom to membrane components, including membrane-associated proteins. Nevertheless, even among glycerophospholipids, species with less fluid saturated chains are relatively abundant. The transition temperature of the lipid chains of complex lipids is one of the most important properties determining the dynamics of biological membranes [53, 54]. In a membrane, lipids that contain saturated alkyl chains characterized by higher transition temperatures are excluded from those that contain unsaturated chains, with a lower transition temperature, determining the separation of an ordered, less fluid, liquid phase. Dipalmitoylphosphatidylcholine is indeed highly enriched in detergent-resistant membrane preparations, putatively corresponding to liquid-ordered membrane domains commonly named lipid rafts [55–58]. On the other hand, gangliosides containing saturated acyl chains such as palmitic and stearic acid are very abundant, especially in the nervous system. In addition, the ceramide moiety is a rigid structure [59]. Thus not surprisingly, gangliosides are highly enriched in detergent-resistant membranes, contributing to the creation of highly organized and rigid membrane microenvironments.

Differential scanning calorimetry has been widely used by several groups in the past to determine used to determine the transition temperatures of complex lipids, including gangliosides and other glycosphingolipids. The transition temperature of complex lipids is dictated mainly by the nature of their acyl chains; however, it does not necessarily correspond to the melting point of the lipid chains. In fact, its value is strictly related to the type and size of aggregate formed by the amphipathic lipid in solution, which in turn depends on their geometrical properties, determined by the features of both the hydrophilic and hydrophobic moieties of the amphiphile. These aggregates range from small, quite spherical micelles that are unfavorable to the chain ordering, to large vesicles that on the other hand favor a high degree of chain ordering. Consequently, significant differences in the transition temperatures of complex lipids can be observed in the case of aggregates characterized by large differences in curvature [60–62]. Ganglioside GM1 containing stearic acid and a sphingosine with 18 carbons shows a transition temperature of 11.7 °C, however the addition of two carbon atoms to the sphingosine moiety determines a shift of the transition temperature to 23.2 °C. Stable removal of water from the hydrophilic head or changes of the oligosaccharide conformation by increasing temperature lead to changes of the surface area occupied by the monomer inserted into the aggregate with consequent changes in surface curvature and in aggregate size

Table 4
Transition temperature of gangliosides [60, 72, 73]

	Tt (°C)
Bovine adrenal medulla GM3 with heterogeneous ceramide	35.3
Bovine brain GM2 with heterogeneous ceramide	29.3
Bovine brain GM1 with heterogeneous ceramide	19.7
Synthetic GM1(d18:1,18:0), previously warmed at 60 °C	11.7
Synthetic GM1(d18:1,18:0), previously warmed at 40 °C	15.0
Synthetic GM1(d18:1,18:0), previously warmed at 25 °C	17.6
Synthetic GM1(d20:1,18:0), previously warmed at 60 °C	23.2
Pig brain FucGM1 with heterogeneous ceramide	13.2
Synthetic FucGM1(d18:1,18:0), previously warmed at 60 °C	10.0
Synthetic FucGM1(d20:1,18:0), previously warmed at 60 °C	18.3
Bovine brain GD1a with heterogeneous ceramide	16.0
Synthetic GD1a(d18:1,18:0), previously warmed at 60 °C	10.1
Synthetic GD1a(d20:1,18:0), previously warmed at 60 °C	19.2
Bovine brain GT1bwith heterogeneous ceramide	7.3

[60–62]. Due to these effects, the transition temperature of the same homogeneous GM1 species is lower (with a difference of 5 °C) when the ganglioside is dissolved at 60 °C instead of 20 °C (Table 4). In addition to this, gangliosides typically show two different transitions, one at lower temperature, related to the hydrophobic moiety, the other, at higher temperature, to the oligosaccharide chains [60–62]. It is important to recall that the transition temperature of the lipid moiety is also dependent on the geometry of the ganglioside monomer, determined the size of the head groups and by their structural differences (as discussed above). In general, the size of the head groups is inversely correlated with the transition temperature. This represents a further effect of the head groups on the fluidity of the hydrophobic moiety, and a further factor contributing to the modulation of the segregation process.

As discussed in detail above, the structure, geometric properties, dynamics and interactions with the water shell of the oligosaccharide head groups of gangliosides are extremely important in dictating the behavior of gangliosides in complex systems, such as biological membranes, and their aggregational properties. In addition, we have recalled the importance of the high transition temperature of the hydrophobic chains of gangliosides in the segregation process. However, the ceramide portion of sphingolipids is contributing with other aspects to the aggregation/segregation of these lipids within membranes. In particular, the contribution of the part of the ceramide molecule sitting at the water–lipid interface seems to be very relevant, as indicated by the observation that sphingolipid-rich domains can be formed in cells that contain sphingomyelin but have low amount or are lacking of glycosphingolipids [63], and in subdomains with sphingomyelin but a very low content of neutral glycosphingolipids and no content of gangliosides [64, 65]. The amide group of ceramide at the water–lipid interface behaves as a rigid system comprising six atoms in a planar conformation, together with the hydroxyl group at position 3. The group has a perpendicular orientation towards the axes of the two hydrocarbon chains, which parallel orientation is stabilized by the presence of double bond at position 4–5 of sphingosine [66, 67]. The availability of an amide nitrogen, of a carbonyl oxygen and of a hydroxyl group, enables sphingolipids to form hydrogen bonds, acting as hydrogen bond donors and acceptors at the same time (a feature lacking in other complex membrane lipids such as glycerophospholipids). Therefore, this feature allows sphingolipids to form a stable net of interactions, which, in the case of gangliosides, further stabilizes the clustering driven by the properties of the hydrophilic head groups. The van der Waals forces between hydrocarbon chains have been estimated to about 2–3 kcal per hydrocarbon chain. The formation of hydrogen bonds at the water–lipid interface contributes with 3–10 kcal to the lipid–lipid interaction, and considering the predicted high number of hydrogen bonds formed in sphingolipid-rich microdomains, this contribution is very significant to the overall stability of the system.

3 Conclusions

The structural complexity of gangliosides has puzzled researchers for decades. When exploring the links between structure and biological functions, much emphasis has been posed on the role of gangliosides oligosaccharide chains. Ganglioside head groups, protruding toward the extracellular space, with a high degree of complexity reached with relatively simple metabolic mechanism, highly dynamic properties and potential for efficient and fast local regulation (e.g., due to remodeling by plasma membrane-associated enzymes [68], or lactonization processes) seem to be functional for complex interactions with molecules sitting on the same cellular membrane (*cis* interactions) or present in the extracellular environment, or on the surface of other cells (*trans* interactions), and there is no doubt that oligosaccharide-dependent interactions are involved in the multifaceted roles of gangliosides as regulators of the “social life” of cells. Nevertheless, it is clear that in some cases the physicochemical, aggregational, and consequently biological properties of these amphipathic membrane lipids are dictated by the properties of the monomers as a whole, which are dependent not only on the composition of the polar head groups. Thus, 55 years after the elucidation of GM1 structure, exploration is still not concluded.

References

1. Carter HE, Glick FJ, Norris WP, Phillips GE (1947) Biochemistry of the sphingolipides: III. Structure of sphingosine. J Biol Chem 170(1):285–294
2. Klenk E, Faillard H, Lempfrid H (1955) Enzymatic effect of the influenza virus. Hoppe Seylers Z Physiol Chem 301(4-6):235–246
3. Kuhn R, Wiegandt H (1963) The constitution of gangliosides G-Ii, G-Iii and G-Iv. Z Naturforsch B 18:541–543
4. Colsch B, Jackson SN, Dutta S, Woods AS (2011) Molecular microscopy of brain gangliosides: illustrating their distribution in hippocampal cell layers. ACS Chem Nerosci 2(4):213–222. https://doi.org/10.1021/cn100096h
5. Hirano-Sakamaki W, Sugiyama E, Hayasaka T, Ravid R, Setou M, Taki T (2015) Alzheimer's disease is associated with disordered localization of ganglioside GM1 molecular species in the human dentate gyrus. FEBS Lett 589(23):3611–3616. https://doi.org/10.1016/j.febslet.2015.09.033. S0014-5793(15)00933-3 [pii]
6. Go S, Go S, Veillon L, Ciampa MG, Mauri L, Sato C, Kitajima K, Prinetti A, Sonnino S, Inokuchi JI (2017) Altered expression of ganglioside GM3 molecular species and a potential regulatory role during myoblast differentiation. J Biol Chem 292(17):7040–7051. https://doi.org/10.1074/jbc.M116.771253
7. Veillon L, Go S, Matsuyama W, Suzuki A, Nagasaki M, Yatomi Y, Inokuchi J (2015) Identification of Ganglioside GM3 Molecular Species in Human Serum Associated with Risk Factors of Metabolic Syndrome. PLoS One 10(6):e0129645. https://doi.org/10.1371/journal.pone.0129645
8. Roisen FJ, Bartfeld H, Nagele R, Yorke G (1981) Ganglioside stimulation of axonal sprouting in vitro. Science 214(4520):577–578
9. Sonnino S, Chigorno V (2000) Ganglioside molecular species containing C18- and C20-sphingosine in mammalian nervous tissues and neuronal cell cultures. Biochim Biophys Acta 1469(2):63–77. doi:S0005-2736(00)00210-8 [pii]
10. Sastry PS (1985) Lipids of nervous tissue: composition and metabolism. Prog Lipid Res 24(2):69–176. doi:0163-7827(85)90011-6 [pii]
11. Riboni L, Acquotti D, Casellato R, Ghidoni R, Montagnolo G, Benevento A, Zecca L, Rubino F, Sonnino S (1992) Changes of the human liver GM3 ganglioside molecular species during aging. Eur J Biochem 203(1-2):107–113
12. Farooqui AA (1981) Metabolism of sulfolipids in mammalian tissues. Adv Lipid Res 18:159–202
13. Yu RK, Ledeen RW (1972) Gangliosides of human, bovine, and rabbit plasma. J Lipid Res 13(5):680–686
14. Kamerling JP, Vliegenthart JF (1975) Isolation and identification of 2-deoxy-2,3-dehydro-N-acetylneuraminic acid from the urine of a patient with sialuria. Eur J Biochem 56(1): 253–258
15. Ghidoni R, Sonnino S, Tettamanti G, Baumann N, Reuter G, Schauer R (1980) Isolation and characterization of a trisialoganglioside from mouse brain, containing 9-O-acetyl-N-acetylneuraminic acid. J Biol Chem 255(14): 6990–6995
16. Chigorno V, Sonnino S, Ghidoni R, Tettamanti G (1982) Isolation and characterization of a tetrasialoganglioside from mouse brain, containing 9-O-acetyl,N-acetylneuraminic acid. Neurochem Int 4(6):531–539. doi:0197-0186(82)90042-0 [pii]
17. Glebov OO, Nichols BJ (2004) Lipid raft proteins have a random distribution during localized activation of the T-cell receptor. Nat Cell Biol 6(3):238–243
18. Lin J, Shaw AS (2005) Getting downstream without a Raft. Cell 121(6):815–816
19. Karlsson KA (1970) On the chemistry and occurrence of sphingolipid long-chain bases. Chem Phys Lipids 5(1):6–43
20. Okerblom J, Varki A (2017) Biochemical, cellular, physiological and pathological consequences of human loss of N-glycolylneuraminic acid. Chembiochem. https://doi.org/10.1002/cbic.201700077

21. Marquina G, Waki H, Fernandez LE, Kon K, Carr A, Valiente O, Perez R, Ando S (1996) Gangliosides expressed in human breast cancer. Cancer Res 56(22):5165–5171
22. Riboni L, Sonnino S, Acquotti D, Malesci A, Ghidoni R, Egge H, Mingrino S, Tettamanti G (1986) Natural occurrence of ganglioside lactones. Isolation and characterization of GD1b inner ester from adult human brain. J Biol Chem 261(18):8514–8519
23. Avrova NF, Ghidoni R, Karpova OB, Nalivayeva NN, Malesci A, Tettamanti G (1986) Systematic position of fish species and ganglioside composition and content. Comp Biochem Physiol B 83(3):669–676
24. Terabayashi T, Kawanishi Y (1998) Naturally occurring ganglioside lactones in Minke whale brain. Carbohydr Res 307(3-4):281–290
25. Song Y, Kitajima K, Inoue S, Khoo KH, Morris HR, Dell A, Inoue Y (1995) Expression of new KDN-gangliosides in rainbow trout testis during spermatogenesis and their structural identification. Glycobiology 5(2):207–218
26. Sonnino S, Prinetti A, Mauri L, Chigorno V, Tettamanti G (2006) Dynamic and structural properties of sphingolipids as driving forces for the formation of membrane domains. Chem Rev 106(6):2111–2125. https://doi.org/10.1021/cr0100446
27. Sonnino S, Kirschner G, Ghidoni R, Acquotti D, Tettamanti G (1985) Preparation of GM1 ganglioside molecular species having homogeneous fatty acid and long chain base moieties. J Lipid Res 26(2):248–257
28. Masserini M, Freire E (1986) Thermotropic characterization of phosphatidylcholine vesicles containing ganglioside GM1 with homogeneous ceramide chain length. Biochemistry 25(5):1043–1049
29. Masserini M, Palestini P, Venerando B, Fiorilli A, Acquotti D, Tettamanti G (1988) Interactions of proteins with ganglioside-enriched microdomains on the membrane: the lateral phase separation of molecular species of GD1a ganglioside, having homogeneous long-chain base composition, is recognized by Vibrio cholerae sialidase. Biochemistry 27(20):7973–7978
30. Masserini M, Palestini P, Freire E (1989) Influence of glycolipid oligosaccharide and long-chain base composition on the thermotropic properties of dipalmitoylphosphatidylcholine large unilamellar vesicles containing gangliosides. Biochemistry 28(12):5029–5034
31. Terzaghi A, Tettamanti G, Masserini M (1993) Interaction of glycosphingolipids and glycoproteins: thermotropic properties of model membranes containing GM1 ganglioside and glycophorin. Biochemistry 32(37):9722–9725
32. Palestini P, Allietta M, Sonnino S, Tettamanti G, Thompson TE, Tillack TW (1995) Gel phase preference of ganglioside GM1 at low concentration in two-component, two-phase phosphatidylcholine bilayers depends upon the ceramide moiety. Biochim Biophys Acta 1235(2):221–230. doi:0005-2736(95)80008-4 [pii]
33. Goins B, Masserini M, Barisas BG, Freire E (1986) Lateral diffusion of ganglioside GM1 in phospholipid bilayer membranes. Biophys J 49(4):849–856. https://doi.org/10.1016/S0006-3495(86)83714-6. S0006-3495(86)83714-6 [pii]
34. Palestini P, Masserini M, Tettamanti G (1994) Exposure to galactose oxidase of GM1 ganglioside molecular species embedded into phospholipid vesicles. FEBS Lett 350(2-3):219–222. doi:0014-5793(94)00765-9 [pii]
35. Ferraretto A, Pitto M, Palestini P, Masserini M (1997) Lipid domains in the membrane: thermotropic properties of sphingomyelin vesicles containing GM1 ganglioside and cholesterol. Biochemistry 36(30):9232–9236. https://doi.org/10.1021/bi970428j. bi970428j [pii]
36. Acquotti D, Poppe L, Dabrowski J, Vonderlieth CW, Sonnino S, Tettamanti G (1990) 3-Dimensional structure of the oligosaccharide chain of Gm1 ganglioside revealed by a distance-mapping procedure – a rotating and laboratory frame nuclear overhauser enhancement investigation of native glycolipid in dimethyl-sulfoxide and in water dodecylphosphocholine solutions. J Am Chem Soc 112(21):7772–7778. https://doi.org/10.1021/Ja00177a043
37. Siebert HC, Reuter G, Schauer R, von der Lieth CW, Dabrowski J (1992) Solution conformations of GM3 gangliosides containing different sialic acid residues as revealed by NOE-based distance mapping, molecular mechanics, and molecular dynamics calculations. Biochemistry 31(30):6962–6971
38. Poppe L, van Halbeek H, Acquotti D, Sonnino S (1994) Carbohydrate dynamics at a micellar surface: GD1a headgroup transformations revealed by NMR spectroscopy. Biophys J 66(5):1642-1652. doi:https://doi.org/10.1016/S0006-3495(94)80956-7. S0006-3495(94)80956-7 [pii]
39. Jarrell HC, Jovall PA, Giziewicz JB, Turner LA, Smith IC (1987) Determination of conformational properties of glycolipid head groups by 2H NMR of oriented multibilayers. Biochemistry 26(7):1805–1811

40. Skarjune R, Oldfield E (1982) Physical studies of cell surface and cell membrane structure. Deuterium nuclear magnetic resonance studies of N-palmitoylglucosylceramide (cerebroside) head group structure. Biochemistry 21(13):3154–3160
41. Nyholm PG, Pascher I (1993) Orientation of the saccharide chains of glycolipids at the membrane surface: conformational analysis of the glucose-ceramide and the glucose-glyceride linkages using molecular mechanics (MM3). Biochemistry 32(5):1225–1234
42. Brocca P, Berthault P, Sonnino S (1998) Conformation of the oligosaccharide chain of G(M1) ganglioside in a carbohydrate-enriched surface. Biophys J 74(1):309–318. https://doi.org/10.1016/S0006-3495(98)77788-4. S0006-3495(98)77788-4 [pii]
43. Brocca P, Acquotti D, Sonnino S (1996) Nuclear overhauser effect investigation on GM1 ganglioside containing N-glycolyl-neuraminic acid (II3Neu5GcGgOse4Cer). Glycoconj J 13(1):57–62
44. Acquotti D, Cantu L, Ragg E, Sonnino S (1994) Geometrical and conformational properties of ganglioside GalNAc-GD1a, IV4Ga lNAcIV3Neu5AcII3Neu5AcGgOse4Cer. Eur J Biochem 225(1):271–288
45. Christian R, Schulz G, Brandstetter HH, Zbiral E (1987) On the side-chain conformation of N-acetylneuraminic acid and its epimers at C-7, C-8, and C-7,8. Carbohydr Res 162(1):1–11. doi:0008-6215(87)80195-7 [pii]
46. Poppe L, Dabrowski J, von der Lieth CW, Numata M, Ogawa T (1989) Solution conformation of sialosylcerebroside (GM4) and its NeuAc(alpha 2–3)Gal beta sugar component. Eur J Biochem 180(2):337–342
47. Brocca P, Acquotti D, Sonnino S (1993) 1H-NMR study on ganglioside amide protons: evidence that the deuterium exchange kinetics are affected by the preparation of samples. Glycoconj J 10(6):441–446
48. Acquotti D, Fronza G, Ragg E, Sonnino S (1991) Three dimensional structure of GD1b and GD1b-monolactone gangliosides in dimethylsulphoxide: a nuclear Overhauser effect investigation supported by molecular dynamics calculations. Chem Phys Lipids 59(2):107–125. doi:0009-3084(91)90001-R [pii]
49. Ha JH, Spolar RS, Record MT Jr (1989) Role of the hydrophobic effect in stability of site-specific protein-DNA complexes. J Mol Biol 209(4):801–816. doi:0022-2836(89)90608-6 [pii]
50. Bach D, Sela B, Miller IR (1982) Compositional aspects of lipid hydration. Chem Phys Lipids 31(4):381–394
51. Cantu L, Corti M, Sonnino S, Tettamanti G (1990) Evidence for spontaneous segregation phenomena in mixed micelles of gangliosides. Chem Phys Lipids 55(3):223–229
52. Regina Todeschini A, Hakomori SI (2008) Functional role of glycosphingolipids and gangliosides in control of cell adhesion, motility, and growth, through glycosynaptic microdomains. Biochim Biophys Acta 1780(3):421–433. https://doi.org/10.1016/j.bbagen.2007.10.008. S0304-4165(07)00240-1 [pii]
53. Sonnino S, Prinetti A (2013) Membrane domains and the "lipid raft" concept. Curr Med Chem 20(1):4–21. doi:CMC-EPUB-20121108-2 [pii]
54. Cantu L, Del Favero E, Sonnino S, Prinetti A (2011) Gangliosides and the multiscale modulation of membrane structure. Chem Phys Lipids 164(8):796-810. doi:https://doi.org/10.1016/j.chemphyslip.2011.09.005. S0009-3084(11)00330-6 [pii]
55. Prinetti A, Chigorno V, Prioni S, Loberto N, Marano N, Tettamanti G, Sonnino S (2001) Changes in the lipid turnover, composition, and organization, as sphingolipid-enriched membrane domains, in rat cerebellar granule cells developing in vitro. J Biol Chem 276(24):21136–21145. https://doi.org/10.1074/jbc.M010666200. M010666200 [pii]
56. Pitto M, Parenti M, Guzzi F, Magni F, Palestini P, Ravasi D, Masserini M (2002) Palmitic is the main fatty acid carried by lipids of detergent-resistant membrane fractions from neural and non-neural cells. Neurochem Res 27(7-8):729–734
57. Sonnino S, Prinetti A (2010) Lipids and membrane lateral organization. Front Physiol 1:153. https://doi.org/10.3389/fphys.2010.00153
58. Sonnino S, Prinetti A (2008) Membrane lipid domains and membrane lipid domain preparations: are they the same thing? Trends Glycosci Glycotechnol 20(116):315–340. https://doi.org/10.4052/Tigg.20.315
59. Bertoli E, Masserini M, Sonnino S, Ghidoni R, Cestaro B, Tettamanti G (1981) Electron paramagnetic resonance studies on the fluidity and surface dynamics of egg phosphatidylcholine vesicles containing gangliosides. Biochim Biophys Acta 647(2):196–202
60. Maggio B, Ariga T, Sturtevant JM, Yu RK (1985) Thermotropic behavior of glycosphingolipids in aqueous dispersions. Biochemistry 24(5):1084–1092
61. Cantu L, Corti M, Del Favero E, Muller E, Raudino A, Sonnino S (1999) Thermal hysteresis in ganglioside micelles investigated by differential scanning calorimetry and light-

scattering. Langmuir 15(15):4975–4980. https://doi.org/10.1021/La981355n

62. Koynova R, Caffrey M (1995) Phases and phase transitions of the sphingolipids. Biochim Biophys Acta 1255(3):213–236
63. Ostermeyer AG, Beckrich BT, Ivarson KA, Grove KE, Brown DA (1999) Glycosphingolipids are not essential for formation of detergent-resistant membrane rafts in melanoma cells. Methyl-beta-cyclodextrin does not affect cell surface transport of a GPI-anchored protein. J Biol Chem 274(48): 34459–34466
64. Iwabuchi K, Yamamura S, Prinetti A, Handa K, Hakomori S (1998) GM3-enriched microdomain involved in cell adhesion and signal transduction through carbohydrate-carbohydrate interaction in mouse melanoma B16 cells. J Biol Chem 273(15):9130–9138
65. Chigorno V, Palestini P, Sciannamblo M, Dolo V, Pavan A, Tettamanti G, Sonnino S (2000) Evidence that ganglioside enriched domains are distinct from caveolae in MDCK II and human fibroblast cells in culture. Eur J Biochem 267(13):4187–4197. doi:ejb1454 [pii]
66. Pascher I (1976) Molecular arrangements in sphingolipids. Conformation and hydrogen bonding of ceramide and their implication on membrane stability and permeability. Biochim Biophys Acta 455(2):433–451. doi:0005-2736(76)90316-3 [pii]
67. Hakomori S, Handa K, Iwabuchi K, Yamamura S, Prinetti A (1998) New insights in glycosphingolipid function: "glycosignaling domain," a cell surface assembly of glycosphingolipids with signal transducer molecules, involved in cell adhesion coupled with signaling. Glycobiology 8(10):xi–xix
68. Sonnino S, Aureli M, Loberto N, Chigorno V, Prinetti AFine tuning of cell functions through remodeling of glycosphingolipids by plasma membrane-associated glycohydrolases. FEBS Lett 584(9):1914–1922. https://doi.org/10.1016/j.febslet.2009.11.020. S0014-5793(09)00913-2 [pii]
69. Svennerholm L (1980) Ganglioside designation. Adv Exp Med Biol 125:11
70. IUPAC-IUB Joint Commission (1999) IUPAC-IUB Joint Commission on biochemical nomenclature (JCBN) nomenclature of glycolipids recommendations 1997. J Mol Biol 286(3):963–970. doi:S0022283698924858 [pii]
71. Sonnino S, Cantu L, Corti M, Acquotti D, Venerando B (1994) Aggregative properties of gangliosides in solution. Chem Phys Lipids 71(1):21–45. doi:0009-3084(94)02304-2 [pii]
72. Milhiet PE, Domec C, Giocondi MC, Van Mau N, Heitz F, Le Grimellec C (2001) Domain formation in models of the renal brush border membrane outer leaflet. Biophys J 81(1):547–555
73. Lawrence JC, Saslowsky DE, Edwardson JM, Henderson RM (2003) Real-time analysis of the effects of cholesterol on lipid raft behavior using atomic force microscopy. Biophys J 84(3):1827–1832

Chapter 2

Gangliosides of the Nervous System

Robert Ledeen and Gusheng Wu

Abstract

This review begins by attempting to recount some of the pioneering discoveries that first identified the presence of gangliosides in the nervous system, their structures and topography. This is presented as prelude to the current emphasis on physiological function, about which much has been learned but still remains to be elucidated. These areas include ganglioside roles in nervous system development including stem cell biology, membranes and organelles within neurons and glia, ion transport mechanisms, receptor modulation including neurotrophic factor receptors, and importantly the pathophysiological role of ganglioside aberrations in neurodegenerative disorders. This relates to their potential as therapeutic agents, especially in those conditions characterized by deficiency of one or more specific gangliosides. Finally we attempt to speculate on future directions ganglioside research is likely to take so as to capitalize on the impressive progress to date.

Key words Gangliosides, Ganglio-series, GM1 ganglioside, Ganglioside structures, Ganglioside topography in the nervous system, Gangliosides in nervous system development, Gangliosides in nervous system cells, Gangliosides in subcellular fractions, Ganglioside function in ion transport, Gangliosides in nervous system receptors, Gangliosides in neurodegenerative disorders

1 Early Milestones as Prologue to Function Elucidation

Glycans have come to be recognized as highly efficient vehicles for information storage, purportedly with greater coding capacity than the genetic code itself [1]. Sialic acid-containing glycoconjugates are recognized as major elements within this functional network, sialo-glycolipids, or gangliosides, being the lipid components with special relevance to the nervous system. This is attested in the fact that gangliosides are the major sialo-glycoconjugates of brain and of neurons in particular, as suggested in the high concentration known to exist in gray matter. Indeed in the course of his groundbreaking discoveries, Klenk coined the term "ganglioside" in reference to what appeared to be a special, perhaps unique relation of these substances to "ganglionzellen" (neurons) [2]. Gangliosides were first detected by Klenk in small amounts as a new substance in the brain of a Niemann–Pick

Sandro Sonnino and Alessandro Prinetti (eds.), *Gangliosides: Methods and Protocols*, Methods in Molecular Biology, vol. 1804, https://doi.org/10.1007/978-1-4939-8552-4_2,

© Springer Science+Business Media, LLC, part of Springer Nature 2018

patient [3] and later in much larger amounts from a Tay–Sachs brain where it was shown to be a new and until then unknown glycosphingolipid [4]. Similar substances were obtained in lesser amounts from normal human brain [2] and shown to contain sialic acid. This glycan, also known as neuraminic acid, has the pyranose conformation corresponding to 5-amino-3,5-dideoxy-d-*glycero*-D-*galacto*-non-2-ulopyranosonic acid (Fig. 1); it exists in multiple substituted forms [5] which, in humans, are limited to the 5-N-acetyl- and 5-N-acetyl-9-O-acetyl derivatives but are more diverse in other species. Sialic acid became the identifying unit for this glycosphingolipid category, setting it apart from neutral and sulfated glycosphingolipids which in toto are more numerous. The idea of gangliosides as unique to neurons or even brain was discarded with the discovery of "hematoside," a triglycan structure (II^3Neu5AcLacCer) later named GM3, in horse erythrocytes by Yamakawa and Suzuki [6] followed by numerous reports indicating ganglioside presence in virtually every tissue and body fluid of vertebrates. Sialic acid-containing glycosphingolipids have also been characterized in some invertebrates (e.g., echinoderms) [7] but of significantly different structures from those found in vertebrates.

2 Structures and Topography

Klenk and coworkers reported several structural features of these new substances from human brain but it was Kuhn and Wiegandt [8] who first described the full carbohydrate structure of the major monosialoganglioside of brain (Fig. 1). This ganglioside received several symbolic designations by different laboratories in that period but it was Svennerholm's nomenclature, GM1, that eventually prevailed, as exemplified in approval by IUB-IUPAC [9]. That symbol first appeared in a landmark study [10] in which Svennerholm demonstrated the utility of thin-layer chromatography (TLC) in depicting (and potentially isolating) the four major and two minor gangliosides of brain. Following their initial report Kuhn and Wiegandt described the other three major gangliosides of brain [11] as did Klenk and Gielen in part [12], all belonging to the gangliotetraose (ganglio-)series and subsequently receiving Svennerholm nomenclature (GD1a, GD1b, GT1b) (Fig. 1). This was followed by characterization of a growing list of minor brain gangliosides [13, 14] including GM2, the one originally discovered by Klenk in Tay–Sachs brain and later observed as a minor species in normal brain [15]; this was shown to have the same oligosaccharide structure as GM1 while lacking the terminal galactose (Gal) [16]. The pace of ganglioside discovery then accelerated and became a truly international undertaking which continues to

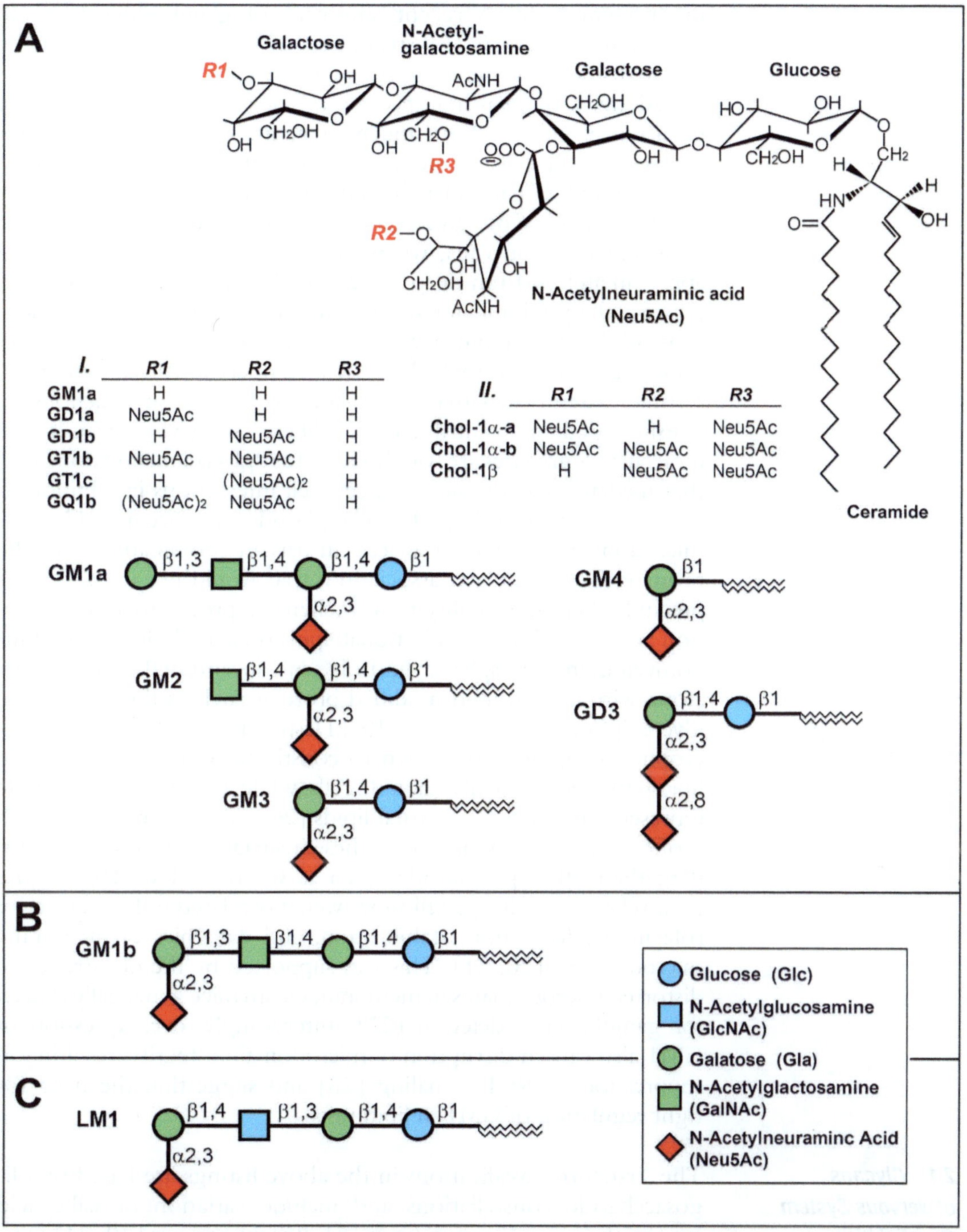

Fig. 1 Structures of gangliosides. (**A**) *I.* Gangliotetrarose (ganglio-series) gangliosides. *II.* Alpha series. Lower portion shows abbreviated depictions of GM1a and four minor gangliosides of brain. (**B**) 0-Series: GM1b. (**C**) PNS ganglioside: LM1

the present time. A recent summary of gangliosides from all vertebrate tissues lists 188 structures [17], a list that will likely be updated since it was compiled approximately a decade ago.

Along with growing recognition of intracellular ganglioside occurrence and function, much of ganglioside research remains focused on the plasma membrane where they are generally most concentrated, in particular the outer half of the bilayer. There they constitute part of the glycan-rich glycocalyx, along with glycoproteins and proteoglycans, where they occupy a shallow region of approximately 40 nm-deep closest to the cell surface. Moreover, they are the predominant sialoglycans, an attribute unique to neurons. Within the lipid bilayer gangliosides occur to a large extent in sphingolipid-rich rafts which are defined operationally as detergent-resistant assemblages that are isolated as low-density fractions by density gradient centrifugation. These somewhat transitional molecular aggregates are widely viewed as "glycosignaling domains" that mediate cell adhesion and cell–cell interactions in addition to transmembrane signaling [18]. Gangliosides that occur within such microdomains likely exist at least in part in combination with the proteins they modulate, associations that would not in most cases preclude their accessibility to histochemical probes used for detection and identification. Additional lipids such as cholesterol, sphingomyelin, and phosphatidylinsitols are concentrated in such rafts, along with certain lipid-modified proteins such as those linked to glycosylphosphatidylinositol [19]. Despite these enrichments glycerophospholipids were shown to constitute over 50% of lipids in lipid rafts from cerebellar granule cells, while gangliosides as a whole represented only 6–7% of total lipids [20]. Moreover GM1, which is often considered a marker for these microdomains, was less abundant than other gangliosides such as GD1a and GT1b. Despite their relative paucity, gangliosides were postulated to have an active role in the formation, stabilization, and dynamics of membrane microdomains [20, 21]. This was supported by the occurrence of distorted microdomains in membranes from mice genetically altered for ganglio-series deletion [22]. Interestingly, overexpression of GM1 also caused disruption of microdomain integrity, resulting in suppression of NGF signaling [23] and suggesting the need for tight regulation of GM1 in lipid rafts.

2.1 Glycans of Nervous System Gangliosides

The structure classifications in the above listings are based on oligosaccharide constellations and include variation in sialic acid among the several known types [5]. Recent reviews [20, 24] indicate some 30 of these gangliosides that have been characterized in mammalian nervous systems. These depictions fail to take into account the ceramide unit which, although less variable than the glycans, does manifest some structural diversity (see below). The above lists reveal that the overwhelming majority of mammalian nervous system gangliosides belong to the ganglio-series (also termed "ganglio-4" series; refs. 20, 24) the prototypic component of which is GM1. The latter is also termed GM1a to denote the a-subseries containing a single sialic acid on the internal Gal and the possibility of one or more on terminal Gal (e.g., GD1a). A disialosyl unit attached to internal Gal denotes the b-subseries and trisialosyl at that position the c-subseries; the latter subseries arose from discovery of polysialo-gangliosides of that type in fish brain [25, 26], some of which are also found in mammalian brains. Attachment of sialic acid to terminal-Gal rather than internal-Gal of the gangliotetraose unit gives rise to GM1b (Fig. 1), prototype of the O series which lacks sialic acid linked to internal Gal. Another isomer of GM1 was identified with sialic acid joined (α2–6) to *N*-acetylgalactosamine (GalNAc) and the designation GM1α [27] to denote its membership in the α-series (sialic acid linked to GalNAc). The other series found in brain include galacto (GM4), lacto (GM3, GD3, O-Acetyl-GD3), and neolacto (e.g., 3′-LM1; Fig. 1). These non-ganglio types constitute minor gangliosides of mammalian nervous systems in most cases, as pertains also to GQ1b, a tetrasialo form (Fig. 1). Penta-, hexa-, and even hepta-sialo gangliosides have been found in lower vertebrates [17].

In relation to carbohydrate structures, nervous system ganglioside patterns are known to vary among vertebrate species as seen in the preponderance of tetrasialogangliosides and pentasialogangliosides in the brains of lower compared to higher vertebrates [25, 26]. There are small pattern variations among mammals which generally display the four major ganglio-series (Fig. 1) as the predominant forms in the CNS. The peripheral nervous system (PNS) contains those in addition to substantial levels of the glucosamine-containing 3′-LM1 (Fig. 1; IV^3nLc_4Cer) [28], most of that occurring in PNS myelin (see below). Minor variations can exist, however, as in the presence of N-glycolylneuraminic acid (Neu5Gc)-containing gangliosides in some mammalian brains but not in human [29]. The sialic acid of GM1 is somewhat unique in being resistant to most forms of mammalian sialidase. This is a device of nature for elevating tissue levels of GM1 when required via endogenous sialidases [30], especially the Neu3 form which is specific for gangliosides and colocalized with substrates (e.g., GD1a) in the membrane [31]. This facilitates the numerous modulatory and signaling reactions which GM1 mediates in the nervous system and elsewhere [24, 32]. The carboxyl group of sialic acid exists predominantly in ionic form in biosystems, which adds to its hydrophilicity and facilitates charge–charge interactions. An interesting departure from the free carboxyl motif was seen in the GD1b lactone in which the carboxyl of the terminal sialic acid forms an inner ester with the C9-hydroxyl of the internal sialic acid, the carboxyl of the latter remaining ionic [33]. This form, whose function is unknown, is present in only trace amounts in infant brain but increases to 3.5% of sialic acid in older subjects. Another example lacking free carboxyl was that of GM2 joined in amide linkage to taurine [34], again of unknown function.

2.2 Ceramide Structures in Nervous System Gangliosides

The presence of ceramide, the hydrophobic component of gangliosides, in conjunction with glycan units accounts for the amphiphilic character of these molecules and their facile solubility in both water and certain organic solvents. Ceramide contains two long-chain hydrophobic units, a long-chain sphingoid base (e.g., C18 sphingosine) and a long-chain fatty acid—predominantly C18 stearoyl—which are joined in amide linkage (Fig. 1). This serves as the anchoring unit of sphingolipids within the lipid bilayer and also contributes to association of gangliosides with specific proteins and other lipids. The double bond of sphingosine has the trans configuration and the asymmetric carbons at C2 and C3 have the 2S, 3R configurations [35]. A unique feature of brain gangliosides is the copresence of C20 sphingosine, this form being absent prior to differentiation and then increasing progressively with development [36]. The functional significance of this change is not known but it was hypothesized that the varying ratio between C18- and C20-sphingosine-containing gangliosides during CNS development can be instrumental in modulating membrane domain organization [36]. The "sphingolipidome" in toto has been considered in relation to metabolic pathways "...that account for the tens of thousands of molecular subspecies (and perhaps more) that appear in nature" [37]. This astonishing assessment includes consideration of both ceramide and glycan variations and suggests the need for recognition of a possible lipidic code along with the sugar code for a more complete understanding of ganglioside reactivities. Semisynthetic GM1 analogs with altered ceramide structures have helped to reveal the functional participation of that unit (see below).

3 Overview on Functional Roles of Nervous System Gangliosides

One can view this large and growing array of nervous system ganglioside structures in the context of the enormous informational power of the sugar code and its major role in regulating what is likely the most complex biological system in nature. The requirement that macromolecules assume and maintain their correct tertiary and quaternary structures becomes evident in the prevalence of pathologies based on misfolded proteins, especially in the realm of neurological disorders (see below). The multifaceted roles of glycosphingolipids in promoting the correct conformation of individual proteins or aggregates is becoming well recognized, a notable example being the role of GM1 in maintaining the helical, non-aggregating conformation of α-synuclein, a protein that suffers misfolding in the absence of sufficient GM1—as occurs in Parkinson's disease [38, 39] (see below). Some protein components of lipid rafts have glycolipid-binding domains to which GM1 and other glycosphingolipids attach [40] and such domains for GM1 have been identified in β-amyloid peptide [41] and α-synuclein [42], proteins of primary importance in neurological diseases (see below). An example of ganglioside modulation of quaternary structure is GM1 promotion of Ret and GFRα1 coalescence, the protein components of the receptor for glial cell-line-derived neurotrophic factor (GDNF) [43]. Such stereochemical modulations might well involve charge–charge interactions as would be supplied by the sialic acid(s) of gangliosides, as seen in the ability of GM1 to modulate a delta-opioid receptor with positively charged amino acid in a key position but failure to do so in the absence of that charge [44]. These and other examples point to another form of functional diversity: the ability of a single ganglioside—e.g., GM1—to interact with and modulate an array of neural proteins [24, 32]. Gangliosides might themselves be the target of reactive proteins, as in the cross-linking of membrane GM1 by galectin-1 to induce Ca influx via TRPC5 channels ([45] and below). The numerous gangliosides in the nervous system and elsewhere attest to the extent of structural tailoring and fine tuning required for the broad array of glycolipid–protein interactions in nature. The majority of such interactions to be reviewed here are in the cis-mode between gangliosides and target proteins in the same membrane, but the same general principles apply to trans-interactions such as those involving ganglioside mediation of cell–cell interactions [18, 46].

The introduction of semisynthetic GM1 analogs with altered ceramide structure by Costa and coworkers proved informative in showing that such alteration did not diminish GM1-mediated activity; in fact replacement of the stearoyl unit with either acetyl (LIGA4) or dichloroacetyl (LIGA20) produced analogs that were more potent than GM1 itself in protecting primary cultures of cerebellar granule neurons from glutamate-induced cell death [47]. Similarly LIGA20 proved considerably more potent than GM1 in blocking kainate-induced seizure activity [48] and Parkinsonian symptoms [49] in ganglio-series-deficient mice (see below). This enhanced neuroprotective activity was initially attributed to prevention of protein kinase C translocation to the membrane [47] while a later interpretation cited greater membrane permeability of the semisynthetic analogs [48]. This was seen as providing greater penetration into brain and access to intraneuronal loci functionally associated with GM1 (e.g., Ca regulation—see below). The increased membrane permeability of these analogs suggested one function of ceramide to be facilitation of compact topology among membrane-associated sphingolipids which is diminished by such structural changes. The results also suggest that at least in these cases the correct ceramide structure was not critical in the related GM1 modulatory effects and the altered structure proved adequate; however, examples where it is critical are also known (see below). The importance of long-chain aliphatic chains in GM1 ceramide has been described in relation to transmembrane interdigitation and alteration of membrane leaflet phase behavior [50].

4 Gangliosides in Nervous System Development and Stem Cell Biology

A detailed study of ganglioside changes in embryonic rat brain revealed predominance of GM3 and GD3 at E14, appearance of b-series gangliosides at E16 and a-series at E18 [51]. A detailed study of glycosphingolipids and glycogenes in embryonic and postnatal mouse brain provided more detail including the transitory appearance of GT3 and GQ1c in embryo [52]; GM3 and GD3 were barely detectable after E16 while expression of GD1a and eventually GM1 gradually became more evident leading into early postnatal development. The complex b-series gangliosides (GD1b, GT1b) followed a similar course. Interestingly, Chol-1-α-a, a cholinergic-specific ganglioside (see below), was found only after birth. These dramatic changes in qualitative and quantitative expression of brain gangliosides are regulated through stage-specific expression of glycosyltransferase genes and these in turn are strongly influenced by histone acetylation. As a form of epigenetic activation, acetylation of histones H3 and H4 of two genes encoding regulatory glycosyltransferases caused gradual enhancement of transferase activities involved in the synthesis of GM2 and GD3 [53]. Activation of *B4galnt1*, the gene encoding GalNAcT (GM2/GD2 synthase) in this manner promoted neural differentiation in primary neuroepithelium culture following exogenous supplementation with GM1. This led to the significant discovery that nuclear GM1 is associated with gene regulation by interacting with chromatin via acetylated histones [54]. Once in the membrane GM1 associates with α5β1 integrin to promote axonogenesis in neuroblastoma cells and cerebellar granule neurons when cross-linked with galectin-1, a process that induces intracellular signaling and transient increase in intracellular Ca by activation of TRPC5 channels [45]. This mechanism is shown in Fig. 2 including depiction of astrocytes as source of Gal-1. Other work has documented galectin-1 as a major receptor for GM1 in neuronal differentiation and indicated the role of sialidase in elevating membrane GM1 to the level required for effective cross-linking [55].

The accumulation of unmetabolized ganglioside in lysosomal storage disorders such as Tay-Sachs disease and GM1 gangliosido-

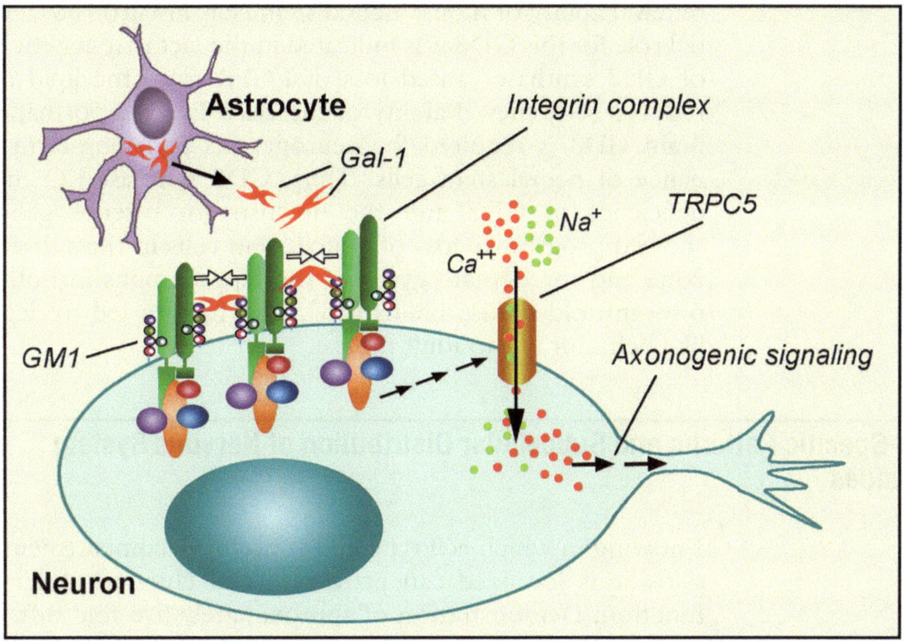

Fig. 2 Galectin-1 (Gal-1) was shown to be an effector of axonogenesis in cerebellar granule neurons (CGNs) and NG108–15 cells by cross-linking GM1 ganglioside and its associated glycoprotein $\alpha_5\beta_1$-integrin. The resulting signaling led to a transient increase in intracellular Ca^{2+} by opening TRPC5 channels. CGNs deficient in GM1 showed retarded axonogenesis, underscoring the relevance of GM1 as functional counterreceptor for Gal-1 in this process. This Gal-1/GM1-induced signaling was manifest at the earliest, initiating stage of axon development. Reproduced from ref. 45 with permission from the Journal of Neurochemistry

sis was shown to induce ectopic neurite outgrowth [56, 57], leading to the hypothesis that gangliosides can serve as a molecular switch for neuronal differentiation. This appeared to receive support in the neuritogenesis that was triggered by GM1 in primary neurons and neuro-2A neuroblastoma cells [58, 59]. However, this idea required reconsideration with the demonstration of nonspecificity in a study showing that all of 11 gangliosides tested proved neuritogenic [60] as did a pseudo-ganglioside (sialic acid attached to diglyceride) [61]. Working with neuroblastoma cells it was shown that neurites induced by exogenous GM1 were dendritic in character while projection of axon-like neurites was dependent on Ca influx [62]; an example of the latter is the physiologically relevant process triggered by Gal-1-induced cross-linking of GM1 and resultant TRPC5 channel activation [45].

Identification of ganglioside markers in stem cells indicated their involvement in this earliest stage of neural development. Several such gangliosides have been detected in both embryonic and somatic stem cells including GM3, GD3 and several members of the ganglio-series [63, 64]. GD3 constitutes 80% or more of total gangliosides in mouse neural stem cells and was shown to colocalize and interact with the EGF receptor to sustain the self-renewal ability of mouse neural stem cells in vitro [65]. An essential role for this GD3 was indicated in the fact that genetic deletion of GD3 synthase caused loss of EGFR from the lipid rafts and reduced self-renewal ability of the cells. Even in postnatal mouse brain GD3 is required for neurogenesis and long-term maintenance of neural stem cells. Thus, GD3 synthase-KO mice with deficiency in GD3 and the downstream b-series gangliosides showed progressive loss of neural stem cells in the subventricular zone and the dentate gyrus of the hippocampus in both 1- and 6-month-old mouse brains [66]. This in turn led to depression-like behavior in the adult mouse.

5 Cell-Specific Patterns and Subcellular Distribution of Nervous System Gangliosides

Knowing in which cell type and subcellular compartments a ganglioside is localized can provide useful clues in understanding function. Demonstration of approximately five-fold difference in dry weight concentration between cerebral cortex and white matter of human adults [67] pointed to regional and therefore cellular differences in quantitative and qualitative occurrence of nervous system gangliosides. The multiplicity of cell types in the nervous system indicated the need for cell fractionation procedures which, together with cell culture methodologies permitted identification of distinctive cellular patterns. Cellular detection of gangliosides has been aided by immunohistochemistry employing ganglioside-specific antibodies [68] or, in the case of GM1, CtxB which binds with high affinity and relative specificity to this ganglioside [69]. Problems with the use of antibodies include ambiguous antibody specificities, disruption via tissue fixation, detergent-mediated redistribution, etc. An effort to overcome at least some of these problems gave rise to a study of differential distribution of major brain gangliosides in mouse brain employing new IgG-class specific monoclonal antibodies [70]. The results were interpreted in part in relation to the hypothesis that GD1a and GT1b act as receptors for the axon-myelin recognition protein, myelin-associated glycoprotein (MAG). A relatively new development is imaging mass spectrometry which, among other applications, was used to detect subtle ceramide-based differences in the distribution of ganglioside subspecies in the hippocampal cell layers [71].

5.1 Comparison of Ganglioside Patterns in CNS Neurons and Glia

Initial studies with primary cultures of rat embryo CNS containing both neuronal and glial elements revealed synthesis of monosialogangliosides and polysialogangliosides [72, 73]. Use of a culturing method specific for neurons demonstrated development of a pattern that included hexosamine-containing polysialogangliosides similar to that of adult brain with predominance of the four major ganglio-series species [74]. This was in contrast to cultured astroglia from rat and chick brain which contained primarily GM3 and GD3 and relatively little hexosamine-containing species, consistent with a very low level of GalNAc-transferase (GM2/GD2 synthase) [75]. Similar study of short-term and long-term cultured rat brain astrocytes revealed predominantly GM3 with no detectable GM1 [76]. These results appeared to conflict with a common finding in other studies based on bulk isolation that astrocytes and neurons displayed similar ganglioside patterns resembling that of whole brain [77–80]. A possible explanation of this discrepancy was suggested in a study employing in vivo labeling via intracerebral injection of [^{3}H]hexosamine followed by cell isolation in which only the neuronal gangliosides were rapidly labeled; those of the astrocytes acquired radioactivity slowly over several hours [81]. This suggested selective synthesis of ganglio-series gangliosides in neurons followed by gradual transfer to astrocytes. This hypothesis was consistent with the reported isolation of glycolipid transfer proteins from brain with activity toward GM1 and the other gangliotetraose species [82]. The significantly higher level of gangliosides (relative to protein) in bulk-isolated astrocytes was attributed to the larger surface area in those cells compared to bulk-isolated neuronal perikarya.

5.2 Gangliosides in Specific CNS Neurons

Although all CNS neurons appear to possess the same major (and to some extent minor) gangliosides, variation in relative concentrations have been noted. Early on, different mouse mutants varying in loss of specific cerebellar neurons suggested that GD1a was enriched in granule cells and GT1a in Purkinje cells [83]. A, striking example of pattern specificity was identification of gangliosides specific to cholinergic neurons, one designated Chol-1β detected in both the electric organ of *Torpedo marmorata* and mammalian brain and found to be a trisialoganglioside similar to GD1b with the third Neu5Ac postulated to occur on GalNAc [84]. A second trisialo member of the Chol-1α family, Chol-1α-a, was later identified as GD1a with the third Neu5Ac attached at C6 of GalNAc [85]. Finally a third member of the Chol-1α family, designated Chol-1α-b, was obtained in 1.3 mg quantity from 5 g of total bovine brain ganglioside and shown to be a tetrasialo-ganglioside with Neu5Ac linked to C6 of GalNAc of GT1b [86]. Isolation of this succeeded through use of a new procedure employing a Q-Sepharose column, shown useful for isolating extremely minor gangliosides such as this and GM1b from brain [87]. A pivotal role for Chol-1α was indicated in the inhibition of acetylcholine release by antibody to this ganglioside which also suppressed memory and learning ability in rats [88]. Conceivably additional gangliosides specific to other neurotransmitter-associated neurons will be detected in due course.

5.3 Gangliosides in CNS Myelin and Oligodendrocytes

The glial cells whose primary function is myelinogenesis are the oligodendroglia in the CNS and Schwann cells in the PNS. Myelin produced by these cells in the CNS contain distinct albeit differing ganglioside patterns among vertebrate species. In general, GM1 is the major ganglioside of mammalian CNS myelin as seen first in the rat [89] and subsequently in humans [90] and several other mammals [91]. GM4 (sialosylgalactosyl ceramide) was identified as a specific marker and major ganglioside of human CNS myelin [90, 92], while aquatic vertebrates had virtually no detectable GM4 and only limited levels of GM1 [91]. Somewhat unexpectedly, avian CNS myelin, like human, had both GM1 and GM4 as major gangliosides [91]. On the other hand, bulk-isolated rat and bovine oligodendroglia cell bodies had minor or trace amounts of GM1 and GM4 [93] in contrast to those of humans [92]. An NMR study revealed interaction of GM1 with myelin basic protein (MBP), specifically with the unique N^G-methylarginine and associated amino acids in a manner suggesting determination of MBP conformation [94]. GM4 can also interact with MBP [95], pointing to the rough equivalence of total monosialogangliosides with MBPs in CNS myelin as indication of conformational interaction. This suggests the myelin occurrence of GM1 alone vs GM1 + GM4 is determined by the specific MBP isoform(s) that is (are) present with which they interact. This is somewhat analogous to the approximate molar equivalence of monosialogangliosides and MBP in PNS myelin (see below).

5.4 Gangliosides of Microglia

Microglia constitute 10–15% of human CNS cells and have major roles in disease processes especially those involving CNS inflammation. Gangliosides have not as yet been well characterized in such cells to our knowledge, although GD3 levels were shown to increase upon inflammation-induced activation [96]. Release of such GD3 was shown to induce apoptosis in oligodendrocytes [97], cells heavily affected by neuroinflammation.

5.5 Gangliosides of the PNS

These have been subjected to less intensive scrutiny than those of the CNS and in some cases neuronal and glial patterns were analyzed jointly. One study of human motor and sensory nerves revealed whole nerves to contain relatively high levels of GM3, GD3, and LM1 in addition to GM1 and other ganglio-series species [28]. The neolacto forms (e.g., LM1) are localized mainly in myelin and PNS myelin-producing (Schwann) cells from motor and sensory nerves; the whole nerves also contain substantial amounts of GM1 and other ganglio-series forms [98]. Ganglio-series gangliosides predominate in PNS axons which contain substantially less LM1, suggesting a somewhat similar pattern in PNS neurons as for CNS neurons. Dorsal root ganglia of the rat contain in addition to the 4 major species of brain, substantial levels of

GM3 and LM1; the latter was a minor ganglioside at birth and increased with maturation [99]. Myelin from both human and bovine peripheral nerves contained less than half the ganglioside levels found in CNS myelin, while this too was predominantly GM1; in bovine this was a mixture of GM1a(Neu5Ac) and GM1a(Neu5Gc) while in human it was GM1a(Neu5Ac) alone [100]. No GM4 was detected in PNS myelin. The molar concentration of GM1 in both types of PNS myelin was found to be roughly equivalent to that of the P_1 form of MBP, both of which are substantially reduced in quantity compared to their levels in CNS myelin [100]. As with CNS myelin, this equivalence supports association of GM1 with MBP.

5.6 Gangliosides in the Neuronal Membrane

An early fractionation study of whole rat brain reported the outer membrane of synaptosomes to contain the highest concentrations of gangliosides, particularly disialo and trisialo forms, with somewhat less in microsomes and least in mitochondria [101]. Additional studies reported ganglioside enrichment in synaptic plasma membranes (SPM) with polysialogangliosides predominating over GM1 [102, 103]. Such findings tended to reinforce a widespread belief of that time that gangliosides are concentrated at the nerve ending, an idea that was disputed by the calculated contributions of gangliosides measured in isolated synaptosomes [104]. The similar ganglioside concentrations and patterns observed in microsomes and SPM [105] supported the concept that gangliosides are distributed over large portions of the neuron. This was supported by studies with peroxidase-linked cholera toxin subunit [106] and observation of a ganglioside pattern in an axolemma-enriched fraction roughly equivalent to that of SPM [107]. The origin of such gangliosides in both axon and nerve ending was shown to be rapid axonal transport [108] and an analogous mechanism—retrograde transport—accounted for their return to the cell body for reutilization or metabolic processing [109]. Study of synaptic membranes isolated from Torpedo electric organ revealed a clear difference in gangliosides between the presynaptic and postsynaptic elements, the former being much higher in total ganglioside content than the latter. The presynaptic membranes contained a paucity of monosialogangliosides and the virtual absence of GM1 while the postsynaptic membranes did contain these [110]. During development gangliosides were detected at high levels in isolated growth cone membranes in a pattern similar to that of mature SPM [111].

5.7 Gangliosides in Synaptic Vesicles

Related to the above findings was the controversial issue of ganglioside presence in synaptic vesicles (SV), some studies reporting positive findings and others suggesting their absence. This matter addressed the question of an exocytosis-endocytosis cycle for the SV during neurotransmitter release which would be indicated by similar lipid compositions in the SV and SPM. One of the more extensive studies of fractions from rat brain showed similar lipid compositions between SV [112] and SPM [102] but not in the case of gangliosides which were reportedly much lower in the SV. Given the problems involved at that time in isolating and quantifying small quantities of gangliosides, the issue appeared to be unresolved with respect to mammalian vesicles. However, improved methods for glycolipid isolation and quantification plus the ability to obtain SV in high quantity and purity from elasmobranch electric organs led to demonstration of high levels of gangliosides in SV of both *Torpedo californica* and *Torpedo marmorata* [113]. Thin-layer chromatography revealed a predominance of polysialogangliosides in keeping with the patterns for lower vertebrates [25, 26]. These findings pertained to cholinergic SV while bovine noradrenergic SV were also shown to contain measurable ganglioside [114] and a high level was found in bovine chromaffin granules [115].

5.8 Gangliosides in the Neuronal Nucleus

Adding to the growing awareness of ganglioside presence and function in loci other than the plasma membrane were studies directed to the nucleus, particularly of neurons. Study of two distinct fractions of large (N1—presumably neuronal) and small (N2—presumably glial) nuclear populations from rat brain revealed close to four times higher levels of gangliosides in the former with identification of both a- and b-series and lesser amounts of GM3 and c-series gangliosides [116]. Specific focus on the ganglio-series revealed GM1 and GD1a to undergo marked elevation in the neuronal membrane during differentiation [117] together with development of a Ca transport system involving a Na/Ca exchanger (Fig. 3); the same was observed in astroglial nuclei (see below). A new dimension to the nuclear ganglioside story appeared with the revelation that GM1 occurs in the chromatin of neurons where it influences gene expression in relation to promotion of GM1 synthesis [54]. This interaction results in epigenetic activation of GalNAcT (GM2/GD2 synthase) which occurs through binding of GM1 to acetylated histones on the promoters of the *B4galnt1* and *NeuroD1* genes. Interestingly, GM1 was bound to those promoters in differentiated neurons but not in undifferentiated neuronal precursor cells. This epigenetic function of GM1 may eventually explain some of the neuroprotective properties attributed to this ganglioside. It remains to be determined whether this chromatin function of GM1 occurs in glial cells.

5.9 Gangliosides in Mitochondria and Endoplasmic Reticulum

As mentioned, fractionation studies based on differential centrifugation revealed low levels of gangliosides in brain mitochondria [101] whereas an advanced isolation procedure employing ficoll and continuous sucrose gradient yielded mitochondria reportedly

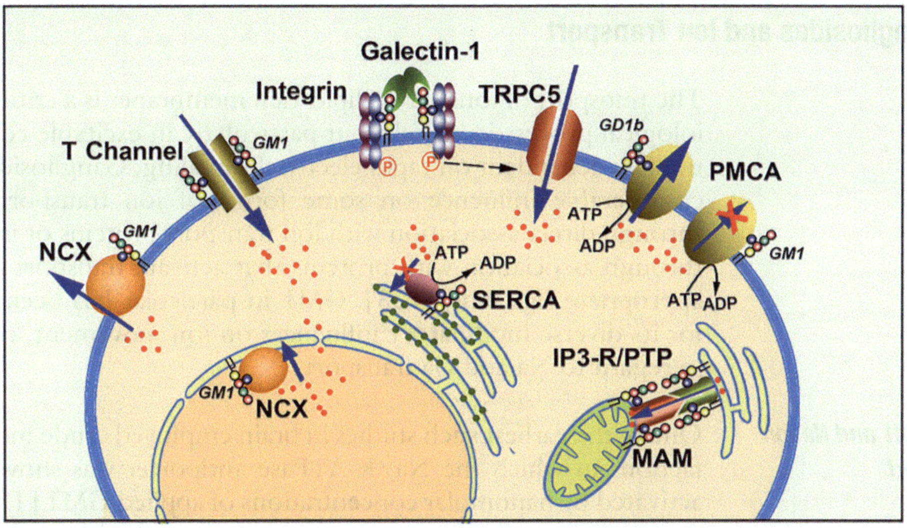

Fig. 3 Summary of ganglioside modulatory roles for neuronal Ca^{2+}. For the plasma membrane, Ca^{2+} influx is promoted by GM1 via T type channels and by GM1 association with alpha5-beta1 integrin, cross-linking of which by galectin-1 leads to TRPC5 channel activation. Calcium efflux is influenced positively in the plasma membrane by GM1 association with Na^{+}/Ca^{2+} exchanger (NCX) and negatively by GM1 association with plasma membrane Ca^{2+}-ATPase (PMCA); Ca^{2+} efflux via PMCA is promoted by GD1b. Intracellular mechanisms include GM1 association with NCX in the inner nuclear membrane that mediates transfer of Ca^{2+} from nucleoplasm to the nuclear envelope (and hence the ER); GM1 also inhibits the SERCA pump. GM1 in the ER influences Ca^{2+} flux from that organelle to mitochondria via the mitochondrial associated membrane (MAM) with binding to the IP3-receptor

devoid of gangliosides [118]. However, an important interaction has been well demonstrated as part of an apoptotic mechanism in which newly synthesized GD3 that accumulates in cells undergoing ceramide-dependent apoptosis causes dissipation of the mitochondrial transmembrane potential due to opening of the permeability transition pore; this in turn causes release of cytochrome c, caspase 9, and apoptosis-inducing factor [119]. The endoplasmic reticulum (ER) is the locus of early stages of ganglioside synthesis (*see* Sandhoff Chapter 5) and is continuous with the outer membrane of the nuclear envelope which has been shown to contain gangliosides [120]. This observation plus the fact that GM1 interacts with the phosphorylated form of inositol 1,4,5-triphosphate (IP3) receptor-1 [121, 122] suggested ganglioside presence throughout the ER membrane though this remains to be more fully substantiated. GM1 was shown to accumulate in the mitochondria-associated ER membranes in a mouse model of GM1-gangliosidosis [122], raising the possibility the same may apply in normal brain. GM1 appears to be involved in Ca transport as part of the ER stress mechanism (see below).

6 Gangliosides and Ion Transport

The transport of ions across lipid-rich membranes is a critical physiological process in all cells but particularly in excitable cells functionally dependent on rapid electrical signaling. Gangliosides exert considerable influence on some forms of ion transport, either through direct association with ion transport proteins or indirectly through association with proteins that activate transport through appropriate signaling [123]. GM1 in particular has been studied for its diverse modulatory influences on ion movement, especially in regard to Na and Ca transport.

6.1 GM1 and Na Ion Transport

One of the earliest such studies in brain employed crude membrane factions in which the Na/K-ATPase antiporter was shown to be activated by nanomolar concentrations of applied GM1 [124]. The fact that this modulatory property was shared with several other ganglio-series gangliosides suggested this might be pharmacological rather than physiological in nature, or possibly the effect of sialidase in the membrane that converted the polysialo-gangliosides to GM1. The authors suggested the multimolecular effect was due to modifications of the membrane lipid environment surrounding the enzyme. Voltage-gated Na channels in nodes of Ranvier were shown to suffer damage by complement-mediated disruption through anti-GM1 antibodies [125], while more direct evidence based on association of GM1 with Na_v channels was suggested in the observation that current densities of both tetrodotoxin-sensitive and insensitive Na channels were significantly decreased by CtxB [126]. Evidence also came from observation of broadened Na_v channel clusters in genetically altered mice lacking GM1 and the ganglio-series [127].

6.2 GM1 and Ca Regulation at the Plasma Membrane

The numerous developmental and functional roles of Ca within neural cells, including its pivotal role in neurotransmission, require precise regulation of this ion so as to maintain its intracellular concentration at one ten-thousandth or less of the extracellular level. GM1 through association with a diversity of regulatory proteins has a major role in these processes including controlled Ca influx and extrusion across the membrane as well as movement between intracellular loci (Fig. 3). The importance of GM1 in regulating neuronal Ca homeostasis was revealed by in vitro studies of cerebellar granule cells exposed to toxic doses of glutamate [128] and similar cells from mice lacking ganglio-series gangliosides subjected to depolarizing levels of K [129]. Calcium homeostasis was restored by application of GM1 and/or membrane-permeable analogs of GM1 (LIGA4, LIGA20) [128]. One influx mechanism involves GM1 activation of low-threshold voltage-dependent T type channels, a process activated by ganglioside-specific sialidase, Neu3, by virtue of its membrane proximity to GD1a which it converts to

GM1 [30, 130]. This was demonstrated in neuroblastoma cells [131, 132] and also primary cultures of hippocampal neurons in which the resulting process outgrowth was shown to be axonal [133]. Calcium influx promoted by Neu3 elevation of GM1 was shown to promote regeneration of injured peripheral (sensory) nerve but not CNS nerve (retinal ganglion cell axons) despite the presence of Neu3 in the latter tissue [134]. This difference was suggested to account for the failure of injured CNS axons to regenerate in contrast to PNS axons which have that ability. GM1 also mediates Ca influx when exposed to crosslinking agents such as CtxB in primary neurons of the CNS [135] and PNS [136]. This mechanism involves the transient receptor potential 5 (TRPC5) channel whose natural activation is triggered by galectin-1-induced crosslinking of GM1 in neurons [137] as well T cells [138, 139]. Such crosslinking of GM1 induces co-crosslinking of $\alpha5\beta1$ integrin heterodimer with which GM1 is associated, which in turn induces tyrosine autophosphorylation of an associated focal adhesion kinase followed by activation of phospholipase Cγ and phosophoinosiide-3 kinase [138]. This occurs at an early stage of neurite outgrowth during which TRPC5 is transiently expressed in the neuronal membrane.

Calcium extrusion from neuronal cytoplasm is also subject to GM1 regulation via at least two mechanisms: plasma membrane calcium ATPase (PMCA) and Na/Ca exchanger (NCX). The former is a high affinity low capacity system while the latter is the opposite. Na/Ca exchanger of the plasma membrane was shown to be associated with GM1 although not as tightly as for NCX of the nuclear membrane ([140]; see below], and loss of this association might well have accounted for at least part of the failed Ca regulation in neurons with reduced ganglio-series gangliosides [129]. PMCAs are P-type ATPases and the isoform 2 is primarily neuronal in distribution [141]. That isoform in porcine synaptosomes was stimulated by GD1b and slightly inhibited by GM1, while GM2 and GM3 were even more inhibitory [142]. A subsequent study showed that inhibition of ganglioside synthesis in neurons or treatment with sialidase both inhibited PMCA activity, further confirming the stimulatory effect of polysialo- and inhibitory effect of monosialogangliosides in this system [143].

6.3 Gangliosides and Ca Modulation in the Cytoplasm

In a mouse model of GM1-gangliosidosis the elevated ganglioside was shown to accumulate in the mitochondria-associated ER membranes where it interacts with the phosphorylated form of IP3 receptor-1; the Ca released from the ER through activation of this channel is taken up by mitochondria causing opening of the permeability transition pore and activation of the mitochondrial apoptotic pathway [122]. Gangliosides are thus proposed to have a role in ER stress mechanisms such as the unfolded protein response [144, 145]. In another mouse model of ganglioside storage disease

GM2 was shown to inhibit Ca transfer from cytosol to the ER via the sarcoplasmic/endoplasmic reticulum Ca-ATPase (SERCA) mechanism [146]. The latter study also demonstrated reduction in the rate of Ca uptake in normal brain microsomes by exogenous GM2, to a lesser extent by GM1, and least by GM3. Similar studies on skeletal muscle sarcoplasmic reticulum also reported inhibition by GM1 on SERCA activity but a positive effect by GM3 [147]. The fact that the above SERCA studies were carried out with applied (exogenous) gangliosides, some in pathological cells, underscores the desirability of ascertaining whether endogenous ER gangliosides have similar modulatory roles in normal and non-pathological systems.

6.4 GM1 and Ca Regulation in the Nucleus

Upregulation of a-series gangliosides (GM1, GD1a) was shown to coincide with appearance of a major Ca regulatory mechanism of the nucleus involving GM1 in association with a Na/Ca exchanger (NCX) [140]. Importantly, the efficiency of this exchanger function depended on high affinity association with GM1, as seen in its functional demise in neurons from mice lacking this ganglioside and restoration with GM1 and even more effectively with LIGA20, a membrane-permeable GM1 analog [148]. Location of this GM1-NCX complex in the inner membrane of the double-layered nuclear envelope places it in position to mediate transfer of nuclear Ca from nucleoplasm into the luminal space of the nuclear envelope and hence the ER with which it is continuous. Such transfer was shown to occur as an alternative to the SERCA pump for transferring cytoplasmic Ca to the ER [149]. This accords with the fact that nucleoplasmic Ca is substantially in equilibrium with cytosolic free Ca via the nuclear pores. The driving force for this gating mechanism is the Na gradient created by Na-K-ATPase, which was shown to also occur in the nuclear membrane [150] suggesting concerted physiological coupling between these transporters. A key topological feature proposed to explain the tight association of GM1 with NCX was charge–charge interaction of negatively charged Neu5Ac of GM1 with the alternative splice region of the NCX loop containing positively charged amino acids [151]. For the plasma membrane this NCX loop is seen as residing on the opposite side of the bilayer as GM1, thus accounting for the lower affinity of that association; nevertheless Western blot analysis revealed GM1 to also be associated with this NCX [140]. GD1a, a metabolic precursor to GM1, is upregulated in the nucleus along with GM1 during development [117] and coexists with the latter in both membranes of the nuclear envelope. The function of GD1a as storage reservoir for GM1 is suggested in the presence of sialidase in both membranes of the nuclear envelope that convert endogenous GD1a to GM1 [120]. The observed NCX-GM1 complex in the nucleus is not limited to neural cells but also occurs in astrocytes [152] and certain other cell types [151].

7 Gangliosides as Receptor Modulators

Many studies over the years have demonstrated ganglioside modulation of growth factor and other receptors with varying results depending on ganglioside species, cell type, and experimental conditions; a methodological approach to resolving observed opposite effects is described [153]. A case in point is fibroblast growth factor 2 (FGF2) which was inhibited from binding to its high-affinity tyrosine-kinase receptor when treated exogenously with GM1 (and other gangliosides) [154], but was also depicted as utilizing GM1 as a functional coreceptor when the latter is cell associated [155]. This is of possible relevance to the ability of FGF2 to stimulate axonal growth in cultured hippocampal neurons with promotion of cerebroside (and consequently ganglioside) synthesis [156]. While GM1 has been a subject of focused interest in regard to receptor interaction [24, 32] it is by no means the only ganglioside to function in this manner, as in the well-known example of GM3 modulation of the insulin receptor [157, 158]. In what follows we attempt to review two areas in which the role of GM1 has been well defined: G-protein coupled receptors and neuroprotective signaling by neurotrophic factors. The latter has special relevance in regard to neurological disorders where such signaling has failed. Additional evidence on the role of GM1 interaction with a variety of receptors is available in recent reviews [24, 32, 159].

7.1 GM1 and Modulation of G-Protein Coupled Receptors

As prototypic G protein coupled receptor (GPCR) opioid-induced activity has been shown to be highly influenced by GM1. This relates to the ability of opioids to assert excitatory when GM1 is associated with this GPCR or inhibitory activity when it is not. This dual activity was demonstrated with such phenomena as action potential duration of sensory neurons [160] and calcium influx in NG108-15 cells [161]. The role of GM1 in facilitating the excitatory response was indicated in blockade of Ca influx by CtxB [162] as well as conversion from inhibitory to excitatory mode by bath application of GM1 in CHO cells expressing the δ-opioid receptor [163]. Opioid receptors of the μ- and kappa-type showed similar dual responses. This reactivity was highly specific for GM1 at nM concentrations, consistent with a true high affinity physiological mechanism. Moreover, the carboxyl group was essential suggesting interaction of GM1 with a positively charged amino acid of the receptor. This was supported by site-directed mutagenesis of the δ-opioid receptor in which arginine-192 was replaced with alanine [44]. GM1 has been shown to modulate G proteins that influence adenylate cyclase activity, such as the β1-adrenergic receptor in Sf9 cells which inhibited isoproterenol-induced cAMP formation [164] and stimulated prostaglandin E1-induced cAMP formation in Neuro-2a cells [165]. A GPCR of relevance to Parkinson's disease is GPR37 which, when expressed on the surface of catecholaminergic cells is neuroprotective against toxins implicated in Parkinson's disease, such as 6-hydroxydopamine, rotenone, and MPTP [166]. This receptor was shown to partition into GM1-containing lipid rafts, a process that required extracellular prosaposin and led to functional association between GPR37, prosaposin, and GM1 [167].

7.2 GM1 and Neurotrophic Factor Receptors

Neurotrophic factors such as nerve growth factor (NGF) are well recognized for their role in neuronal differentiation and also their essentiality in regard to survival of mature neurons. Lifelong survival, a unique feature of neurons in contrast to most other mammalian cells, depends on continuous support of this kind of neuroprotective signaling, and gradual failure with age is coming to be recognized as an important contributor to neurodegenerative processes. GM1 has been shown to assert a special role in this regard, occurring in high affinity association with TrkA, the tyrosine kinase receptor for NGF which serves to activate the receptor [168, 169]. The same has been observed for GM1 and TrkB, the receptor for brain-derived neurotrophic factor (BDNF), the activity of which is modulated by the amount of GM1 in the membrane environment [170]. Of interest was the finding that unglycosylated Trk protein does not colocalize nor associate with GM1 [171]. Expression of TrkA in cells lacking GM1 elicited failure of this receptor to appear on the cell surface, a defect that was corrected by stable transfection of GM1 synthase [172]. This suggested among other things a trafficking role involving intracellular association of GM1 with TrkA during raft assembly. A similar finding was reported for the receptor for GDNF, a neurotrophic factor essential for preserving the viability of catecholaminergic (and perhaps additional) neuronal types (see below). GM1 was shown to associate with the GDNF receptor complex comprised of Ret, the tyrosine kinase component, and GFRα, a GPI-anchored coreceptor; Ret association with GFRα was severely impaired in neurons totally or even partially devoid of GM1 [43]. This too suggested a trafficking role with respect to Ret, which is recruited to lipid rafts by GFRα1 following GDNF binding.

8 Gangliosides and Neurodegenerative Diseases

Following Klenk's initial discoveries [3, 4] gangliosides were first viewed in relation to neurological disorders, which led to more intensive studies of GM2- and GM1-gangliosidoses ([173]; *see* also

Chapter 5 and ref. [174]). An early indication that gangliosides might also serve a therapeutic role came with the observation of improved functional recovery of denervated cat nictitating membrane following application of brain ganglioside mixture [175], leading to further animal studies and eventually clinical trials [176]. Neither GM1 nor other gangliosides are employed therapeutically in the US or Europe at this time, although some promising clinical trials with GM1 suggest that may yet occur (see below). A potentially significant development was introduction of the LIGA series of "semisynthetic" derivatives of GM1 in which the stearoyl fatty acid of ceramide was replaced by shorter chain fatty acids [47]. These GM1 analogs turned out to differ importantly from GM1 in having enhanced membrane permeability and therefore greater ability to penetrate the blood brain barrier and neuronal plasma membrane. One such analog, LIGA20 for example, proved neuroprotective in reducing ethanol-induced apoptosis in rat cerebellar granule neurons [177] and affording in vivo protection against enhanced kainate-induced seizures in ganglio-series-deficient mice [48]. While gangliosides have now been intensively studied in relation to several neurological disorders the following three conditions have in common one key feature: significant involvement of gangliosides, especially the a-series (GM1, GD1a). Considering that GD1a functions as metabolic precursor to GM1 via membrane-associated sialidase [30], this puts the primary focus on GM1 as the principal active agent. The conditions in which it is in deficit suggest the possibility of special vulnerability of the genes expressing GM!- and GD1a-synthase, leading to diverse neuropathological lesions depending on other genetic and environmental factors. A preliminary report [178] suggested a genetic (or epigenetic) approach may prove fruitful.

8.1 Gangliosides and Parkinson's Disease (PD)

This year marks the 200th anniversary of Parkinson's first description of the "shaking palsy" syndrome, subsequently named after him and viewed initially as a prototypic movement disorder. However, as in-depth studies progressed it gradually became clear that PD is a "progressive multi-system disease with variegated motor and non-motor deficiencies" [179, 180]. Its systemic manifestations include impaired olfaction, gastrointestinal, cardiovascular and respiratory functions as well as sleep, sensory and neuropsychiatric disorders. Included among the systemic symptoms are non-dopamine lesions [181] along with the classical L-dopa-responsive symptoms. Well before any thought was given to possible ganglioside involvement in PD etiology, ganglioside therapy was shown to exert significant benefit in rat and mouse PD models involving damage to the nigrostriatal dopamine (DA) system. These lesions included surgical transection [182] and neurotoxin-induced degeneration via 1-methyl-4-phenyl,1,2,3,6-tetrahydropyridine (MPTP) [183, 184] and 6-hydroxydopamine [185]. Systemic treatment with GM1 in most cases resulted in significant restoration of depleted striatal DA and nigrostriatal neuron recovery. GM1-induced improvements were also seen in subhuman primate models of PD [186, 187]. LIGA20, the above-mentioned membrane permeable analog of GM1, proved superior to GM1 in the mouse MPTP model and had the important advantage of oral bioavailability [188].

These promising preclinical results led to clinical trials by Schneider's group beginning with small scale studies that established safety and the likelihood of benefit with GM1. These were followed by a 5-year open-label study in which PD subjects were given two subcutaneous injections of 100 mg GM1 per day, resulting in reduced, slow linear projection of motor dysfunction (UPDRS motor score and activities of daily living scores); significantly, after five years the motor disability score was still lower than that at baseline [189]. The same investigators then carried out a randomized, controlled, delayed-start phase II trial that demonstrated GM1 superiority to placebo in reducing motor impairment and slowing symptom progression over a 2-year period; as before, UPDRS motor scores showed improvement at trial end (120 weeks) compared to base line, again suggesting that "...GM1 may have symptomatic and potentially disease modifying effects" on PD [190].

The mechanisms underlying those clinical benefits were largely unknown but the possibility that they might represent a form of replacement therapy in PD patients deficient in a-series gangliosides was suggested in studies of ganglio-series-deficient *B4galnt1*$^{-/-}$ mice which manifested impaired movement and virtually all the neuropathological symptoms of PD; these included depletion of striatal DA, elevation of aggregated α-synuclein, loss of tyrosine hydroxylase (TH)-expressing DA neurons in the substantia nigra pars compacta (SNpc), and diminished pRet expression in those neurons [43, 191]. Importantly, heterozygotes (*B4galnt1*$^{+/-}$) with roughly half the normal level of GM1 showed virtually the same PD symptoms [43, 49]; what highlighted this result was the parallel findings of diminished GM1 in PD tissues, including the nigrostriatal neurons of the SNpc [49] and the occipital cortex of PD brains [43]. This suggested the possibility of systemic GM1 deficiency in PD subjects, an hypothesis that has gained support in the observation of subnormal GM1 (and GD1a) in a number of non-CNS tissues of PD subjects. It was noteworthy that these deficiencies in PD tissues (34–47%) approximated those of the heterozygous mice. An important consequence of such deficiency was shown to be impaired GDNF signaling, as revealed in reduced expression of phosphorylated Ret (pRet) and pMAPK in the SNpc of the GM1-deficient mice as well as PD subjects [43]. As mentioned, striatal tissues of these mice showed impaired formation of the Ret/GFRα1 receptor complex for GDNF, the extent of this failed cohesion being proportional to the GM1 deficit. GDNF is a member of the transforming growth factor superfamily and functions to preserve the viability of catecholaminergic neurons—such as the DA neurons at risk in PD [192] along with other catecholaminergic neurons in the CNS and PNS. This GDNF receptor lesion and the defective pRet signaling were corrected with LIGA20, again pointing to GM1 as the ganglioside involved in GDNF-induced neuroprotection [43].

An additional function of GM1, likely of critical relevance to PD pathophysiology, is its association with α-synuclein, an intracellular protein localized at the synapse and nucleus which, under adverse conditions, aggregates to form Lewy bodies and neurites. These inclusions constitute a hallmark of PD pathology [193] and sequential appearance of such aggregates served as the major guide to Braak's 6-stage hypothesis demonstrating an early prodromal phase preceding overt motor impairment by many years and a final stage often characterized by cognitive dysfunction [194, 195]. It was noteworthy that α-synuclein fibrillations developed spontaneously over time in the above ganglio-series-deficient mice and largely disappeared upon administration of LIGA20 at low dose [49] or GM1 at higher dose (Wu and Ledeen, in preparation). This very likely relates to the described association of GM1 with α-synuclein in a manner that retains the nonaggregating helical conformation of this protein [38]. Such association is enhanced by N-terminal acetylation of α-synuclein, which significantly increases GM1-binding specificity and resistance to fibrillation [39]. The glycolipid-binding domain reported to exist within α-synuclein which may be responsible for this effect was described as a structurally related loop centered on tyrosine (Y39) with favored binding to GM1 and GM3 [42]. It is conceivable that α-synuclein aggregation results primarily from the observed GM1 deficiency in tissues of the above mice and PD subjects, while impaired GDNF signaling might also contribute. Yet another aspect of GM1 involvement in PD is the above mentioned GPR37 G-protein which is neuroprotective when expressed on the surface of catecholaminergic neurons; a functional association of GPR37, prosaposin, and GM1 has been described [166, 167] which, like the many other neuronal GM1 functions, likely depends on an adequate level of this ganglioside. For more detail on ganglioside involvement in PD, please *see* review [196].

Finally an especially puzzling aspect of PD involving glycosphingolipids is its association with loss of glucocerebrosidase (GBA) activity as occurs in Gaucher's disease. This association applies as well to carriers of GBA mutations without actual Gaucher symptoms. All patients with PD or those with sporadic PD had significantly lower GBA activity than controls, leading to the conclusion that lower GBA activity contributes to the pathogenesis of PD [197]. This applied to both GBA mutation carriers and PD patients without GBA mutations. Interestingly, an extensive literature search revealed that additional lysosomal storage disorders might be associated with PD [198], including GM1 gangliosidosis [199]. How such changes impinge on the vulnerable genes responsible for GM1 and GD1a synthesis remains to be determined.

8.2 Gangliosides and Alzheimer's disease (AD)

Selective loss of cholinergic neurons, particularly in the basal forebrain [200], pointed to NGF involvement in AD [201] and accordingly GM1, owing to its essential association with the TrkA NGF receptor [168, 169]. Despite the seeming paradox of AD brain containing increased pro-NGF, precursor to NGF, together with impaired NGF signaling, this was resolved through demonstration of diminished conversion of pro-NGF to its mature form and of greater NGF degradation in AD [201]. Owing to the dependence of NGF activity on GM1 association with TrkA it has been of interest to know whether GM1 is deficient in AD brain.as it is in PD brain (see above). One report showed some ganglio-series gangliosides to be decreased in the early-onset form of AD, primarily the b-series (GD1b, GT1b) but not GM1; that study showed little ganglioside change in the late-onset form with the exception of the temporal lobe which manifested mildly depleted a-series (GM1, GD1a) [202]. Another study similarly found a relatively normal level of total gangliosides in AD frontal cortex but with some decrease in the b-series [203]. On the other hand two studies reported a general decrease in ganglio-series gangliosides, one a study of 9 brain regions involved in AD in which the b-series was preferentially affected [204] and the other a study employing a novel quantitation method which similarly found decreases in the b-series [205] Finally one other study reported all ganglio-series to be decreased in the temporal and frontal cortex and nucleus basalis of Meynert [206]. Hence there is no clear consensus as yet on possible changes in GM1 levels, while the evidence to date would tend to suggest preferential loss of b-series gangliosides in AD-relevant regions. If verified that would accord with the well-established loss of axon terminals in AD and the predominance of GD1b and GT1b in those structures [101, 102].

Nevertheless a number of studies have linked AD with gangliosides, especially GM1, in both positive and negative modes [207]. The negative role assigned GM1 in AD was based on its ability when bound to Aβ to act as seed for fibrillogenesis of soluble Aβ [208, 209]. GM1 combined with Aβ(1–40) also proved cytotoxic to mouse embryonic neural stem cells [210], of potential relevance to efforts for AD regenerative therapy. On the other hand GM1 ganglioside has shown indications of therapeutic efficacy as in the study of Svennerholm et al. [211] in which five patients with the early-onset form of AD received continuous infusion of GM1 into the frontal horns of the lateral ventricles via a programmed minipump over a period of 1 year; all five patients were reported to show significant physical, mental, and emotional improvement. Also in organotypic hippocampal slice cultures, pretreatment with

GM1 prevented the toxicity triggered by fibrillar Aβ(25–35) [212]. In contrast to the toxicity of GM1 when associated with Aβ, GM1 alone proved beneficial by inhibiting amyloid beta-protein-induced release of proinflammatory cytokines such as IL-1β, IL-6, and TNF-α [213]. Similarly, GM1 increased the viability of PC12 cells transfected with mutant gene expressing amyloid precursor protein $(APP)_{SW}$ exposed to Aβ(25–35) (or hydrogen peroxide) [214].

A notable example of therapeutic benefit was seen in elevation of endogenous GM1 and GD1a genetically—through deletion of GD3-synthase in the double transgenic APP/PSEN1 mouse model of AD; this resulted in almost complete elimination of both aggregated and unaggregated Aβ and associated neuropathology [215]. Remarkably, in contrast to the double-transgenic APP/PSEN1 mice which exhibited robust impairments on a number of reference-memory tasks, the triple mutant APP/PSEN1/$GD3S^{-/-}$ mice performed as well as wild-type mice. The authors suggested these improvements might be attributed to elevation of GM1 and GD1a but also noted possible benefit from removal of the b-series gangliosides, especially GD3 considering its prominent role in apoptotic pathways including Aβ-induced cell death [216].

These apparent therapeutic benefits of GM1 are somewhat perplexing in view of the above mentioned absence of clear evidence of a-series ganglioside deficiency in key AD brain regions. In that regard it is well to consider that total tissue levels may be less meaningful than the amount of functionally available GM1, considering the numerous essential roles of this ganglioside in the nervous system [24, 32]. Functionally available GM1 would admittedly be difficult to determine but it seems reasonable to assume that association of GM1 with Aβ would remove it from its normal functional associations. This might, for example, explain the common occurrence in AD brain of α-synuclein aggregates (Lewy bodies/neurites) [217] that are a cardinal feature of PD arising through GM1 deficiency [43, 49]. As mentioned, this association constitutes an essential intraneuronal function of GM1 that serves to maintain α-synuclein in nonaggregating conformation. Such deficit could in principle be restored through GM1 replacement therapy of the kind attempted with initial promise in PD and possibly in effect in the above Svennerholm study [211]. For an informative review of ganglioside involvement in AD and other neurodegenerative disorders please *see* [159, 207].

8.3 Gangliosides and Huntington's Disease (HD)

This condition is caused by an autosomal dominant mutation in one allele of the huntingtin (Htt) gene. This leads to expansion of a CAG triplet repeat which in turn causes expansion of a polyglutamine sequence in the N-terminal region of the Htt protein. The primary symptoms are chorea, loss of muscle coordination and eventually psychiatric and cognitive decline. Depressed expression of certain genes encoding glycosyltransferases and sialyltransferases involved in ganglioside synthesis were observed in the striatum of HD transgenic (R6/1) mice consistent with a depressed level of GM1 in the striatum [218]. Significantly, the same was found in postmortem caudate nucleus tissue from HD subjects; the depressed genes included B4galnt1, St3gal5, St8sia3, and St3gal2. That tissue showed 39% less GM1 compared to control caudate, but that difference did not reach significance due to high variability. However, GD3 was significantly elevated (162%). Another study employing the YAC128 mouse model of HD showed GM1 reduction in the striatum and cortex, two regions involved in HD, and down-regulation of B3gGalT4 (GM1/GD1b synthase) in the YAC128 cortex [219]. GM1 was shown to be reduced in fibroblasts from HD subjects, suggesting the possibility of a systemic epigenetic defect. Application of GM1 to a knock-in cell model of HD restored GM1 levels as well as cell survival, apparently through promotion of Akt activation with resultant phosphorylation of the mutant Htt protein [220]. It was perhaps surprising that LIGA20, the above mentioned GM1 analog that proved more effective than GM1 itself in most such experiments, was ineffective in this in vitro test. This suggested an active role for the ceramide unit in this signaling process. GM1 application proved therapeutic in vivo as well, intraventricular infusion restoring normal motor function in already symptomatic HD mice [220]. This was apparently achieved, as with the above HD knock-in cells, through induced phosphorylation of the Htt protein at the critical serine residues that attenuated toxicity of the mutant protein. These studies provide another informative example of ganglioside synthase suppression leading to neurodegeneration, caused in this case by expression of an unrelated misfolded mutant protein. This appears to be a common mechanism for suppression of ganglioside synthesizing enzymes.

9 Concluding Remarks and Future Prospects

The foregoing review has attempted to outline the impressive progress to date of numerous groups in defining the basic biochemical properties of nervous system gangliosides leading to a more profound understanding of their essential roles in the complex functioning of the nervous system. Looking to the future, efforts to heighten our understanding in these basic areas have as an ultimate goal alleviation and if possible actual cure of neurodegenerative diseases that increasingly devastate the world's aging populations. That gangliosides are essential in maintaining a healthy, homeostatic nervous system is well illustrated in the disastrous consequences of genetic deletion of the ganglio-series gangliosides in humans through disruption of the B4galnt1

(GM2/GD2 synthase) gene, leading to early onset spastic paraplegia [221]; mice with this lesion are similarly though less drastically affected [191, 222]. Deletion of these gangliosides along with GM3 and GD3 by mutation of the ST3GAL5 (GM3 synthase) gene was even more devastating to humans [223] and mice [224, 225]. The latter condition is widespread in the Amish religious community [226], which also manifests the world's highest prevalence of PD—approximately three times the prevalence of the USA, despite the USA having the highest PD prevalence of any country. Animal models of these and other types have proved invaluable in defining ganglioside functions and their roles in particular disorders [227]. Such models have revealed the important finding that total deletion is not necessary to induce neurodegenerative pathology, partial reduction of GM1, for example, being sufficient to suppress critical GDNF neuroprotective signaling thereby inducing PD symptoms [43]. The essential nature of GM1 interaction with neurotrophic factor receptors will need to be more fully defined, perhaps elucidating additional examples of GM1 acting to induce cohesion of receptor protein components—as with the GDNF receptor [43]. This alludes to the possibility that other neurotrophic factors dependent on GM1 can become operative, likely in the later (cognitive) stages of PD. The above findings suggest that genetic and epigenetic considerations will command increasing attention in future studies including the impact of other genetic changes on the expression patterns of ganglioside synthase genes. This could conceivably include genetic changes that are therapeutically beneficial, such as deletion of GD3 synthase which elevated GM1 and GD1a in a mouse model of AD and resolved the neuropathological and behavioral disorders [212]. Another genetic mechanism for elevating endogenous gangliosides was revealed in the finding that GM1 occurs in the chromatin of neurons where it promotes its own further synthesis through epigenetic activation of B4galnt1 (GM2/GD2 synthase) [54]. Other epigenetic regulatory mechanisms of ganglioside expression [228] are likely to receive attention including the fascinating new area of gut microbiota which have already been studied in relation to a model of PD [229]. Efforts to clinically remedy the various forms of GM1 deficiency perceived in the three neurodegenerative conditions reviewed above will likely continue apace in the pharmacological mode following the pioneering efforts devised for PD [190] and AD [208]. In that realm the use of membrane-permeable ganglioside analogs which are better able to access intraneuronal GM1-associated sites may hold considerable therapeutic promise.

References

1. Gabius H-J (ed) (2009) The sugar code. Fundamenals of glycosciences. Weinheim, Wiley-Blackwell
2. Klenk E (1942) Uber die Ganglioside, eine neue Gruppe von zuckerhaltigen Gehirnlipoiden. Hoppe Seylers Z Physiol Chem 273:76–86
3. Klenk E (1935) Uber die Natur der Phosphatide und anderer Lipoid des Gehirns und der Leber bei der Niemann-Pickschen Krankheit. Z Physiol Chem 235:24–36
4. Klenk E (1937) Die Fettstoffe des Gehirns bei amaurotischer Idiotie und Niemann-Pick'scher Krankheit. Ber Physiol 96:659–660
5. Varki A, Schauer R (2009) Sialic acids. In: Varki A et al (eds) Essentials of glycobiology, 2nd edn. Cold Spring Harbor Laboratory Press, Cold Spring Harbor
6. Yamakawa T, Suzuki S (1951) The chemistry of the lipids of posthemolytic residue or stroma of erythrocytes. I Concerning the ether-insoluble lipids of lyophilized horse blood stroma. J Biochem 38:199–212
7. Kawano Y, Higuchi R, Komori T (1990) Isolation and structure of five new gangliosides. Liebig's Ann Chem 1990:43–50
8. Kuhn R, Wiegandt H (1963) Die Konstitutionder Ganglio-N-tetraose und des gangliosids GI. Chem Ber 96:866–880
9. Chester MA (1998) IUPAC-IUB Joint Commission on Biochemical Nomenclature (JCBN). Nomenclature of glycolipids—recommendations 1997. Eur J Biochem 257:293–298
10. Svennerholm L (1963) Chromatographic separation of human bran gangliosides. J Neurochem 10:613–623
11. Kuhn R, Wiegandt H (1963) Die Konstitution der Ganglioside G_{II}, G_{III}, und G_{IV}. Z Naturforsch 18b:541–543
12. Klenk E, Gielen W (1963) Uber ein zweites hexosaminhaltiges Gangliosid aus Menschengehirn. Z Physiol Chem 330:218–226
13. Kuhn R, Wiegandt H (1964) Weitere Ganglioside aus Menschenhirn. Z Naturforsch 19b:256–257
14. Klenk E (1970) On the discovery and chemistry of neuraminic acid and gangliosides. Chem Phys Lipids 5:193–197
15. Klenk E, Liedtke U, Gielen W (1963) Das Gangliosid des Gehirns bei der intantilen amaurotischen idiote von Typ Tay-Sachs. Z Physiol Chem 334:186–192
16. Ledeen R, Salsman K (1965) Structure of the Tay-Sachs ganglioside. Biochemist 4:2225–2233
17. Yu RK, Yanagisawa M, Ariga T (2007) Glycosphingolipid sructures. In: Kamerling JP (ed) Comprehensive glycoscience. Elsevier, Oxford, pp 73–122
18. Hakomori S, Handa K, Iwabuchi K, Yasmamura S, Prinetti A (1998) New insights in glycosphingolipid function: "glycosignaling domain," a cell surface assembly of glycosphingolipids with signal transducer molecules, involved in cell adhesion coupled with signaling. Glycobiol 8:xi–xix
19. Simons K, Sampaio JL (2011) Membrane organization and lipid rafts. Cold Spring Harb Perspect Biol 3:a004697
20. Sonnino S, Mauri L, Chigorno V, Prinetti A (2007) Gangliosides as components of lipid membrane domains. Glycobiology 17:1R–13R
21. Furukawa K, Ohmi Y, Ohkawa Y, Tokuda N, Kondo Y, Tajima O, Furukawa K (2011) Regulatory mechanisms of nervous systems with glycoosphingolipids. Neurochem Res 36:1578–1586
22. Ohmi Y, Ohkawa Y, Yamauchi Y, Tajima O, Furukawa K, Furukawa K (2012) Essential roles of gangliosides in the formation and maintenance of membrane microdomains in brain tissues. Neurochem Res 37:1185–1191
23. Nishio M, Fukumoto S, Furukawa K, Ichimura A, Miyazaki H, Kusunoki S, Urano T, Furukawa K (2004) Overexpressed GM1 suppresses nerve growth factor (NGF) signals by modulating the intracellular localization of NGF receptors and membrane fluidity in PC12 cells. J Biol Chem 279:33368–33378
24. Aureli M, Mauri L, Grazia Ciampa M, Prinetti A, Toffano G, Secchieri C, Sonnino S (2016) GM1 ganglioside: past studies and future potential. Mol Neurobiol 53:1824–1842
25. Ishizuka I, Wiegandt H (1972) An isomer of trisialoganglioside and the structure of tetra- and pentasialogangliosides from fish brain. Biochim Biophys Acta 260:279–289
26. Yu RK, Ando S (1980) Structure of some new complex gangliosides of fish brain. Adv Exp Med Biol 125:33–45
27. Irie F, Kurono S, Li Y-T, Seyama Y, Hirabayashi Y (1996) Isolation of three novel cholinergic neuron-specific gangliosides from bovine brain and their inn vitro syntheses. Glycoconj J 13:177–186
28. Svennerholm L, Bostrom K, Fredman P, Jungbjer B, Lekman A, Mansson JE, Rynmark BM (1994) Gangliosides and allied glycosphingolipids in human peripheral

nerve and spinal cord. Biochim Biophys Acta 1214:115–123

29. Yu RK, Ledeen R (1970) Gas-liquid chromatographic assay of lipid-bound sialic acids: measurement of gangliosides in brain of several species. J Lipid Res 11:506–516
30. Miyagi T, Yamaguchi K (2012) Mammalian sialidases: physiological and pathological roles in cellular functions. Glycobiology 22:880–896
31. Wang Y, Yamaguchi K, Wada T, Hata K, Zhao X, Fujimoto T, Miyagi T (2002) A close association of the ganglioside-specific sialidase Neu3 with caveolin in membrane microdomains. J Biol Chem 277:26252–26259
32. Ledeen R, Wu G (2015) The multi-tasked life of GM1 ganglioside, a true factotum of nature. Trends Biochem Sci 40:407–418
33. Riboni L, Sonnino S, Acquotti D, Malesci A, Ghidoni R, Egge H et al (1986) Natural occurrence of ganglioside lactones. Isolation and characterization of GD1b inner ester from adult human brain. J Biol Chem 261:8514–8519
34. Li YT, Maskos K, Chou CW, Cole RB, Li SC (2003) Presence of an unusual GM2 derivative, taurine-conjugated GM2, in Tay-Sachs brain. J Biol Chem 278:35286–35291
35. Carter HE, Glick FJ, Norris WP, Phillips GE (1947) Biochemistry of sphingolipids. III. Structure of sphingosine. J Biol Chem 170:285–294
36. Sonnino S, Chigorno V (2000) Ganglioside molecular species containing C18- and C20-sphingosine in mammalian nervous tissues and neuronal cell cultures. Biochim Biophys Acta 1469(2):63–77
37. Merrill AH Jr (2011) Sphingolipid and glycosphingoliipid metabolic pathways in the era of sphingolipidomics. Chem Rev 111:6387–6422
38. Martinez Z, Zhu M, Han S, Fink AL (2007) GM1 specifically interacts with α-synuclein and inhibits fibrillation. Biochemist 46:1868–1877
39. Bartels T, Kim NC, Luth ES, Selkoe DJ (2014) N-alpha-acetylation of alpha-synuclein increases its helical folding propensity, GM1 binding specificity and resistance to aggregation. PLoS One 9:e103727
40. Fantini J (2007) Interaction of proteins with lipid rafts through glycolipid-binding domains: biochemical background and potential therapeutic applications. Curr Med Chem 14:2911–2917
41. Fantini J, Yahi N, Garmy N (2013) Cholesterol accelerates the binding of Alzheimer's β-amyloid peptide to ganglioside GM1 through a universal hydrogen-bond-dependent sterol tuning of glycolipid conformation. Front Physiol 4(120). https://doi.org/10.3389/fphys.2013.00120
42. Fantini J, Yahi N (2011) Molecular basis for the glycosphingolipid-binding specificity of α-synuclein: key role of tyrosine 39 in membrane insertion. J Mol Biol 408:654–669
43. Hadaczek P, Wu G, Sharma N, Ciesielska A, Bankiewicz K, Davidow AL et al (2015) GDNF signaling implemented by GM1 ganglioside; failure in Parkinson's disease and GM1-deficient murine model. Exp Neurol 263:177–189
44. Wu G, Lu Z-H, Wei TJ, Howells RD, Christoffers K, Ledeen RW (1998) The role of GM1 ganglioside in regulating excitatory opioid effects. Ann N Y Acad Sci 845:126–138
45. Wu G, Lu Z-H, Andre S, Gabius H J, Ledeen RW (2016) Functional interplay between ganglioside GM1 and cross-linking galectin-1 induces axon-like neuritogenesis via integrin-based signaling and TRPC5-dependent Ca influx. J Neurochem 136:550–563
46. Lopez PH, Schnaar RL (2009) Gangliosides in cell recognition and membrane protein regulation. Curr Opin Struct Biol 19:549–557
47. Manev H, Favaron M, Vicini S, Guidotti A, Costa E (1990) Glutamate-induced neuronal death in primary cultures of cerebellar granule cells: protection by synthetic derivatives of endogenous sphingolipids. J Pharmacol Exp Ther 252:419–527
48. Wu G, Lu Z-H, Wang J, Wang Y, Xie X, Meyenhofer MF, Ledeen RW (2005) Enhanced susceptibility to kainate-induced seizures, neuronal apoptosis, and death in mice lacking gangliotetraose gangliosides: protection with LIGA20, a membrane-permeant analog of GM1. J Neurosci 25:11014–11022
49. Wu G, Lu Z-H, Kulkarni N, Ledeen RW (2012) Deficiency of ganglioside GM1 correlates with Parkinson's disease in mice and humans. J Neurosci Res 90:1997–2008
50. Manna M, Javanainen M, Monne HM-S, Gabius H-J, Rog T, Vattulainen I (2017) Long-chain GM1 gangliosides alter transmembrane domain registration through interdigitation. Biochim Biophys Acta Biomembr 1859:870–878
51. Yu RK, Macala LJ, Taki T, Weinfeld HM, Yu FS (1988) Developmental changes in ganglioside composition and synthesis in embryonic rat brain. J Neurochem 50:1825–1829

52. Ngamukote S, Yanagisawa M, Ariga T, Ando S, Yu RK (2007) Developmental changes of glycosphingolipids and expression of glycogenes in mouse brains. J Neurochem 103:2327–2341
53. Tsai Y-T, Yu RK (2014) Epigenetic activation of mouse ganglioside synthase genes: implications for neurogenesis. J Neurochem 128:101–110
54. Tsai Y-T, Itokazu Y, Yu RK (2016) GM1 ganglioside is involved in epigenetic activation loci of neuronal cells. Neurochem Res 41:107–115
55. Kopitz J, von Reitzenstein C, Burchert M, Cantz M, Gabius H-J (1998) Galectin-1 is a major receptor for ganglioside GM1, a product of the growth controlling activity of a cell surface ganglioside sialidase on human neuroblastoma cells in culture. J Biol Chem 273:11205–11211
56. Purpura DP, Suzuki K (1976) Distortion of neuronal geometry and formation of aberrant synapses in neuronal storage disease. Brain Res 116:1–21
57. Walkley SU, Wurzelmann S, Purpura DP (1981) Ultrastructure of neuritis and meganeurites of cortical pyramidal neurons in feline gangliosidosis as revealed by the combined Golgi-EM technique. Brain Res 211:393–398
58. Roisen FJ, Bartfeld H, Nagele R, Yorke G (1981) Ganglioside stimulation of axonal sprouting in vitro. Science 214:577–578
59. Facci L, Leon A, Toffano G, Sonnino S, Ghidoni R, Tettamanti G (1984) Promotion of neuritogenesis in mouse neuroblastoma cells by exogenous gangliosides. Relationship between the effect and the cell association of ganglioside GM1. J Neurochem 42:299–305
60. Byrne MC, Ledeen RW, Roisen FJ, Yorke G, Sclafani JR (1983) Ganglioside-induced neuritogenesis: verification that gangliosides are the active agents, and comparison of molecular species. J Neurochem 41(5):1214–1222
61. Cannella MS, Acher AJ, Ledeen RW (1998) Stimulation of neurite outgrowth in vitro by a glycerol-gangliioside. Int J Dev Neurosci 6:319–326
62. Wu G, Fang Y, Lu Z-H, Ledeen RW (1998) Induction of axon-like and dendrite-like processes in neuroblastoma cells. J Neurocytol 27:1–14
63. Yu RK, Suzuki Y, Yanagisawa M (2010) Membrane glycolipids in stem cells. FEBS Lett 584:1694–1699
64. Yanagisawa M (2011) Stem cell glycolipids. Neurochem Res 36:1623–1635
65. Wang J, Yu RK (2013) Interaction of ganglioside GD3 with an EGF receptor sustains the self-renewal ability of mouse neural stem cells in vitro. Proc Nat Acad Sci 110:19137–19142
66. Wang J, Cheng A, Wakade C, Yu RK (2014) Ganglioside GD3 is required for neurogenesis and long-term maintenance of neural stem cells in the postnatal mouse brain. J Neurosci 34:13790–13800
67. Suzuki K (1965) The pattern of mammalian brain gangliosides – III. J Neurochem 12:969–979
68. Schwarz A, Futerman AH (1996) The localization of gangliosides in neurons of the central nervous system: the use of anti-ganglioside antibodies. Biochim Biophys Acta 1286:247–267
69. Schengrund C-L, Ringler NJ (1989) Binding of Vibrio cholera toxin and the heat labile enterotoxin of Escherichia coli to GM1 and derivatives of GM1. J Biol Chem 264:13233–13237
70. Vajn K, Viljetic B, Degmecic I, Schnaar RL, Heffer M (2013) Differential distribution of major brain gangliosides in the adult mouse central nervous system. PLoS One 8:e75720
71. Colsch B, Jackson SN, Dutta S, Woods AS (2011) Molecular microscopy of brain gangliosides: illustrating their distribution in hippocampal cell layers. ACS Chem Neurosci 2:213–222
72. Schengrund C-L, Repman MA (1977) Cell culture of sixteen-day-old rat embryo cerebra and associated changes in ganglioside pattern. J Neurochem 29:923–927
73. Yavin E, Yavin Z (1979) Ganglioside profiles during neural tissue development. Dev Neurosci 2:25–37
74. Dreyfus H, Louis JC, Harth S, Mandel P (1980) Gangliosides in cultured neurons. Neuroscience 5:1647–1655
75. Sbaschnig-Agler M, Dreyfus H, Norton WT, Sensenbrenner M, Farooq M, Byrne MC, Ledeen RW (1988) Gangliosides of cultured astroglia. Brain Res 461:98–106
76. Asou H, Brunngraber EG (1983) Absence of ganglioside GM1 in astroglial cells from 21-day old rat brain: immunohistochemical, histochemical, and biochemical studies. Nerochem Res 8:1045–1057
77. Abe T, Norton WT (1974) The characterization of sphingolipids from neurons and astroglia of immature rat brain. J Neurochem 23:1025–1036

78. Norton WT, Abe T, Poduslo SE, DeVries GH (1975) The lipid composition of isolated brain cells and axons. J Neurosci Res 1:57–75
79. Hamburger A, Svennerholm L (1971) Composition of gangliosides and phospholipids of neuronal and glial cell enriched fractions. J Neurochem 18:1821–1829
80. Robert J, Freysz L, Sensenbrenner M, Mandel P, Rebel G (1975) Gangliiosides of glial cells: a comparative study of normal astroblasts in tissue culture and glial cells isolated on sucrose-ficoll gradients. FEBS Lett 50:144–146
81. Byrne MC, Farooq M, Sbaschnig-Agler M, Norton WT, Ledeen RW (1988) Ganglioside content of astroglia and neurons isolated from maturing rat brain: consideration of the source of astroglial gangliosides. Brain Res 461:87–97
82. Gammon CM, Vaswani KK, Ledeen RW (1987) Isolation of two glycolipid transfer proteins from bovine brain: reactivity toward gangliosides and neutral glycosphingolipids. Biochemist 26:6239–6243
83. Seyfried TN, Yu RK, Miyazawa N (1982) Differential cellular enrichment of gangliosides in the mouse cerebellum: analysis using neurological mutants. J Neurochem 38:551–559
84. Giuliani A, Calappi E, Borroni E, Whittaker VP, Sonnino S, Tettamanti G (1990) Further studies on the gangliosidic nature of the cholinergic-specific antigen, Chol-1. Arch Biochem Biophys 280:211–216
85. Ando S, Hirabayashi Y, Kon K, Inagaki F, Tate S, Whittaker VP (1992) A trisialoganglioside containing a sialyl α2-6 N-acetylgalactosamine residue is a cholinergic specific antigen, Choll-1α. J Biochem 111:287–290
86. Hirabayashi Y, Nakao T, Irie F, Whittaker VP, Kon K, Ando S (1992) Structural characterization of a novel cholinergic neuron-specific ganglioside in bovine brain. J Biol Chem 267:12973–12978
87. Hirabayashi Y, Hyogo A, Nakao T, Tsuchiya K, Suzuki Y, Matsumoto M, Kon K, Ando S (1990) Isolation and characterization of extremely minor gangliosides, GM1b and GD1α, in adult bovine brains as developmentally regulated antigens. J Biol Chem 265:8144–8151
88. Ando S, Tanaka Y, Kobayashi S, Fukui F, Iwamoto M, Waki H, Tai T, Hirabayashi Y (2004) Synaptic function of cholinergic-specific Chol-1alpha ganglioside. Neurochem Res 29:857–867
89. Suzuki K, Poduslo SE, Norton WT (1967) Gangliosides in the myelin fraction of developing rats. Biochim Biophys Acta 144:375–381
90. Ledeen RW, Yu RK, Eng LF (1973) Gangliiosides of human myelin: sialosylgalactosylceramide (G7) as a major component. J Neurochem 21:829–839
91. Cochran FB Jr, Yu RK, Ledeen RW (1982) Myelin gangliosides in vertebrates. J Neurochem 39:773–779
92. Yu RK, Iqbal K (1979) Sialosylgalactosyl ceramide as a specific marker for human myelin and oligodendroglial perikarya: gangliosides of human myelin, oligodendroglia and neurons. J Neurochem 32:293–300
93. Yu RK, Macala LJ, Farooq M, Sbaschnig-Agler NWT, Ledeen RW (1989) Ganglioside and lipid composition of bulk-isolated rat and bovine oligodendroglia. J Neurosci Res 23:136–141
94. Littlemore LAT, Ledeen RW (1979) N.M.R. studies of myelin basic protein. III. Interactions of the protein with lipid micelles by ^{1}H and ^{31}P N.M.R. Aust J Chem 32:2631–2636
95. Mullin BR, Decandis FX, Montanaro AJ, Reid JD (1981) Myelin basic protein interacts with the myelin-specific ganglioside GM4. Brain Res 222:218–221
96. Andersson LM, Fredman P, Lekman A, Rosengren L, Gisslen M (1998) Increased cerebrospinal fluid ganglioside GD3 concentrations as a marker of microglial activation in HIV type 1 infection. AIDS Res Hum Retrovir 14:1065–1969
97. Simon BM, Malisan F, Testi R, Nicotera P, Leist M (2002) Disialoganglioside GD3 is released by microglia and induces oligodendrocyte apoptosis. Cell Death Differ 9:758–767
98. Ogawa-Goto K, Abe T (1998) Gangliosides and glycosophingolipids of peripheral nervous system myelins—a minireview. Neurochem Res 23:305–310
99. Ohsawa T (1990) Rat dorsal root ganglion gangliosides during early development until senescence. Mech Aging Dev 53:259–266
100. Fong JW, Ledeen RW, Kundu SK, Brostoff SW (1976) Gangliosides of peripheral nerve myelin. J Neurochem 26:157–162
101. Avrova NF, Chenykaeva EY, Obukhova EL (1973) Ganglioside composition and content of rat-brain subcellular fractions. J Neurochem 20:997–1004
102. Breckenridge WC, Gombos G, Morgan IG (1972) The lipid composition of adult rat brain synaptosomal plasma membranes. Biochim Biophys Acta 266:695–707
103. Lapetina EG, Soto EF, De Robertis E (1968) Lipids and proteolipids in isolated subcellular

membranes of rat brain cortex. J Neurochem 15:437–445
104. Ledeen RW (1978) Ganglioside structures and distribution: are they localized at the nerve ending? J Supramol Struct 8:1–17
105. Skrivanek JA, Ledeen RW, Margolis RU, Margolis RK (1982) Gangliosides associated with microsomal subfractions of brain: comparison with synaptic plasma membranes. J Neurobiol 13:95–106
106. Hansson HA, Holmgren J, Svennerholm L (1977) Ultrastructural localization of cell membrane GM1 ganglioside by cholera toxin. Proc Natl Acad Sci U S A 74:3782–3786
107. Devries GH, Zmachinski CJ (1980) The lipid composition of rat CNS axolemma-enriched fractions. J Neurochem 34:424–430
108. Forman DS, Ledeen RW (1972) Axonal transport of gangliosides in the goldfish optic nerve. Science 177:630–633
109. Aquino DA, Bisby MA, Ledeen RW (1987) Bidirectional transport of gangliosides, glycoproteins, and neutral glycosphingoliipids in the sensory neurons of rat sciatic nerve. Neuroscience 20:1023–1029
110. Ledeen RW, Diebler MF, Wu G, Lu ZH, Varoqui H (1993) Ganglioside composition of subcellular fractions, including pre- and postsynaptic membranes, from Torpedo electric organ. Neurochem Res 18:1151–1155
111. Sbaschnig-Agler M, Pfenninger KH, Ledeen RW (1988) Gangliosides and other lipids of the growth cone membrane. J Neurochem 51:212–220
112. Breckenridge WC, Morgan IG, Zanetta JP, Vincendon G (1973) Adult rat brain synaptic vesicles II. Lipid composition. Biochim Biophys Acta 320:681–686
113. Ledeen RW, Parsons SM, Diebler MF, Sbaschnig-Agler M, Lazereg S (1988) Ganglioside composition of synaptic vesicles from Torpedo electric organ. J Neurochem 51:1465–1469
114. Margolis RU, Ledeen RW, Sbaschnig-Agler M, Byrne MC, Kline RL, Douglas II, Margolis RK (1987) Complex carbohydrate composition of large dense-cored vesicles from sympathetic nerve. J Neurochem 49:1839–1844
115. Geisler D, Martinek A, Margolis RU, Margolis RK, Skrivanek JA, Ledeen R, Konig P, Winkler H (1977) Composition and biogenesis of complex carbohydrates of ox adrenal chromaffin granules. Neuroscience 2:685–693
116. Saito M, Sugiyama K (2002) Characterization of nuclear gangliosides in rat brain: concentration, composition, and developmental changes. Arch Biochem Biophys 398:153–159
117. Wu G, Lu Z-H, Ledeen RW (1995) Induced and spontaneous neritogenesis are associated with enhanced expression of ganglioside GM1 in the nuclear membrane. J Neurosci 15:3739–3746
118. Kiebish MA, Han X, Cheng H, Lunceford A, Clarke CF, Moon H et al (2008) Lipidomic analysis and electron transport chain activities in C57BL/6J mouse brain mitochondria. J Neurochem 106:299–312
119. Rippo MR, Malisan F, Ravagnan L, Tomassini B, Condo I, Costantini P et al (2000) GD3 ganglioside directly targets mitochondria in a bcl-2-controlled fashion. FASEB J 14:2047–2054
120. Wang J, Wu G, Miyagi T, Lu Z-H, Ledeen RW (2009) Sialidase occurs in both membranes of the nuclear envelope and hydrolyzes endogenous GD1a. J Neurochem 111:547–554
121. Ravichandra B, Joshi PG (1999) Regulation of transmembrane signaling by ganglioside GM1: interaction of anti-GM1 with neuro2a cells. J Neurochem 73:557–567
122. Sano R, Annunziata I, Patterson A, Moshiach S, Gomero E, Opferman J et al (2009) GM1-ganglioside accumulation at the mitochondria-associated ER membrane links ER stress to Ca-dependent mitochondrial apoptosis. Mol Cell 36:500–511
123. Nowycky MC, Wu G, Ledeen RW (2014) Glycobiology of ion transport in the nervous system. In: Yu RK, Schengrund C-L (eds) Glycobiology of the Nervous System. Springer, New York, pp 321–342
124. Leon A, Facci L, Toffano G, Sonnino S, Tettamanti G (1981) Activation of (Na+, K+)-ATPase by nanomolar concentations of GM1 ganglioside. J Nerochem 37:350–357
125. Susuki K, Rasband MN, Tohyama K, Koibuchi K, Okamoto S, Funakoshi K et al (2007) Anti-GM1 antibodies cause complement-mediated disruption of sodium channel clusters in peripheral motor nerve fibers. J Neurosci 27:3956–3967
126. Qiao GF, Cheng ZF, Huo R, Sui XH, Lu YJ, Li BY (2008) GM1 ganglioside contributes to retain the neuronal conduction and neuronal excitability in visceral and baroreceptor afferents. J Neurochem 106:1637–1645
127. Susuki K, Baba H, Tohyama K, Kanai K, Kuwabara S, Hirata K et al (2007) Gangliosides contribute to stability of paranodal junctions and ion channel clusters in myelinated nerve fibers. Glia 55:746–757

128. de Erausquin GA, Manev H, Guidotti A, Costa E, Brooker G (1990) Gangliosides normalize distorted single-cell intracellular free Ca dynamics after toxic doses of glutamate in cerebellar granule cells. Proc Natl Acad Sci U S A 87:8017–8021
129. Wu G, Xie X, Lu Z-H, Ledeen RW (2001) Cerebellar neurons lacking complex gangliosides degenerate in the presence of depolarizing levels of potassium. Proc Natl Acad Sci U S A 98:307–312
130. Monti E, Bassi MT, Papini N, Riboni M, Manzoni M, Venerando B et al (2000) Identification and expression of NEU3, a novel human sialidase associated to the plasma membrane. Biochem J 349(1):343–351
131. Wu G, Ledeen RW (1991) Stimulation of neurite outgrowth in neuroblastoma cells by neuraminidase: putative role of GM1 ganglioside in differentiation. J Neurochem 56:95–104
132. Fang Y, Wu G, Xie X, Lu Z-H, Ledeen RW (2000) Endogenous GM1 gangliioside of the plasma membrane promotes neuritogenesis by two mechanisms. Neurochem Res 25:931–940
133. Rodriguez JA, Piddini E, Hasegawa T, Miyagi T, Dotti CG (2001) Plasma membrane ganglioside sialidase regulates axonal growth and regeneration in hippocampal neurons in culture. J Neurosci 21:8387–8395
134. Kappagantula S, Andrews MR, Cheah M, Abad-Rodriguez J, Dotti CG, Fawcett JW (2014) Neu3 sialidase-mediated ganglioside conversion is necessary for axon regeneration and is blocked in CNS axons. J Neurosci 34:2477–2492
135. Wu G, Lu Z-H, Nakamura K, Spray DC, Ledeen RW (1996) Trophic effect of cholera toxin B subunit in cultured cerebellar granule neurons: modulation of intracellular calcium by GM1 ganglioside. J Neurosci Res 44:243–254
136. Milani D, Minozzi MC, Petrelli L, Guidolin D, Skaper SD, Spoerri PE (1992) Interaction of ganglioside GM1 with the B subunit of cholera toxin modulates intracellular free calcium in sensory neurons. J Neurosci Res 33:466–475
137. Wu G, Lu ZH, Obukhov AG, Nowycky MC, Ledeen RW (2007) Induction of calcium influx through TRPC5 channels by cross-linking of GM1 ganglioside associated with alpha5beta1 integrin initiates neurite outgrowth. J Neurosci 27:7447–7458
138. Wang J, Lu Z-H, Gabius H-J, Rohowsky-Kochan C, Ledeen RW, Wu G (2009) Cross-linking of GM1 ganglioside by galectin-1 mediates regulatory T cell activity involving TRPC5 channel activation: possible role in suppressing experimental autoimmune encephalomyelitis. J Immunol 182:4036–4045
139. Ledeen RW, Wu G, Andre S, Bleich D, Huet G, Kaltner H et al (2012) Beyond glycoproteins as galectin-counterreceptors: tumor-effector cell growth control via ganglioside GM1. Ann N Y Acad Sci 1253:206–221
140. Xie X, Wu G, Lu Z-H, Ledeen RW (2002) Potentiation of a sodium-calcium exchanger in the nuclear envelope by nuclear GM1 ganglioside. J Neurochem 81:1185–1195
141. Burette A, Rockwood JM, Strehler EE, Weinberg RJ (2003) Isoform-specific distribution of the plasma membrane Ca2+ ATPase in the rat brain. J Comp Neurol 467:464–476
142. Zhao Y, Fan X, Yang F, Zhang X (2004) Gangliosides modulate the activity of the plasma membrane Ca(2+) ATPase from porcine brain synaptosomes. Arch Biochem Biophys 427:204–212
143. Jiang L, Bechtel MD, Bean JL, Winefield R, Williams TD, Zaidi A et al (2014) Effects of gangliosides on the activity of the plasma membrane Ca2+-ATPase. Biochim Biophys Acta Biomembr 1838:1255–1265
144. Tessitore A, del P Martin M, Sano R, Ma Y, Mann L, Ingrassia A et al (2004) GM1-ganglioside-mediated activation of the unfolded protein response causes neuronal death in a neurodegenerative gangliosidosis. Mol Cell 15:753–766
145. d'Azzo A, Tessitore A, Sano R (2006) Gangliosides as apoptotic signals in ER stress response. Cell Death Diff 13:404–414
146. Pelled D, Lloyd-Evans E, Riebeling C, Jeyakumar M, Platt FM, Futerman AH (2003) Inhibition of calcium uptake via the sarco/endoplasmic reticulum Ca2+-ATPase in a mouse model of Sandhoff disease and prevention by treatment with butyldeoxynojirimycin. J Biol Chem 278(32):29496–29501
147. Wang Y, Tsui Z, Yang F (1999) Antagonistic effect of ganglioside GM1 and GM3 on the activity and conformation of sarcoplasmic reticulum Ca^{2+}-ATPase. FEBS Lett 457:144–148
148. Wu G, Lu Z-H, Xie X, Ledeen RW (2004) Susceptibility of cerebellar granule neurons from GM2/GD2 synthase-null mice to apoptosis induced by glutamate excitotoxicity and elevated KCl: rescue by GM1 and LIGA20. Glycoconj J 21:305–313
149. Wu G, Xie X, Lu Z-H, Ledeen RW (2009) Sodium-calcium exchanger complexed with GM1 ganglioside in the nuclear membrane

transfers calcium from nucleoplasm to endoplasmic reticulum. Proc Natl Acad Sci U S A 106:10829–10834
150. Galva C, Artigas P, Gatto C (2012) Nuclear Na^+/K^+-ATPase plays an active role in nucleoplasmic Ca^{2+} homeostasis. J Cell Sci 125:6137–6147
151. Xie X, Wu G, Lu Z-H, Rohowsky-Kochan C, Ledeen RW (2004) Presence of sodium-calcium exchanger/GM1 complex in the nuclear envelope of non-neural cells: nature of the exchanger-GM1 interaction. Neurochem Res 29:2135–2146
152. Xie X, Wu G, Ledeen RW (2004) C6 cells express a sodium-calcium exchanger/GM1 complex in the nuclear enveklope but have no exchanger in the plasma membrane: comparison to astrocytes. J Neurosci Res 76:363–357
153. Kaucic K, Liu Y, Ladisch S (2006) Modulation of growth factor signaling by gangliosides: positive or negative? Methods Enzymol 417:168–185
154. Rusnati M, Tanghetti E, Urbinati C, Tulipano G, Marchesini S, Ziche M, Presta M (1999) Interaction of fibroblast growth factor-2 (FGF-2) with free gsangliosides: biochemical characterization and biological consequences in endothelial cell cultures. Mol Biol Cell 10:313–327
155. Rusnati M, Urbinati C, Tanghetti E, Dell'Era P, Lortat-Jacob H, Presta M (2002) Cell membrane GM1 ganglioside is a functional coreceptor for fibroblast growth factor 2. Proc Natl Acad Sci U S A 99:4367–4372
156. Boldin SA, Futerman ASH (2000) Up-reguloation of glucosylceramide synthesis upon stimulation of axonal growth by basic fibroblast growth factor. Evidence for post-translational modification of glucosylceramide synthase. J Biol Chem 275:9905–9909
157. Nojiri H, Stroud M, Hakomori S-I (1991) A specific type of ganglioside as a modulator of insulin-dependent cell growth and insulin receptor tyrosine kinase activity. J Biol Chem 266:4531–4537
158. Kabayama K, Sato T, Saito K, Loberto N, Prinetti A, Sonnino S et al (2007) Dissociation of the insulin receptor and caveolin-1 complex by ganglioside GM3 in the state of insulin resistance. Proc Natl Acad Sci U S A 104:13678–13683
159. Schengrund C-L (2015) Gangliosides: glycosphingolipids essential for normal neural development and function. Trends Biochem Sci 40:397–406
160. Shen KF, Crain SM (1989) Dual modulation of the action potential duration of mouse dorsal root ganglion neurons in culture. Brain Res 491:227–242
161. Jin W, Lee NM, Loh HH, Thayer SA (1992) Dual excitatory and inhibitory effects of opioids on intracellular calcium in neuroblastoma x glioma hybrid NG108-15 cells. Mol Pharm 42:1083–1089
162. Shen KF, Crain SM (1990) Cholera toxin B-subunit blocks opioid excitatory effects on sensory neuron action potentials indicating the GM1 ganglioside may regulate G_s-linked opioid receptor functions. Brain Res 531:1–7
163. Wu G, Lu Z-H, Ledeen RW (1997) Interaction of δ-opioid receptor with GM1 ganglioside: conversion from inhibitory to excitatory mode. Mol Brain Res 44:341–346
164. Saito M, Frielle T, Benovic JL, Ledeen RW (1995) Modulation by GM1 ganglioside of β1-adrenergic receptor-induced cyclic AMP formation in Sf9 cells. Biochim Biophys Acta 1267:1–5
165. Wu G, Lu Z-H, Ledeen RW (1996) GM1 ganglioside modulates prostaglandin E1 stimulated adenylyl cyclase in Neuro-2A cells. Glycoconj J 13:235–239
166. Lundius EG, Stroth N, Vukojevic V, Terenius L, Svenningsson P (2012) Functional GPR37 trafficking protects against toxicity induced by 6-OHDA, MPP+ or rotenone in a catecholaminergic cell line. J Neurochem 124:410–417
167. Lundius EG, Vukojevic V, Hertz E, Stroth N, Cederlund A, Hiraiwa M et al (2014) GPR37 protein trafficking to the plasma membrane regulated by prosaposin and GM1 gangliosides promotes cell viability. J Biol Chem 289:4660–4673
168. Mutoh T, Tokuda A, Miyadai T, Hamaguchi M, Fujiki N (1995) Ganglioside GM1 binds to the Trk protein and regulates receptor function. Proc Natl Acad Sci U S A 92:5087–5091
169. Rabin SJ, Mocchetti I (1995) GM1 ganglioside activates the high-affinity nerve growth factor receptor TrkA. J Neurochem 65:347–354
170. Pitto M, Mutoh T, Kuriyama M, Ferraretto A, Palestini P, Masserini M (1998) Influence of endogenous GM1 ganglioside on TrkB activity in cultured neurons. FEBS Lett 439:93–96
171. Mutoh T, Hamano T, Tokuda A, Kuriyama M (2000) Unglycosylated Trk protein does not co-localize nor associate with ganglioside GM1 in stable clone of PC12 cells overexpressing Trk (PCtrk cells). Glycoconj J 17:233–237
172. Mutoh T, Hamano T, Yano S, Koga H, Yamamoto H, Furukawa K, Ledeen RW

(2002) Stable transfection of GM1 synthase gene into GM1-deficient NG108-15 cells, CR-72 cells, rescues the responsiveness of Trk-neurotrophin receptor to its ligand, NGF. Neurochem Res 27:801–806
173. Volk BW, Aronson SM (eds) (1972) Sphingolipids, sphingolipidoses and allied disorders. Advances in experimental medicine and biology, vol 19. Plenum, London
174. Kolter T (2012) Ganglioside Biochemistry. ISRN Biochem 2012(506160):1–36
175. Ceccarelli B, Aporti F, Finesso M (1976) Effects of brain gangliosides on functional recovery in experimental regeneration and reinnervation. Adv Exp Med Biol 71:275–293
176. Ledeen RW, Hogan EL, Tettamanti G, Yates AJ, Yu RK (eds) (1988) New trends in ganglioside research. Neurochemical and neuroregenerative aspects. Fidia research series, vol 14. Liviana Press, Padova
177. Saito M, Saito M, Berg MJ, Guidotti A, Marks N (1999) Gangliosides attenuate ethanol-induced apoptosis in rat cerebellar granule neurons. Neurochem Res 24:1107–1115
178. Kidd SK, Mettil W, Anderson DW, Schneider JS (2012) Ganglioside regulation in human substantia nigra and its relation to Parkinson's disease. Program no. 754.16. 2012 Neuroscience meeting planner. Society for Neuroscience, New Orleans. Online: 2012
179. Jellinger KA (2011) Synuclein deposition and non-motor symptoms in Parkinson disease. J Neurol Sci 310:107–111
180. Lang AE (2011) A critical appraisal of the premotor symptoms of Parkinson's disease: clinical features and detection strategies. Mov Disord 26:775–783
181. Halliday GM, Barker RA, Rowe DB (eds) (2011) Non-dopamine Lesions in Parkinson's disease. Oxford University Press, Oxford
182. Toffano G, Savoini G, Moroni F, Lombardi G, Calza L, Agnati LF (1983) GM1 ganglioside stimulates the regeneration of dopaminergic neurons in the central nervous system. Brain Res 261:163–166
183. Hadjiconstantinou M, Rossetti ZL, Paxton RC, Neff NH (1986) Administration of GM1 ganglioside restores the dopamine content in striatum after chronic treatment with MPTP. Neuropharmacology 25:1075–1077
184. Schneider JS, Yuwiler A (1989) GM1 ganglioside treatment promotes recovery of striatal dopamine concentrations in the mouse model of MPTP-induced parkinsonism. Exp Neurol 105:177–183
185. Tilson HA, Harry GJ, Nanry K, Hudson PM, Hong JS (1988) Ganglioside interactions with the dopaminergic system of rats. J Neurosci Res 19:88–93
186. Schneider JS, Pope A, Simpson K, Taggart J, Smith MG, DiStefano L (1992) Recovery from experimental parkinsonism in primates with GM1 ganglioside treatment. Science 256:843–846
187. Herrero MT, Perez-Otano I, Oset C, Kastner A, Hirsch EC, Agid Y et al (1993) GM-1 ganglioside promotes the recovery of surviving midbrain dopaminergic neurons in MPTP-treated monkeys. Neuroscience 56:965–972
188. Schneider JS, DiStefano L (1995) Response of the damaged dopamine system to GM1 and semisynthetic gangliosides: effects of dose and extent of lesion. Neuropharmacology 34:489–493
189. Schneider JS, Sendek S, Daskalakis C, Cambi F (2010) GM1 ganglioside in Parkinson's disease: results of a five-year open study. J Neurol Sci 292:45–51
190. Schneider JS, Gollomp SM, Sendek S, Colcher A, Cambi F, Du W (2013) A randomized, controlled, delayed start trial of GM1 ganglioside in treated Parkinson's disease patients. J Neurol Sci 324:140–148
191. Wu G, Lu Z-H, Kulkarni N, Amin R, Ledeen RW (2011) Mice lacking major brain gangliosides develop parkinsonism. Neurochem Res 36:1706–1714
192. Pascual A, Hidalgo-Figueroa M, Piruat JI, Pintado CO, Gomez-Diaz R, Lopez-Barneo J (2008) Absollute requirement of GDNF for adult catecholaminergic neuron survival. Nat Neurosci 11:755–761
193. Spillantini MG, Schmidt ML, Lee VM, Trojanowski JQ, Jakes R, Goedert M (1997) Alpha-synuclein in Lewy bodies. Nature 388:839–840
194. Braak H, de Vos RAI, Bohl J, Del Tredici K (2006) Gastric α-synuclein immunoreactive inclusions in Meissner's and Auerbach's plexuses in cases staged for Parkinson's disease-related brain pathology. Neurosci Lett 396:67–72
195. Braak H, Del Tredici K, Rub U, de Vos RAI, Steur ENHJ, Braak E (2003) Staging of brain pathology related to sporadic Parkinson's disease. Neurobiol Aging 24:197–211
196. Schneider JS (2014) Gangliosides and glycolipids in neurodegenerative disorders. In: Yu RK, Schengrund C-L (eds) Glycobiology of the nervous system. Advances in neurobiology, vol 9. Springer, New York, pp 449–464

197. Alcalay RN, Levy OA, Waters CC, Fahn S, Ford B, Kuo S-H et al (2015) Glucocerebrosidase activity in Parkinson's disease with and without GBA mutation. Brain 138:2648–2658
198. Shachar T, Lo Bianco C, Recchia A, Wiessner C, Raas-Rothschild A, Futerman AH (2011) Lysosomal storage disorders and Parkinson's disease: Gaucher disease and beyond. Mov Disord 26:1593–1604
199. Roze E, Paschke E, Lopez N, Eck T, Yoshida K, Maurel-Olivier A et al (2005) Dystonia and parkinsonism in GM1 type 3 gangliosidosis. Mov Disord 20:1366–1369
200. Davies P, Maloney AJ (1976) Selective loss of central cholinergic neurons in Alzheimer's disease. Lancet 2:1403
201. Lulita MF, Cuello AC (2014) Nerve growth factor metabolic dysfunction in Alzheimer's disease and Down syndrome. Trends Pharm Sci 35:338–348
202. Svennerholm L, Gottfries C-G (1994) Membrane lipids, selectively diminished in Alzheimer brains, suggest synapse loss as a primary event in early onset (type I) and demyelination in late-onset form (type II). J Neurochem 62:1039–1047
203. Brooksbank BW, McGovern J (1989) Gangliosides in the brain in adult Down's syndrome and Alzheimer's disease. Mol Chem Neuropathol 11:143–156
204. Crino PB, Ullman MD, Vogt BA, Bird ED, Volicer L (1989) Brain gangliosides in dementia of the Alzheimer type. Arch Neurol 46:398–401
205. Valdes-Gonzalez T, Goto-Inoue N, Hirano W, Ishiyama H, Hayasaka T, Setou M, Taki T (2011) New approach for glyco-and lipidomics – molecular scanning of human brain gangliosides by TLC-Blot and MALDI-QIT-TOF MS. J Neurochem 116:678–683
206. Kracun I, Rosner H, Drnovsek V, Heffer-Lauc M, Cosovic C, Lauc G (1991) Human brain gangliosides in development, aging and disease. Int J Dev Biol 35:289–295
207. Ariga T, McDonald MP, Yu RK (2008) Role of ganglioside metabolism in the pathogenesis of Alzheimer's disease – a review. J Lipid Res 49:1157–1175
208. Hayashi H, Kimura N, Yamaguchi H, Hasegawa K, Yokoseki T, Shibata M et al (2004) A seed for Alzheimer amyloid in the brain. J Neurosci 24:4894–4902
209. Yanagisawa K (2007) Role of gangliosides in Alzheimer's disease. Biochim Biophys Acta 1768:1943–1951
210. Yanagisawa M, Ariga T, Yu RK (2010) Cytotoxic effects of GM1 ganglioside and amyloid β-peptide on mouse embryonic neural stem cells. ASN Neuro 2:e00029
211. Svennerholm L, Brane G, Karlsson I, Lekman A, Ramstrom I, Wikkelso C (2002) Alzheimer disease – effect of continuous intracerebroventricular treatment with GM1 ganglioside and a systematic activation programme. Dement Geriatr Cogn Disord 14:128–136
212. Kreutz F, Frozza RL, Breier AC, de Oliveira VA, Horn AP, Pettenuzzo LF et al (2011) Amyloid-β induced toxicity involves ganglioside expression and is sensitive to GM1 neuroprotective action. Neurochem Int 59:648–655
213. Ariga T, Yu RK (1999) GM1 inhibits amyloid beta-protein-induced cytokine release. Neurochem Res 24:219–226
214. Sokolova TV, Zakharova IO, Furaev VV, Rychkova MP, Avrova NF (2007) Neuroprotective effect of ganglioside GM1 on the cytotoxic action of hydrogen perkoxide and amyloid β-peptide in PC12 cells. Neurochem Res 32:1302–1313
215. Bernardo A, Harrison FE, McCord M, Zhao J, Bruchey A, Davies SS et al (2009) Elimination of GD3 synthase improves memory and reduces amyloid-β beta plaque load in transgenic mice. Neurobiol Aging 30:1777–1791
216. Copani A, Melchiorri D, Caricasole A, Martini F, Sale P, Carnevale R et al (2002) β-amyloid-induced synthesis of the ganglioside GD3 is a requisite for cell cycle reactivation and apoptosis in neurons. J Neurosci 22:3963–3968
217. Galpern WSR, Lang AE (2006) Interface between tauopathies and synucleinopathies: a tale of two proteins. Ann Neurol 59:449–458
218. Desplats PA, Denny CA, Kass KE, Gilmartin T, Head SR, Sutcliffe JG et al (2007) Glycolipid and ganglioside metabolism imbalances in Huntington's disease. Neurobiol Dis 27:265–277
219. Maglione V, Marchi P, Di Pardo A, Lingrell S, Horkey M, Tidmarsh E, Sipione S (2010) Impaired ganglioside metabolism in Huntington's disease and neuroprotective role of GM1. J Neurosci 30:4072–4080
220. Di Pardo A, Maglione V, Alpaugh M, Horkey M, Atwal RS, Sassone J et al (2012) Ganglioside GM1 induces phosphorylation of mutant huntingtin and restores normal motor behavior in Huntington disease mice. Proc Natl Acad Sci U S A 109:3528–3533
221. Boukhris A, Schule R, Loureiro JL, Lourenco CM, Mundwiller E, Gonzalwz MA et al (2013) Alteration of ganglioside biosynthesis responsible for complex hereditary spastic paraplegia. Am J Hum Gen 93:118–123

222. Chiavegatto S, Sun J, Nelson RJ, Schnaar RL (2000) A functional role for complex gangliosides: motor deficits GM2/GD2 synthase knockout mice. Exp Neurol 166:227–234
223. Simpson MA, Cross H, Proukakis C, Priestman DA, Neville DCA, Reinkensmeier G (2004) Infantile-onset symptomatic epilepsy syndrome caused by a homozygous loss-of-function mutation of GM3 synthase. Nat Gen 36:1225–1229
224. Yoshikawa M, Go S, Takasaki K, Kakazu Y, Ohashi M, Nagafuku M et al (2009) Mice lacking ganglioside GM3 synthase exhibit complete hearing loss due to selective degeneration of the organ of Corti. Proc Natl Acad Sci U S A 106:9483–9488
225. Niimi K, Nishioka C, Miyamoto T, Takahashi E, Miyoshi I, Itakura C, Yamashita T (2011) Impairment of neuropsychological behaviors in ganglioside GM3-knockout mice. Biochem Biophys Res Commun 406:524–528
226. Tell S (2012) Medical genetic research on the Amish: from genetic tourism to community health centers. Einstein J Biol Med 2012:18–24
227. Allende ML, Proia RL (2014) Simplifying complexity: genetically resculpting glycosphingolipid synthesis pathways in mice to reveal function. Glyconj J 31:613–622
228. Itokazu Y, Tsai Y-T, Yu RK (2016) Epigenetic regulation of ganglioside expression in neural stem cells and neuronal cells. Glycoconj J. https://doi.org/10.1007/s10719-016-9719-6
229. Sampson TR, Debellus JW, Thron T, Wittung-Stafshede P, Knight R, Mazmanian SK (2016) Gut microbiota regulate motor deficits and neuroinflammation in a model of Parkinson's disease. Cell 167:1469–1480

Chapter 3

Scattering Techniques and Ganglioside Aggregates: Laser Light, Neutron, and X-Ray Scattering

Elena Del Favero, Paola Brocca, and Laura Cantù

Abstract

Scattering techniques are applied to studying the structural features of ganglioside aggregates in solution. Here it is described how different probing radiations allow to access different structural and dynamical parameters on different lengthscales. Besides a brief but comprehensive description of the scattering measurements, several practical suggestions are given concerning the experiments and the data analysis.

Key words Aggregation, Packing parameter, Specific volume, Contrast, Refractive index, Scattering length density, Form factor, Structure factor, Interacting particles, Hydrodynamic radius, Effective charge

1 General Concepts on Scattering of Radiation

Scattering of radiation is caused by dishomogeneities in the medium it crosses. Pure solvents themselves scatter radiation, due to dishomogeneities on different lengthscales, from molecules to density fluctuations. In a solution or suspension, particles generate an excess scattering with respect to the solvent, according to their concentration, their mass and their contrast against the hosting medium. Generally speaking, scattering increases linearly with the number concentration of same particles, quadratically with their mass given the same number concentration, and quadratically with their contrast. Moreover, scattering of radiation is not isotropic, but it is modulated in space, depending on the relative size of the dishomogeneities and of the wavelength of the radiation. Then, such modulation is connected to the shape of the individual scatterers, to their internal contrast profile and also to their spatial arrangement, if not randomly distributed. General and more specific references on scattering techniques applied to colloids solutions can be found in reference [1].

Sandro Sonnino and Alessandro Prinetti (eds.), *Gangliosides: Methods and Protocols*, Methods in Molecular Biology, vol. 1804, https://doi.org/10.1007/978-1-4939-8552-4_3, © Springer Science+Business Media, LLC, part of Springer Nature 2018

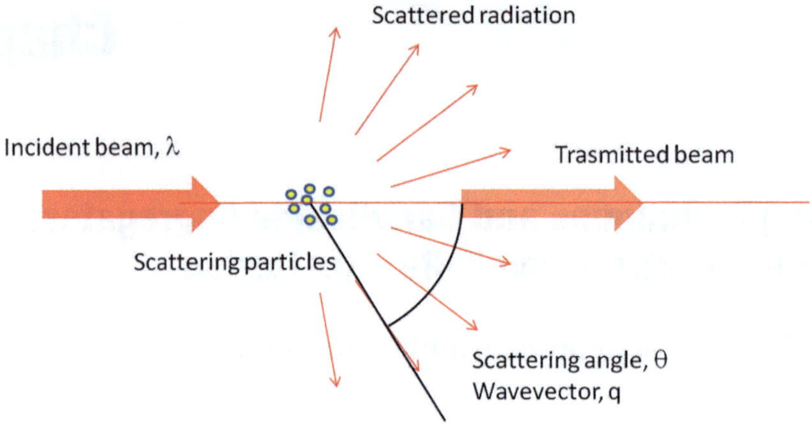

Fig. 1 Scheme of a scattering experiment. An incident beam, with wavelength λ, traverses a solution hosting *N* nanoparticles per unit volume, with mass *M*. Part of the beam is scattered in directions different from that of the trasmitted beam, at scattering angles θ's, provided that the particles have some contrast with the hosting medium. $q = 2\pi/\lambda \sin(\theta/2)$ is the wavevector

The general scheme of a scattering experiment is sketched in Fig. 1.

The general simplified form for the excess scattered intensity can be written as:

$$I(q) = A\,N\,M^2\,(\text{contrast})^2\,P(q)\,S(q),$$

where A is a constant accounting for specific experimental details, N is the number density of particles with mass M, $P(q)$ is the form factor, an oscillating function describing the shape of the particles, $S(q)$ is the structure factor, depending on the particles arrangement in space. $q = 2\pi/\lambda \sin(\theta/2)$ is the wavevector, where λ is the wavelength of the incident radiation and θ the scattering angle. The form factor $P(q) = 1$ in the limit of $q = 0$. $S(q)$ is an oscillating function as well, that flattens to 1 for randomly distributed non-interacting particles in dilute solution. This simple general expression for the excess scattered intensity can be enriched to account for other parameters, like particle polydispersity for example, or for nonhomogeneous particles.

Then, scattering techniques are well suited for studying the physicochemical properties of nanoparticles in solution, meanwhile being largely noninvasive. In particular, we will deal with aggregates of gangliosides in aqueous solution, mainly micelles with size in the range 10–15 nm. Such structures consist of a core region hosting the hydrophobic chains of gangliosides, shielded by a hydrophilic shell constituted by their sugar moiety. We will focus on scattering of the following radiations: laser light ($\lambda \approx 500$ nm, larger than the ganglioside micelles), X-ray, and neutrons ($\lambda \approx 0.1$ nm and $\lambda \approx 1$ nm, respectively, both smaller than the ganglioside micelles). Their combined use allows investigating the colloidal particles in solution at different level of detail.

2 Contrast and the Aggregate Visibility

In order to generate scattering, particles have to contrast with the hosting medium. The definition of contrast depends on the radiation, and moreover the physical parameter of interest depends on the relative size of the nanoparticle and the radiation wavelength.

In fact, the wavelength of the visible laser light is in the range of 500 nm, larger (or similar) than the size of nanoparticles, which are seen mostly as globular uniform particles, contributing to the overall refractive index of the solution. So the contrast term for laser light scattering is given by the rate of refractive index increment with solute concentration, $\mathrm{d}n/\mathrm{d}c$. This parameter has to be measured experimentally, with a refractometer, by determining the slope of the linear dependence of the refractive index of solutions at different dilution as a function of the concentration of solute,

$$n(c) = n_{\text{solvent}} + \mathrm{d}n / \mathrm{d}c \times c.$$

Within the class of gangliosides, rescaling with the mol fraction of head/tail balance allows to nicely reproduce the $\mathrm{d}n/\mathrm{d}c$ values, as seen in Table 1, by comparing the experimental $(\mathrm{d}n/\mathrm{d}c)_{\text{exp}}$ and calculated $(\mathrm{d}n/\mathrm{d}c)_{\text{calc}}$ values [2].

Conversely, the wavelength of X-ray radiation, typically in the range of 0.1 nm for scattering experiments, is smaller than the size of aggregates. Their internal structure is accessible with such radiation, provided that contrast exists between the different regions.

Table 1
Partial specific volume, refractive index increment, neutron scattering length density, and X-ray electron density for different gangliosides. The experimental values of the specific volume, determined by densitometry, are compared with the calculated ones obtained with $\langle v\rangle_{calc} = \langle v\rangle_t X_t + \langle v\rangle_h X_h$, where $\langle v\rangle_t = 1.1437 \pm 0.015$ cm^3/g, $\langle v\rangle_h = 0.6043 \pm 0.0065$ cm^3/g and X_t, X_h are the weight fractions of the tails and head moieties of the molecule, respectively. The experimental values of the refractive index increment are compared with the calculated ones obtained with $dn/dc = \langle dn/dc\rangle_t X_t + \langle dn/dc\rangle_h X_h$, where $\langle dn/dc\rangle_t = 0.08987 \pm 0.00764$ cm^3/g and $\langle dn/dc\rangle_h = 0.1783 \pm 0.00357$ cm^3/g are the individual contributions of heads and tails

	X_h: X_t (w:w)	$\langle v\rangle_{exp}$ (cm^3/g)	$\mathrm{d}n/\mathrm{d}c_{exp}$ (cm^3/g)	$\langle v\rangle_{calc}$ (cm^3/g)	$\mathrm{d}n/\mathrm{d}c_{calc}$ (cm^3/g)	Neutron SLD in D_2O (10^{10} cm^{-2})	X-ray electron density (el A^{-3})
Head		0.6043	0.1783			4.15 (2.29 in H_2O)	0.54
Tails		1.1437	0.08987			−0.09	0.3
GT1b	26:74	0.744	0.155	0.7445	0.1553	2.46	0.44
GD1a	30:70	0.7665	0.152	0.7661	0.1518	2.25	0.43
GM1	36:64	0.7976	0.146	0.7985	0.1465	2.05	0.42
GM2	40:60			0.8201	0.1429	1.78	0.41

The contrast term for X-rays is usually expressed as a function of the electron density, namely, the difference in the electron density between the adjacent regions. This parameter can be calculated for different molecules, starting from their chemical formula, thus the number of electrons, and molecular volume. For ganglioside micelles, the electron density of the sugar shell is higher than both those of the solvent water and the hydrophobic core, which are quite similar. The X-ray investigation is then particularly sensitive to the properties of the hydrophilic region of the ganglioside aggregates.

Also neutron beams are used for scattering experiments, with an equivalent wavelength of the order of 1 nm, smaller than the size of aggregates. The contrast for neutrons is expressed in terms of the scattering length density (SLD). Also this parameter can be calculated for different molecules, starting from their chemical formula, thus the number of specific atoms, and molecular volume [3]. Again, the difference between the SLDs of adjacent regions of the nanoparticle determines the visibility of its internal structure. In standard neutron scattering experiments on particles in solution, D_2O is used as a solvent instead of H_2O, as the two give rise to a very different amount and type of the neutron scattering. This is a very interesting feature of the neutron scattering investigation, connected to an H-D isotopic substitution that can enhance the visibility of particles (or of regions of particles) without significative chemical drawback. H-nuclei originate high incoherent scattering, while D-nuclei give low mostly coherent scattering [3].

In Fig. 2, the contrast profiles of a globular core–shell particle sketching a ganglioside micelle are drawn, for the three considered

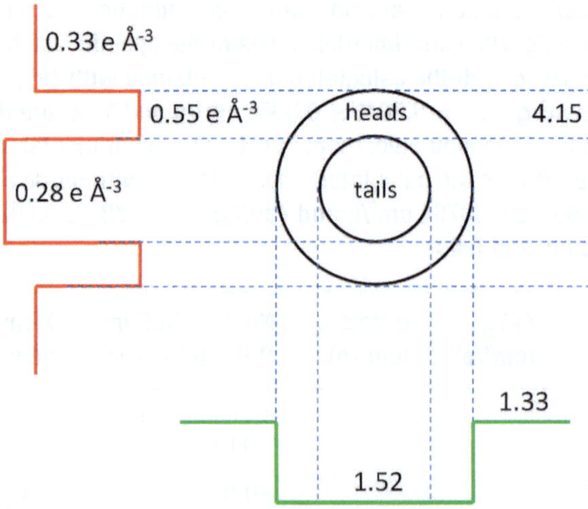

Fig. 2 Contrast profiles of the same globular core–shell particle, sketching a ganglioside micelle in aqueous solution, for the three considered radiations. The inner core hosts the hydrophobic chains (tails), the shell is made by the sugar groups (heads). *Left side, red.* For X-rays, the electron density profile (electrons/Å^3). *Right side, blue.* For neutrons, the scattering length density (SLD, cm^{-2}). The solvent is D_2O. Exchangeable protons in the sugar headgroup have been replaced by deuterons. *Bottom, green.* For laser visible light, the refractive index

radiations. In a real aggregate, the contrast profile is not step-like, hedges being smeared by the interfacial rugosity. Nonetheless, often the simple step-like profile is enough to fit the experimental data. Moreover, the outer shell contrast is affected by water penetration. In this respect, the assessment of the effective contrast of this region from experimental data allows to give an estimate of the number of water molecules per ganglioside headgroup entering the sugar layer of the aggregate. In Table 1, parameters used to define the contrast terms are reported for some gangliosides.

3 The Aggregates Mass

For dilute solution of noninteracting ($S(q) = 1$) small homogeneous particles and in the limit of small scattering wavevector ($P(q) = 1$) the expression of the excess scattering becomes very simple. Then, the so-called *static* laser light scattering is useful to determine the mass of ganglioside micelles in dilute solution (typically of the order of mM), provided that some salt (typically NaCl, of the order of 100 mM) is added in order to screen the electrostatic interactions originated by ganglioside dissociation, as we will point out in a following section. In this case of small non interacting micelles, measurements can be done at the convenient 90° scattering angle. Once the excess scattered intensity has been normalized by that of the solvent with known Rayleigh ratio (water for aqueous solutions), [4] it can be expressed as:

$$I_{\text{norm}} = A\,c\,M\left(\mathrm{d}n / \mathrm{d}c\right)^2,$$

where $c = NM$ is the w/v concentration of solute, gangliosides in our case, and A is an instrumental constant. Loss of accuracy in this measurement occurs if dust is present in the solution, as dust particles originate strong scattering. For this reason, solutions have to be carefully and repeatedly filtered on appropriate membranes before measurement. For gangliosides, polycarbonate membranes with 0.2 μm pore size are best suited. Moreover, measuring cells made of optical glass, optically polished, are to be used, in order to reduce flare from the walls. Polydispersity influences of course the average value of M. So carefully purified gangliosides have to be used, in order to avoid the presence of lipid impurities, readily solubilized by gangliosides in swollen micelles, or nanoemulsion particles. In pure and dust free condition, a 1 mM GM1aqueous solution (100 mM NaCl) gives an excess scattered intensity 100 times higher than that of the solvent. We point out that, as the intrinsic weighting function in light scattering is the scattered intensity, the so-determined weight-average value of M is shifted to higher values with respect to the number-average. Of course, if the particles in solution have a non-monomodal distribution, M cannot be simply deduced by the *static* laser light scattering technique. In the case of accurately purified gangliosides, forming nicely monodisperse micelles in solution, the weigh-average aggregate mass was determined by laser light scattering, as reported in Table 2 [5–10]. The light scattering apparati used are noncommercial versatile instruments, each one equipped with a thermostatted cell with a filtration inlet line, an Argon-ion or a Nd-Yag laser, vertically polarized, operating on green line and a photomultiplier detection system on the equatorial plane, described in detail in [11].

As another example connected to the influence of ganglioside aggregates mass to the scattering, Fig. 3 shows the ongoing reduction of scattered intensity of X-rays from a dilute micellar solution of GD1a, due to the progressive decrease of the aggregate mass following the removal of the external sialic acid of GD1a operated by sialidase. A 3 mM GD1a solution, 100 mM NaCl, was carefully dissolved at room temperature, then put in a polycarbonate capillary, 2 mm diameter, closed on both sides with silicon caps and placed in front of the beam. Just before shining the X-ray beam, sialidase was injected in the solution, and the evolution of the scattered intensity was followed in time. The effect is clearly visible in Fig. 3. In the right panel, the reduction of extrapolated intensity at q = 0 as a function of time is reported, corresponding to a reduction

Table 2
Some micellar parameters of various gangliosides, deduced by static laser light scattering. The micellar mass *M* and aggregation number *N*. The molecular surface area a_0 and the packing parameter $P = V/a_0l_c$. Values are given for the *standing* conformation. In (parentheses) the parameters corresponding to the *laying* conformation are also given for GM2, GM1, GD1a, and GD1b gangliosides. The gangliosides GM3 and GM4 do not form oblate micelles in solution, but bigger aggregates, with bilayer type structure

	Mass of the aggregate (kDa)	Aggregation number	Surface area per monomer (Å^2)	Packing parameter
GM2	740 (365)	529 (261)	92.0 (96.5)	0.440 (0.423)
GM1$_{acetyl}$	102	76	67.0	0.37
GM1	470 (320)	301 (205)	95.4 (99.5)	0.428 (0.411)
GD1a	418 (318)	226 (172)	98.1 (101.7)	0.416 (0.402)
GalNac-GD1a	509	246	97.0	0.421
GD1b	311 (280)	168 (151)	100.8 (104.6)	0.405 (0.391)
GD1b-L	424	229	97.6	0.418
GT1b	378	176	100.8	0.405
GM4	18,300	18,000	~ 80	> 0.5
GM3	16,700	14,000	~ 80	> 0.5

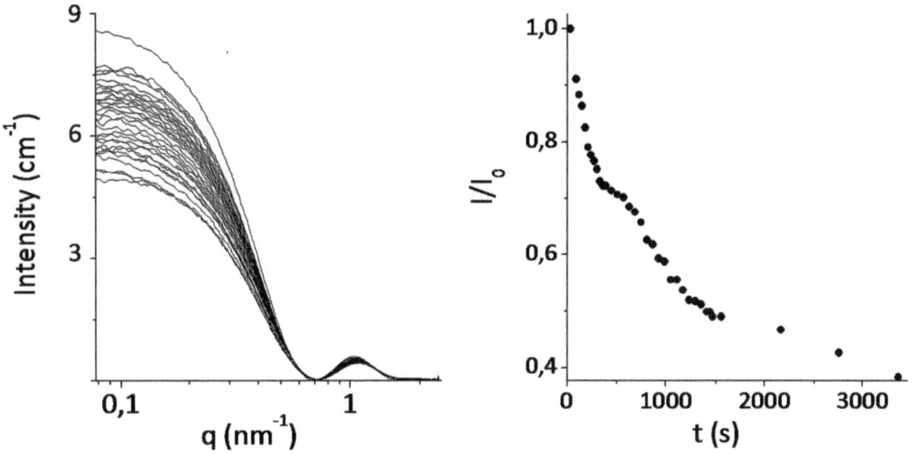

Fig. 3 *Left panel.* X-rays scattered intensity from a micellar solution of GD1a, submitted to the removal of the external sialic acid by sialidase. The decrease of the scattered intensity is connected to the reduction of aggregate mass. Data are reported in lin-log scale. *Right panel.* The normalized extrapolated intensity at $q = 0$ as a function of time displays a steplike reduction, normalization is towards the initial value, I_0

in the aggregate molecular mass (from 418 to 353 kDa), showing a step-like behavior. On this basis, we could outline a model for enzymatic digestion of aggregated gangliosides, the enzyme scooting all over one micelle surface, and removing all external sialic groups of the GD1a monomers there condensed, before addressing a next one [12, 13]. The same experiment could hardly be done by laser light scattering, as the solution containing the enzyme both could not be efficiently filtered (for dust removal) without loss of activity and it was itself contributing a too high scattered intensity in the region of low q, overwhelming with respect to the dilute micelles. Conversely, its contribution in the q region accessible to X-rays was negligible, and the kinetics of micelles digestion could be clearly observed.

As a further application to the behavior of ganglioside micelles, Fig. 4 shows the variation in the neutron scattering from a 1 mM micellar solution of GM1, in NaCl 100 mM, following a collective surface rearrangement of ganglioside headgroups, passing from a *standing* to a *laying* conformation with respect to the aggregate surface. This conformational transition can be induced by different external solicitations, like thermal annealing or crowding or enzymatic action [14–22], and is common to most gangliosides. In the case presented in Fig. 4, the solution was investigated at room temperature, both just after careful dissolution and after heating at 60 °C for 30 min, directly in the measuring cell. As thermal treatment is widely used to accelerate the dissolution of solute, one should be aware that this can induce a modification in the surface organization of gangliosides, influencing the exposure of headgroup portions to the solvent. The *laying* conformation requires a

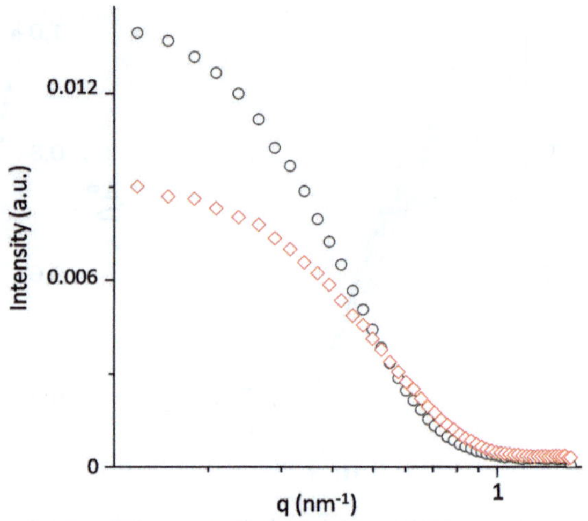

Fig. 4 Neutron scattered intensity from a 1 mM micellar solution of GM1, NaCl 100 mM, upon conversion from the *standing* (black upper curve) to the *laying* (red lower curve) conformation. Lowering of the scattered intensity is connected to the reduction of aggregate mass following the increase in the required surface area per monomer a_0. Data are reported in lin-log scale

wider surface area per headgroup, then, a smaller and more curved micelle is formed. The variation in the aggregate mass is readily detected by scattering, and the molecular rearrangement at the basis of this phenomenon, although small, is clearly revealed through the amplifying response of micellar aggregation. The *laying*-conformation micellar parameters for GM2, GM1, GD1a, and GD1b are reported (in parentheses) in Table 2.

4 The Aggregates Shape

As already evident from the scattered intensity profiles of Figs. 3 and 4, scattering of radiation by noninteracting dilute particles in solution is not isotropic, but it is modulated in space. Modulation depends on the relative size of the particles and on the wavelength of the radiation. In fact, the wavelength of the radiation dictates the range of accessible wavevectors q's in correspondence to the geometrical (0–π) scattering angles θ's. This is illustrated in Fig. 5, where the form factor $P(q)$ of homogeneous spheres, 6 nm in radius, similar in size to ganglioside micelles, is reported as a function of the wavevector q. The analytical form of the form factor $P(q)$ of the homogeneous sphere can be found, for example in [23, 24]. The q-range accessible to visible laser light is confined to below $q_{\mathrm{maxLS}} \approx 3 \times 10^{-2}\ \mathrm{nm}^{-1}$ much shallower than for X-ray or neutrons, due to the large difference in wavelength. In that shallow range, the variation of scattered intensity is very small,

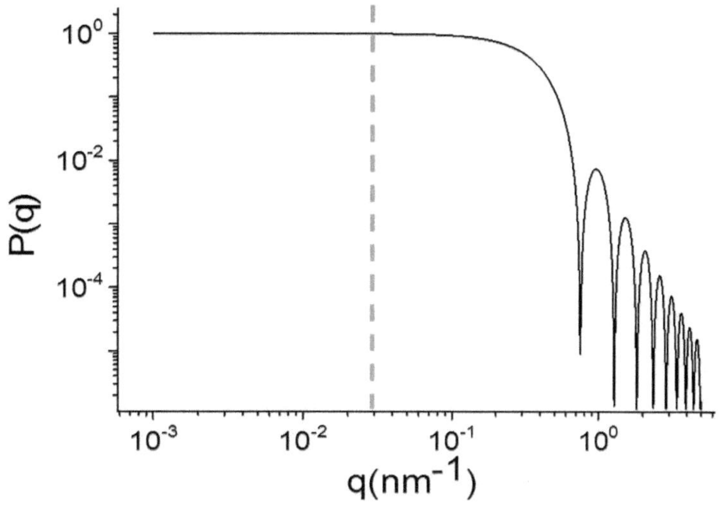

Fig. 5 The calculated form factor $P(q)$ of a homogeneous sphere, 6 nm in radius, similar size to ganglioside micelles [23]. The dashed vertical line marks the upper limit of accessibility for visible laser light scattering. A much wider range is accessible to X-rays and neutrons, allowing for clear detection of spatial anisotropy of the scattered intensity. Plot is in log-log scale

becoming less for smaller particles. As already noted, in fact, in the case of light scattering from small diluted non interacting particles, $P(q) = 1$ (and $S(q) = 1$). The extended q-range is accessible by X-rays and neutrons, being the corresponding wavelengths shorter by more than two orders of magnitude, then accessing the anisotropy and modulation of the scattered intensity.

On the basis of their mass (then volume) and monomer extension, it can be deduced that ganglioside micelles are not spherical. Table 2 reports the values of the packing parameter, $P = V/a_0l_c$ (V and l_c are the monomer hydrophobic volume and length, a_0 is area required by monomer at the hydrophobic/hydrophilic interface), expressing the geometry of packing of ganglioside molecules within the aggregate, and connected to the aggregate shape [25]. Shortly, if monomers pack with a conical geometry, then $P = 1/3$ and they give rise to small spherical aggregates. If they pack with cylindrical geometry, then $P = 1$ and they form a flat bilayer. Molecular geometries in between the cone and the cylinder, namely truncated cones, have intermediate values of P and originate non-spherical micelles, flat or elongated (modeled as oblate or prolate ellipsoids), as sketched in Fig. 6.

The angular dependence of the scattering anisotropy depends on the particle shape. Figure 7 reports the form factor for oblate (disk-like) and prolate (rodlike) homogeneous nanoparticles [23, 24], obtained by deforming the 6 nm sphere of Fig. 5, that is, keeping the same volume. The analytical form of the form factor

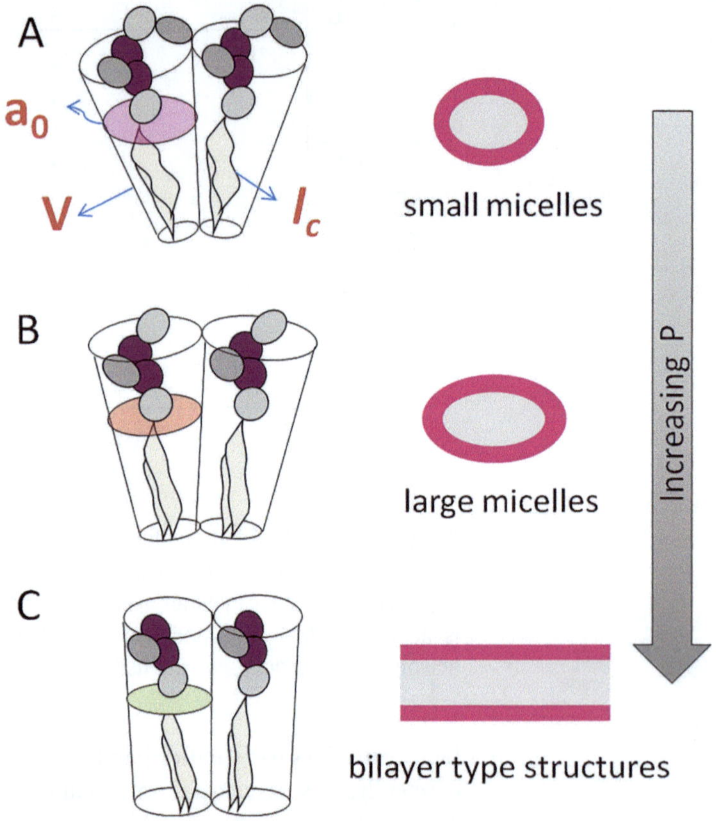

Fig. 6 Sketch of couples of gangliosides highlighting the packing parameter (*P*) determinants: l_c, the chain length, *V*, the hydrophobic volume, and a_0, the area per headgroup. A, B, and C correspond to *P* values 0.41, 0.43, and >0.5, respectively. On the right side the corresponding aggregate shape

$P(q)$ of the homogeneous ellipsoids of revolution can be found, for example, in [23, 24]. The deformation was constrained to obey the micelle packing criterion "no water in the hydrophobic core," then fixing one semiaxis to the length of the ganglioside molecule. Of course, the axial ratios of the two deformed particles of same volume are very different, being ≈2 for the oblate and ≈6 for the prolate.

The form factor of the ellipsoidal particles is different from that of the sphere in the *q*-range accessible to radiation of short wavelength, X-rays and neutrons. Nonetheless, a similar effect is obtained by considering size polydispersity, so it is not always easy to determine the shape of particles of small size, as ganglioside micelles, by just looking at the q-dependence of the scattered radiation. We will see in the following how this information can be obtained by combining *static* and *dynamic* laser light scattering, giving independent but contemporary access to the absolute scattered intensity (then mass and volume of the nanoparticles) and the diffusion coefficient. By this experimental procedure, the ganglioside micelles could be assessed to be globular aggregates of the oblate ellipsoidal type (disk-like), although the visible laser light wavelength is too long to allow for the q-region of scattered intensity modulation. The inspection of the anisotropy of the scattered light is highly informative and detailed for larger particles and in the study of the internal structure of nonhomogeneous particles.

5 The Internal Description of Aggregates

If nanoparticles are non-homogeneous for the incident radiation, that is, if their contrast profile is not flat to the incident radiation, then the internal structure can be accessed. This happens for ganglioside aggregates, as pointed out in Fig. 2. Then, the form factor $P(q)$ besides the "geometrical" shape accounts also for the "optical" shape of the aggregates [23]. This is seen in Fig. 8. The homogeneous oblate ellipsoidal particle of Fig. 7, has been replaced by a particle with identical geometrical external features, but with the contrast profiles for gangliosides illustrated in Fig. 2, for neutrons (black line) and X-rays (gray line). The scattering aggregates are core–shell particles, with the same geometrical parameters but with different "optical" profiles. The X-rays scattered intensity graph of Fig. 8 displays the contrast-minimum, typical for optically hollow

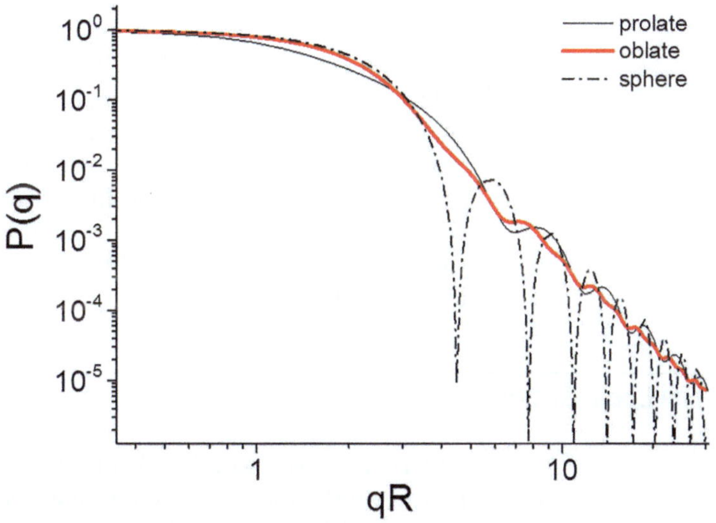

Fig. 7 The calculated form factor of monodisperse spheres, and of particles obtained by oblate and prolate elllipsoidal deformations at constant volume. The axial ratio of the two deformations is such that the minor axis is kept to the monomer length, thus obeying the general rule of micelle aggregation. Form factors are drawn as a function of the dimensionless variable qR, R being the radius of the sphere with the same volume, then showing the shape dependence

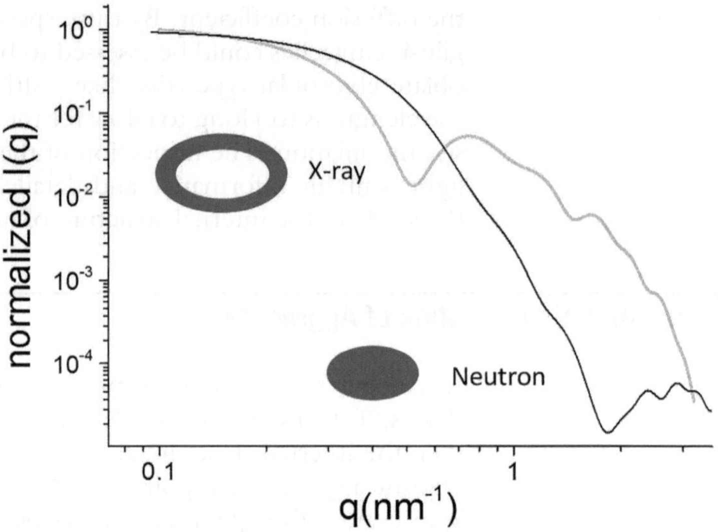

Fig. 8 Calculated normalized scattered intensity for the oblate ellipsoidal particle of Fig. 7, where the internal homogeneity has been replaced by the contrast profiles for gangliosides illustrated in Fig. 2, for neutrons (black line) and X-rays (gray line). The scattering aggregates are core–shell particles, with the same geometrical parameters but with different "optical" profiles

particles, like ganglioside micelles, clearly visible also in the two-dimensional X-ray scattering pattern shown in Fig. 9.

It can be realized, then, that the X-ray scattered intensity distribution is sensitive to changes in the shell region of the ganglioside aggregates. As an example, in Fig. 10, the first and last spectra of Fig. 3 (*left panel*), relative to sialidase digestion of GD1a, are zoomed in the region of the contrast-minimum. In this case, the core is the same, and in the final situation one sugar group (one sialic acid residue) out of six of each of the 226 micellized GD1a molecules has been removed from the periphery of the shell, with clear visibility.

6 The Local Structure

At larger *q*'s, the so-called Wide-Angle region, the organization of matter on the local scale can be assessed. This is the *q*-region where the typical interatomic distances are determined, for example in a crystallized protein. This is also the *q*-region where pure fluids, like water or alkanes, have a nonisotropic scattering, connected to the spatial distribution of molecules or molecular groups. In this

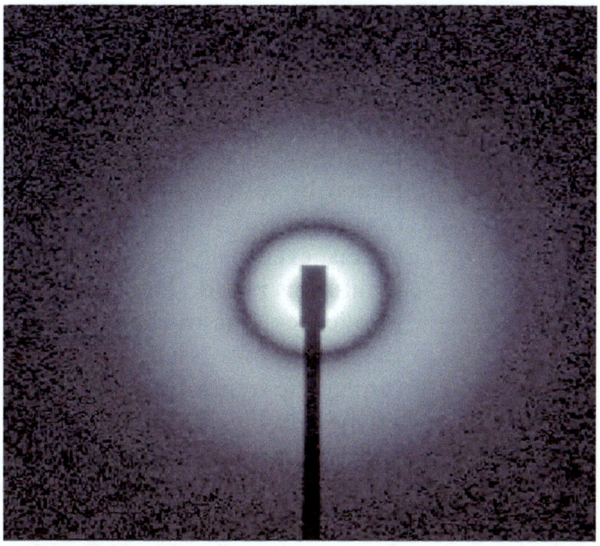

Fig. 9 2D X-ray scattering pattern from a GM1 ganglioside micellar solution. The contrast-minimum, typical for optically hollow particles is clearly seen as a dark ring. The intensity profiles are then constructed by angular regrouping of the scattered intensity

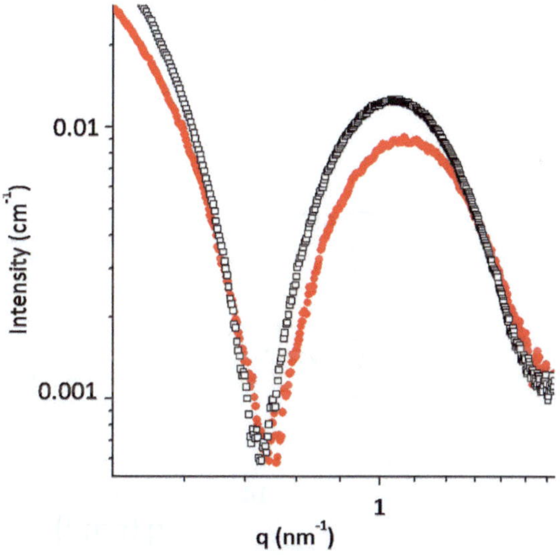

Fig. 10 The first and last spectra of Fig. 3 (left panel), zoomed in the region of the contrast-minimum. Data are presented in log-log scale

q-region the intensity peaks of the lipid chains of phospholipids arranged into bilayers can be found. These peaks move in position and width when the phospholipids chains undergo the well-known order–disorder transition, from the gel to the fluid phase, upon heating, with an associated enthalpy of transition that can be determined by differential scanning calorimetry (DSC). The transition occurs provided that an "ordered" state can be assumed by the chains, eventually transforming in a "disordered" state when temperature is raised. This "ordered" state cannot usually be assumed in micelles, due to the curvature and density constraints. Usually, the micelle core is a fluid drop, and the calorimetry profile does not display the transition enthalpy peak. Conversely, the disk-like shape of the ganglioside micelles allows for some ordering in the core and a peak can be seen [26]. This can be appreciated in Fig. 11, where the left panel shows the local-arrangement peak at $q = 15\ \text{nm}^{-1}$, although with very low intensity. Despite coming from a dilute 3 mM GM1 solution, the peak was still detectable with the high-brilliance X-ray beam of the used ESRF synchrotron source and ID02 beamline. This chain-peak is much more visible in phospholipid liposomes or lamellar phases, and possible modification in the local-structure of the bilayer induced by the presence of gangliosides can be appreciated. In the right panel of Fig. 11, the DSC calorigram obtained for a GM1 aqueous solution shows the existence of a transition enthalpy peak centered at 20 °C,

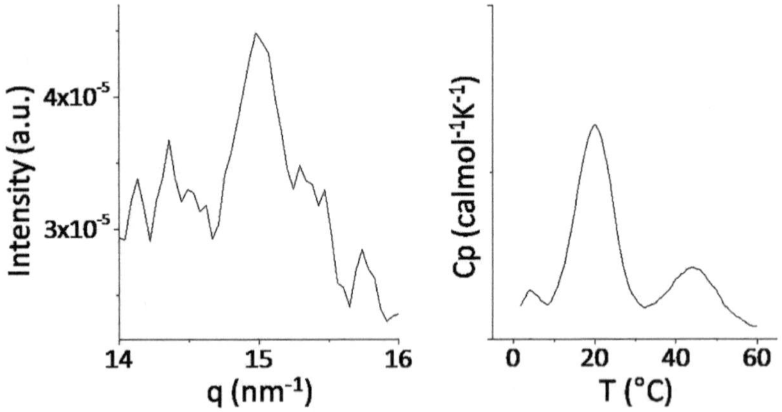

Fig. 11 *Left panel.* The local-order peak at $q = 15\ \text{nm}^{-1}$ originated by a 3 mM solution of GM1. In real space, this value corresponds to a distance of $2\pi/q \sim 0.4$ nm, typical for hydrocarbon chains [27]. Data are in lin-lin scale. *Right panel.* Differential scanning calorimetry (DSC) profile relative to a 3 mM GM1 aqueous solution. The wide enthalpy peak centered at $T = 20$ °C corresponds to the chain melting transition of the GM1 ceramide moiety

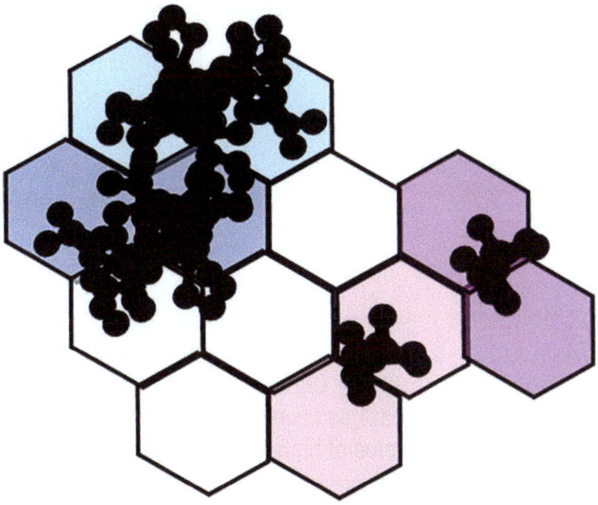

Fig. 12 Top view of the headgroup crowding of GM3 (upper left), as compared to DMPC (lower right) on the same triangular lattice. Hexagons correspond to the Voronoi cell occupied by one lipid chain. Headgroup size is properly rescaled

although very large (due to the small number of monomers in a micelle as compared to the number of phospholipids into a liposome), corresponding to a gel phase melting into a liquid phase.

In this q-region, some additional features were observed for ganglioside GM3, revealing that some additional order exists, besides in the aggregate core. By observing these additional features, we could infer that structuring occurs on the surface of GM3 aggregates. The headgroups of GM3 are too small to give rise to small micellar aggregates, but too big to arrange freely on the aggregate surface, originating trapped structures where an organized sugar shell hangs over a liquid core. This is an interesting feature, as it reveals the ability of ganglioside headgroups of keeping an organized structure on the surface of the aggregate where they are embedded, meanwhile allowing for core fluidity. This is described in [28, 29]. In Fig. 12 a sketch of the crowded packing of gangliosides headgroups on the aggregate surface is compared to the loose arrangement of phosphocholines in phospholipid membranes.

7 The Organization of Aggregates in Space

If micelles are not randomly distributed in space, then the scattered radiation is affected. In fact, in this case the structure factor $S(q)$ is no more a flat function ($S(q) = 1$ for all q) but it comes out to be oscillating around unity. Simply speaking, it represents an interference pattern, reflecting the organization of aggregates in space. The amplitude and disposition of the oscillations depend on how

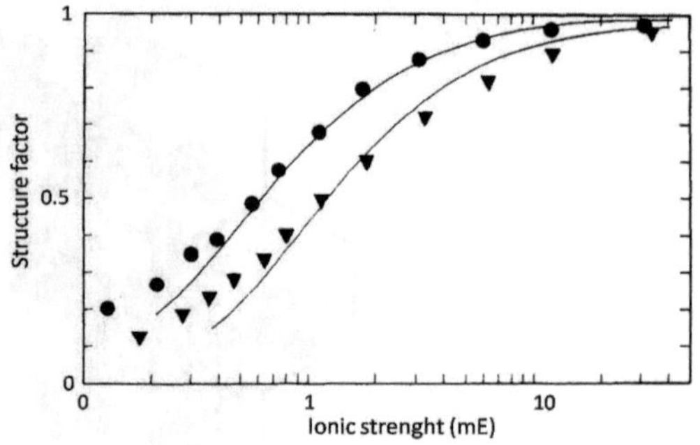

Fig. 13 Dependence of the excess normalized laser light intensity scattered at 90° by GM1 solutions at 0.5 mM (lower curve) and 1 mM (upper curve) upon NaCl addition from 0 to 100 mM. Normalization has been made to the limiting value at high ionic strength. Data are presented in lin-log scale

nonrandomly distributed and how far from each other are the aggregates in solution. Nonrandom distribution occurs either because the number concentration is high or because the particles are interacting. Theoretical models, with different approximations, allow to predict the expression of $S(q)$ starting from particles charge, number concentration and ionic strength, and shape, as can be found, for example, in [30]. Then, the effective particle charge can be inferred from the scattering pattern. Being charged, gangliosides have repulsive interactions among each other. So if interactions are not screened by a suitable ionic strength of added salt, they repel each other and assume a nonrandom distribution, so $S(q) \neq 1$, with a minimum in the low-q-region accessible to laser light scattering. If this effect is not accounted for, an apparent lower value is determined for the aggregate mass. This effect is strong in ganglioside solutions. As an example, Fig. 13 shows the variation in the excess normalized laser light intensity scattered at 90° by 0.5 mM and 1 mM GM1 solutions upon NaCl addition from 0 to 100 mM. The high-ionic-strength limit is the one corresponding to $S(q) = 1$, the one that allows the micellar mass to be calculated, as described above, by *static* laser light scattering. If no salt is added to the solution, a depletion of the laser light scattered intensity by one order of magnitude is seen, a huge reduction. If not properly accounted for, this would induce to estimate an apparent micellar mass even ten times lower than real. From the behavior of the scattered light intensity as a function of the ionic strength of the solution, the effective charge of the micelles could be determined [31–33], being much lower than the structural charge, $48/301 \approx 16\%$, a well-known effect in dissociable polyions.

The effect of interparticle interaction shows up clearly in the extended q-range accessible to X-ray and neutrons. In Fig. 14 the experimental X-ray intensity scattered by a GM1 solution with no added salt displays a clear oscillation due to the structure factor

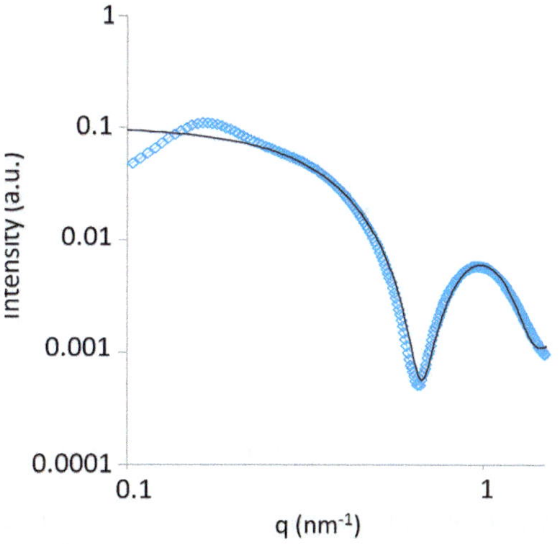

Fig. 14 Experimental X-ray intensity scattered by a GM1 solution with no added salt (blue diamonds). The experimental spectrum displays a clear oscillation due to the structure factor $S(q)$, superimposed to the optically hollow-particle profile (dark line). Scale is log-log

$S(q)$, superimposed to the optically hollow-particle profile described before (shown as a dark line). The depletion of the intensity seen to the left extends also to smaller q-values, accessible by laser light scattering, as just described. In the extended q-range of Fig. 14, also the positive oscillation (higher than the dark line) can be seen. The position of this structure factor peak depends on micelle concentration, as it is connected to the distance of nonrandomly distributed interacting particles.

In fact, when concentration is increased and particles come closer to each other, the position of the peak moves to higher q. The evolution of the neutron scattering profile of GM1 micellar solutions upon the concentration increase in the range 5–25% (w/w) is shown in the left panel of Fig. 15. GM1 solutions were prepared in pure D_2O solvent, with increasing concentration, put in quartz cells, 2 mm thickness, then placed in front of the neutron beam, and the scattered neutrons intensity was collected and regrouped as a function of q, as shown in the figure. The contribution of the form factor $P(q)$, still visible for the lowest concentration, is progressively eaten up by the initial depletion and the first peak of the structure factor $S(q)$, increasing in intensity and moving to higher q, corresponding to shorter distances. In the right panel of Fig. 15, the structure factor has been isolated, and shown independently, from a series of X-ray scattering spectra generated by GM1 solutions of increasing concentration in pure water [21].

If concentration is higher than a limiting value, the ganglioside micelles assume an ordered disposition in space, and the scattered intensity shows the Bragg peaks, typical of the symmetry assumed

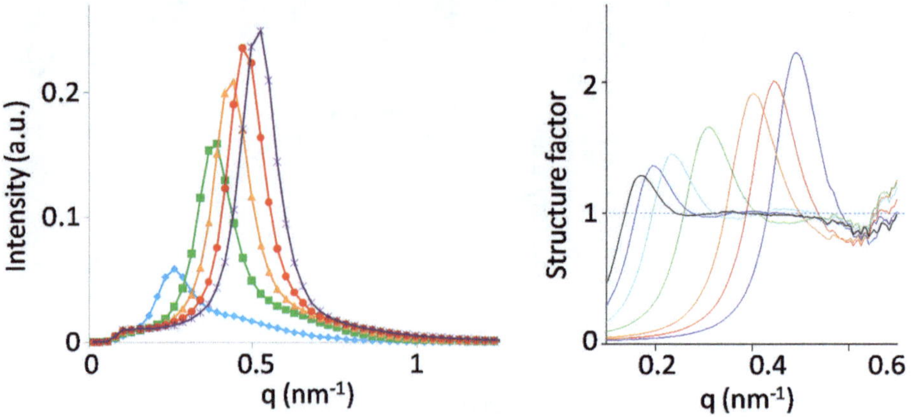

Fig. 15 *Left panel.* The evolution of the neutron scattering profile of GM1 micellar solution upon concentration increase in the range (5–25% w/w) in pure deuterated water. The contribution of the form factor $P(q)$, still visible for the lowest concentration, is progressively eaten up by the first peak of the structure factor $S(q)$, increasing in intensity and moving to higher q, corresponding to shorter distances. *Right panel.* The structure factor has been isolated, and shown independently, from a series of X-ray scattering spectra generated by GM1 solutions of increasing concentration in pure water. Data are shown in lin-lin scale

by the colloidal crystal. This is immediately seen by X-rays. As an example, Fig. 16 shows the 2D pattern of scattered X-rays corresponding to the most evident lamellar phase, found in all lipid systems even at low concentration, with the 1–2–3 repetitions.

For GM1, the lamellar phase exists in the high concentration regime, above 60% [34]. For intermediate concentrations, more complicate spatial arrangements are assumed [35]. As the concentration increases, micellar crystals are formed, with cubic symmetry (Pm3n) in the range 30–45%. The corresponding crystal parameters are reported in Table 3. The micelles keep the same mass all over the concentration region, despite the growing interparticle interactions, just crowding in space. This allows to conclude that ganglioside patches can be very resistant to external forces. For higher concentration, 45–55%, also bicontinuous phases with cubic symmetry (Pn3m and Ia3d) are found, evidencing the ability of GM1 to confer saddle curvature to the aggregated structure where they are embedded, like in bud necks or invaginations.

8 The Aggregates Brownian Motion

The most known and widely used scattering technique is the *dynamic* laser light scattering, DLS. Many laboratories have a commercial push-button DLS to assess the size of particles in solution. It is based on the fact that the laser light scattered at a given angle

Fig. 16 2D pattern of scattered X-rays originated by a lamellar phase. The 1–2–3 repetitions are clearly visible as bright rings

Table 3
Micellar parameters in the Pm3n cubic micellar phase. The lattice parameter a (Å) and the aggregation number N (*italics*)

GM1 volume fraction	Temperature				
	25 °C	30 °C	40 °C	43 °C	60 °C
0.26			220, *170*		214, *157*
0.32		215, *192*		212, *184*	
0.37					206, *194*
0.40	200, *191*	202, *197*			

by a collection of particles is not constant, but fluctuating around an average value. These fluctuations are originated by the fact that particles are not fixed in space, but in Brownian motion. The characteristic time of fluctuation of the scattered intensity at a given angle is connected to the characteristic time of the collective translational diffusion of the scattering particles, that is, to their translational diffusion coefficient D. Then, D is related to the hydrodynamic radius of the particles R_H, via the Stokes–Einstein relation:

$$D = kT / (6\pi\eta R_{\mathrm{H}}).$$

where k is the Boltzmann constant, T the absolute temperature, and η the viscosity of the solvent.

So the hydrodynamic radius of the solute particles can be determined provided that D can be extracted from the features of the fluctuating scattered intensity. This is typically done by building the time-correlation function of the laser light scattered intensity, by means of a correlator [36]. The time-correlation function for a random distribution of particles in Brownian motion contains a decaying contribution, of the negative-exponential type as a function of time. If the particles are monodisperse, such contribution is:

$$G_2(t) \approx \exp\left(-2Dq^2t\right).$$

so that in a log-lin plot it becomes a linear function, with negative slope proportional to the translational diffusion coefficient. If particles in solution are polydisperse, the time-correlation function contains different contributions of the same type, and can be analyzed by different approaches. The most common is the cumulant analysis, [37]. This analysis gives an estimate of the weight-average hydrodynamic size of the particles, $\langle R_H \rangle = \langle 1/D \rangle\ kT/(6\pi\eta)$, and their polydispersity, $\langle D^2 \rangle / \langle D \rangle^2$. This kind of analysis is suitable for monomodal distributions of particles. If the distribution is multimodal, more sophisticated data analysis routines are used, like NNLS or CONTIN [38, 39] that give a frequency plot of particles as a function of size. Usually, those frequency plots can be obtained as intensity-distribution, or volume distribution, or number distribution. Due to the dependency of the scattered light on particles mass and concentration, the three types of distribution are unbalanced with respect to each other, the intensity-distribution privileging bigger particles. The average estimated size is highest for the intensity-distribution, smaller for the volume-distribution and smallest for the number-distribution. In the case of the cumulant analysis, the scattered light intensity is the weighting function, so the estimated average size is higher than for number average, opposite to what one would estimate by microscopy, for example.

Carefully purified gangliosides give rise to monomodal low-polydispersity micellar solutions. If properly filtered, the cumulant analysis of the laser light they scatter give stable and very reproducible results, allowing to calculate the average hydrodynamic radius of micelles in solution.

We note that for DLS, careful filtration is required in order to remove spurious correlated signal coming from dust, but disposable plastic transparent cells can be used, with no particular optical quality. In fact, flare from cell walls constitutes only an uncorrelated background that does not affect the result of the measurement.

Rather, interactions among particles in solution affect the result. In fact, particle Brownian motion is affected by interactions, leading to determine an apparent size of the particles that is higher or lower than true according to whether interactions are attractive or repulsive. So not only the intensity of the scattered radiation is affected by interactions (through the structure factor $S(q)$) but also it correlation function. For a GM1 1 mM solution in the absence of added salt, micelles undergo a Brownian motion ten times faster than when interactions are screened, presenting an apparent hydrodynamic radius one order of magnitude smaller than real [31]. So also for DLS, interactions have to be properly accounted for.

DLS measurements on carefully purified gangliosides solutions, in 100 mM NaCl, properly filtered, allow to calculate the average hydrodynamic radius of micelles. Results are reported in Table 4. For all gangliosides, the polydispersity is low, of the order of 5%. Then, their hydrodynamic radius can be estimated with confidence.

The hydrodynamic radius is an indication of how fast particles move, and depends on particle shape. For a given mass, then a given volume, ellipsoidal particles diffuse slower than spheres, and rods slower than discs. The hydrodynamic radius and the equivalent radius (the radius of the sphere with the same volume) are connected by the Perrin factor, that depends on shape and on axial ratio [40]. So simultaneous measurements of the average mass and the average hydrodynamic radius of micelles allows to infer their shape. For gangliosides, the shape was determined to be of the

Table 4
Hydrodynamic radius, axial ratio and electric charge (electron units) of ganglioside micelles

	Hydrodynamic radius (nm)	Axial ratio	Electric charge
GM2	6.60	3.1	100
GM1	5.87	2.3	48
$GM1_{acetyl}$	6.00	1.1	
GD1a	5.80	2.0	60
GalNac-GD1a	5.00	2.1	
GD1b	5.10	1.8	
GD1b-L	5.95	2.1	
GT1b	5.32	1.8	
GM4	27		
GM3	24		

oblate ellipsoidal type (disklike) with an axial ratio around 2, higher for bigger micelles formed by smaller-headgroup gangliosides. If modeled as rodlike, given their mass, estimated by *static* laser light scattering, their hydrodynamic radius would have been clearly higher than assessed by DLS.

9 General Guidelines to Operation

The most known and widely used scattering technique is the *dynamic* laser light scattering, DLS. Many laboratories have a commercial push-button DLS to assess the size of particles in solution, sometimes associated to the Z-potential functionality, addressing their charge.

As already noted, for gangliosides, bearing a net negative charge, care has to be taken regarding the ionic strength of the solution, in order to screen repulsive electrostatic interactions. If not properly accounted for, one would estimate an apparent size of the micelles lower than real, as well as an apparent mass lower than real would be assessed by *static* laser light scattering.

Careful filtration is necessary in order to remove dust from the solution, as dust gives strong "forward" scattering, in the *q*-region accessible with laser light. Dust has to be removed from the solution, but also from the measuring cell. So the cell itself has to be rinsed with the filtered solution, repeatedly cycled into the cell.

The cells for DLS can be transparent plastic disposable cells, compatible with the solvent. Conversely, for *static* laser light scattering, cells made of optical glass, optically polished are required, and carefully cleaned also on the external walls, in order to reduce flare. A standard for cell thickness is 10 mm, and few ml of solution are needed to operate a careful filtration and rinsing of the cell.

Other probes than light are needed to have access to higher *q* range, that means to investigate the small distances. X-ray and neutron spectroscopy are applied to unravel particle structure at the angstrom scale and up to the hundred nanometer scale, by performing wide and small angle scattering.

A large number of instrumental setups exist to perform small angle scattering measurements, from conventional instrument to more advanced and powerful ones, and we briefly describe the criteria at the basis of the best experimental approach. The choice of the instrument depends on the sample to be investigated, in terms of (1) the contrast for X-ray (governed by the atomic number) or neutron (depending on the nuclei variety in the sample in a non-ruled way), (2) the amount of material and/or the concentration of interest, and (3) the expected range of the distances to look for. Investigating a *q* range as wide as possible is of course the best choice. However, it should be considered that the lowest *q* required in order to detect the overall particles structure corresponds to few times the inverse of the particle size. In the case of gangliosides, the largest micelles are about 13 nm (*see* Table 2) that requires a minimum q of about 0.1 nm^{-1}. Of course, if the $S(q)$ contribution to the scattered intensity is relevant, the range should be decreased to lower q. Considering the first point (1), in the case of ganglioside micelles, similarly to other systems of aggregates in solution and to microemulsion, any instrument can be used because contrast is present for both neutron and X-ray. The neutron probe gives the opportunity to play on the contrast variation of the solvent. In fact, lipid tails of the core, sugars of the heads, and the solvent change their contrast differences by varying the isotopic D_2O/H_2O proportion in the solvent. This can be applied to highlight one component over the other. A parallel and complementary use of SANS and SAXS on the same sample is the optimum approach to exploit at the best the contrast variation opportunity. Point (2) can pose limitation to the use of neutron scattering in biological application due to the often low amount of molecules available. Volume of each sample for SANS must be at least 300–400 μl. The concentration required in the case of ganglioside molecules at the best neutron sources, such as, for example, the lines of ILL in Grenoble (Fr) [41], is around 5 mg/ml. Other instruments, as those at LLB in Scalay (Fr) [42], where some of the data reported in the previous sections were acquired, need more material. For SAXS investigation, a large choice of conventional instruments is available that allow for in-house measurements (for a review *see* Pedersen 1995 in Modern Aspect of Small Angle Scattering [43]). However, besides the common drawback of the smearing of the scattering profile due to reduced beam collimation, the flux attainable from conventional sources is much lower than that provided by a synchrotron. In particular, the X-ray scattering data reported in this chapter where acquired at the ID02 beamline of ESRF in Grenoble (Fr) [44] . The extremely high brilliance of the ESRF synchrotron radiation allows to investigate very small sample volumes, down to a couple of tens of microliters with concentration down to 1 mg/ml. An extremely large q-range is attainable at the ID02 beamline using different sample to detector distances from very low q's (interesting for long-correlated structures) to wide q's for WAXS measurements detecting the intermolecular distance range.

The cells for neutron scattering are made in transparent quartz, optically polished, with well-defined thickness, usually 1 or 2 mm. There is a sort of standard accepted by all facilities providing neutron beams, so the same cells can be used worldwide, 40 mm in height, 10 mm wide. Typically, 0.5 ml of solution is needed per sample, depending on the neutron beamsize, in order to shine all incoming neutrons onto the sample. Sample filtration is not needed, but for typical neutron scattering experiments, D_2O has to be used as solvent, added with ionic strength. Care has to be taken that D_2O does not contain H_2O. For specific neutron contrast variation experiments, exact mixtures of the two solvents can also be used.

The cells for X-rays must provide high transmission without causing spurious scattering themselves in *q*-regions of interest. Usually, capillaries are used, 2 mm thickness. High-transmission Lindemann glass capillaries can be used, but they are very fragile, and their thickness is not constant all over the single capillary and from one capillary to the other. Plastic capillaries can also be used, but not all the plastics are suitable. Kapton is often used, and also some polycarbonates. For strong scatterers, also mica can be used, for the windows of flat cells. Anyway, all materials give some scattering when shined with X-rays, so it is important that their contribution can be carefully evaluated and subtracted from the scattering spectra. Sample filtration is not needed, and low amounts of sample are used, typically of the order of 0.1 ml.

For updated information about instrument characteristic and improvements is essential to refer to the homepages of the different instruments.

References

1. Lindner P, Zemb T (eds) (2002) Neutrons, X-rays and light. Elsevier Science B.V., Amsterdam
2. Corti M, Cantu' L, Salina P (1991) Aggregation properties of biological amphiphiles. Adv Coll Int Sci 36:153–171
3. Neutron News (1992) 3(3):29–37
4. Degiorgio V, Corti M, Minero C (1984) Laser light scattering in micellar solutions. Il Nuovo Cimento D 3:44–61
5. Corti M, Cantu' L, Sonnino S, Tettamanti G (1988) Aggregation properties of gangliosides in aqueous solution. In: Ledeen RW, Hogan EL, Tettamanti G, Yates AJ, Yu RK (eds) New trends in ganglioside research: neurochemical and neuroregenerative aspects, Fidia research. Series, vol 14. Liviana Press, Padova
6. Cantu' L, Corti M, Acquotti D, Sonnino S (1993) Aggregation properties of gangliosides: influence of the primary and secondary structure of the headgroup. J Phys IV 3:357–364
7. Sonnino S, Acquotti D, Brocca P, Cantu' L, Corti M (1993) Influence of primary and secondary structures of oligosaccharide chain on the aggregative and geometrical properties of gangliosides. Indian J Biochem Biophys 30:370–375
8. Sonnino S, Cantu' L, Corti M, Acquotti D, Venerando B (1994) Aggregative properties of gangliosides in solution. Chem Phys Lipids 71:21–45
9. Corti M, Cantu' L, Del Favero E (1994) Peculiarities in the Aggregative Behaviour of Gangliosides, Glucosidic Surfactants of Biological Origin. Nuovo Cimento D 9:1391–1400
10. Corti M, Cantu' L, Del Favero E (1996) Selfaggregation properties of gangliosides, glycosidic surfactants of biological origin. Tens Surf Detergents 33:214–219
11. Lago P, Rovati L, Cantù L, Corti M (1993) A quasielastic light scattering detector for chromatographic analysis. Rev Sci Instrum 64:1797–1802
12. Del Favero E, Brocca P, Motta S, Rondelli V, Sonnino S, Cantu' L (2011) Nanoscale structural response of ganglioside-containing aggregates to the interaction with sialidase. J Neurochem 116:833–839
13. Del Favero E, Raudino A, Pannuzzo M, Brocca P, Motta S, Cantu' L (2012) Transient step-like kinetics of enzyme reaction on fragmented-condensed micellar substrates. J Phys Chem B 116:9570–9579
14. Cantu' L, Corti M, Del Favero E, Digirolamo E, Sonnino S, Tettamanti G (1996) Experimental evidence of a temperature-related conformational change of the hydrophilic

portion of gangliosides. Chem Phys Lipids 79:137–145

15. Cantu' L, Corti M, Del Favero E, Digirolamo E, Raudino A (1996) Thermal Hysteresis Phenomena in Micellar Solutions of Gangliosides: Theory and Experiments. J Phys France 6:1067–1090
16. Cantu' L, Corti M, Del Favero E, Muller E, Raudino A, Sonnino S (1999) Thermal Hysteresis in Ganglioside Micelles Investigated by Calorimetry and Light Scattering. Langmuir 15:4975–4980
17. Cantu' L, Corti M, Del Favero E, Raudino A (2000) Phase Transition at the Surface of Mixed Micelles of the Ganglioside GM1 and Dodecylphosphocholine. J Phys: Condensed Matter 12:A321–A325
18. Raudino A, Cantu' L, Corti M, Del Favero E (2000) Bistable molecular selfassembly. Curr Opin Colloid Interface Sci 5:13–18
19. Brocca P, Rondelli V, Mallamace F, Di Bari MT, Deriu A, Lohstroh W, Del Favero E, Corti M, Cantu' L (2017) Water response to ganglioside GM1 surface remodelling. Biochim Biophys Acta 1861:3573–3580
20. Brocca P, Cantu' L, Corti M, Del Favero E, Raudino A (2007) Intermicellar interactions may induce anomalous size behaviour in micelles carrying out bulky heads with multiple spatial arrangements. Langmuir 23(6):3067–3074
21. Del Favero E, Brocca P, Rondelli V, Motta S, Raudino A, Cantu' L (2014) Optimizing the crowding strategy: sugar-based ionic micelles in the dilute-to-condensed regime. Langmuir 30(30):9157–9164
22. Rondelli V, Brocca P, Fragneto G, Daillant J, Tringali C, Cantu' L, Del Favero E (2017) In situ digestion of gangliosides by sialidase induces surface restructuring in raft-mime membranes: a Synchrotron X-Ray reflectivity insight. Biochim Biophys Acta 1859(5): 845–851
23. Pedersen JS (1997) Analysis of small-angle scattering data from colloids and polymer solutions: modeling and least-squares fitting. Adv Colloid Interf Sci 70:171–210
24. Pedersen JS (2002) Modelling of small-angle scattering data from colloids and polymer systems. In: Lindner P, Lindner T (eds) Neutrons, X-rays and light. Elsevier Science B.V., Amsterdam, pp 391–420
25. Israelachvili J, Mitchell DJ, Ninham BW (1976) Theory of self-assembly of hydrocarbon amphiphiles into micelles and bilayers. J Chem Soc Faraday Trans 2(72):1525–1568
26. Cantu' L, Corti M, Del Favero E, Muller E, Raudino A, Sonnino S (1999) A calorimetric study of thermal hysteresis effects in ganglioside micelles. Progr Colloid Polym Sci 112:5–10
27. Luzzati V, Mustacchi H, Skoulios A, Husson F (1960) La structure des colloïdes d'association. I. Les phases liquide–cristallines des systèmes amphiphile–eau. Acta Cryst 13:660–667
28. Brocca P, Cantu' L, Del Favero E, Dubois M, Motta S, Tunesi S, Zemb T (2005) Headgroup and chain melting transition in dispersed bilayers of GM3 ganglioside. Coll Surf A 259:125–133
29. Sega M, Vallauri R, Brocca P, Cantu' L, Melchionna S (2007) Short range structure of a GM3 ganglioside membrane: comparison between experimental WAXS and computer simulation results. J Phys Chem B 111:10965–10969
30. Klein R (2002) Interacting colloidal suspensions. In: Lindner P, Zemb T (eds) Neutrons, X-rays and light. Elsevier Science B.V., Amsterdam, p 351
31. Cantu' L, Corti M, Degiorgio V (1987) Static and dynamic light scattering study of solutions of strongly interacting ionic micelles. Faraday Discuss Chem Soc 83(20):287–295
32. Cantu' L, Corti M, Degiorgio V, Piazza R, Rennie A (1988) Neutron scattering from ganglioside micelles. Progr Colloid Polym Sci 76:216–220
33. Degiorgio V, Cantu' L, Corti M, Piazza R, Rennie A (1989) Light and neutron scattering study of strongly interacting ionic micelles. Coll Surf 38:169–178
34. Cantu' L, Corti M, Del Favero E, Raudino A (2000) Tightly packed lipid lamellae with large conformational flexibility in the interfacial region may exhibit multiple periodicity in their repeat distance. A theoretical analysis and X-ray verification. Langmuir 16(23):8903–8911
35. Boretta M, Cantu' L, Corti M, Del Favero E (1997) Cubic phases of gangliosides in water: possible role of the conformational bistability of the headgroup. Phys A 236:162–176
36. Cummins HZ, Pike ER (eds) (1977) Photon correlation spectroscopy and velocimetry. Plenum, New York
37. Koppel DE (1972) Analysis of macromolecular polydispersity in intensity correlation spectroscopy: the method of cumulants. J Chem Phys 57:4814–4820
38. Lawson CL, Hanson RJ (1995) Solving least squares problems, vol 15. SIAM, Philadelphia

39. Provencher SW (1982) A constrained regularization method for inverting data represented by linear algebraic or integral equations. Comput Phys Comm 27:213–227
40. Perrin F (1936) Mouvement brownien d'un ellipsoide. Le J. de. Physique 7(1):3–11
41. *D11, D22, D33 instruments, see ILL website:* www.ill.eu
42. PACE, PAXE, *PAXY instruments, see LLB website: www*-llb.cea.fr
43. Pedersen JS (1995) Instrumentation for small-angle scattering. In: Brumberger H (ed) Modern aspect of small angle scattering. Springer, New York
44. *ID02, BM26, BM29, BM02, ID13 instruments, see ESRF website:* www.esrf.eu

Chapter 4

Gangliosides in the Immune System: Role of Glycosphingolipids and Glycosphingolipid-Enriched Lipid Rafts in Immunological Functions

Kazuhisa Iwabuchi

Abstract

Although individuals are constantly exposed to infectious agents, these agents are generally resisted by the innate and acquired immune systems. Both the innate and acquired immune systems protect against invading organisms, but they differ functionally in several ways. The innate immune system is the body's inborn defense mechanism and the first line of defense against invading organisms, such as bacteria, fungi, and viruses. Glycosphingolipids (GSLs), which are expressed on the outer leaflet of plasma membranes (Murate et al., J Cell Sci 128(8):1627–1638, 2015), are involved in both innate and acquired immunity (Inokuchi et al., Biochim Biophys Acta 1851(1):98–106, 2015; Nakayama et al., Arch Immunol Ther Exp (Warsz) 61(3):217–228, 2013; Rueda, Br J Nutr 98(Suppl 1):S68–73, 2007; Popa and Portoukalian, Pathol Biol (Paris) 51(5):253–255, 2003).

Recent studies have indicated that innate immunity is not a "nonspecific" immune system. Large numbers of viruses, bacteria, and bacterial toxins have been reported to bind to host surface carbohydrates, a number of which are components of GSLs (Schengrund, Biochem Pharmacol 65(5):699–707, 2003). Binding studies have also demonstrated that some glycolipids function as receptors for microorganisms and bacterial toxins (Yates and Rampersaud, Ann N Y Acad Sci 845:57–71, 1998). These findings clearly indicate that GSLs are involved in host–pathogen interactions.

GSLs are composed of hydrophobic ceramide and hydrophilic sugar moieties (Hakomori, Annu Rev Biochem 50:733–764, 1980). The ceramide moiety of sphingolipids and the cholesterol sterol-ring system are thought to interact via hydrogen bonds and hydrophobic van der Waal's forces (Mukherjee and Maxfield, Annu Rev Cell Dev Biol 20:839–866, 2004). Additional hydrophilic *cis* interactions among GSL headgroups have been found to promote their lateral associations with surrounding lipid and protein membrane components. These interactions result in the separation in cell membranes of lipid rafts, which are lipid domains rich in GSLs, cholesterol, glycosylphosphatidylinositol (GPI)-anchored proteins and membrane-anchored signaling molecules (Pike, J Lipid Res 47(7):1597–1598, 2006). These GSL-enriched lipid rafts play important roles in immunological functions (Inokuchi et al., Biochim Biophys Acta 1851(1):98–106, 2015; Iwabuchi et al., Mediators Inflamm 2015:120748, 2015; Anderson and Roche, Biochim Biophys Acta 1853(4):775–780, 2015; Zuidscherwoude et al., J Leukoc Biol 95(2):251–263, 2014; Dykstra et al., Annu Rev Immunol 21:457–481, 2003). This introductory chapter describes the roles of GSLs and their lipid rafts in the immune system.

Key words Glycosphingolipid, Ganglioside, Innate immunity, Acquired immunity, Lipid rafts, Pattern recognition receptor, Intracellular pathogens

Sandro Sonnino and Alessandro Prinetti (eds.), *Gangliosides: Methods and Protocols*, Methods in Molecular Biology, vol. 1804, https://doi.org/10.1007/978-1-4939-8552-4_4, © Springer Science+Business Media, LLC, part of Springer Nature 2018

1 GSLs Are Immune Modulator Molecules

Gangliosides are acid GSLs widely distributed in most vertebrate tissues and fluids. Gangliosides have been found to function as unintentional target binders for bacterial adhesion in specific tissues, with several gangliosides binding to microorganisms and toxins. For example, polyoma virus (Py) was shown to bind to gangliosides GD1a and GT1b expressed on human erythrocytes [15]. In addition, the ganglioside GM1 was found to be involved in the recognition of viruses, such as simian virus 40 (SV40) [15], and bacteria, such as *Brucella suis* [16]. GM1 is also specifically bound by cholera toxin (CT) B subunit [17, 18]. CT and *Escherichia coli* (*E. coli*) enterotoxin bind to GM1 expressed on epithelial cells [19]. Human milk gangliosides were shown to be involved in the inhibition of *E. coli* and *Vibrio cholerae* enterotoxins [20], an inhibitory action later attributed to GM1. The contents and distribution of individual gangliosides in human milk change during lactation. For example, GD3 is the most abundant ganglioside in colostrum, whereas GM3 is the major ganglioside in mature milk. Orally administered gangliosides may act as putative traps that interfere with pathogen binding in the intestine, thereby preventing infection.

The effects of dietary gangliosides on parameters associated with the development of the intestinal immune system, such as cytokine and intestinal IgA production, have been evaluated in animal models [4]. GM3 and GD3 were shown to inhibit the lipopolysaccharide (LPS)-induced maturation of murine dendritic cells and to modulate LPS-induced cytokine production by these cells [21]. However, there is no evidence that dietary gangliosides activate immune system cells in the intestine.

Evaluations of neutral GSLs have shown that globotriaosyl ceramide (Gb3) can bind to Shiga toxin (Stx) and verotoxin B-subunit. Binding sites for Stx have been detected on the surfaces of human epithelial and endothelial cells [22–25], and the B-subunit of Stx was shown to bind to globotriaosyl ceramide (Gb3) on the surfaces of these cells [22–24]. Asialo GM1 (GA1) expressed on epithelial cells was shown to bind *Pseudomonas aeruginosa*, *Bifidobacterium bifidum*, and *Lactobacillus* [26, 27].

LacCer, which is also expressed on intestinal epithelial cells, was shown to bind specifically to various pathogenic microorganisms, including *E. coli*, *Bordetella pertussis*, *Bacillus dysenteriae*, *Propionibacterium freudenreichii*, *Candida albicans*, *Mycobacterium avium-intracellulare* complex, and *M. tuberculosis* [28–42]. LacCer is also the major GSL component of human and bovine milk [43], suggesting that milk-derived LacCer can act as putative traps that interfere with pathogen binding in the intestines, thereby preventing infection. *Pneumocystis carinii*-derived β-glucan has been found to stimulate the production by rat alveolar epithelial cells of significant amounts of MIP-2 via LacCer expressed on these cells [30]. CSBG was shown to induce the migration of human neutrophils, a migration completely inhibited by LacCer liposomes or the Src family kinase inhibitor PP1 [44]. Taken together, these observations clearly suggest that LacCer is a key molecule in intestinal infection and immunity.

2 Gangliosides Mediate Immunological Functions of Lymphocytes

The best studied ganglioside-related immune function of lymphocytes is the involvement of GM1-enriched lipid rafts in B cell antigen receptor (BCR) signaling [45–47]. In resting B cells, IgM-BCR is excluded from GM1-enriched lipid rafts, while IgD-BCR is close to these domains [45]. Following stimulation of BCRs, IgM-BCRs colocalize with GM1-enriched lipid rafts in a caveolin-1-dependent manner [48], while IgD-BCRs diffuse from these domains, such that nanoclusters of IgM BCRs cluster with nanoclusters of IgD BCRs [49].

The initial event in T-cell activation involves the movement of T-cell receptors (TCRs) into lipid rafts [50]. However, the roles of lipid rafts in TCR signaling remain unclear [51, 52], primarily because of confusion about the binding specificity of the CT B subunit and its use as a lipid raft marker. The CT B subunit is able to bind not only GM1 but other gangliosides, especially fucosylated GM1 (Fuc-GM1), but not neutral GSLs [53, 54]. The association constants of CT B subunit with GM1 and Fuc-GM1 are comparable. Thus, antibody to CT B subunit alone would be inconclusive for visualization of lipid rafts [55]. Mouse $CD4^+$ T cells mainly express GM1a and GD1b. Thus, cells from mice lacking GM3 synthase (GM3-null mice) exhibit severe impairments in TCR-mediated immune responses, including cytokine production and clonal expansion [52]. These defects were rescued by preincubation of T cells from GM3-null mice with GM3 and GM1a. In contrast, mouse $CD8^+$ T cells express GA1, GM1b, GalNAc-GM1b, and extended-GM1b. Cells from GM2/GD2S-null mice also exhibited severe impairments in TCR-mediated cytokine production and clonal expansion, impairments rescued by preincubation with GA1 and GM1b. Therefore, it is likely that TCR-dependent activation of $CD4^+$ T cells selectively requires a-series gangliosides, whereas activation of $CD8^+$ T cells requires o-series gangliosides [52].

Ganglioside GQ1b was found to enhance both IL-6 and IL-10 production by T cells, without affecting cytokine production by monocytes and B cells [56]. Preincubation of T cells with GQ1b, followed by their coculture with B cells, also enhanced Ig production. Exogenous IL-6 and IL-10 each enhanced Ig production by B cells alone, with the two showing additive effects. These results suggest that GQ1b induces IL-6 and IL-10 production by T cells, resulting in Ig production by B cells [56]. Extracellular gangliosides may therefore act as modulators of immune functions.

3 GSLs Work as Pattern Recognition Receptors in Innate Immunity

The innate immune system recognizes pathogen-associated molecular patterns (PAMPs) shared by groups of related microorganisms, components of which are essential for the survival of these organisms; PAMPs, however, are not present on mammalian cells [57]. Pattern recognition receptors (PRRs) expressed on innate immune cells play a central role in the proper functions of the innate immune system. The binding of PRRs to PAMPs is the initial step in activating the innate immune system. PAMPs can also be identified by GSLs, which organize specialized lipid rafts. GD1a on human monocytes binds to a subunit of type IIb *Escherichia coli* enterotoxin, promoting its interaction with the TLR2/TLR1 signaling complex and activating NF-κB [58]. The binding of bacterial flagellin to asialoGM1 and TLR5 has been found to induce the autocrine release of ATP from human lung epithelial cells [59]. These observations suggest that the cooperation of GSLs and PRRs is one of the key steps in innate immune responses.

The killing of microorganisms by phagocytosis is one of the most important functions of the innate immune system. Phagocytes express and utilize PRRs to recognize pathogens directly or indirectly. PAMPs are directly recognized by several types of PRR, such as Toll-like receptors (TLRs) [60, 61], mannose receptor [62], the LPS-binding receptor CD14 [63], scavenger receptors [64, 65], the C-type lectin receptor (CLR) dectin-1 [66], and αMβ2-integrin [67]. PAMPs can be sensed not only by glycoproteins and C-type lectin receptors but by GSLs such as LacCer [28, 44, 68, 69] and GM1 [15, 16], which constitute specialized lipid rafts. These types of PRR act as sensors for nonopsonic infectious microorganisms in host phagocytes, contributing to the nonopsonic recognition of infectious microorganisms by phagocytes. These cells efficiently engulf microorganisms opsonized with complement component C3bi and IgG. αMβ2-Integrin is involved in the adhesion, migration, chemotaxis, phagocytosis of opsonized and nonopsonized microorganisms, as well as respiratory bursts and the degranulation of phagocytes [70]. Although αMβ2-integrin-induced outside-in signaling is highly dependent on Src family kinases [71, 72], αMβ2-integrin has short cytoplasmic domains and is devoid of catalytic activities responsible for signaling inside the cells [73, 74]. Thus, adaptor or mediator molecules are required for αMβ2-integrin-mediated outside-in signaling. The αM subunit has a unique structure, containing not only a binding site for ligands, such as ICAM-1 and C3bi, but a spatially separated carbohydrate binding domain that serves as a PRR for β-glucan and LPS [75, 76]. αMβ2-Integrin and LacCer localize close to each other on the neutrophil plasma membrane (Fig. 1a). In addition to activating αMβ2-integrin, integrin molecules move to LacCer-enriched domains (Fig. 1b). Indeed, the β2 subunit was immunoprecipitated with anti-LacCer antibody from the lipid raft fraction of zymosan-engulfed phagosomal membranes but not from plasma membranes of resting neutrophils [69]. LacCer forms Lyn-coupled lipid rafts, which mediate neutrophil chemotaxis, phagocytosis and superoxide generation

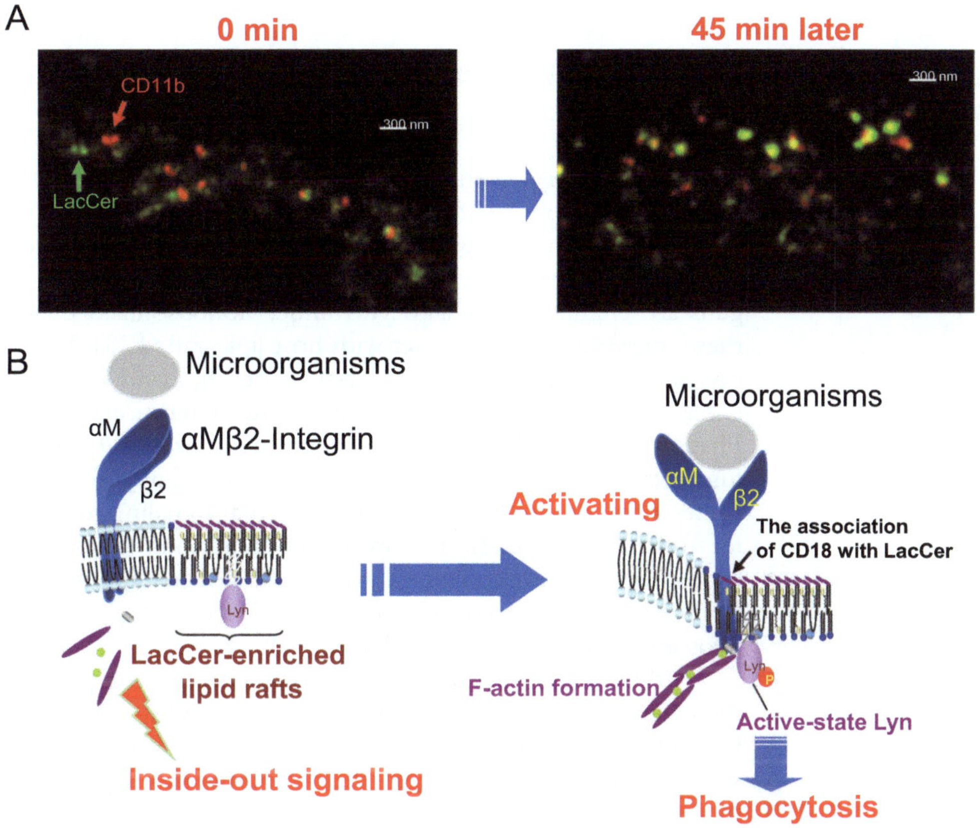

Fig. 1 Lyn-coupled LacCer-enriched lipid rafts are essential for αMβ2-integrin-mediated phagocytosis of nonopsonized microorganisms. **(A)** Colocalization of αMβ2-integrin and LacCer on the plasma membranes of neutrophils. Neutrophils were labeled with anti-LacCer IgM T5A7 (green) and anti-αM subunit IgG ICRF44 (red) on ice. After incubation at room temperature for 45 min, the cells were viewed with a Gated STED microscope equipped with a Plan-Apochromat 100× oil immersion DIC objective. **(B)** αMβ2-integrin-dependent neutrophil phagocytosis is mediated by LacCer-enriched lipid rafts. Upon binding to nonopsonized microorganisms, αMβ2-integrins are activated and undergo a conformational change, resulting in the rearrangement of cytoskeletal proteins, including β-actinin. Inside-out signaling via specific cytokine or chemoattractant receptors initiates the conversion of αMβ2-integrin from the resting to the activated state. Following rearrangement of cytoskeletal proteins, αMβ2-integrin translocates into LacCer-enriched lipid rafts, allowing αMβ2-integrin to transmit stimulatory signals to the Src family kinase Lyn through LacCer-enriched lipid rafts. These signaling cascades lead to the formation of actin-enriched phagocytic cups, resulting in phagosome formation

[28, 44, 69]. Experiments using si-RNA demonstrated that Lyn-coupled LacCer-enriched domains are indispensable for αMβ2-integrin mediated neutrophil activation by nonopsonized microorganisms [68]. The main molecular species of LacCer in human neutrophil plasma membranes contain very long C24 fatty acid chains. Experiments using azide-photoactivatable tritium-labeled C24- and C18-LacCer showed that C24- but not C16-LacCer was directly associated with Lyn and Gαi [77]. These results suggest that the interactions of C24 fatty acid chains of LacCer with the palmitic chains of Lyn and Gαi were essential for the outside-in signal transduction of LacCer-enriched lipid rafts.

4 Intracellular Pathogens Target GSL-Enriched Lipid Rafts to Escape from Host Immunity

Listeria, *Salmonella*, and *Mycobacteria* are intracellular pathogens that target lipid rafts [78] in the plasma membranes of host cells, as well as proteinaceous PRRs. These organisms can survive in host cells by manipulating various signaling pathways. If intracellular pathogens are engulfed by phagocytes under nonopsonized conditions, these microorganisms interact with host lipid rafts [78]. Pathogenic mycobacteria also use both cholesterol-rich lipid rafts and PRR to enter phagocytes under nonopsonized conditions [79, 80]. Phagosomes containing serum-opsonized *M. tuberculosis* effectively fuse with lysosomes, whereas phagosomes containing nonopsonized bacteria do not fuse [81, 82], suggesting that signaling responses differ under opsonized and nonopsonized conditions. Once intracellular pathogens are internalized within the phagosome, they escape from the vacuolar compartment to reach the cytosol or arrest the lysosome–phagosome fusion [80, 83–90].

Pathogenic mycobacteria have been shown to use effector molecules, including lipoarabinomannan (LAM), to manipulate host immune systems [91, 92]. LAMs, which are classified into several families according to their capping motifs, constitute a major glycolipid component of all mycobacterial species [93]. Pathogenic mycobacteria contain LAMs with a terminal mannose cap (ManLAM), whereas nonpathogenic mycobacteria including *M. smegmatis* contain LAM with a terminal phospho-myoinositol cap (PILAM) or no cap [91]. ManLAMs have been associated with several PRRs and binding molecules on phagocytes, including the αMβ2 integrin, macrophage mannose receptor (CD206), and dendritic cell-specific, intercellular adhesion molecule-3 (ICAM-3)-grabbing nonintegrin (DC-SIGN) [94–96]. Deletion of the gene encoding lipoarabinomannan carrier protein (LprG), a cell envelope lipoprotein that is essential for the virulence of *M. tuberculosis*, results in the defective surface exposure of ManLAM and impaired inhibition of phagosome maturation in mouse macrophages [97, 98]. These and other findings support the hypothesis, that ManLAM plays an important role in the process by which pathogenic mycobacteria prevent phagosome-lysosome fusion [99].

Human neutrophils engulf both pathogenic and nonpathogenic mycobacteria through Lyn-coupled LacCer-enriched lipid rafts and αMβ2 integrin [42]. Regardless of their pathogenicity, LAMs have a mannan core structure, which consists of a 21–34 residue α(1 → 6)-mannopyranose backbone and 5–10 α(1 → 2)-mannopyranose side chains [100]. Among several GSLs, *M. tuberculosis*- and *M. smegmatis*-derived LAMs bind to LacCer [42]. Importantly, PILAM from an *M. smegmatis* α1,2-mannosyltransferase deletion mutant (*ΔMSMEG_4247*), which lacks the α1,2-monomannose side branches of the LAM mannan core, did not bind to LacCer nor induce phagocytosis. These findings indicate that LacCer recognizes the α1,2-monomannose side branches of LAM, which are expressed by both nonpathogenic and pathogenic mycobacteria.

LacCer clusters bind to the β-1,6 long glucopyranose side chains branched with a β-1,3 glucopyranose of β-glucan [28]. LacCer apparently recognizes the common structural patterns of fungal β-glucans and mycobacterial LAMs. Although human neutrophils phagocytose pathogenic and nonpathogenic mycobacteria via the same mechanism, the fates of these engulfed mycobacteria differ. Neutrophil engulfment of nonpathogenic mycobacteria results in lysosome fusion with nascent bacteria-containing phagosomes [101]. In contrast, engulfed pathogenic mycobacteria arrest the fusion of lysosomes to phagosomes, resulting in bacterial survival in host phagocytes. About 90% of LacCer molecules localize to the granular membranes of human neutrophils [44, 102], forming lipid rafts not only on granular membranes but on phagosomes [69]. The Src family kinase Hck, which is present in the granular fractions of human neutrophils [103], has been reported essential for the fusion of lysosomes to phagosomes [104]. LacCer-enriched lipid rafts were found to associate with Hck on phagosomes containing PILAM- but not ManLAM-coated beads [42], suggesting that the mannose cap of ManLAM inhibits the Hck-associated formation of such rafts. Hck has been associated with the maturation of mycobacteria-containing phagosomes in human neutrophils (41). These findings suggest that ManLAMs derived from pathogenic mycobacteria affects the integrity of lipid rafts reorganized by LacCer with Hck in phagosomal membranes, allowing bacterial escape from killing by neutrophils.

5 Perspective

This section has reviewed the roles of GSLs in immune functions. Although gangliosides have been shown to act as modulators of the immune system, neutral GSLs on host cells have been shown important in host defense mechanisms. Several GSLs, including GQ1b and α-GalCer, have been identified as specific ligands for receptors [56, 105, 106]. α-GalCer, which was originally isolated from a marine sponge, was identified as a lipid antigen that activates NKT cells [105] and to enhance immune responses to a range of infectious organisms and some types of cancer, as well as suppressing autoimmune diseases, allograft rejection and graft-versus-host diseases [107]. These observations strongly indicate that glycolipid mediated NKT activation is a potentially important immunotherapeutic target with widespread clinical potential. Recently, α-glycosylceramides were identified as the major endogenous ligands of NKT cells [108]. Moreover, β-GlcCer was identified as an endogenous ligand for Mincle and shown to possess immunostimulatory activity [109]. Although the mechanism by which GSL molecules specifically bind to their specific receptors in vivo remains unclear, other GSLs have been identified as ligands for important immune receptors. Unknown GSL-mediated immunological functions will be elucidated in the future.

References

1. Murate M, Abe M, Kasahara K, Iwabuchi K, Umeda M, Kobayashi T (2015) Transbilayer distribution of lipids at nano scale. J Cell Sci 128(8):1627–1638. https://doi.org/10.1242/jcs.163105
2. Inokuchi J, Nagafuku M, Ohno I, Suzuki A (2015) Distinct selectivity of gangliosides required for CD4(+) T and CD8(+) T cell activation. Biochim Biophys Acta 1851(1):98–106. https://doi.org/10.1016/j.bbalip.2014.07.013
3. Nakayama H, Ogawa H, Takamori K, Iwabuchi K (2013) GSL-enriched membrane microdomains in innate immune responses. Arch Immunol Ther Exp 61(3):217–228. https://doi.org/10.1007/s00005-013-0221-6
4. Rueda R (2007) The role of dietary gangliosides on immunity and the prevention of infection. Br J Nutr 98(Suppl 1):S68–S73. https://doi.org/10.1017/s0007114507832946
5. Popa I, Portoukalian J (2003) Relationship between lipids and cutaneous immunity: example of the gangliosides. Pathol Biol (Paris) 51(5):253–255
6. Schengrund CL (2003) "Multivalent" saccharides: development of new approaches for inhibiting the effects of glycosphingolipid-binding pathogens. Biochem Pharmacol 65(5):699–707
7. Yates AJ, Rampersaud A (1998) Sphingolipids as receptor modulators. An overview. Ann N Y Acad Sci 845:57–71
8. Hakomori S (1981) Glycosphingolipids in cellular interaction, differentiation, and oncogenesis. Annu Rev Biochem 50:733–764. https://doi.org/10.1146/annurev.bi.50.070181.003505
9. Mukherjee S, Maxfield FR (2004) Membrane domains. Annu Rev Cell Dev Biol 20:839–866. https://doi.org/10.1146/annurev.cellbio.20.010403.095451
10. Pike LJ (2006) Rafts defined: a report on the keystone symposium on lipid rafts and cell function. J Lipid Res 47(7):1597–1598. https://doi.org/10.1194/jlr.E600002-JLR200
11. Iwabuchi K, Nakayama H, Oizumi A, Suga Y, Ogawa H, Takamori K (2015) Role of ceramide from glycosphingolipids and its metabolites in

immunological and inflammatory responses in humans. Mediat Inflamm 2015:120748. https://doi.org/10.1155/2015/120748
12. Anderson HA, Roche PA (2015) MHC class II association with lipid rafts on the antigen presenting cell surface. Biochim Biophys Acta 1853(4):775–780. https://doi.org/10.1016/j.bbamcr.2014.09.019
13. Zuidscherwoude M, de Winde CM, Cambi A, van Spriel AB (2014) Microdomains in the membrane landscape shape antigen-presenting cell function. J Leukoc Biol 95(2):251–263. https://doi.org/10.1189/jlb.0813440
14. Dykstra M, Cherukuri A, Sohn HW, Tzeng SJ, Pierce SK (2003) Location is everything: lipid rafts and immune cell signaling. Annu Rev Immunol 21:457–481. https://doi.org/10.1146/annurev.immunol.21.120601.141021
15. Tsai B, Gilbert JM, Stehle T, Lencer W, Benjamin TL, Rapoport TA (2003) Gangliosides are receptors for murine polyoma virus and SV40. EMBO J 22(17):4346–4355
16. Naroeni A, Porte F (2002) Role of cholesterol and the ganglioside GM(1) in entry and short-term survival of Brucella suis in murine macrophages. Infect Immun 70(3):1640–1644
17. Cuatrecasas P (1973) Vibrio cholerae choleragenoid. Mechanism of inhibition of cholera toxin action. Biochemistry 12(18):3577–3581
18. Cuatrecasas P (1973) Gangliosides and membrane receptors for cholera toxin. Biochemistry 12(18):3558–3566
19. Hyun CS, Kimmich GA (1984) Interaction of cholera toxin and Escherichia coli enterotoxin with isolated intestinal epithelial cells. Am J Phys 247(6 Pt 1):G623–G631
20. Otnaess AB, Laegreid A, Ertresvag K (1983) Inhibition of enterotoxin from Escherichia coli and vibrio cholerae by gangliosides from human milk. Infect Immun 40(2):563–569
21. Bronnum H, Seested T, Hellgren LI, Brix S, Frokiaer H (2005) Milk-derived GM(3) and GD(3) differentially inhibit dendritic cell maturation and effector functionalities. Scand J Immunol 61(6):551–557. https://doi.org/10.1111/j.1365-3083.2005.01566.x
22. Takenouchi H, Kiyokawa N, Taguchi T, Matsui J, Katagiri YU, Okita H, Okuda K, Fujimoto J (2004) Shiga toxin binding to globotriaosyl ceramide induces intracellular signals that mediate cytoskeleton remodeling in human renal carcinoma-derived cells. J Cell Sci 117(Pt 17):3911–3922
23. Lingwood CA (1996) Role of verotoxin receptors in pathogenesis. Trends Microbiol 4(4):147–153
24. Louise CB, Kaye SA, Boyd B, Lingwood CA, Obrig TG (1995) Shiga toxin-associated hemolytic uremic syndrome: effect of sodium butyrate on sensitivity of human umbilical vein endothelial cells to Shiga toxin. Infect Immun 63(7):2766–2769
25. Tyrrell GJ, Ramotar K, Toye B, Boyd B, Lingwood CA, Brunton JL (1992) Alteration of the carbohydrate binding specificity of verotoxins from gal alpha 1-4Gal to GalNAc beta 1-3Gal alpha 1-4Gal and vice versa by site-directed mutagenesis of the binding subunit. Proc Natl Acad Sci U S A 89(2):524–528
26. Mukai T, Kaneko S, Matsumoto M, Ohori H (2004) Binding of Bifidobacterium bifidum and lactobacillus reuteri to the carbohydrate moieties of intestinal glycolipids recognized by peanut agglutinin. Int J Food Microbiol 90(3):357–362
27. de Bentzmann S, Roger P, Dupuit F, Bajolet-Laudinat O, Fuchey C, Plotkowski MC, Puchelle E (1996) Asialo GM1 is a receptor for Pseudomonas aeruginosa adherence to regenerating respiratory epithelial cells. Infect Immun 64(5):1582–1588
28. Sato T, Iwabuchi K, Nagaoka I, Adachi Y, Ohno N, Tamura H, Seyama K, Fukuchi Y, Nakayama H, Yoshizaki F, Takamori K, Ogawa H (2006) Induction of human neutrophil chemotaxis by Candida albicans-derived beta-1,6-long glycoside side-chain-branched beta-glucan. J Leukoc Biol 80(1):204–211. https://doi.org/10.1189/jlb.0106069
29. Angstrom J, Teneberg S, Milh MA, Larsson T, Leonardsson I, Olsson BM, Halvarsson MO, Danielsson D, Naslund I, Ljungh A, Wadstrom T, Karlsson KA (1998) The lactosylceramide binding specificity of helicobacter pylori. Glycobiology 8(4):297–309
30. Hahn PY, Evans SE, Kottom TJ, Standing JE, Pagano RE, Limper AH (2003) Pneumocystis carinii cell wall beta-glucan induces release of macrophage inflammatory protein-2 from alveolar epithelial cells via a lactosylceramide-mediated mechanism. J Biol Chem 278(3):2043–2050
31. Karlsson KA (1986) Animal glycolipids as attachment sites for microbes. Chem Phys Lipids 42(1–3):153–172
32. Saukkonen K, Burnette WN, Mar VL, Masure HR, Tuomanen EI (1992) Pertussis toxin has eukaryotic-like carbohydrate recognition domains. Proc Natl Acad Sci U S A 89(1):118–122

33. Zimmerman JW, Lindermuth J, Fish PA, Palace GP, Stevenson TT, DeMong DE (1998) A novel carbohydrate-glycosphingolipid interaction between a beta-(1-3)-glucan immunomodulator, PGG-glucan, and lactosylceramide of human leukocytes. J Biol Chem 273(34):22014–22020
34. Fukuda MN, Dell A, Oates JE, Wu P, Klock JC, Fukuda M (1985) Structures of glycosphingolipids isolated from human granulocytes. The presence of a series of linear poly-N-acetyllactosaminylceramide and its significance in glycolipids of whole blood cells. J Biol Chem 260(2):1067–1082
35. Teneberg S, Angstrom J, Ljungh A (2004) Carbohydrate recognition by enterohemorrhagic Escherichia coli: characterization of a novel glycosphingolipid from cat small intestine. Glycobiology 14(2):187–196
36. Stromberg N, Karlsson KA (1990) Characterization of the binding of propionibacterium granulosum to glycosphingolipids adsorbed on surfaces. An apparent recognition of lactose which is dependent on the ceramide structure. J Biol Chem 265(19):11244–11250
37. Stromberg N, Ryd M, Lindberg AA, Karlsson KA (1988) Studies on the binding of bacteria to glycolipids. Two species of Propionibacterium apparently recognize separate epitopes on lactose of lactosylceramide. FEBS Lett 232(1):193–198
38. Payne D, Tatham D, Williamson ED, Titball RW (1998) The pH 6 antigen of Yersinia pestis binds to beta1-linked galactosyl residues in glycosphingolipids. Infect Immun 66(9):4545–4548
39. Jansson L, Tobias J, Lebens M, Svennerholm AM, Teneberg S (2006) The major subunit, CfaB, of colonization factor antigen i from enterotoxigenic Escherichia coli is a glycosphingolipid binding protein. Infect Immun 74(6):3488–3497
40. Stromberg N, Deal C, Nyberg G, Normark S, So M, Karlsson KA (1988) Identification of carbohydrate structures that are possible receptors for Neisseria gonorrhoeae. Proc Natl Acad Sci U S A 85(13):4902–4906
41. Backenson PB, Coleman JL, Benach JL (1995) Borrelia burgdorferi shows specificity of binding to glycosphingolipids. Infect Immun 63(8):2811–2817
42. Nakayama H, Kurihara H, Morita YS, Kinoshita T, Mauri L, Prinetti A, Sonnino S, Yokoyama N, Ogawa H, Takamori K, Iwabuchi K (2016) Lipoarabinomannan binding to lactosylceramide in lipid rafts is essential for the phagocytosis of mycobacteria by human neutrophils. Sci Signal 9(449):ra101. https://doi.org/10.1126/scisignal.aaf1585
43. Newburg DS, Chaturvedi P (1992) Neutral glycolipids of human and bovine milk. Lipids 27(11):923–927
44. Iwabuchi K, Nagaoka I (2002) Lactosylceramide-enriched glycosphingolipid signaling domain mediates superoxide generation from human neutrophils. Blood 100(4):1454–1464
45. Minguet S, Klasener K, Schaffer AM, Fiala GJ, Osteso-Ibanez T, Raute K, Navarro-Lerida I, Hartl FA, Seidl M, Reth M, Del Pozo MA (2017) Caveolin-1-dependent nanoscale organization of the BCR regulates B cell tolerance. Nat Immunol 18:1150. https://doi.org/10.1038/ni.3813
46. Shrestha D, Exley MA, Vereb G, Szollosi J, Jenei A (2014) CD1d favors MHC neighborhood, GM1 ganglioside proximity and low detergent sensitive membrane regions on the surface of B lymphocytes. Biochim Biophys Acta 1840(1):667–680
47. Kimata H (1994) GM1, a ganglioside that specifically enhances immunoglobulin production and proliferation in human plasma cells. Eur J Immunol 24(11):2910–2913. https://doi.org/10.1002/eji.1830241149
48. Gupta N, DeFranco AL (2007) Lipid rafts and B cell signaling. Semin Cell Dev Biol 18(5):616–626. https://doi.org/10.1016/j.semcdb.2007.07.009
49. Maity PC, Blount A, Jumaa H, Ronneberger O, Lillemeier BF, Reth M (2015) B cell antigen receptors of the IgM and IgD classes are clustered in different protein islands that are altered during B cell activation. Sci Signal 8(394):ra93. https://doi.org/10.1126/scisignal.2005887
50. Harder T, Rentero C, Zech T, Gaus K (2007) Plasma membrane segregation during T cell activation: probing the order of domains. Curr Opin Immunol 19(4):470–475. https://doi.org/10.1016/j.coi.2007.05.002
51. Popik W, Alce TM (2004) CD4 receptor localized to non-raft membrane microdomains supports HIV-1 entry. Identification of a novel raft localization marker in CD4. J Biol Chem 279(1):704–712. https://doi.org/10.1074/jbc.M306380200
52. Nagafuku M, Okuyama K, Onimaru Y, Suzuki A, Odagiri Y, Yamashita T, Iwasaki K, Fujiwara M, Takayanagi M, Ohno I, Inokuchi JI (2012) CD4 and CD8 T cells require different membrane gangliosides for activation. Proc Natl Acad Sci U S A 109(6):E336–E342. https://doi.org/10.1073/pnas.1114965109

53. Kuziemko GM, Stroh M, Stevens RC (1996) Cholera toxin binding affinity and specificity for gangliosides determined by surface plasmon resonance. Biochemistry 35(20):6375–6384. https://doi.org/10.1021/bi952314i
54. Masserini M, Freire E, Palestini P, Calappi E, Tettamanti G (1992) Fuc-GM1 ganglioside mimics the receptor function of GM1 for cholera toxin. Biochemistry 31(8):2422–2426
55. Yanagisawa M, Ariga T, Yu RK (2006) Cholera toxin B subunit binding does not correlate with GM1 expression: a study using mouse embryonic neural precursor cells. Glycobiology 16(9):19G–22G
56. Kanda N, Tamaki K (1998) Ganglioside GQ1b enhances Ig production by human PBMCs. J Allergy Clin Immunol 102(5):813–820
57. Medzhitov R, Janeway C Jr (2000) Innate immune recognition: mechanisms and pathways. Immunol Rev 173:89–97
58. Liang S, Wang M, Tapping RI, Stepensky V, Nawar HF, Triantafilou M, Triantafilou K, Connell TD, Hajishengallis G (2007) Ganglioside GD1a is an essential coreceptor for toll-like receptor 2 signaling in response to the B subunit of type IIb enterotoxin. J Biol Chem 282(10):7532–7542. https://doi.org/10.1074/jbc.M611722200
59. McNamara N, Gallup M, Sucher A, Maltseva I, McKemy D, Basbaum C (2006) AsialoGM1 and TLR5 cooperate in flagellin-induced nucleotide signaling to activate Erk1/2. Am J Respir Cell Mol Biol 34(6):653–660. https://doi.org/10.1165/rcmb.2005-0441OC
60. Blander JM, Medzhitov R (2004) Regulation of phagosome maturation by signals from toll-like receptors. Science 304(5673):1014–1018
61. Doyle SE, O'Connell RM, Miranda GA, Vaidya SA, Chow EK, Liu PT, Suzuki S, Suzuki N, Modlin RL, Yeh WC, Lane TF, Cheng G (2004) Toll-like receptors induce a phagocytic gene program through p38. J Exp Med 199(1):81–90
62. Ezekowitz RA, Sastry K, Bailly P, Warner A (1990) Molecular characterization of the human macrophage mannose receptor: demonstration of multiple carbohydrate recognition-like domains and phagocytosis of yeasts in Cos-1 cells. J Exp Med 172(6):1785–1794
63. Jiang Z, Georgel P, Du X, Shamel L, Sovath S, Mudd S, Huber M, Kalis C, Keck S, Galanos C, Freudenberg M, Beutler B (2005) CD14 is required for MyD88-independent LPS signaling. Nat Immunol 6(6):565–570
64. Elomaa O, Kangas M, Sahlberg C, Tuukkanen J, Sormunen R, Liakka A, Thesleff I, Kraal G, Tryggvason K (1995) Cloning of a novel bacteria-binding receptor structurally related to scavenger receptors and expressed in a subset of macrophages. Cell 80(4):603–609
65. Peiser L, Gough PJ, Kodama T, Gordon S (2000) Macrophage class a scavenger receptor-mediated phagocytosis of Escherichia coli: role of cell heterogeneity, microbial strain, and culture conditions in vitro. Infect Immun 68(4):1953–1963
66. Herre J, Marshall AS, Caron E, Edwards AD, Williams DL, Schweighoffer E, Tybulewicz V, Reis e Sousa C, Gordon S, Brown GD (2004) Dectin-1 uses novel mechanisms for yeast phagocytosis in macrophages. Blood 104(13):4038–4045
67. Brown EJ (1991) Complement receptors and phagocytosis. Curr Opin Immunol 3(1):76–82
68. Iwabuchi K, Prinetti A, Sonnino S, Mauri L, Kobayashi T, Ishii K, Kaga N, Murayama K, Kurihara H, Nakayama H, Yoshizaki F, Takamori K, Ogawa H, Nagaoka I (2008) Involvement of very long fatty acid-containing lactosylceramide in lactosylceramide-mediated superoxide generation and migration in neutrophils. Glycoconj J 25(4):357–374. https://doi.org/10.1007/s10719-007-9084-6
69. Nakayama H, Yoshizaki F, Prinetti A, Sonnino S, Mauri L, Takamori K, Ogawa H, Iwabuchi K (2008) Lyn-coupled LacCer-enriched lipid rafts are required for CD11b/CD18-mediated neutrophil phagocytosis of nonopsonized microorganisms. J Leukoc Biol 83(3):728–741. https://doi.org/10.1189/jlb.0707478
70. Arnaout MA (1990) Structure and function of the leukocyte adhesion molecules CD11/CD18. Blood 75(5):1037–1050
71. Evangelista V, Pamuklar Z, Piccoli A, Manarini S, Dell'elba G, Pecce R, Martelli N, Federico L, Rojas M, Berton G, Lowell CA, Totani L, Smyth SS (2007) Src family kinases mediate neutrophil adhesion to adherent platelets. Blood 109(6):2461–2469
72. Piccardoni P, Manarini S, Federico L, Bagoly Z, Pecce R, Martelli N, Piccoli A, Totani L, Cerletti C, Evangelista V (2004) SRC-dependent outside-in signalling is a key step in the process of autoregulation of beta2 integrins in polymorphonuclear cells. Biochem J 380(Pt 1):57–65
73. Rabb H, Michishita M, Sharma CP, Brown D, Arnaout MA (1993) Cytoplasmic tails of human complement receptor type 3 (CR3, CD11b/CD18) regulate ligand avidity and the internalization of occupied receptors. J Immunol 151(2):990–1002

74. Dedhar S, Hannigan GE (1996) Integrin cytoplasmic interactions and bidirectional transmembrane signalling. Curr Opin Cell Biol 8(5):657–669
75. Vetvicka V, Thornton BP, Ross GD (1996) Soluble beta-glucan polysaccharide binding to the lectin site of neutrophil or natural killer cell complement receptor type 3 (CD11b/CD18) generates a primed state of the receptor capable of mediating cytotoxicity of iC3b-opsonized target cells. J Clin Invest 98(1):50–61. https://doi.org/10.1172/JCI118777
76. Wright SD, Levin SM, Jong MT, Chad Z, Kabbash LG (1989) CR3 (CD11b/CD18) expresses one binding site for Arg-Gly-asp-containing peptides and a second site for bacterial lipopolysaccharide. J Exp Med 169(1):175–183
77. Chiricozzi E, Ciampa MG, Brasile G, Compostella F, Prinetti A, Nakayama H, Ekyalongo RC, Iwabuchi K, Sonnino S, Mauri L (2015) Direct interaction, instrumental for signaling processes, between LacCer and Lyn in the lipid rafts of neutrophil-like cells. J Lipid Res 56(1):129–141. https://doi.org/10.1194/jlr.M055319
78. Manes S, del Real G, Martinez AC (2003) Pathogens: raft hijackers. Nat Rev Immunol 3(7):557–568
79. Peyron P, Bordier C, N'Diaye EN, Maridonneau-Parini I (2000) Nonopsonic phagocytosis of mycobacterium kansasii by human neutrophils depends on cholesterol and is mediated by CR3 associated with glycosylphosphatidylinositol-anchored proteins. J Immunol 165(9):5186–5191
80. Gatfield J, Pieters J (2000) Essential role for cholesterol in entry of mycobacteria into macrophages. Science 288(5471):1647–1650
81. Vieira OV, Botelho RJ, Grinstein S (2002) Phagosome maturation: aging gracefully. Biochem J 366(Pt 3):689–704. https://doi.org/10.1042/BJ20020691
82. Armstrong JA, Hart PD (1975) Phagosome-lysosome interactions in cultured macrophages infected with virulent tubercle bacilli. Reversal of the usual nonfusion pattern and observations on bacterial survival. J Exp Med 142(1):1–16
83. Stuart ES, Webley WC, Norkin LC (2003) Lipid rafts, caveolae, caveolin-1, and entry by Chlamydiae into host cells. Exp Cell Res 287(1):67–78
84. Baorto DM, Gao Z, Malaviya R, Dustin ML, van der Merwe A, Lublin DM, Abraham SN (1997) Survival of FimH-expressing enterobacteria in macrophages relies on glycolipid traffic. Nature 389(6651):636–639
85. Watarai M, Makino S, Fujii Y, Okamoto K, Shirahata T (2002) Modulation of Brucella-induced macropinocytosis by lipid rafts mediates intracellular replication. Cell Microbiol 4(6):341–355
86. Tamilselvam B, Daefler S (2008) Francisella targets cholesterol-rich host cell membrane domains for entry into macrophages. J Immunol 180(12):8262–8271
87. Harrison RE, Brumell JH, Khandani A, Bucci C, Scott CC, Jiang X, Finlay BB, Grinstein S (2004) Salmonella impairs RILP recruitment to Rab7 during maturation of invasion vacuoles. Mol Biol Cell 15(7):3146–3154. https://doi.org/10.1091/mbc.E04-02-0092
88. Gekara NO, Weiss S (2004) Lipid rafts clustering and signalling by listeriolysin O. Biochem Soc Trans 32(Pt 5):712–714
89. Knodler LA, Vallance BA, Hensel M, Jackel D, Finlay BB, Steele-Mortimer O (2003) Salmonella type III effectors PipB and PipB2 are targeted to detergent-resistant microdomains on internal host cell membranes. Mol Microbiol 49(3):685–704
90. Brumell JH, Kujat-Choy S, Brown NF, Vallance BA, Knodler LA, Finlay BB (2003) SopD2 is a novel type III secreted effector of salmonella typhimurium that targets late endocytic compartments upon delivery into host cells. Traffic 4(1):36–48
91. Kaur D, Obregon-Henao A, Pham H, Chatterjee D, Brennan PJ, Jackson M (2008) Lipoarabinomannan of mycobacterium: mannose capping by a multifunctional terminal mannosyltransferase. Proc Natl Acad Sci U S A 105(46):17973–17977. https://doi.org/10.1073/pnas.0807761105
92. Fratti RA, Chua J, Vergne I, Deretic V (2003) Mycobacterium tuberculosis glycosylated phosphatidylinositol causes phagosome maturation arrest. Proc Natl Acad Sci U S A 100(9):5437–5442. https://doi.org/10.1073/pnas.0737613100
93. Briken V, Porcelli SA, Besra GS, Kremer L (2004) Mycobacterial lipoarabinomannan and related lipoglycans: from biogenesis to modulation of the immune response. Mol Microbiol 53(2):391–403. https://doi.org/10.1111/j.1365-2958.2004.04183.x
94. Schlesinger LS (1993) Macrophage phagocytosis of virulent but not attenuated strains of mycobacterium tuberculosis is mediated by mannose receptors in addition to complement receptors. J Immunol 150(7):2920–2930
95. Geijtenbeek TB, Van Vliet SJ, Koppel EA, Sanchez-Hernandez M, Vandenbroucke-Grauls CM, Appelmelk B, Van Kooyk Y

(2003) Mycobacteria target DC-SIGN to suppress dendritic cell function. J Exp Med 197(1):7–17

96. Kang PB, Azad AK, Torrelles JB, Kaufman TM, Beharka A, Tibesar E, DesJardin LE, Schlesinger LS (2005) The human macrophage mannose receptor directs mycobacterium tuberculosis lipoarabinomannan-mediated phagosome biogenesis. J Exp Med 202(7):987–999. https://doi.org/10.1084/jem.20051239
97. Shukla S, Richardson ET, Athman JJ, Shi L, Wearsch PA, McDonald D, Banaei N, Boom WH, Jackson M, Harding CV (2014) Mycobacterium tuberculosis lipoprotein LprG binds lipoarabinomannan and determines its cell envelope localization to control phagolysosomal fusion. PLoS Pathog 10(10):e1004471. https://doi.org/10.1371/journal.ppat.1004471
98. Gaur RL, Ren K, Blumenthal A, Bhamidi S, Gonzalez-Nilo FD, Jackson M, Zare RN, Ehrt S, Ernst JD, Banaei N (2014) LprG-mediated surface expression of lipoarabinomannan is essential for virulence of mycobacterium tuberculosis. PLoS Pathog 10(9):e1004376. https://doi.org/10.1371/journal.ppat.1004376
99. Diacovich L, Gorvel JP (2010) Bacterial manipulation of innate immunity to promote infection. Nat Rev Microbiol 8(2):117–128. https://doi.org/10.1038/nrmicro2295
100. Mishra AK, Driessen NN, Appelmelk BJ, Besra GS (2011) Lipoarabinomannan and related glycoconjugates: structure, biogenesis and role in mycobacterium tuberculosis physiology and host-pathogen interaction. FEMS Microbiol Rev 35(6):1126–1157. https://doi.org/10.1111/j.1574-6976.2011.00276.x
101. Dell'Angelica EC, Mullins C, Caplan S, Bonifacino JS (2000) Lysosome-related organelles. FASEB J 14(10):1265–1278
102. Kniep B, Skubitz KM (1998) Subcellular localization of glycosphingolipids in human neutrophils. J Leukoc Biol 63(1):83–88
103. Mohn H, Le Cabec V, Fischer S, Maridonneau-Parini I (1995) The src-family protein-tyrosine kinase p59hck is located on the secretory granules in human neutrophils and translocates towards the phagosome during cell activation. Biochem J 309(Pt 2):657–665
104. Peyron P, Maridonneau-Parini I, Stegmann T (2001) Fusion of human neutrophil phagosomes with lysosomes in vitro: involvement of tyrosine kinases of the Src family and inhibition by mycobacteria. J Biol Chem 276(38):35512–35517. https://doi.org/10.1074/jbc.M104399200
105. Kawano T, Cui J, Koezuka Y, Toura I, Kaneko Y, Sato H, Kondo E, Harada M, Koseki H, Nakayama T, Tanaka Y, Taniguchi M (1998) Natural killer-like nonspecific tumor cell lysis mediated by specific ligand-activated Valpha14 NKT cells. Proc Natl Acad Sci U S A 95(10):5690–5693
106. Kawano T, Cui J, Koezuka Y, Toura I, Kaneko Y, Motoki K, Ueno H, Nakagawa R, Sato H, Kondo E, Koseki H, Taniguchi M (1997) CD1d-restricted and TCR-mediated activation of valpha14 NKT cells by glycosylceramides. Science 278(5343):1626–1629
107. Rossjohn J, Pellicci DG, Patel O, Gapin L, Godfrey DI (2012) Recognition of CD1d-restricted antigens by natural killer T cells. Nat Rev Immunol 12(12):845–857
108. Kain L, Webb B, Anderson BL, Deng S, Holt M, Costanzo A, Zhao M, Self K, Teyton A, Everett C, Kronenberg M, Zajonc DM, Bendelac A, Savage PB, Teyton L (2014) The identification of the endogenous ligands of natural killer T cells reveals the presence of mammalian alpha-linked glycosylceramides. Immunity 41(4):543–554
109. Nagata M, Izumi Y, Ishikawa E, Kiyotake R, Doi R, Iwai S, Omahdi Z, Yamaji T, Miyamoto T, Bamba T, Yamasaki S (2017) Intracellular metabolite beta-glucosylceramide is an endogenous Mincle ligand possessing immunostimulatory activity. Proc Natl Acad Sci U S A 114(16):E3285–E3294

Chapter 5

Ganglioside Metabolism and Its Inherited Diseases

Bernadette Breiden and Konrad Sandhoff

Abstract

Gangliosides are sialic acid containing glycosphingolipids, which are abundant in mammalian brain tissue. Several fatal human diseases are caused by defects in glycolipid metabolism. Defects in their degradation lead to an accumulation of metabolites upstream of the defective reactions, whereas defects in their biosynthesis lead to diverse problems in a large number of organs.

Gangliosides are primarily positioned with their ceramide anchor in the neuronal plasma membrane and the glycan head group exposed on the cell surface. Their biosynthesis starts in the endoplasmic reticulum with the formation of the ceramide anchor, followed by sequential glycosylation reactions, mainly at the luminal surface of Golgi and TGN membranes, a combinatorial process, which is catalyzed by often promiscuous membrane-bound glycosyltransferases.

Thereafter, the gangliosides are transported to the plasma membrane by exocytotic membrane flow. After endocytosis, they are degraded within the endolysosomal compartments by a complex machinery of degrading enzymes, lipid-binding activator proteins, and negatively charged lipids.

Key words Glycosphingolipids, Biosynthesis, Catabolism, Bis(monoacylglycero)phosphate, Sialic acid, Membranes, Luminal vesicles, Ceramide, Gangliosidosis

1 Introduction

Sphingolipids were discovered by the German physician Johann L.W. Thudichum in 1884, when he analyzed alcoholic brain extracts after acid hydrolysis [1]. He named the so far unknown nitrogen containing organic base sphingosine, now known as the main backbone of most sphingolipids. Gangliosides were first described by Ernst Klenk, when he analyzed acidic glycolipids of postmortem brain tissue of infantile patients with amaurotic idiocy, a rare inherited disease. He named the *N*-acetyl-neuraminic acid containing glycolipids gangliosides, as they appeared to be enriched in ganglion cells [2, 3]. In 1963, Kuhn and Wiegandt published the first correct chemical structure of a complex brain ganglioside, the ganglioside GM1 (Fig. 1a). This was the starting point for the deduction of the structure of its degradation products, which were generated by acid hydrolysis [4], a prerequisite for the characterization of several

Sandro Sonnino and Alessandro Prinetti (eds.), *Gangliosides: Methods and Protocols*, Methods in Molecular Biology, vol. 1804, https://doi.org/10.1007/978-1-4939-8552-4_5, © Springer Science+Business Media, LLC, part of Springer Nature 2018

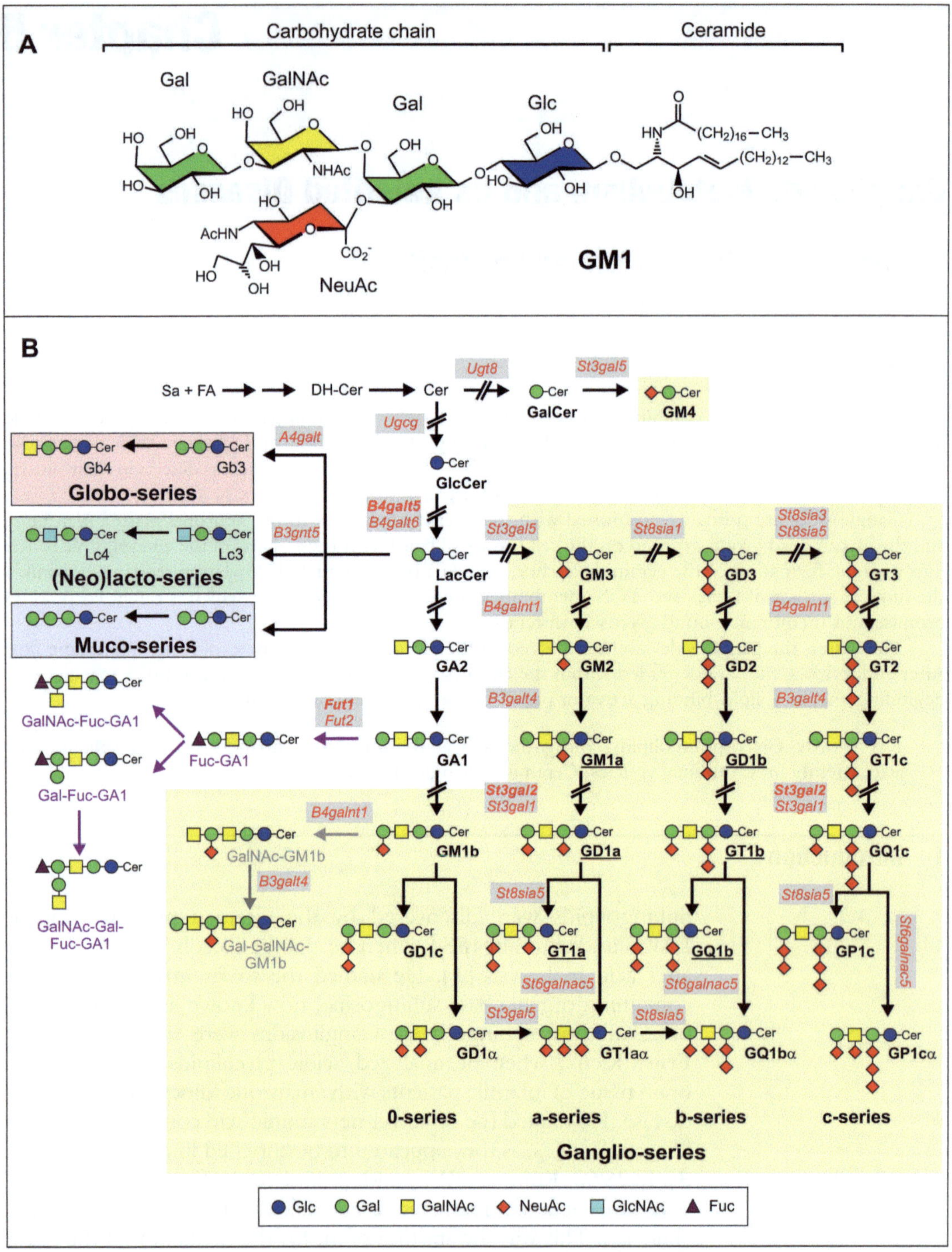

Fig. 1 Biosynthesis of mammalian gangliosides. Structure of ganglioside GM1 (**A**) [4] consisting of a ceramide anchor and a carbohydrate head group containing glucose (blue), galactose (green), *N*-acetylgalactosamine (yellow), and *N*-acetylneuraminic acid (red). Pathways of the ganglioside biosynthesis are depicted in (**B**). Major gangliosides occurring on neurons in adult mammalian brain are underlined (GM1a, GD1a, GT1a, GD1b, GT1b,

materials accumulating in tissues of patients with Tay–Sachs disease and other forms of amaurotic idiocy and the investigation of their metabolic relationship [5, 6]. Early metabolic studies with radiolabeled precursors unraveled a slow turnover of rat brain gangliosides in vivo with a half-life of about 10 days [7], much longer than turnover rates of glycolipids in cultured cells [8–10]. The investigation of ganglioside and sphingolipid metabolism was facilitated by autopsy material and cultured fibroblasts obtained from patients with rare metabolic defects, in which sphingolipids accumulate due to mutations inactivating proteins of their lysosomal degradation. Analysis of these genetic diseases led to the identification of several molecular defects in gangliosidoses and sphingolipidoses and to the hypothesis on the topology of endocytosis and lysosomal digestion of lipids and membranes [11]. It also allowed to analyze the function of activator proteins that are needed to facilitate hydrolysis of membrane-bound sphingolipids (SL) by water-soluble hydrolases. Thus, the analysis of metachromatic leukodystrophy, a sulfatide storage disease, and the AB-variant of GM2-gangliosidosis led to the identification of sulfatide activator protein (now called saposin B (Sap B)) [12] and the GM2 activator protein (GM2AP) [13], respectively. The discovery of patients with a Sap-precursor (now prosaposin) deficiency enabled the analysis of the in vivo functions of Sap A–D.

Glycosphingolipids (GSL) had been identified early on as major components of mammalian plasma membranes residing almost exclusively on the cellular surface [14]. Comparative studies of Klenk, Yamakawa, and Hakomori led to the present concept that the pattern, distribution and location of GSLs is species, cell type and organelle specific (Fig. 2). It is changing with differentiation, which indicates adaption of GSLs to cell function.

Neuronal cells primarily synthesize gangliosides of the ganglio-series (Fig. 2, *lane 1*) [15, 16], whereas oligodendrocytes of the brain predominantly produce myelin forming GSLs, galactosylceramide, ganglioside GM4 and sulfatides (Fig. 2, *lane 2*). Fibroblasts of the skin, however, mainly generate GSLs of the globo-series (Fig. 2, *lane 3*), whereas for example male germ cells, upon differentiation, switch between gangliosides of the a- and 0-series when crossing the blood–testis barrier [17].

Gangliosides contribute to 10–12% of the lipid content in the nervous system [18], where they form cell type-specific patterns on neuronal cell surfaces that change with differentiation. Although the exact biological functions of gangliosides and GSLs are still unclear, we know that they stabilize neuronal plasma membranes,

Fig. 1 (continued) GQ1b). Enzyme defects in knockout mice are depicted by arrows with slashes through them. Fucosylation and addition of GalNAc to higher gangliosides are depicted by purple and gray arrows. *Cer* ceramide, *DH-Cer* dihydroceramide, *FA* fatty acid, *GalCer* galactosylceramide, *GlcCer* glucosylceramide, *Sa* sphinganin. Ganglioside names are abbreviated according to Svennerholm [246]

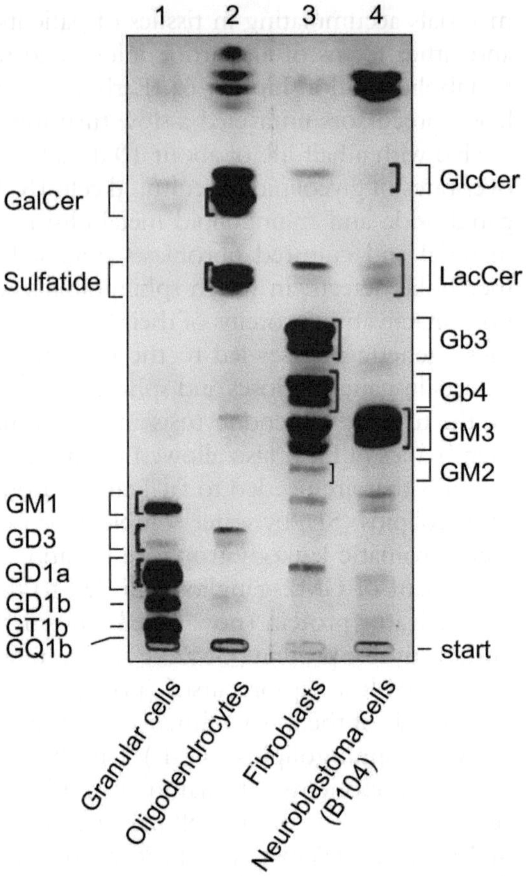

Fig. 2 Cell type specific expression of glycosphingolipids. Primary cultured cerebellar neurons (granule cells) (*1*), oligodendrocytes (*2*), fibroblasts (*3*), and neuroblastoma cells (B 104) (*4*) were metabolically labeled with [^{14}C] galactose for 48 h and the radiolabeled lipids were separated by thin layer chromatography, and visualized by fluorography [247]

together with sphingomyelin and a high cholesterol level [19, 20]. Furthermore, GSLs are involved in cell adhesion processes [21]: Whereas cells devoid of GSLs [22] can survive in culture, GSLs are required for the development of multicellular organisms [23]. They can mediate cell-to-cell adhesion processes, for example the myelin associated glycoprotein recognizes terminal carbohydrates of the oligosaccharide chains of gangliosides, an essential process for axon myelin stability [24].

2 Pathways of Ganglioside Biosynthesis

The main pathway of ganglioside biosynthesis starts at the endoplasmic reticulum (ER) with the formation of ceramide, the hydrophobic backbone and membrane anchor of many sphingolipids and GSLs. Its de novo synthesis is catalyzed in four sequential steps by membrane-bound enzymes at the cytosolic leaflet of ER membranes [25, 26] (Fig. 3).

In a first step, serine palmitoyltransferase (SPT) catalyzes the pyridoxal phosphate mediated condensation of L-serine and a coenzyme A (CoA) activated fatty acid (usually palmitoyl-CoA) to form 3-ketosphinganine with a chain length of 18 carbon atoms, which is reduced to sphinganine by NADPH in a second step. Sphinganine is acylated by a coenzyme A activated fatty acid to form dihydroceramides, which can be dehydrogenated by a desaturase to generate ceramide [27, 28].

Genetic ablation of dihydroceramide desaturase 1 (Des1) strongly alters sphingolipid (SL) levels and glucose homeostasis [29]. In addition, Des1 knockout mice revealed an incompletely penetrant lethality. Surviving animals had a growth defect with a complex phenotype with tremors, scaly skin and sparse hair.

SPT catalyzes a rate limiting step in SL and ganglioside biosynthesis [30] usually generating sphingoid bases/long chain bases with 18 carbon atoms, which are the main components of all mammalian SLs. Gangliosides of the central nervous system additionally contain significant amounts of long chain sphingoid bases with 20 carbon atoms [31, 32]. These sphingoid bases are largely absent in gangliosides of undifferentiated cells, but their molar percentage increases continuously with differentiation and aging of the neurons to reach substantial concentrations. Apparently, the substrate specificity of SPT for acyl-CoA, which is directed by the expression of its two smallest subunits, must change in order to accept stearoyl-CoA as substrate in ageing neurons [33].

Mutations in the genes *SPTLC1* and *SPTLC2*, which code for two major subunits of SPT, lead to a less strict acceptor specificity of SPT for serine. These mutations cause a rare inherited sensory and autonomic neuropathy (HSAN), in which patients suffer from a progressive loss of pain and temperature sensation [34, 35]. The mutant SPTs can also accept L-alanine and glycine as substrates besides serine, thereby generating atypical neurotoxic 1-deoxyceramides. Since they lost the hydroxyl group at the C1-position, these deoxyceramides cannot be converted to more complex SLs, nor can they enter the canonical catabolic pathway through sphingosine-1-phosphate [36, 37]. An oral supplementation of HSAN1 patients with L-serine normalized the deoxyceramide levels in the blood and improved sensation, wound healing and several autonomous functions [38]. The cytotoxic 1-deoxysphingolipids are, nevertheless, metabolized by a rather slow cytochrome P450-dependent pathway. Increased 1-deoxysphingolipid levels found in patients with metabolic syndrome, and type 2 diabetes might result from downregulation of its enzymes [39].

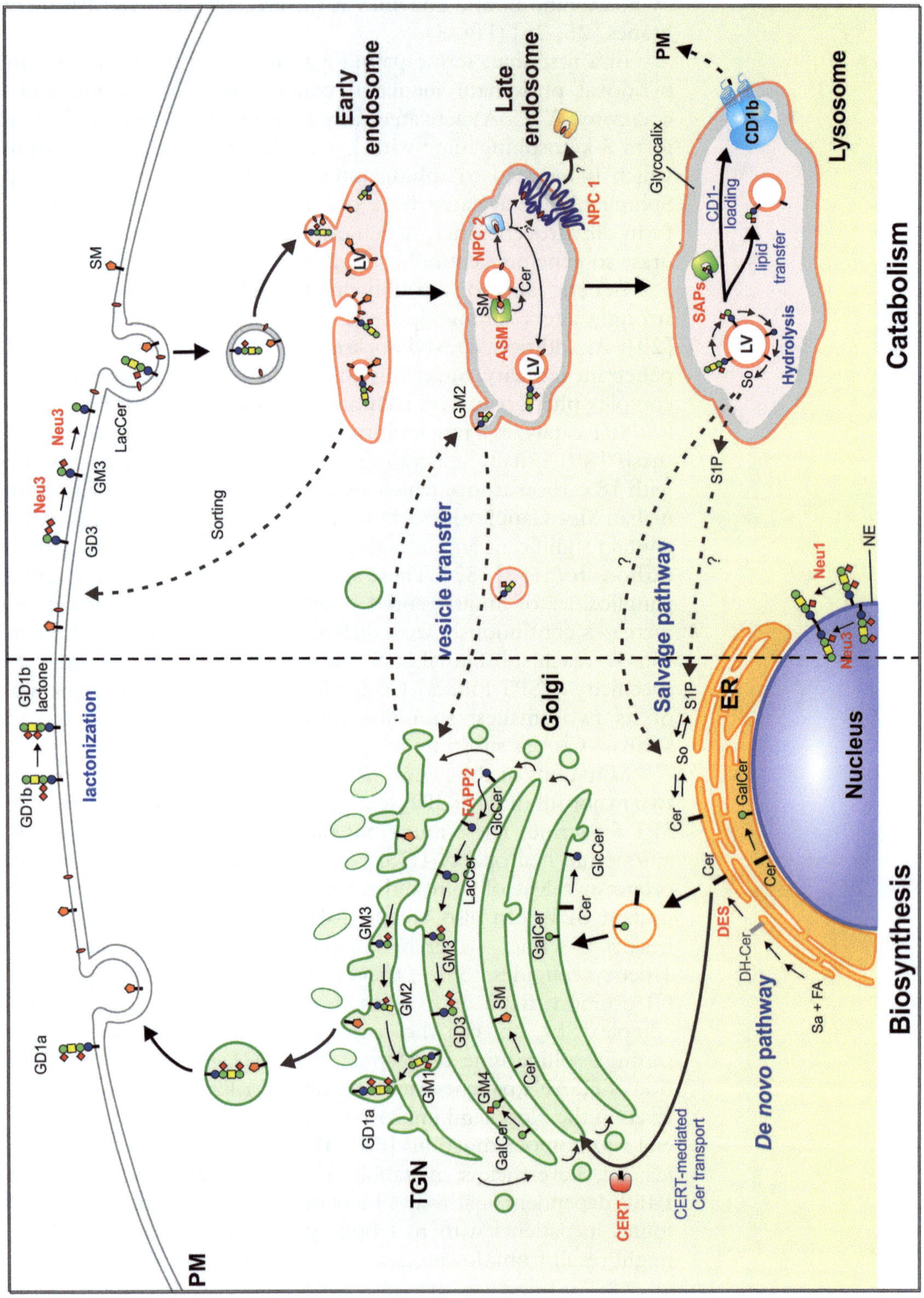
PM
GD1a
GD1b
GD1b lactone
lactonization
GD3
Neu3
GM3
LacCer
SM
Sorting
Early endosome
LV
Late endosome
GM2
ASM
SM
Cer
NPC 2
NPC 1
vesicle transfer
Lysosome
PM
CD1b
Glycocalix
SAPs
CD1-loading
lipid transfer
Hydrolysis
So
S1P
Salvage pathway
Catabolism
TGN
Golgi
GD1a
GM1
GM2
GM3
GD3
GM4
GalCer
LacCer
GlcCer
FAPP2
Cer
CERT
CERT-mediated Cer transport
De novo pathway
DES
DH-Cer
Sa + FA
ER
Nucleus
Neu1
NE
Biosynthesis

Fig. 3 Major pathways of ganglioside metabolism, its subcellular compartmentalization and intracellular trafficking. The left side indicates the biosynthetic/secretory pathway and the right side the endocytic, recycling, and catabolic pathways. The schematic representation and colors of ganglioside structures are the same as indicated in Fig. 1 and of sphingomyelin as in Fig. 4. *ASM* acid sphingomyelinase, *Cer* ceramide, *CERT* ceramide transfer protein, *DES* desaturase, *DH-Cer* dihydroceramide, *FA* fatty acid, *GlcCer* glucosylceramide, *ER* endoplasmatic reticulum, *LacCer* lactosylceramide, *LV* intraendolysosomal luminal vesicles, *NE* nuclear envelope, *NEU* neuraminidase, *NPC* Niemann–Pick protein, *PM* plasma membrane, *Sa* sphinganine, *SAP* Sphingolipid activator protein, *SM* sphingomyelin, *So* sphingosine, *S1P* sphingosine-1-phosphate, *TGN* trans Golgi network

2.1 Ceramide Synthases

Ceramide synthases (CerSs) of animal cells are a group of membrane-bound acyltransferases that catalyze the *N*-acylation of sphingoid bases, usually at the cytosolic surface of the ER [40] (Fig. 3). In mammalian tissues, six CerSs (CerS1-CerS6) have been discovered to date. They have different substrate specificities, with a wide range of acyl-chain distributions and expression patterns and are expressed in a cell type specific manner, which changes with cellular differentiation. Some functions of ceramides depend on their acyl-chain length, e.g., their impact on inflammation, autophagy, cancer, and Alzheimer's disease [41].

As critical regulators of SL metabolism, CerSs control many aspects and functions of cellular development, programmed cell death, cellular signaling, cancer and the formation of complex SLs. They synthesize predominantly dihydroceramides, but also ceramides and phytoceramides, mostly with acyl chains of C18 (CerS1), C22–26 (CerS2), C18 and C22–38 (CerS3), C18–24 (CerS4), C14–18 (CerS5) and C14–18 (CerS6).

Besides the regulation of CerS activity by phosphorylation [42], we do not have a clear picture of the regulation of individual CerSs yet, neither by membrane lipids or hormones, nor at the transcriptional and posttranscriptional levels [43, 44]. Downregulation of an individual CerS by siRNAs can trigger an upregulation of other CerSs [45], e.g., downregulation of CerS2 decreased levels of very long chain ceramides and upregulated shorter chain ceramides by increased expression of CerS4, 5 and 6. A large deletion affecting the gene of CerS2 caused a progressive myoclonic epilepsy [46].

In mice, CerS6 is mainly expressed in kidney, small and large intestine, and in the brain [47]. Its knockout showed behavioral abnormalities and habitational defects. The content of C16-sphingolipids decreased substantially in several organs (by 60–70%), whereas the ganglioside pattern and content in the brain were hardly affected. Under high fat diet, however, CerS6-deficient mice were protected from glucose intolerance [48], indicating a functional importance of the acyl chain length of free ceramides and/or ceramides as membrane anchors of SLs.

However, when CerS1 activity is lost, the brain is heavily affected. CerS1 is mainly expressed in neurons and catalyzes the formation of C18 ceramides. Its knockout induced a reduction of C18 ceramides and an increase of free and toxic sphingoid bases in mice. They suffer from a foliation defect in the brain, brain shrinkage and neuronal apoptosis in the cerebellum [49], whereas human patients suffer from loss of cerebellar Purkinje cells, progressive ataxia, and myoclonic epilepsy with dementia [49, 50]. Ganglioside levels and the expression of the myelin-associated glycoprotein are substantially reduced, suggesting an impaired stabilization of the oligodentrocyte derived myelin-associated glycoprotein by the lowered interaction with gangliosides of the neuronal surfaces [49].

This is in line with the observation that CerS1 mutant pathology was almost completely suppressed when the level of free sphingoid bases was normalized [51].

CerS5 is mainly expressed in testis, eWAT (epididymal white adipose tissue), lung and thymus and only at a low level in the brain. Although its loss in mice triggered a reduced weight gain compared to wild type animals, no ganglioside data were reported [52].

Loss of CerS3 in genetically engineered mice causes a lethal skin barrier disruption in the outermost epidermal layer, the *stratum corneum*. Unique ceramides with ultralong chain acyl residues are synthesized exclusively by CerS3 in the *stratum granulosum* of the skin and are required to generate the extracellular lipid lamellae, which are essential for the formation of barriers in the skin that protect mammals against water loss and infection [53]. Loss of CerS3 also impairs formation of complex sphingolipids and gangliosides in testis, affecting male fertility.

A novel class of fucosylated GSLs and gangliosides containing polyunsaturated very long chain fatty acid residues in their ceramide residues are synthesized in testicular germ cells and their synthesis depends on CerS3 activity [17, 54]. Mutations in human CerS3 cause a congenital ichthyosis with patients' skin appearing as a collodion membrane [55, 56].

CerS2 is widely expressed and generates ceramides with very long chain fatty acid residues (C22–26). In organs of adult CerS2 knockout mice, these ceramides are present in much lower levels than in wild type animals. This is also true for downstream sphingolipids and gangliosides that also contain long acyl chains. These mice lose myelin and myelin-associated glycoprotein, show neuronal degeneration and develop hepatocarcinoma, suggesting that ceramides with long acyl chains stabilize myelin, cerebellar architecture and protect against hepatocarcinomas [57]. The lower levels of very long chain ceramides and downstream SLs was compensated by an increase of C16-ceramide synthesis, keeping total ceramide levels unaltered, however increasing free levels of toxic long chain sphingoid bases [58, 59]. Ablation of CerS2 affects biophysical properties of membranes [60], causes chronic oxidative stress due to disruption of the mitochondrial respiratory chain [61] and encephalopathy [62]. Furthermore, CerS2-deficient mice appear to be more sensitive to develop hepatocellular carcinoma and pheochromocytoma [57, 59, 63].

2.2 Biosynthesis of Ganglioside GM4

Only the ganglioside GM4 is derived from galactosylceramide (Table 1, Fig. 1b) and belongs to the gala-series. GM4 is localized in erythrocytes, kidney, intestine, and, as a minor component in, human brain [4], where it was discovered as a component of myelin [64]. Little is known about the function of GM4, e.g., whether it interacts with the myelin basic protein and whether it has immunosuppressive properties [65].

Table 1
Schematic formulas of ganglioside in the mammalian nervous system

	Cer	Ceramide, *N*-acylsphingosine
	GalCer	Galβ1→1Cer
	GM4	Neu5Acα2→3Galβ1→1Cer
	GlcCer	Glcβ1→1Cer
	LacCer	Galβ1→4Glcβ1→1Cer
0-Series	GA2	GalNAcβ1→4Galβ1→4Glcβ1→1Cer
	GA1	Galβ1→3GalNAcβ1→4Galβ1→4Glcβ1→1Cer
	GM1b	Neu5Acα2→3Galβ1→3GalNAcβ1→4Galβ1→4Glcβ1→1Cer
	GD1c	Neu5Acα2→8Neu5Acα2→3Galβ1→3GalNAcβ1→4Galβ1→4Glcβ1→1Cer
	GD1α	Neu5Acα2→3Galβ1→3(Neu5Acα2→6)GalNAcβ1→4Galβ1→4Glcβ1→1Cer
a-Series	GM3	Neu5Acα2→3Galβ1→4Glcβ1→1Cer
	GM2	GalNAcβ1→4(Neu5Acα2→3)Galβ1→4Glcβ1→1Cer
	GM1a	Galβ1→3GalNAcβ1→4(Neu5Acα2→3)Galβ1→4Glcβ1→1Cer
	GD1a	Neu5Acα2→3Galβ1→3GalNAcβ1→4(Neu5Acα2→3)Galβ1→4Glcβ1→1Cer
	GT1a	Neu5Acα2→8Neu5Acα2→3Galβ1→3GalNAcβ1→4(Neu5Acα2→3)Galβ1→4Glcβ1→1Cer
	GT1α	Neu5Acα2→3Galβ1→3(Neu5Acα2→6)GalNAcβ1→4(Neu5Acα2→3)Galβ1→4Glcβ1→1Cer
b-Series	GD3	Neu5Acα2→8Neu5Acα2→3Galβ1→4Glcβ1→1Cer
	GD2	GalNAcβ1→4(Neu5Acα2→8Neu5Acα2→3)Galβ1→4Glcβ1→1Cer
	GD1b	Galβ1→3GalNAcβ1→4(Neu5Acα2→8Neu5Acα2→3)Galβ1→4Glcβ1→1Cer
	GT1b	Neu5Acα2→3Galβ1→3GalNAcβ1→4(Neu5Acα2→8Neu5Acα2→3)Galβ1→4Glcβ1→1Cer
	GQ1b	Neu5Acα2→8Neu5Acα2→3Galβ1→3GalNAcβ1→4(Neu5Acα2→8Neu5Acα2→3)Galβ1→4Glcβ1→1Cer
	GQ1bα	Neu5Acα2→3Galβ1→3(Neu5Acα2→6)GalNAcβ1→4(Neu5Acα2→8Neu5Acα2→3)Galβ1→4Glcβ1→1Cer

c-Series	GT3	Neu5Acα2→8Neu5Acα2→8Neu5Acα2→3Galβ1→4Glcβ1→1Cer
	GT2	GalNAcβ1→4(Neu5Acα2→8Neu5Acα2→8Neu5Acα2→3)Galβ1→4Glcβ1→1Cer
	GT1c	Galβ1→3GalNAcβ1→4(Neu5Acα2→8Neu5Acα2→8Neu5Acα2→3)Galβ1→4Glcβ1→1Cer
	GQ1	Neu5Acα2→3Galβ1→3GalNAcβ1→4(Neu5Acα2→8Neu5Acα2→8Neu5Acα2→3) Galβ1→4Glcβ1→1Cer
	GP1c	Neu5Acα2→8Neu5Acα2→3Galβ1→3GalNAcβ1→4(Neu5Acα2→8Neu5Acα2→8Neu5Acα2→3) Galβ1→4Glcβ1→1Cer
	GP1cα	Neu5Acα2→3Galβ1→3(Neu5Acα2→6)GalNAcβ1→4(Neu5Acα2→8Neu5Acα2→8Neu5Acα2→3) Galβ1→4Glcβ1→1Cer
Fucosylated	Fuc-GM1	Fucα1→2Galβ1→3GalNAcβ1→4(Neu5Acα2→3)Galβ1→4Glcβ1→1Cer
	Fuc-Gal-GM1	Fucα1→2Galα1→3Galβ1→3GalNAcβ1→4(Neu5Acα2→3)Galβ1→4Glcβ1→1Cer
	Fuc-GD1b	Fucα1→2Galβ1→3GalNAcβ1→4(Neu5Acα2→8Neu5Acα2→3)Galβ1→4Glcβ1→1Cer
	Fuc-Gal-GD1b	Fucα1→2Galα1→4Galβ1→3GalNAcβ1→4(Neu5Acα2→8Neu5Acα2→3)Galβ1→4Glcβ1→1Cer
Galactosylated	Gal-GD1b	Galα1→4Galβ1→3GalNAcβ1→4(Neu5Acα2→8Neu5Acα2→3)Galβ1→4Glcβ1→1Cer
	Gal-GalNAc-GM1b	Galβ1→3GalNAcβ1→4(Neu5Acα2→3)Galβ1→3GalNAcβ1→4Galβ1→4Glcβ1→1Cer
	GalNAc-GM1	GalNAcβ1→4Galβ1→3GalNAcβ1→4(Neu5Acα2→3)Galβ1→4Glcβ1→1Cer
	GalNAc-GM1b	GalNAcβ1→4(Neu5Acα2→3)Galβ1→3GalNAcβ1→4Galβ1→4Glcβ1→1Cer
	GalNAc-GD1b	GalNAcβ1→4Galβ1→3GalNAcβ1→4(Neu5Acα2→8Neu5Acα2→3)Galβ1→4Glcβ1→1Cer
Acetylated	9-*O*-Ac-GD3	Neu5,9Ac_2α2→8Neu5Acα2→3Galβ1→4Glcβ1→1Cer
	9-*O*-Ac-GD1b	Galβ1→3GalNAcβ1→4(Neu5,9Ac_2α2→8Neu5Acα2→3)Galβ1→4Glcβ1→1Cer
	9-*O*-Ac-GT1b	Neu5Acα2→3Galβ1→3GalNAcβ1→4(Neu5,9Ac_2α2→8Neu5Acα2→3)Galβ1→4Glcβ1→1Cer
	9-*O*-Ac-GQ1b	Neu5Acα2→8Neu5Acα2→3Galβ1→3GalNAcβ1→4(Neu5,9Ac_2α2→8Neu5Acα2→3)Galβ1→4Glcβ1→1Cer
	9-*O*-Ac-GT3	Neu5,9Ac_2α2→8Neu5Acα2→8Neu5Acα2→3Galβ1→4Glcβ1→1Cer
	9-*O*-Ac-GT2	GalNAcβ1→4(Neu5,9Ac_2α2→8Neu5Acα2→8Neu5Acα2→3)Galβ1→4Glcβ1→1Cer

(continued)

Table 1 (continued)

Lactone	GD1b-lactone	Galβ1→3GalNAcβ1→4(Neu5Acα2→8, 1→9 Neu5Acα2→3)Galβ1→4Glcβ1→1Cer
	3′isoLM1	Neu5Acα2→3Galβ1→3GlcNAcβ1→3Galβ1→4Glcβ1→1Cer
	6′LM1	Neu5Acα2→6Galβ1→4GlcNAcβ1→3Galβ1→4Glcβ1→1Cer
	3′LM1	Neu5Acα2→3Galβ1→4GlcNAcβ1→3Galβ1→4Glcβ1→1Cer
	3′LD1	Neu5Acα2→8Neu5Acα2→3Glcβ1→4GalNAcβ1→3Galβ1→4Glcβ1→1Cer
	GbOse3Cer	Galβ1→4Galβ1→4Glcβ1→1Cer
	GbOse4Cer	GalNAcβ1→3-Galβ1→4Galβ1→4Glcβ1→1Cer

Sialic acid containing glycosphingolipids are underlined in yellow
Cer ceramide, *Neu5Ac N*-acetylneuraminic acid, *Glc* glucose, *Gal* galactose, *GalNAc N*-acetylgalactosamine

The biosynthesis of galactosylceramide is performed on the luminal side of the ER (Fig. 3), which is opposite to glucosylceramide (GlcCer) biosynthesis (*see* Subheading 2.3). Galactosylation of ceramide is catalyzed by UDP-galactose:ceramide galactosyltransferase (*Ugt8*, also known as GalT3, EC 2.4.1.45). The further steps of the biosynthesis of gala-series GSL take place in the lumen of the Golgi apparatus (Fig. 3). GM4 is also formed by GM3-synthase (CMP-NeuAc:lactosylceramide-α2,3-sialyltransferase (*St3gal5*), EC 2.4.99.9)[1] (Fig. 1b).

2.3 Glucosylceramide Biosynthesis

Ceramide glucosyltransferase (*Ugcg*, UDP-glucose:ceramide glucosyltransferase, EC 2.4.1.80) is a type III transmembrane enzyme, which is resident in the Golgi [66, 67], with its catalytic domain facing the cytosol [68, 69]. The dimeric enzyme [70] transfers glucose from UDP-glucose to membrane-bound ceramides and dihydroceramides, which reach the Golgi membrane by vesicular transport from the ER [71] as precursors of ganglioside biosynthesis (Fig. 3). Ceramides can also reach trans-Golgi network (TGN) membranes by the ceramide transfer protein CERT [72, 73], where they are involved in sphingomyelin (SM) and globo-series GSL biosynthesis (Fig. 3). A fraction of the GlcCer pool reaches the luminal side of Golgi membranes and is glycosylated in a stepwise procedure to form complex GSL and gangliosides of the ganglio-series [74]. It has been reported that GlcCer of the Golgi membranes undergoes rapid transbilayer movement [75, 76]. At the luminal leaflet of trans-Golgi and TGN membranes (Fig. 3) [77], lactosylceramide (LacCer) is formed by galactosylation of GlcCer, mainly by the LacCer synthase β4-galactosyl-transferase 5 (*B4galt5*) and to some extent also by β4-galactosyl-transferase 6 (*B4galt6*) [78] (Fig. 1b). Synthesis of complex GSL of the globo-series furthermore requires the GlcCer transfer protein FAPP2 [79].

2.4 The Assembly Line for Gangliosides in the Golgi Apparatus

As suggested by Saul Roseman, synthesis of complex glycolipids and carbohydrates occurs by the multiglycosyltransferase systems [80–82]. Indeed, glycolipid glycosyltransferases (GGTs) that use GlcCer and complex glycolipids as acceptors show a modular organization in homomultienzyme and heteromultienzyme complexes in the Golgi membranes [83–85]. For example, in cultured CHO-K1 cells, LacCer synthase, GM3-synthase (CMP-NeuAc:lactosylceramide-α2,3-sialyltransferase (*St3gal5*), EC 2.4.99.9[2]) and GD3-synthase (*St8sia1*, EC 2.4.99.8[3]) participate in a heterocomplex in proximal Golgi membranes, as well as in TGN membranes [86–88].

[1] Also known as ST3Gal V or sialyltransferase 9 (*Siat 9*).
[2] Also known as ST3Gal V or sialyltransferase 9 (*Siat 9*) or SAT-I.
[3] Also known as α-2,8-sialyltransferase (*Siat 8*) or SAT II.

The biosynthesis of higher gangliosides by stepwise glycosylation reactions occurs at the luminal face of the Golgi apparatus [89]. The growing ceramide anchored oligosaccharide chains are oriented luminally, an orientation topologically equivalent to exposure of the complex carbohydrates of GSLs on the cell surface.

Enzymatic steps for the biosynthesis of a- and b-series gangliosides have already been identified in the 1960s and 1970s [89, 90]. Transferases catalyzing the formation of LacCer and its sialylated charged derivatives, the small gangliosides GM3 and GD3, are highly specific for their glycolipid substrates. Together with GT3, they serve as precursors for the 0- [91], a-, b-, and c-series of gangliosides (Fig. 1b), and differ only in the number of sialic acid residues bound to the inner galactose.

Early in vitro studies that reproduced the *N*-acetylgalactosamine (GalNAc)-transferase activities of isolated Golgi membranes catalyzing the formation of GM2 and GD2 from their precursors GM3 and GD3 needed the addition of detergents or of an anionic lipid such as phosphatidylglycerol to the incubation mixture to achieve reasonable biosynthetic rates [81, 82]. It was expected that the assays would allow the identification of specific GalNAc-transferases, one for the biosynthesis of a-series and another one for the formation of b-series gangliosides. The use of different lipid substrates, GM3 and GD3, however, failed to detect specific GalNAc-transferases, one for the biosynthesis of GM2 (a-series), and another one for the formation of GD2 (b-series). On the contrary, competition experiments between the lipid substrates GM3 and GD3 proved that both lipid acceptors competed for one and the same active site of a promiscuous GalNAc-transferase with limited lipid substrate specificity [92].

Further sugar residues are later added to the growing oligosaccharide chain in a stepwise manner. All glycosyltransferases involved in the elongation of the growing oligosaccharide chains studied so far turned out to be quite unspecific for their lipid substrates [93–95].

Surprisingly, the stepwise glycosylation of the precursors is catalyzed by only few glycosyltransferases with limited acceptor specificity, which was completely unexpected at that time. Acceptor competition experiments and analysis of enzyme kinetics revealed that in each case only one promiscuous transferase can catalyze the incorporation of *N*-acetylgalactosamine and galactose, respectively, into the main mammalian gangliosides of the 0-, a-, b-, and c-series: there is a single β1,4-GalNAc-transferase (*B4galnt1*, EC 2.4.1.92[4]), as well as a single β1,3-galactosyl-transferase (*B3galt4*, EC 2.4.1.62[5]). However, several sialyl-transferases exist, (e.g., *St3gal2*,

[4] Synonyms are GalNAcT, GM2/GD2 synthase.
[5] Also known as GalT II.

EC 2.4.99.4[6] > *St3gal1*, EC 2.4.99.6[7]) along with another sialyl-transferase (*St8sia5*[8]) with limited specificity [92, 95, 96]. In the same manner, GA2 can also be metabolized to GM2α by the enzyme St6galnac6, in a next step to GM1α by β1,4-GalNAc-transferase (*B4galnt1*) and to GD1β by a sialyl-transferase (*St8sia5*).

GD1α synthase (coded by *St6GalNAc 3–6*, EC 2.4.99.7[9]) is involved in the synthesis of α-series gangliosides [97]. The enzyme also has activity toward GD1a, GT1b, GM1b and GQ1c (Fig. 1b). Furthermore, GQ1bα can be generated from GD1α over GT1aα by GM3-synthase (*St3gal5*) followed by a sialyl-transferase (*St8sia5*) (Fig. 1b).

All these analyses resulted in the assembly line given in Fig. 1b. It shows the major de novo biosynthetic routes of mammalian brain- and liver gangliosides. The concept has been confirmed by analysis of genetically modified mice [98, 99].

2.5 Functions Learned from Mouse Models with Genetically Engineered Blocks in Ganglioside Biosynthesis

The scheme of combinatorial ganglioside biosynthesis of mammalian tissues [99], presented in Fig. 1b, allows for the formation of different ganglioside patterns on individual neuronal cell surfaces. Mammalian epidermal keratinocytes and testicular germ cells also express SLs and GSLs with ultralong chain acyl moieties (with 26–36 carbon atoms), depending on their CerS3 activity [100]. The essential role of polyunsaturated ultra-long chain SLs, GSLs and gangliosides in differentiating spermatocytes/spermatids is still poorly characterized [100], however they are required for proper meiosis and stabilization of intercellular bridges formed by clonal spermatids [54]. Although it is not clear how the individual GSL patterns are generated on individual cell surfaces, there is evidence for transcriptional and posttranscriptional control of glycosyl transferases. The complex regulation patterns identified so far have been reviewed by [43, 88, 99, 101, 102]. Inhibitors of GSL metabolism have also been summarized [103–105].

Similarly, the kinetic parameters of transferases in their special lipid environment in the Golgi compartment are currently unknown, as is the flow rate of individual metabolites through the secretory pathway. Biosynthetic rates for individual gangliosides depend heavily on the lipid environment of the reaction [82, 91] and on the formation of multienzyme complexes at the Golgi and TGN membranes [106–109].

We furthermore suspect that lipid substrate and enzyme interactions proceed in the 2-dimensional plane of the membrane and, therefore, should follow two-dimensional Michaelis–Menten

---

[6] Synonyms are SAT IVB, GD1a/GT1b synthase.
[7] Synonyms St3Gal I, SAT IV.
[8] Synonyms are SAT V, GT1a/GQ1b synthase.
[9] Also known as α-2,6-sialyltransferase, SAT VIIc-f.

kinetics, as it has been described for the ganglioside–sialidase interaction at neuronal plasma membranes [110].

The development of mice with distinct blocks in ganglioside biosynthesis shed some light on the formation of ganglioside patterns found on eukaryotic cells. This has already been reviewed [98, 99], so only a few remarks will be made here. Blocks in the early steps of the combinatorial pathway are indicated in Fig. 1b. Formation of GlcCer is required for the development of a multicellular organism but not for growth and reproduction of GSL-free cells in culture [22]. Mice with a disruption of the ceramide glucosyltransferase gene (Fig. 1b) are not viable and die in early embryonic development. Apparently, the formation of GlcCer and subsequent higher glycolipids, e.g., the stage-specific embryonic antigens [21], is essential for normal differentiation and development [111]. Mice with an engineered block in galactosylceramide formation, a precursor of negatively charged myelin components, sulfatides and ganglioside GM4 as well as the neutral galabiaosylceramide in kidney, still form myelin but incorporate GlcCer with hydroxylated acylchains instead of galactosylceramide into their myelin sheets. Mice with such instable and functionally impaired myelin suffer from generalized tremor, ataxia and conduction deficiency [112, 113]. Surprisingly, mice defective in the formation of major brain gangliosides like GM2, GD2, GM1a, GD1b, GD1a, GT1b, GT1a, and GQ1b due to a loss of GalNAc-transferase activity encoded by *B4galnt1*, and mice deficient in b-series gangliosides like GD3, GD2, GD1b, GT1b, and GQ1b, due to a loss of sialyltransferase activity encoded by *St8sia1*, show only subtle impairment of brain stability and functions [114–117], which, however, progressively increases with age [118]. In contrast, double knockout mice with deficiencies in both genes *B4galnt1* and *St8sia1*, which express only GM3 as their major ganglioside [117], are extremely susceptible to induction of lethal seizures. On the other hand, mice which cannot express ganglioside GM3 any more due to a knockout of GM3-synthase (*St3gal5*) cannot generate gangliosides of the a-, b-, and c-series anymore. Instead, they express gangliosides of the 0-series (GM1b, GD1c and GD1α), which are not found in healthy brain tissue [119]. The GM3-free mice have an altered glucose homeostasis with an accelerated insulin receptor signaling pathway. The data indicate that GM3 is an important inhibitor of the insulin receptor [120] and also of the EGF receptor [121]. Modulation of membrane receptor activities by gangliosides appears to be a more general aspect of ganglioside function. Selective deletion of the *Ugcg gene* (GlcCer synthase) in hypothalamic neurons resulted in obese mice and revealed an activating interaction of gangliosides, likely GD1a, with the leptin receptor [122].

A deficiency of both GalNAc-transferase (*B4galnt1*) and GM3-synthase (*St3gal5*) activities in mice makes the formation of all major gangliosides of the brain impossible. The knockout mice suffer from a severe and fatal neurodegeneration [20] (Fig. 1b). Their LacCer levels in the brain are elevated up to 100-fold, and a new GSL, LacCer-sulfate, which had never been found before in mammalian brains, was detected. LacCer-sulfate may compensate at least in part the missing negative surface charges lost with the disappearance of the gangliosides. Such, at least the loss of glycolipid-bound sialic acid residues at the cell surfaces is compensated to a large extent [11]. Neuronal deletion of the glucosylceramide synthase also eliminates all neuronal gangliosides with no compensation of negative lipid-bound charge. These mice are born with an apparently normal brain as seen by light microscopy, but they die even earlier than the above mentioned double knockout mice [19]. Gangliosides are apparently not needed to generate a mammalian brain, but they are needed to stabilize it [123]. Patients with defects in ganglioside biosynthesis found so far suffer mainly from epilepsy, ataxia and paraplegia [35, 124]. An inherited GM3-synthase activity defect causes a symptomatic epilepsy syndrome [124] with seizures and intellectual disability and a neurocutaneous syndrome [125]. Mutations in the *B4galnt1* gene (GM2-synthase) induce a recessive form of spastic paraplegia accompanied by cortical atrophy and cognitive impairment [126, 127].

2.6 Recycling Pathways in Ganglioside Formation

Besides de novo biosynthetic pathways discussed so far, few salvage pathways have been observed in GSL biosynthesis.

In order to get some insight into the metabolic routes, the cellular metabolism of exogenously added gangliosides was studied in cultured fibroblasts and neurons. Radiotracer techniques and ESR spectroscopy showed that [^{3}H]- and nitroxide-labeled gangliosides bind in micellar form to the surface of cultured cells, a binding that can be removed by trypsinization. A trypsin resistant component of about 10–15 nmol labeled ganglioside per mg of cellular protein, however, displayed characteristic anisotropic ESR spectra and order parameters showing that here, their ceramide moieties were fully incorporated into the plasma membrane [128]. Incorporated ganglioside GM2 that had been radiolabeled in its sphingoid base generated the expected lysosomal degradation products, including the release of radiolabeled sphingoid bases, which apparently escaped the lysosomes and reached biosynthetic compartments where they were incorporated into sphingomyelin, and the gangliosides GM1 and GD1a. In GM2 gangliosidosis cells, which have a block in GM2 degradation, however, incorporated ganglioside GM2 produced no lysosomal degradation products and no sphingomyelin, but still formed radiolabeled gangliosides GM1 and GD1a by direct glycosylation of the internalized GM2, presumably by trafficking from endosomes to the TGN compartment [129] mediated by the retromer complex [130] (Fig. 3).

Recycling from the endosomal compartments to the biosynthetic compartment in the Golgi and the TGN was also demonstrated after injection of radiolabeled ganglioside GM1 into rats. They formed GD1b and GT1b by direct glycosylation, whereas the released sialic acid was incorporated into newly generated glycoconjugates [131, 132].

These findings are in agreement the observation that GM1 containing unsaturated acyl chains efficiently pull cholera toxin from the plasma membrane through endosomes to the TGN and ER [133]. The ceramide anchor of GM1 also seems to play a critical role for its potential of transcytosis [134].

Internalization of a fluorescent and nondegradable analog of glucosylceramide was followed either by its recycling to the plasma membrane [135] or by its direct glycosylation in cultured cerebellar granule cells to form complex gangliosides, again suggesting a traffic of SL intermediates from the endocytotic to the biosynthetic compartment [136].

Metabolic studies support a recycling of SL metabolites from the endosomes to the biosynthetic compartments (Golgi, TGN) in cultured human epithelium cells. They also demonstrate a salvage pathway of sphingoid bases from the lysosomal to the ER compartment for ceramide formation [137–139].

2.7 Metabolic Modification of Gangliosides

As components of cellular surfaces, gangliosides play a crucial role in the maintenance of tissue integrity and function [140]. Metabolic modifications of gangliosides and change in ganglioside pattern have been observed in several tumors and cancer cells and contribute to malignancy and cancer progression [141].

An interesting example is the *N*-glycolyl-neuraminic acid (NeuGc), which differs only by one additional oxygen atom from the C5 *N*-acetyl group of *N*-acetyl-neuraminic acid (NeuAc). NeuGc is found in several tumor types, such as melanoma and breast cancer [141], but rarely in normal human tissue. NeuGc, however, is a normal component of gangliosides from visceral organs of most animal species, including mice, horse, sheep, and goats. It cannot be synthesized in human tissues due to the absence or irreversible inactivation of relevant genes. Its incorporation into brain gangliosides disrupts the interaction between gangliosides of neuronal axons and the myelin associated glycoprotein of myelin sheets surrounding the axons. NeuGc overexpression in mice brain resulted abnormal locomotor activity, impaired object recognition memory, and abnormal axon myelination [142]. It is presumed that incorporation from diet, hypoxic conditions, and endogenous metabolic mechanisms support the generation of NeuGc and its incorporation into human gangliosides, especially in fetal tissues and some cancers.

O-acetylation of hydroxyl groups is one of the most frequent modifications of sialic acid [143]. It has been observed at the C4,

C7, C8, and C9 hydroxyl groups [144] and occurs in many *O*-acetylated gangliosides (Table 1) [145]. In mammals, two enzyme systems have been described for *O*-acetylation: (a) AcCoA:sialate-4-*O*-acetyltransferase (EC 2.3.1.44) and (b) an AcCoA:sialate-7(9)-*O*-acetyltransferase (EC 2.3.1.45). The acetylation is a small modification of the ganglioside structure but causes a major change of the physiological properties. For example, 9-*O*-acteylated sialic acid mediates the adhesion of influenza C virus to the cell surfaces and is resistant to neuraminidase. Increased synthesis of 9-*O*-acetyl-GD3 occurs in malignant melanomas and apparently limits apoptosis. On the other hand, metastatic melanoma cells are enriched in GD3 in contrast to poorly metastatic and healthy cells, suggesting that GD3 may promote metastasis possibly by suppressing the antitumor immune response.

Similarly, the 5-*N*-deacetylated form of GM3 is expressed in metastatic melanomas, but not in healthy tissue or even in primary melanomas; it is considered to be a specific marker for the metastatic condition and a target for potential therapy.

The fucosyl (Fuc)-gangliosides Fuc-GM1 and Fuc-GD1b (Table 1) are gangliosides with an α1,2-fucosylated galactose at the nonreducing end [146] and are found in a number of cancers, but rarely in normal tissue. They are also considered to be potential markers for cancer and candidates for immunotherapy. Fuc-GM1 and Fuc-GD1b are generated in the last step of their biosynthetic pathway, by adding fucose with the help of an α1,2-fucosyltransferase (*Fut1*, EC 2.4.1.69 > *Fut2*) [147] (analogously to fucosylation of GA1 in Fig. 1b).

Furthermore, galactose can be added to GD1b by α1,3-galactosyltransferase 2 (*A3galt2*, EC 2.4.1.87) (Table 1) and GalNAc can be added to GM1b (Fig. 1b), GM1 or GD1b by β1,4-*N*-acetyl-galactosaminyltransferase 1 (*B4galnt1*, EC 2.4.1.92) (Table 1).

3 Ganglioside Remodeling at or Near Cell Surfaces

Brain gangliosides are highly concentrated on the surface of neuronal plasma membranes, especially at the membranes of nerve endings [148]. Their functional and molecular plasticity is considered to be of great importance for memory, behavior, and learning functions. Remodeling of sphingolipid and ganglioside pattern at neuronal and synaptic surfaces may well participate in and contribute to the synaptic plasticity and may affect LTPs (long term potentials) in the hippocampus [149, 150]. Impairment of ganglioside biosynthesis affects cognitive and learning abilities in mice [151, 152]. A drastic ganglioside loss has been identified in grey and white areas of an early-onset form of Alzheimer brains, suggesting a loss of synapses as a primary event [153]. Synaptosomal membranes carry both a sialidase, Neu3, and a sialyltransferase activity. Phosphatidic acid binds, activates, and translocates Neu3 to the plasma membrane [154], where it contributes to the desialylation of oligosialyl gangliosides and glycoproteins (Fig. 3). It is significantly upregulated in tumor tissues, triggers EGF receptor signaling [155], and contributes neoplastic potential on colon cancer cells by regulating Wnt signaling [156]. Selective ganglioside desialylation at the plasma membrane promotes growth and differentiation of human neuroblastoma cells [157]. Expression and activity of Neu3 are increased during pharmacologically induced differentiation of neuroblastoma cells, and *Neu3* gene transfection triggered neurite growth. Although it was also reported that overexpressed Neu3 in COS-7 cells can hydrolyze ganglioside substrate from neighboring cells [158, 159], it remains unclear if the Neu3 reactions occur exclusively at the highly glycosylated and glycocalyx-covered cell surface or in part also during the fast cycling of plasma membrane components through early endosomes back to the plasma membrane.

Radiolabeled ganglioside GD1a inserted into isolated neuronal PMs is hydrolyzed by plasma membrane-bound sialidase. The interaction between plasma membrane-bound substrate and enzyme obviously occurs by diffusion within the plane of the plasma membrane and follows two-dimensional Michaelis–Menten kinetics [110, 160, 161]. The enzymatic reaction rate is strongly stimulated by an increase of plasma membrane fluidity activated either by addition of general anesthetics like Xenon gas or halothane or by the incorporation of long chain alcohols [110, 160, 161].

In cultured cerebellar granula cells, surface hydrolases like sialidase Neu3, plasma membrane-bound β-galactosidase, β-glucosidase, and β-hexosaminidase could contribute to the remodeling of ganglioside pattern during differentiation beside the major biosynthetic machinery along the secretory pathway in the Golgi and TGN compartments and the catabolic pathway in the endolysosomal system [10]. Their role in intact brain tissue, however, still has to be established. As discussed below (*see* Subheading 5.7), important factors in the microenvironment of a catabolic reaction are needed to allow significant hydrolytic rates and GSL degradation. For ganglioside GM2 and GlcCer hydrolysis by hexosaminidase A (Hex A) and glucocerebrosidase (GBA1), respectively, several essential conditions have been identified: the presence of appropriate sphingolipid activator proteins (SAPs) (GM2AP and Sap C, respectively), a negative surface charge and a stimulatory lipid composition of the GSL-substrate carrying membranes (e.g., anionic lipids at the cellular surface, low level of inhibiting lipids like SM), an appropriately low pH value (pH 3.8–4.6 for GM2 hydrolysis) and a low ionic strength [162, 163]. All these conditions can be achieved within the lysosomal microenvironment to reach relevant catabolic rates, but are hardly reached at the surface of cellular plasma membranes in mammalian tissues. Some of the plasma membrane enzymes, however, differ from intracellular enzymes and seem to play a specific role in the brain tissue [9].

Another fascinating mechanism at the plasma membrane is the lactonization of gangliosides containing a disialosyl residue, e.g., GD1b (Fig. 3). Ganglioside lactones are minor components of the mammalian brain [164]. The mechanism of their formation is not completely understood, but it has been described that the formation of GD1b-lactone is influenced by conformational, aggregational, and biological properties. It seems that lactonization/delactonization triggers gangliosides-mediated cellular processes [159].

4 Nuclear and Mitochondrial Gangliosides

Gangliosides are localized mainly in the plasma membrane, the TGN membranes and the endolysosomal perimeter membranes, and in small amounts also in mitochondria, nucleus and ER. Little is known about their origin and metabolism in these organelles. It might be possible that they are synthesized in the Golgi and TGN and reach these organelles by membrane flow.

Membranes of the nuclear envelope also contain the gangliosides GM1 and GD1a, together with two sialidases, Neu3 in the inner membrane of the nuclear envelope and Neu1 in the outer membrane (Fig. 3) [165]. GM1 of the inner membrane is associated with the Na^{+}–Ca^{2+} exchanger and potentiates its activity. Absence of GM1 in *B4galnt1* knockout mice renders cells susceptible to apoptotic elimination.

GD3 is a minor ganglioside in most normal tissues [166], but upregulated in many tumors [167]. It induces a marked loss of mitochondrial transmembrane potential, besides DNA fragmentation, and apoptosis [18].

Other sphingolipids also occur in the nucleus and the mitochondria, e.g., sphingomyelin. It can be degraded to ceramide to trigger apoptosis, and furthermore to sphingosine and free fatty acids [168].

5 Pathways of Ganglioside Catabolism

5.1 Topology of Ganglioside Internalization by Endocytosis

The major and constitutive pathway of ganglioside and SL catabolism occurs in the digestive compartment of the cell, the lysosomes.

It was described more than 40 years ago that ganglioside GM3 incorporated into the limiting lysosomal membrane is protected from degradation [169], whereas, it is easily degraded in the lysosomes during the normal membrane turnover when it is endocytosed by cultured cells [138]. Therefore, it is highly unlikely that sphingolipids and the other membrane lipids reach the lysosomal perimeter membrane for lysosomal digestion. Data collected on endocytosis of plasma membrane-bound radiolabeled and biotin-labeled lipids, analyzing GSL and SL degradation by immunoelectron microscopy in wild type and mutant cells obtained from patients with inherited blocks in sphingolipid and membrane digestion, indicate that intraluminal vesicles (LVs) inside the late endosomal and lysosomal compartment are the main platforms for SL and lipid catabolism [170–172]. They are also called ILVs (intralysosomal luminal vesicels) and are the major sites of membrane-lipid and membrane digestion in the cell [173]. The occurrence of LVs is not only observed in cells with impaired membrane digestion, but also in cultured wild type cells, where they are visible under the electron microscope as multivesicular bodies. They are formed by an inward budding of the limiting endosomal membrane and budding off as catalyzed by ESCRT proteins [174]. Lipids and proteins are sorted either to the limiting or internal membranes. As components of the internal membranes, they are easy prey for the attack by the lysosomal digestive juice.

5.2 Lysosomes as Intracellular Stomachs for the Digestion of Macromolecules and Membranes

Lysosomes are intracellular stomachs. They degrade a great variety of macromolecules and release their components as nutrients into the cytosol for use in biosynthetic salvage pathways and energy metabolism. Macromolecules, membrane components and complex lipids reach LVs and intralysosomal membranes of the lysosomal compartment for digestion by autophagy [175], phagocytosis [175] and endocytic pathways [173]. An overview on the topology of ganglioside digestion along the endocytic pathways is given in Fig. 3. The LVs are prepared for lysosomal digestion by a lipid sorting process beginning at the level of endosomes [173, 176] by removal of membrane stabilizing lipids. First, SM is hydrolyzed by acid sphingomyelinase (ASM) to generate ceramide and phosphocholine [177]. The decrease of inhibiting SM levels and the concomitant increase of stimulating ceramide levels then facilitate cholesterol secretion from the LVs by the sterol-binding NPC2 glycoprotein in the lysosol [178]. Cholesterol is secreted from the late endosomes mainly by two sterol-binding proteins, NPC2 and NPC1, a transmembrane spanning glycoprotein of the perimeter membrane. Anionic bis(monoacylglycero)phosphate (BMP), which substantially stimulates cholesterol transfer by NPC2 and several steps in GSL and SL catabolism, is formed in the LVs as an intermediate of phosphatidylglycerol degradation [178, 179].

Gangliosides and SLs of the LVs are degraded in a stepwise manner, mainly by water-soluble but often protein and membrane associated hydrolases with the help of lipid-binding proteins, the SAPs (Fig. 4). SAPs are small lysosomal lipid-binding glycoproteins with different properties and functions [180]. Degradation of gangliosides, GSLs, SLs, and SM proceeds at the surface of

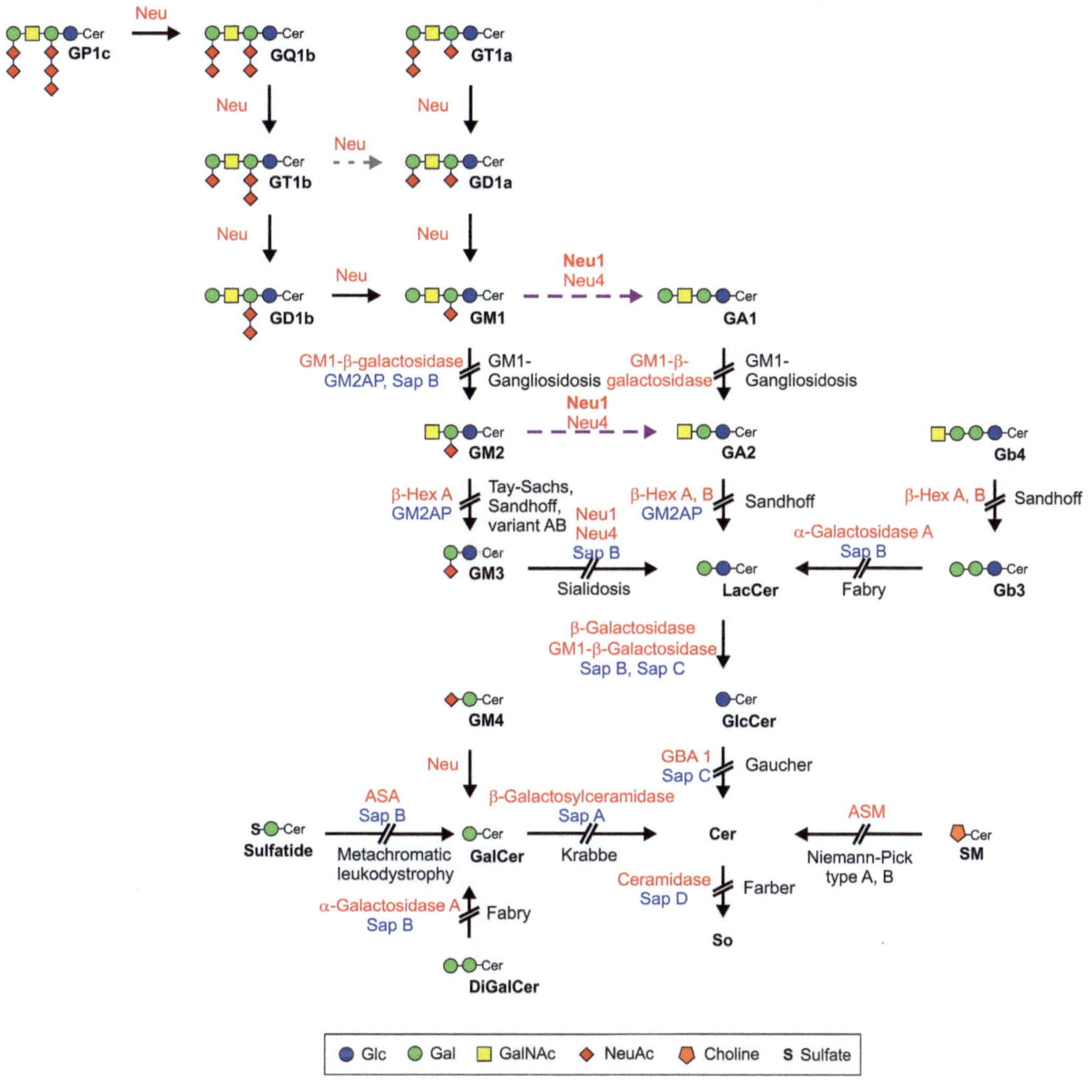

Fig. 4 Scheme of lysosomal ganglioside catabolism. The eponyms of known metabolic diseases and those of hydrolases (red) and sphingolipid activator proteins (blue) necessary for in vivo degradation are indicated. Broken purple arrows indicate the bypass pathway of the ganglioside GM1 degradation in mice. The schematic representation and colors of ganglioside structures are the same as indicated in Fig. 1. *ASA* arylsulfatase A, *ASM* acid sphingomyelinase, *GBA1* lysosomal β-glucocerebrosidase, *Hex A* β-hexosaminidase, *Neu* neuraminidase, *Sap* saposin, *variant AB* variant AB of GM2-gangliosidosis (deficiency of GM2AP)

intralysosomal vesicles (LVs) and membrane structures (Fig. 3) [173]. The perimeter membrane is protected by a luminally oriented thick glycocalyx, rich in digestion resistant polylactosamine structures, and also by having high cholesterol content [181, 182].

Water-soluble lysosomal hydrolases can attack water-soluble macromolecules such as oligosaccharides or glycoproteins, directly, but they can hardly digest lipophilic membrane-embedded gangliosides, GSLs and other lipophilic membrane components. Their digestion at the interphase needs a more complex cooperation between soluble hydrolases and lipid-binding transfer proteins at the surface of LVs [173, 183, 184]. Due to the lipid phase problem, soluble glycosidases hardly attack amphiphilic GSL embedded in lipid bilayer or micelle structures directly, but need the help of bilayer perturbing and lipid-binding proteins, the SAPs. Their inherited defects cause rare and fatal, often neurodegenerative diseases [180, 185]. Their clinical picture is often quite similar to that of the respective hydrolase deficiency [11, 186, 187]. A deficiency of prosaposin, the common precursor of all four saposins, causes an early postnatal fatal disease with accumulation of LVs, ceramide and several GSLs in the endolysosomal compartment, and the loss of the water permeability barrier in the skin [173, 188, 189].

5.3 Ganglioside Digestion at Luminal Vesicles

Desialylation of oligosialogangliosides and glycoproteins in the endosomal and lysosomal compartment is a crucial step for constitutional ganglioside degradation in mammalian tissues [190]. The sialidases Neu1, Neu4 and also Neu3 are the major mammalian hydrolases that cleave off terminal sialyl residues from oligosialylgangliosides to generate ganglioside GM1 [191, 192]. The lysosomal sialidase Neu1 is complexed and stabilized by a protective protein, cathepsin A (PPCA), and the GM1 cleaving β-galactosidase. Whereas inherited defects of Neu1 trigger the lysosomal accumulation of sialylated metabolites in sialidosis, inherited defects of PPCA cause a galacatosialidosis with an additional storage of oligosaccharides and glycolipids [193, 194].

Using electron dense endocytotic tracers and immunoelectron microscopy, the uptake of biotin-GM1 was investigated along the endocytotic pathway [171]. Tagged ganglioside GM1, inserted into the plasma membrane of cultured human fibroblasts, was mainly endocytosed to the LVs for constitutive degradation [195]. In cultured fibroblasts and neuroblastoma cells, catabolism of GM1 analogs, carrying a bulky digoxigenin or biotin residue at the neuraminyl residue, however, was blocked at the level of GM2 and GM3 analogs. Due to their bulky tag at the sialic acid residues, they were not recognized by sialidases and accumulated. Therefore, they seem to be suitable to study their distribution in endocytotic pathways and LVs [196].

Exogenously added radiolabeled biotin-GM1 binds to patches (0.1 μm in diameter), caveolae-like structures and coated pits of the plasma membrane. During endocytosis, biotin-GM1 reached late endosomes and lysosomes of EGF stimulated fibroblasts. Both biotin-GM1 and the EGF receptor, are then transported to LVs, indicating that both membrane constituents followed the same endocytotic pathway [195].

The digestion of SLs taken up by endocytosis occurs at the LVs. It is blocked by inherited defects of lysosomal proteins (Fig. 4) [173]. Patients deficient in all four saposins due to an inherited prosaposin deficiency suffer from an intraendolysosomal accumulation of GSLs, SLs, and LVs. LVs with an average diameter of 90 nm have been identified by electron microscopy in patient skin samples and in patients' cultured fibroblasts [172, 197–199]. In fibroblasts they serve as platforms of SL and membrane digestion [195]. Feeding nanomolar concentrations of the missing protein, prosaposin (Sap precursor), to the patients' mutant fibroblasts restored their capability to degrade the accumulated GSLs (ceramide, GlcCer, LacCer, ganglioside GM3) completely [172] and reversed the adherent accumulation of SLs and vesicular structures. In healthy cells, LVs are obviously the major location of SL- and membrane catabolism [171, 172].

Whereas most membrane lipids are rapidly degraded in healthy lysosomes, the degradation of phosphatidylglycerol results in a substantial accumulation of one of its intermediates the anionic lysolipid BMP, which is only slowly catabolized and therefore increases up to 40–70 mol% of LVs phospholipids in the lysosomes [176, 200, 201].

Like other phospholipids (phosphatidic acid, phosphatidylglycerol, phosphatidylinositol), BMP is negatively charged at pH values down to pH 4, and is expected to generate a negative surface potential on LVs [202].

In an acidic lysosomal environment of around pH 4–5, however, most lysosomal enzymes are protonated and therefore bind electrostatically and concentrate on the surface of the LVs [202, 203].

Surface-bound lysosomal enzymes like ASM [204], acid ceramidase [205], Hex A, and other hydrolases are apparently stabilized against premature proteolytic digestion in the lysosomes. Uptake of cationic amphiphilic drugs by cultured fibroblasts and their concentration in negatively charged membranes like LVs compensates the negative surface charge of LVs, reduces the electrostatic binding of hydrolases to the LV surface and favors the proteolytic digestion of released hydrolases, triggering an induced phospholipidosis [203, 204, 206].

Taken together, these observations show that SL, GSL and membrane digestion occurs at LVs within the late endosomal and lysosomal compartment. In healthy and young mammalian tissue, LVs are readily degraded, together with their content of lipids and proteins, whereas in cells cultured in a rich medium, when uptake probably exceeds turnover rates, they can be observed. In several SL and GSL storage diseases, they accumulate, e.g., in GM1 gangliosidosis [207] and prosaposin deficient human tissue [172, 197]. In contrast to the almost glycocalyx-free LVs, the lysosomal perimeter membrane is protected against attacks by membrane-degrading enzymes of the lysosol by a thick glycocalyx covering its luminal surface. The glycocalyx is formed by peripheral and integral lysosomal glycoproteins, which are highly *N*-glycosylated with digestion-resistant polylactosamine units [182]. Apparently, the hydrophilic glycocalyx is an efficient barrier that prevents the lysosomal hydrolases and SAPs from reaching and attacking lipid substrates in the lysosomal perimeter membrane. It is also a barrier for cholesterol, phospholipids, GSLs, SLs, and most of their hydrophobic degradation products to escape from the lumen of the lysosome and to reach the lipid bilayer of the perimeter membrane. Defects in the biosynthesis of lysosomal perimeter membrane glycoproteins indeed increase the turnover of the limiting lysosomal membrane [182]. This also implies that degradation products like sphingosine must be exported actively, e.g., through the action of transmembrane transporters or by trapping sphingosine diffusing through the lysosomal perimeter membrane to its surface by phosphorylation [208]. NPC1, which has been in focus, however, seems not to be relevant for lysosomal sphingosine export [209].

5.4 Regulation of Lipid Trafficking and Catabolic Steps by Membrane Lipids of the Late Endosomes

Lipid and protein composition of eukaryotic cellular membranes is organelle-specific [210]. Neuronal plasma membranes are stabilized by a high content of complex gangliosides, SM and cholesterol. A cholesterol content of about 40 mol% of all membrane lipids is essential for the function of many proteins in or at the plasma membrane, e.g., for the optimal function of the Na^+–K^+ ATPase to generate electrochemical gradients and keep the sodium ion permeability low, close to zero [211]. A decrease of the cholesterol level in the plasma membrane would attenuate the generation of action potentials. On the other hand, cholesterol appears to be inhibitory for some steps of lysosomal SL degradation [202] and, therefore, must be removed along the endocytic pathway. At the level of late endosomes, LVs mature by a sorting process, by an efflux of cholesterol, mediated mainly by two sterol-binding and transfer proteins, the small water-soluble glycoprotein NPC2 and the perimeter membrane-spanning NPC1 [212]. A deficiency of either of these proteins causes a modest accumulation of cholesterol in the lysosomes of patients suffering from fatal Niemann–Pick disease type C (NPC), although the overall cholesterol content of the brain tissue appears to be almost unchanged [213]. We could also show that sphingomyelin inhibits the GM2 hydrolysis by Hex A and GM2AP [162], and that of GlcCer by glucocerebrosidase GBA1 [214]. In healthy cells SM of the LVs is efficiently hydrolyzed by ASM [202].

In addition to lowered cholesterol and SM levels, BMP, an anionic lysophospholipid, is generated during endocytosis specifically in the LVs and is quintessential to stimulate several degradative steps in GSL and SL catabolism at the late endosomes and lysosomes [215].

5.5 Threshold Theory and Protracted Clinical Forms of Gangliosidoses

An almost complete loss of catabolic activities usually results in an excessive lysosomal lipid accumulation in infantile forms of diseases, triggering pathologic mechanisms [215] including an accumulation of storage material also in nonlysosomal compartments, which affects cellular functions [216]:

- Inhibition of autophagy, expansion of lysosomal compartments at the expense of cytosol and other organelles.
- Impaired digestion of macromolecules not directly affected by the catabolic defect: Inhibition of catabolic enzymes and proteins (accumulating lipids like SM inhibit NPC2 and lipid sorting, cholesterol inhibits GM2AP, Sap A, Sap B, etc.). Accumulation of cytotoxic cationic lipids: galactosylsphingosine (psychosine hypothesis, Krabbe), glucosylsphingosine, sphingosine, and sphinganine. They inhibit catabolic steps and lysosomal digestion, and trigger proteolysis of ASM and other catabolic hydrolases.
- Release of pro-inflammatory cytokines, activation of microglia, influx of macrophages in Gaucher mice [217], and central nervous system inflammation in mouse models of gangliosidoses [218].

Allelic mutations, however, which produce proteins with residual functions and some catabolic activities left, give rise to protracted milder clinical forms often described as late infantile, juvenile or chronic diseases [11] with a wide range of clinical symptomatology. At the biochemical level, the clinical heterogeneity is paralleled by a variation of the extent and the pattern of glycolipid accumulation and by different levels of residual catabolic activities detected in cultures of patients' fibroblasts. This observation can be modeled very well on the basis of a greatly simplified kinetic theory [11, 219]. A nice correlation was observed between the predicted and the measured lipid substrate turnover, and also between the rate of the ganglioside GM2 turnover and the clinical course of Tay–Sachs disease and its related juvenile and adult forms. 10–20% of normal GM2-cleaving activity appears to be sufficient for normal life, with chronic and juvenile disease forms starting below that threshold. Similar observations were reported for metachromatic leukodystrophy (MLD), Gaucher disease, Sandhoff disease, and ASM-deficient Niemann–Pick disease type A and B [220–222].

5.6 Anionic Lipids Stimulate and High Cholesterol Levels Inhibit Ganglioside GM2 Catabolism

We now know all the major proteins essential for lysosomal degradation of ganglioside GM2 (Fig. 5a), the main storage compound in Tay–Sachs disease [162, 223]. Reconstitution experiments employing GM2-carrying liposomal vesicles (as mimics of LVs of the endolysosomal system), purified HexA and GM2AP (as essential lipid transfer protein) show that in addition to the known proteins, anionic membrane lipids are needed in the vesicular

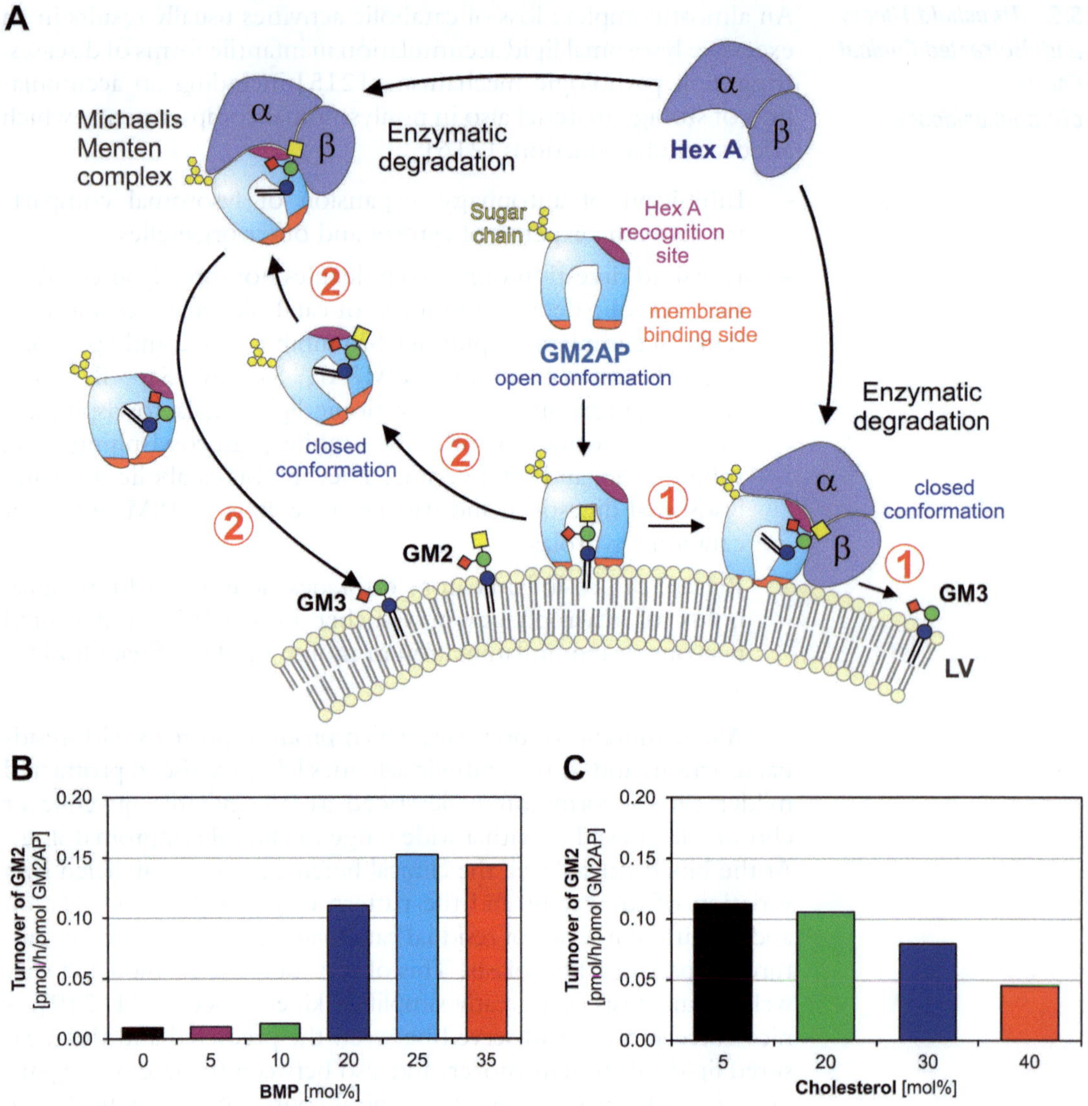

Fig. 5 Scheme of the GM2 degradation by human β-hexosaminidase A (Hex A) (**A**) and their regulation by membrane lipids (**B**, **C**). (**A**) Model for GM2AP-stimulated hydrolysis of GM2 by human Hex A. The digestion of GM2 takes place at intralysosomal vesicles and membranes (LV). GM2AP contains a hydrophobic cavity, with dimensions that can accommodate the ceramide portion of GM2 and other lipids, lined by surface loops and a single short helix. Two conformations of GM2AP are possible: one open and one closed. The open empty activator binds to the membrane using the hydrophobic loops (orange) and penetrates into the hydrophobic region of the bilayer. The lipid recognition site of the activator can interact with the substrate, and its ceramide portion can move inside the hydrophobic cavity. At this point, the conformation of the lipid-loaded activator may change to the closed one, thus the complex becomes more water soluble and can stay (1) or leave the membrane (2), exposing GM2 to the water-soluble Hex A to be degraded. The schematic representation and colors of GM2 and GM3 are the same as indicated in Fig. 1. The GM2 hydrolysis is stimulated by anionic lipids like BMP (**B**) and inhibited by cholesterol (Chol) (**C**) or sphingomyelin. In (**B**) liposomes contain 5 mol% cholesterol, 0–35 mol% BMP, and phosphatidylcholine as host lipid and in (**C**) 5–40 mol% cholesterol, 20 mol% BMP, and phosphatidylcholine as host lipid [162]

GM2-carrying membranes to achieve physiologically relevant turnover rates for GM2 [162] (Fig. 5b).

NPC disease is characterized by a lysosomal accumulation of both, cholesterol and GSLs like GlcCer, LacCer, gangliosides GM3 and GM2 [212, 224]. It had been suggested that GSL and cholesterol accumulation are both triggered by NPC1 deficiency in late endosomes [225]. The identification of the defective proteins NPC1 in NPC type 1 and NPC2 in NPC type 2 as cholesterol-binding and transfer proteins [226–228], however, suggested a primary defect in cholesterol efflux from late endosomes and lysosomes. Therefore, our recent finding that high cholesterol and SM concentrations, as they occur in cellular plasma membranes and as storage material in the lysosomes of NPC patients, inhibit the BMP-enhanced hydrolysis of GM2 significantly (Fig. 5) [162], is important to explain the increased lysosomal levels of GM2 in NPC as a secondary accumulation [223]. The strong cholesterol-induced inhibition of SAPs like Sap A [229], Sap B [230] and GM2AP [162] may well contribute to the additional secondary storage of minor GSLs such as GlcCer, LacCer and ganglioside GM3 as observed in NPC patients [212]. The secondary accumulation of GSLs seems to be important for the molecular pathogenesis of NPC, as miglustat, an inhibitor of sphingolipid biosynthesis, indeed improves the clinical course of the patients [231, 232].

Cholesterol also inhibits other functions of GM2AP, such as the solubilization of membrane lipids and their inter-vesicular transfer, identifying the small glycoprotein GM2AP as an efficient lysosomal lipid transfer protein which, however, is not fusogenic [233]. Some lysosomal SAPs mediate an inter-vesicular lipid transfer and some others can trigger fusion of vesicles at low pH values [180, 234, 235].

5.7 SAPs and Negatively Charged Lipids Stimulate, whereas SM and Sphingoid Bases Inhibit Lysosomal Ganglioside and Glucosylceramide Degradation

All steps in lysosomal sphingolipid digestion studied so far are strongly enhanced by both, lysosomal SAPs and anionic lipids [183, 186, 223]. To demonstrate the extent of regulation, Fig. 6 presents quantitative data on the stimulation of the enzymatic degradation of membrane-bound ganglioside GM1 by lysosomal β-galactosidase in the presence and absence of lysosomal SAPs (GM2AP and Sap B) and anionic lipids.

Membrane-bound complex GSLs like GM1 (Fig. 6) and GM2 (Fig. 5) are hardly hydrolyzed by their respective hydrolases in the absence of anionic lipids, and even less so in the absence of both, SAPs and anionic lipids, whereas the smaller glucosylceramide is already cleaved by β-glucosidase in the presence of anionic lipids and absence of SAPs [214, 236]. Hydrolysis of membrane-bound glucosylceramide by β-glucocerebrosidase (GBA1) is stimulated at low pH up to over 100-fold by insertion of anionic lipids into the GlcCer carrying membranes, depending on the nature and position of the negative charge in

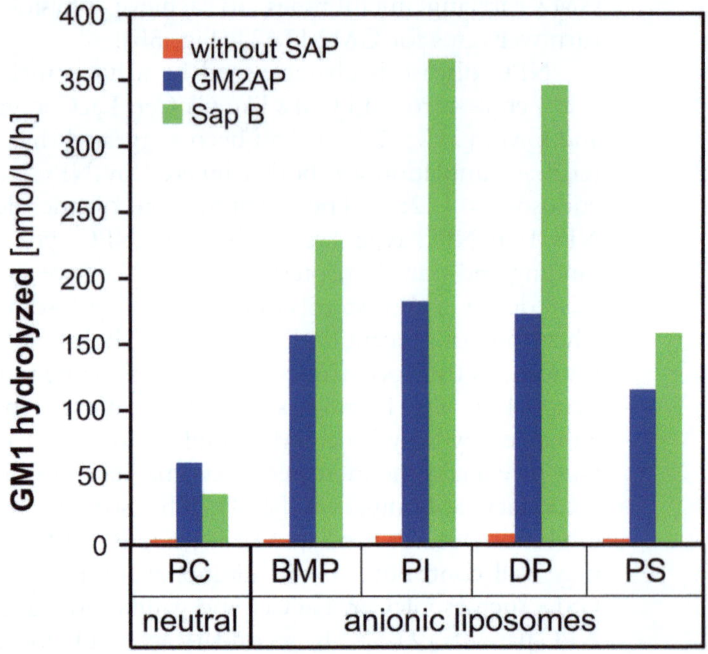

Fig. 6 SAPs and anionic lipids stimulate hydrolysis of ganglioside GM1 by β-galactosidase. Anionic lipids and sphingolipid activator proteins enhance the degradation of ganglioside GM1 by β-galactosidase up to 100-fold. Using phosphatidylcholine (PC) containing neutral liposomes in the absence of anionic lipids and SAPs, β-galactosidase hardly hydrolyzes any detectable amounts of ganglioside GM1 (column far left). However, incorporation of anionic lipids (BMP, phosphatidylinositol (PI), dolicholphosphate (DP), phosphatidylserine (PS)) into the GM1-carrying liposomes and addition of a SAP, either GM2AP or Sap B, stimulated the GM1 hydrolysis up to 100-fold. As indicated, lysosomal anionic lipids (10 mol%) were incorporated in lipid unilamellar vesicles, composed of 10 mol% ganglioside GM1, 20 mol% cholesterol, and 60 mol% PC. Assays were carried out in the absence of an activator protein and in the presence of 5 mM of GM2AP or 5 mM of Sap B, both isolated from human spleen. Mean error was determined to be less than 10%. Figure and legend were modified after [248]

their head groups (Fig. 7), while cationic lipids strongly inhibit the reaction, especially the minor cytotoxic storage compound glucosylsphingosine [237], a biomarker for Gaucher disease [238]. Also, lysosomal lipid degradation products (ceramide, fatty acids, monoacylglycerol, and diacylglycerol) stimulate GlcCer hydrolysis, whereas SM (Fig. 7), sphingosine, and sphinganine strongly attenuate GlcCer degradation, thereby explaining the secondary storage of GlcCer in Niemann–Pick diseases. Surprisingly, cholesterol enhances the GlcCer hydrolysis rate strongly in the presence of BMP, whereas Sap C is only a weak stimulator in its presence; Sap C enhances GlcCer hydrolysis strongly in the absence of anionic lipids, that is, in neutral membranes [214].

Carbohydrate-free sphingolipids, like SM [239] and Cer [240] are, however, already slowly cleaved by their respective hydrolases in the absence of SAPs and anionic lipids, but they reach physiological

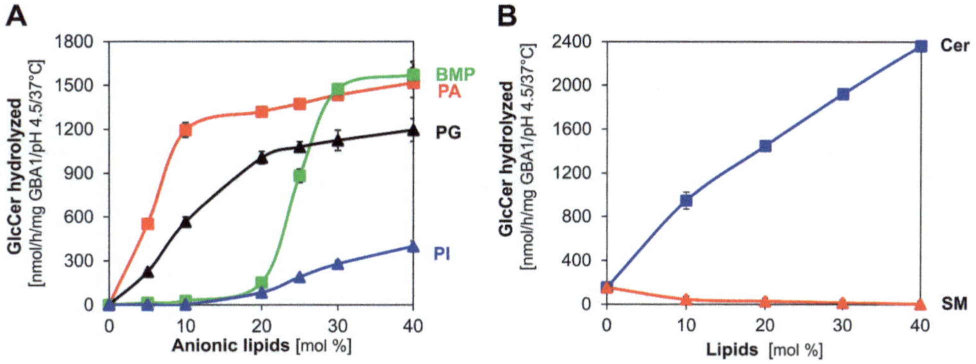

Fig. 7 Membrane lipids regulate GlcCer degradation by GBA1 [214]. Anionic lipids (**A**) and ceramide (**B**) enhances and sphingomyelin (**B**) inhibits the hydrolysis of GlcCer by GBA1. Liposomes in (**A**) contain 5 mol% cholesterol, 0–40 mol% anionic lipids and phosphatidylcholine as host lipid and in (**B**) 5 mol% cholesterol, 20 mol% anionic lipids, ceramide and sphingomyelin as indicated and phosphatidylcholine as host lipid. *BMP* bis(monoacylglycero)phosphate, *Cer* ceramide, *PA* phosphatidic acid, *PG* phosphatidylglycerol, *PI* phosphatidylinositol, *PS* phosphatidylserine, *SM* sphingomyelin

rates of degradation only in the presence of SAPs and anionic lipids. The *N*-glycan chains of SAPs are functionally important though structurally heterogeneous, due to an inherent heterogeneity in *N*-glycan biosynthesis. An inherited glycosylation defect in Sap B was fatal in patients with metachromatic leukodystrophy, despite the presence of the unglycosylated protein in the lysosomal compartment and its ability to stimulate the enzymatic hydrolysis of sulfatides in vitro even better than the glycosylated wild type Sap B [230]. However, the unglycosylated Sap B of the patient lost the ability of the glycosylated Sap B to rescue Sap B deficient cultured patient fibroblasts and to mediate an inter-vesicular lipid transfer in vitro [230]. Hence, glycosylation is required for lipid-loaded Sap B to efficiently take-off again from LVs. A recent investigation on homogeneous, synthetic glycoforms of human Sap D revealed a strong carbohydrate dependence of its bioactivity [241]. Although all chemically synthesized Sap D glycoforms stimulated ceramide cleavage by acid ceramidase to a similar extent, their membrane binding and fusion capacity increased substantially by shortening their *N*-glycan moiety. As expected, however, their ability to form soluble ceramide-Sap D complexes increased with increasing oligosaccharide side chains.

The absence of all four Saps A–D caused by the inherited absence of their precursor protein prosaposin [199] triggers a fatal perinatal disease, mainly induced by the loss of the water permeability barrier in the skin [172, 186, 242], which depends on the extracellular degradation of esterified and protein-bound GlcCers to corresponding ceramides. However, the absence of all four Saps causes a massive accumulation of ceramides, GlcCer, ganglioside GM3 and many other lipids [172, 197, 243], despite the presence of anionic lipids and active hydrolases in the tissue of the patients, emphasizing again the essential interplay of SAPs together with anionic lipids and soluble hydrolases for lysosomal turnover of SLs.

5.8 Lipid Modifiers Affect the Catabolic Activity of Lysosomal Hydrolases on Endogenous Lipid Substrates

The lipid composition of eukaryotic organelles and vesicular membranes is particle specific and apparently adjusted to their individual functions. Disturbed lipid composition of organelle membranes may contribute to clinical heterogeneity of lysosomal diseases [223]. Data obtained so far indicate that catabolic activities of lysosomal enzymes, as exemplified above for Hex A and GBA1, are strongly modified by the microenvironment in which the enzymatic reaction occurs and by the nature of the substrate used, e.g., either in free solution acting on a water-soluble synthetic and unnatural substrate like 4-methylumbelliferyl-*N*-acetyl-β-D-glucosaminide in the absence of SAPs and membrane lipids, or on a membrane-bound natural glycolipid in the presence of lipid modifiers and SAPs. Hence, enzyme activity assays with artificial, soluble substrates may not always reveal potential dysfunctions of patient enzymes for their natural substrates since they rarely employ modifiers important for the degradation of lipid substrates. Important modifiers for the catabolism of a lipid substrate like ganglioside GM2 by HexA or GlcCer by GBA1 are:

- *The presence or absence of a lipid-binding and transfer protein* like GM2AP or Sap C [223, 244].
- *The surface charge and lipid composition of membranes* carrying the GM2 or GlcCer [162, 202, 214].
- *The nature of stimulating anionic lipids* in the lipid substrate (GM2 or GlcCer) carrying membrane [162, 202, 214].
- *The presence of inhibiting lipids*, e.g., SM, cholesterol, sphingosine, sphinganine, and glycosylsphingosine in the GM2 or GlcCer carrying membrane [162, 202, 214].
- *The local pH value* (whereas Hex A cleaves soluble synthetic substrates in a wide pH range of 3–7, cleavage of membrane-bound GM2 in the presence of GM2AP occurs only in a narrow pH range of 3.8–4.6) [6, 162, 245].
- *The ionic strength and ion composition* in the microenvironment of the catabolic reaction [162, 202, 214].

The strong and cumulative influence of these modifiers on the rate of enzyme driven lipid cleavage, e.g., of GM2 and GlcCer hydrolysis, may well affect the correlation between genotype and phenotype obtained for patients with lysosomal diseases. Cells generating high levels of storage compounds biosynthetically and/or by cellular uptake mechanisms (e.g., for macrophages) will be the first to suffer from cytotoxic damage and will dominate the clinical picture of the disease.

Acknowledgments

We thank Roger Sandhoff and Christina Schuette for helpful corrections and discussions. This work was supported by German Research Foundation Grants SFB 645 and TRR83.

References

1. Thudichum JLW (1884) A treatise on the chemical constitution of the brain. London, Bailliere, Tindall and Cox
2. Klenk E (1939) Niemann-Pick'sche Krankheit und Amaurotische Idiotie. Hoppe-Seyler's Z Physiol Chem 262:128–143
3. Klenk E (1942) Über die Ganglioside, eine neue Gruppe von zuckerhaltigen Gehirnlipoiden. Hoppe Seyler's Z Physiol Chem 273:76–86
4. Kuhn R, Wiegandt H (1963) Die Konstitution der Ganglio-N-tetraose und des Gangliosids GI. Chem Ber 96:866–880
5. Jatzkewitz H, Sandhoff K (1963) On a biochemically special form of infantile amaurotic idiocy. Biochim Biophys Acta 70:354–356
6. Sandhoff K, Harzer K, Wässle W, Jatzkewitz H (1971) Enzyme alterations and lipid storage in three variants of Tay-Sachs disease. J Neurochem 18(12):2469–2489
7. Harzer K, Jatzkewitz H, Sandhoff K (1969) Incorporation of labelled glucose into the individual major gangliosides of the brain of young rats. J Neurochem 16(8):1279–1282
8. Skotland T, Ekroos K, Kavaliauskiene S, Bergan J, Kauhanen D, Lintonen T, Sandvig K (2016) Determining the turnover of glycosphingolipid species by stable-isotope tracer lipidomics. J Mol Biol 428(24 Pt A):4856–4866. https://doi.org/10.1016/j.jmb.2016.06.013
9. Aureli M, Samarani M, Murdica V, Mauri L, Loberto N, Bassi R, Prinetti A, Sonnino S (2014) Gangliosides and cell surface ganglioside glycohydrolases in the nervous system. Adv Neurobiol 9:223–244. https://doi.org/10.1007/978-1-4939-1154-7_10
10. Aureli M, Loberto N, Lanteri P, Chigorno V, Prinetti A, Sonnino S (2011) Cell surface sphingolipid glycohydrolases in neuronal differentiation and aging in culture. J Neurochem 116(5):891–899. https://doi.org/10.1111/j.1471-4159.2010.07019.x
11. Sandhoff K (2012) My journey into the world of sphingolipids and sphingolipidoses. Proc Jpn Acad Ser B 88:554–582
12. Mehl E, Jatzkewitz H (1964) Eine cerebrosidsulfatase aus Schweineniere. Hoppe-Seylers. Zeitschr Physiol Chem 339(1–6):260–276
13. Conzelmann E, Sandhoff K (1978) AB variant of infantile GM2 gangliosidosis: deficiency of a factor necessary for stimulation of hexosaminidase A-catalyzed degradation of ganglioside GM2 and glycolipid GA2. Proc Natl Acad Sci U S A 75(8):3979–3983
14. Yamakawa T (1996) A reflection on the early history of glycosphingolipids. Glycoconj J 13(2):123–126. https://doi.org/10.1007/bf00731485
15. Wiegandt H (1968) The structure and the function of gangliosides. Angew Chem Int Ed Engl 7(2):87–96. https://doi.org/10.1002/anie.196800871
16. Wiegandt H (1995) The chemical constitution of gangliosides of the vertebrate nervous system. Behav Brain Res 66(1–2):85–97
17. Sandhoff R, Geyer R, Jennemann R, Paret C, Kiss E, Yamashita T, Gorgas K, Sijmonsma TP, Iwamori M, Finaz C, Proia RL, Wiegandt H, Gröne HJ (2005) Novel class of glycosphingolipids involved in male fertility. J Biol Chem 280(29):27310–27318. https://doi.org/10.1074/jbc.M502775200
18. Posse de Chaves E, Sipione S (2010) Sphingolipids and gangliosides of the nervous system in membrane function and dysfunction. FEBS Lett 584(9):1748–1759. https://doi.org/10.1016/j.febslet.2009.12.010
19. Jennemann R, Sandhoff R, Wang S, Kiss E, Gretz N, Zuliani C, Martin-Villalba A, Jäger R, Schorle H, Kenzelmann M, Bonrouhi M, Wiegandt H, Gröne H-J (2005) Cell-specific deletion of glucosylceramide synthase in brain leads to severe neural defects after birth. Proc Natl Acad Sci U S A 102(35):12459–12464
20. Yamashita T, Wu YP, Sandhoff R, Werth N, Mizukami H, Ellis JM, Dupree JL, Geyer R, Sandhoff K, Proia RL (2005) Interruption of ganglioside synthesis produces central nervous system degeneration and altered axon-glial interactions. Proc Natl Acad Sci U S A 102(8):2725–2730
21. Fukushi Y, Hakomori S, Shepard T (1984) Localization and alteration of mono-, di-, and trifucosyl alpha 1–3 type 2 chain structures

during human embryogenesis and in human cancer. J Exp Med 160(2):506–520

22. Ichikawa S, Nakajo N, Sakiyama H, Hirabayashi Y (1994) A mouse B16 melanoma mutant deficient in glycolipids. Proc Natl Acad Sci U S A 91(7):2703–2707
23. Kolter T, Magin TM, Sandhoff K (2000) Biomolecule function: no reliable prediction from cell culture. Traffic 1(10):803–804
24. Schnaar RL (2010) Brain gangliosides in axon-myelin stability and axon regeneration. FEBS Lett 584(9):1741–1747. https://doi.org/10.1016/j.febslet.2009.10.011
25. Merrill AH Jr (2002) De novo sphingolipid biosynthesis: a necessary, but dangerous, pathway. J Biol Chem 277(29):25843–25846. https://doi.org/10.1074/jbc.R200009200
26. Merrill AH Jr (2011) Sphingolipid and glycosphingolipid metabolic pathways in the era of sphingolipidomics. Chem Rev 111(10):6387–6422. https://doi.org/10.1021/cr2002917
27. Rother J, van Echten G, Schwarzmann G, Sandhoff K (1992) Biosynthesis of sphingolipids: dihydroceramide and not sphinganine is desaturated by cultured cells. Biochem Biophys Res Commun 189(1):14–20
28. Michel C, van Echten-Deckert G, Rother J, Sandhoff K, Wang E, Merrill AH Jr (1997) Characterization of ceramide synthesis. A dihydroceramide desaturase introduces the 4,5-trans-double bond of sphingosine at the level of dihydroceramide. J Biol Chem 272(36):22432–22437
29. Holland WL, Brozinick JT, Wang LP, Hawkins ED, Sargent KM, Liu Y, Narra K, Hoehn KL, Knotts TA, Siesky A, Nelson DH, Karathanasis SK, Fontenot GK, Birnbaum MJ, Summers SA (2007) Inhibition of ceramide synthesis ameliorates glucocorticoid-, saturated-fat-, and obesity-induced insulin resistance. Cell Metab 5(3):167–179. https://doi.org/10.1016/j.cmet.2007.01.002
30. Hanada K (2003) Serine palmitoyltransferase, a key enzyme of sphingolipid metabolism. Biochim Biophys Acta 1632(1–3):16–30
31. Sonnino S, Chigorno V (2000) Ganglioside molecular species containing C18- and C20-sphingosine in mammalian nervous tissues and neuronal cell cultures. Biochim Biophys Acta 1469(2):63–77
32. Palestini P, Masserini M, Sonnino S, Giuliani A, Tettamanti G (1990) Changes in the ceramide composition of rat forebrain gangliosides with age. J Neurochem 54(1):230–235
33. Zhao L, Spassieva S, Gable K, Gupta SD, Shi LY, Wang J, Bielawski J, Hicks WL, Krebs MP, Naggert J, Hannun YA, Dunn TM, Nishina PM (2015) Elevation of 20-carbon long chain bases due to a mutation in serine palmitoyltransferase small subunit b results in neurodegeneration. Proc Natl Acad Sci U S A 112(42):12962–12967. https://doi.org/10.1073/pnas.1516733112
34. Davidson G, Murphy S, Polke J, Laura M, Salih M, Muntoni F, Blake J, Brandner S, Davies N, Horvath R, Price S, Donaghy M, Roberts M, Foulds N, Ramdharry G, Soler D, Lunn M, Manji H, Davis M, Houlden H, Reilly M (2012) Frequency of mutations in the genes associated with hereditary sensory and autonomic neuropathy in a UK cohort. Journal of neurology 259(8):1673–1685. https://doi.org/10.1007/s00415-011-6397-y
35. Astudillo L, Sabourdy F, Therville N, Bode H, Segui B, Andrieu-Abadie N, Hornemann T, Levade T (2015) Human genetic disorders of sphingolipid biosynthesis. J Inherit Metab Dis 38(1):65–76. https://doi.org/10.1007/s10545-014-9736-1
36. Penno A, Reilly MM, Houlden H, Laura M, Rentsch K, Niederkofler V, Stoeckli ET, Nicholson G, Eichler F, Brown RH Jr, von Eckardstein A, Hornemann T (2010) Hereditary sensory neuropathy type 1 is caused by the accumulation of two neurotoxic sphingolipids. J Biol Chem 285(15):11178–11187. https://doi.org/10.1074/jbc.M109.092973
37. Zitomer NC, Mitchell T, Voss KA, Bondy GS, Pruett ST, Garnier-Amblard EC, Liebeskind LS, Park H, Wang E, Sullards MC, Merrill AH Jr, Riley RT (2009) Ceramide synthase inhibition by fumonisin B1 causes accumulation of 1-deoxysphinganine: a novel category of bioactive 1-deoxysphingoid bases and 1-deoxydihydroceramides biosynthesized by mammalian cell lines and animals. J Biol Chem 284(8): 4786–4795. https://doi.org/10.1074/jbc.M808798200
38. Garofalo K, Penno A, Schmidt BP, Lee HJ, Frosch MP, von Eckardstein A, Brown RH, Hornemann T, Eichler FS (2011) Oral L-serine supplementation reduces production of neurotoxic deoxysphingolipids in mice and humans with hereditary sensory autonomic neuropathy type 1. J Clin Invest 121(12):4735–4745. https://doi.org/10.1172/JCI57549
39. Alecu I, Othman A, Penno A, Saied EM, Arenz C, von Eckardstein A, Hornemann T (2017) Cytotoxic 1-deoxysphingolipids are metabolized by a cytochrome P450-dependent pathway. J Lipid Res 58(1):60–71. https://doi.org/10.1194/jlr.M072421
40. Mullen TD, Hannun YA, Obeid LM (2012) Ceramide synthases at the centre of sphingolipid

metabolism and biology. Biochem J 441(3):789–802. https://doi.org/10.1042/BJ20111626

41. Grosch S, Schiffmann S, Geisslinger G (2012) Chain length-specific properties of ceramides. Prog Lipid Res 51(1):50–62. https://doi.org/10.1016/j.plipres.2011.11.001
42. Sassa T, Hirayama T, Kihara A (2016) Enzyme activities of the ceramide synthases CERS2-6 are regulated by phosphorylation in the C-terminal region. J Biol Chem 291(14):7477–7487. https://doi.org/10.1074/jbc.M115.695858
43. Yu RK, Bieberich E, Xia T, Zeng G (2004) Regulation of ganglioside biosynthesis in the nervous system. J Lipid Res 45(5):783–793. https://doi.org/10.1194/jlr.R300020-JLR200
44. Wegner MS, Schiffmann S, Parnham MJ, Geisslinger G, Grosch S (2016) The enigma of ceramide synthase regulation in mammalian cells. Prog Lipid Res 63:93–119. https://doi.org/10.1016/j.plipres.2016.03.006
45. Mullen TD, Spassieva S, Jenkins RW, Kitatani K, Bielawski J, Hannun YA, Obeid LM (2011) Selective knockdown of ceramide synthases reveals complex interregulation of sphingolipid metabolism. J Lipid Res 52(1):68–77. https://doi.org/10.1194/jlr.M009142
46. Mosbech MB, Olsen AS, Neess D, Ben-David O, Klitten LL, Larsen J, Sabers A, Vissing J, Nielsen JE, Hasholt L, Klein AD, Tsoory MM, Hjalgrim H, Tommerup N, Futerman AH, Moller RS, Færgeman NJ (2014) Reduced ceramide synthase 2 activity causes progressive myoclonic epilepsy. Ann Clin Trans Neurol 1(2):88–98. https://doi.org/10.1002/acn3.28
47. Ebel P, Vom Dorp K, Petrasch-Parwez E, Zlomuzica A, Kinugawa K, Mariani J, Minich D, Ginkel C, Welcker J, Degen J, Eckhardt M, Dere E, Dörmann P, Willecke K (2013) Inactivation of ceramide synthase 6 in mice results in an altered sphingolipid metabolism and behavioral abnormalities. J Biol Chem 288(29):21433–21447. https://doi.org/10.1074/jbc.M113.479907
48. Turpin SM, Nicholls HT, Willmes DM, Mourier A, Brodesser S, Wunderlich CM, Mauer J, Xu E, Hammerschmidt P, Brönneke HS, Trifunovic A, LoSasso G, Wunderlich FT, Kornfeld JW, Blüher M, Krönke M, Brüning JC (2014) Obesity-induced CerS6-dependent C16:0 ceramide production promotes weight gain and glucose intolerance. Cell Metab 20(4):678–686. https://doi.org/10.1016/j.cmet.2014.08.002
49. Ginkel C, Hartmann D, vom Dorp K, Zlomuzica A, Farwanah H, Eckhardt M, Sandhoff R, Degen J, Rabionet M, Dere E, Dörmann P, Sandhoff K, Willecke K (2012) Ablation of neuronal ceramide synthase 1 in mice decreases ganglioside levels and expression of myelin-associated glycoprotein in oligodendrocytes. J Biol Chem 287(50):41888–41902. https://doi.org/10.1074/jbc.M112.413500
50. Zhao L, Spassieva SD, Jucius TJ, Shultz LD, Shick HE, Macklin WB, Hannun YA, Obeid LM, Ackerman SL (2011) A deficiency of ceramide biosynthesis causes cerebellar purkinje cell neurodegeneration and lipofuscin accumulation. PLoS Genet 7(5):e1002063. https://doi.org/10.1371/journal.pgen.1002063
51. Spassieva SD, Ji X, Liu Y, Gable K, Bielawski J, Dunn TM, Bieberich E, Zhao L (2016) Ectopic expression of ceramide synthase 2 in neurons suppresses neurodegeneration induced by ceramide synthase 1 deficiency. Proc Natl Acad Sci U S A 113(21):5928–5933. https://doi.org/10.1073/pnas.1522071113
52. Gosejacob D, Jäger PS, Vom Dorp K, Frejno M, Carstensen AC, Köhnke M, Degen J, Dörmann P, Hoch M (2016) Ceramide synthase 5 is essential to maintain C16:0-ceramide pools and contributes to the development of diet-induced obesity. J Biol Chem 291(13):6989–7003. https://doi.org/10.1074/jbc.M115.691212
53. Jennemann R, Rabionet M, Gorgas K, Epstein S, Dalpke A, Rothermel U, Bayerle A, van der Hoeven F, Imgrund S, Kirsch J, Nickel W, Willecke K, Riezman H, Gröne HJ, Sandhoff R (2012) Loss of ceramide synthase 3 causes lethal skin barrier disruption. Hum Mol Genet 21(3):586–608. https://doi.org/10.1093/hmg/ddr494
54. Rabionet M, Bayerle A, Jennemann R, Heid H, Fuchser J, Marsching C, Porubsky S, Bolenz C, Guillou F, Grone HJ, Gorgas K, Sandhoff R (2015) Male meiotic cytokinesis requires ceramide synthase 3-dependent sphingolipids with unique membrane anchors. Hum Mol Genet 24(17):4792–4808. https://doi.org/10.1093/hmg/ddv204
55. Eckl KM, Tidhar R, Thiele H, Oji V, Hausser I, Brodesser S, Preil ML, Onal-Akan A, Stock F, Müller D, Becker K, Casper R, Nürnberg G, Altmüller J, Nürnberg P, Traupe H, Futerman AH, Hennies HC (2013) Impaired epidermal ceramide synthesis causes autosomal recessive congenital ichthyosis and reveals the importance of ceramide acyl chain length. J Invest Dermatol 133(9):2202–2211. https://doi.org/10.1038/jid.2013.153

56. Radner FP, Marrakchi S, Kirchmeier P, Kim GJ, Ribierre F, Kamoun B, Abid L, Leipoldt M, Turki H, Schempp W, Heilig R, Lathrop M, Fischer J (2013) Mutations in CERS3 cause autosomal recessive congenital ichthyosis in humans. PLoS Genet 9(6):e1003536. https://doi.org/10.1371/journal.pgen.1003536
57. Imgrund S, Hartmann D, Farwanah H, Eckhardt M, Sandhoff R, Degen J, Gieselmann V, Sandhoff K, Willecke K (2009) Adult ceramide synthase 2 (CERS2)-deficient mice exhibit myelin sheath defects, cerebellar degeneration, and hepatocarcinomas. J Biol Chem 284(48):33549–33560
58. Pewzner-Jung Y, Brenner O, Braun S, Laviad EL, Ben-Dor S, Feldmesser E, Horn-Saban S, Amann-Zalcenstein D, Raanan C, Berkutzki T, Erez-Roman R, Ben-David O, Levy M, Holzman D, Park H, Nyska A, Merrill AH Jr, Futerman AH (2010) A critical role for ceramide synthase 2 in liver homeostasis: II. insights into molecular changes leading to hepatopathy. J Biol Chem 285(14):10911–10923. https://doi.org/10.1074/jbc.M109.077610
59. Pewzner-Jung Y, Park H, Laviad EL, Silva LC, Lahiri S, Stiban J, Erez-Roman R, Brügger B, Sachsenheimer T, Wieland F, Prieto M, Merrill AH Jr, Futerman AH (2010) A critical role for ceramide synthase 2 in liver homeostasis: I. Alterations in lipid metabolic pathways. J Biol Chem 285(14):10902–10910. https://doi.org/10.1074/jbc.M109.077594
60. Silva LC, Ben David O, Pewzner-Jung Y, Laviad EL, Stiban J, Bandyopadhyay S, Merrill AH Jr, Prieto M, Futerman AH (2012) Ablation of ceramide synthase 2 strongly affects biophysical properties of membranes. J Lipid Res 53(3):430–436. https://doi.org/10.1194/jlr.M022715
61. Zigdon H, Kogot-Levin A, Park JW, Goldschmidt R, Kelly S, Merrill AH Jr, Scherz A, Pewzner-Jung Y, Saada A, Futerman AH (2013) Ablation of ceramide synthase 2 causes chronic oxidative stress due to disruption of the mitochondrial respiratory chain. J Biol Chem 288(7):4947–4956. https://doi.org/10.1074/jbc.M112.402719
62. Ben-David O, Pewzner-Jung Y, Brenner O, Laviad EL, Kogot-Levin A, Weissberg I, Biton IE, Pienik R, Wang E, Kelly S, Alroy J, Raas-Rothschild A, Friedman A, Brugger B, Merrill AH Jr, Futerman AH (2011) Encephalopathy caused by ablation of very long acyl chain ceramide synthesis may be largely due to reduced galactosylceramide levels. J Biol Chem 286(34):30022–30033. https://doi.org/10.1074/jbc.M111.261206
63. Park WJ, Brenner O, Kogot-Levin A, Saada A, Merrill AH Jr, Pewzner-Jung Y, Futerman AH (2015) Development of pheochromocytoma in ceramide synthase 2 null mice. Endocr Relat Cancer 22(4):623–632. https://doi.org/10.1530/ERC-15-0058
64. Ledeen RW, Yu RK, Eng LF (1973) Gangliosides of human myelin: sialosylgalactosylceramide (G7) as a major component. J Neurochem 21(4):829–839
65. Mullin BR, Patrick DH, Poore CM, Smith MT (1984) Prevention of experimental allergic encephalomyelitis by ganglioside GM4. Brain Res 296(1):174–176
66. Coste H, Martel MB, Got R (1986) Topology of glucosylceramide synthesis in Golgi membranes from porcine submaxillary glands. Biochim Biophys Acta 858(1):6–12
67. Jeckel D, Karrenbauer A, Burger KN, van Meer G, Wieland F (1992) Glucosylceramide is synthesized at the cytosolic surface of various Golgi subfractions. J Cell Biol 117(2):259–267
68. Ichikawa S, Sakiyama H, Suzuki G, Hidari KI, Hirabayashi Y (1996) Expression cloning of a cDNA for human ceramide glucosyltransferase that catalyzes the first glycosylation step of glycosphingolipid synthesis. Proc Natl Acad Sci U S A 93(10):4638–4643
69. Futerman AH, Pagano RE (1991) Determination of the intracellular sites and topology of glucosylceramide synthesis in rat liver. Biochem J 280(2):295–302
70. Marks DL, Wu K, Paul P, Kamisaka Y, Watanabe R, Pagano RE (1999) Oligomerization and topology of the Golgi membrane protein glucosylceramide synthase. J Biol Chem 274(1): 451–456
71. D'Angelo G, Uemura T, Chuang CC, Polishchuk E, Santoro M, Ohvo-Rekila H, Sato T, Di Tullio G, Varriale A, D'Auria S, Daniele T, Capuani F, Johannes L, Mattjus P, Monti M, Pucci P, Williams RL, Burke JE, Platt FM, Harada A, De Matteis MA (2013) Vesicular and non-vesicular transport feed distinct glycosylation pathways in the Golgi. Nature 501(7465):116–120. https://doi.org/10.1038/nature12423
72. De Matteis MA, Di Campli A, D'Angelo G (2007) Lipid-transfer proteins in membrane trafficking at the Golgi complex. Biochim Biophys Acta 1771(6):761–768. https://doi.org/10.1016/j.bbalip.2007.04.004
73. Hanada K, Kumagai K, Yasuda S, Miura Y, Kawano M, Fukasawa M, Nishijima M (2003)

Molecular machinery for non-vesicular trafficking of ceramide. Nature 426(6968):803–809. https://doi.org/10.1038/nature02188

74. Lannert H, Bünning C, Jeckel D, Wieland FT (1994) Lactosylceramide is synthesized in the lumen of the Golgi apparatus. FEBS Lett 342(1):91–96
75. Buton X, Hervé P, Kubelt J, Tannert A, Burger KN, Fellmann P, Müller P, Herrmann A, Seigneuret M, Devaux PF (2002) Transbilayer movement of monohexosylsphingolipids in endoplasmic reticulum and Golgi membranes. Biochemistry 41(43):13106–13115
76. Eckford PD, Sharom FJ (2005) The reconstituted P-glycoprotein multidrug transporter is a flippase for glucosylceramide and other simple glycosphingolipids. Biochem J 389(Pt 2):517–526. https://doi.org/10.1042/BJ20050047
77. Lannert H, Gorgas K, Meissner I, Wieland FT, Jeckel D (1998) Functional organization of the Golgi apparatus in glycosphingolipid biosynthesis. Lactosylceramide and subsequent glycosphingolipids are formed in the lumen of the late Golgi. J Biol Chem 273(5):2939–2946
78. Tokuda N, Numata S, Li X, Nomura T, Takizawa M, Kondo Y, Yamashita Y, Hashimoto N, Kiyono T, Urano T, Furukawa K, Furukawa K (2013) Beta4GalT6 is involved in the synthesis of lactosylceramide with less intensity than beta4GalT5. Glycobiology 23(10):1175–1183. https://doi.org/10.1093/glycob/cwt054
79. D'Angelo G, Polishchuk E, Di Tullio G, Santoro M, Di Campli A, Godi A, West G, Bielawski J, Chuang CC, van der Spoel AC, Platt FM, Hannun YA, Polishchuk R, Mattjus P, De Matteis MA (2007) Glycosphingolipid synthesis requires FAPP2 transfer of glucosylceramide. Nature 449(7158):62–67
80. Roseman S (1970) The synthesis of complex carbohydrates by multiglycosyltransferase systems and their potential function in intercellular adhesion. Chem Phys Lipids 5(1):270–297
81. Yusuf HK, Pohlentz G, Schwarzmann G, Sandhoff K (1983) Ganglioside biosynthesis in Golgi apparatus of rat liver. Stimulation by phosphatidylglycerol and inhibition by tunicamycin. Eur J Biochem 134(1):47–54
82. Yusuf HK, Pohlentz G, Sandhoff K (1984) Ganglioside biosynthesis in Golgi apparatus: new perspectives on its mechanism. J Neurosci Res 12(2–3):161–178
83. Simons K, Gerl MJ (2010) Revitalizing membrane rafts: new tools and insights. Nat Rev Mol Cell Biol 11(10):688–699. https://doi.org/10.1038/nrm2977
84. Simons K, Toomre D (2000) Lipid rafts and signal transduction. Nat Rev Mol Cell Biol 1(1):31–39. https://doi.org/10.1038/35036052
85. Bagatolli LA, Gratton E (2000) Two photon fluorescence microscopy of coexisting lipid domains in giant unilamellar vesicles of binary phospholipid mixtures. Biophys J 78(1):290–305. https://doi.org/10.1016/S0006-3495(00)76592-1
86. Baumgart T, Hess ST, Webb WW (2003) Imaging coexisting fluid domains in biomembrane models coupling curvature and line tension. Nature 425(6960):821–824. https://doi.org/10.1038/nature02013
87. Mukherjee S, Maxfield FR (2000) Role of membrane organization and membrane domains in endocytic lipid trafficking. Traffic 1(3):203–211
88. Spessott W, Crespo PM, Daniotti JL, Maccioni HJ (2012) Glycosyltransferase complexes improve glycolipid synthesis. FEBS Lett 586(16):2346–2350. https://doi.org/10.1016/j.febslet.2012.05.041
89. Keenan TW, Morre DJ, Basu S (1974) Ganglioside biosynthesis. Concentration of glycosphingolipid glycosyltransferases in Golgi apparatus from rat liver. J Biol Chem 249(1):310–315
90. Kaufman B, Basu S, Roseman S (1968) Enzymatic synthesis of disialogangliosides from monosialogangliosides by sialyltransferases from embryonic chicken brain. J Biol Chem 243(21):5804–5807
91. Pohlentz G, Klein D, Schmitz D, Schwarzmann G, Peter-Katalinic J, Sandhoff K (1988) Biosynthesis of gangliosides from asialogangliosides in rat liver Golgi vesicles. Biol Chem Hoppe Seyler 369(1):55–63
92. Pohlentz G, Klein D, Schwarzmann G, Schmitz D, Sandhoff K (1988) Both GA2, GM2, and GD2 synthases and GM1b, GD1a, and GT1b synthases are single enzymes in Golgi vesicles from rat liver. Proc Natl Acad Sci U S A 85(19):7044–7048
93. Iber H, Zacharias C, Sandhoff K (1992) The c-series gangliosides GT3, GT2 and GP1c are formed in rat liver Golgi by the same set of glycosyltransferases that catalyse the biosynthesis of asialo-, a- and b-series gangliosides. Glycobiology 2(2):137–142
94. Iber H, Sandhoff K (1989) Identity of GD1C, GT1a and GQ1b synthase in Golgi vesicles from rat liver. FEBS Lett 254(1–2):124–128

95. Iber H, van Echten G, Sandhoff K (1991) Substrate specificity of alpha 2–3-sialyltransferases in ganglioside biosynthesis of rat liver golgi. Eur J Biochem 195(1):115–120
96. Iber H, van Echten G, Sandhoff K (1992) Fractionation of primary cultured cerebellar neurons: distribution of sialyltransferases involved in ganglioside biosynthesis. J Neurochem 58(4):1533–1537
97. Irie F, Jwa Hidari KIP, Tai T, Li Y-T, Seyama Y, Hirabayashi Y (1994) Biosynthetic pathway for a new series of gangliosides, GT1aα and GQ1bα. FEBS Lett 351(2):291–294. https://doi.org/10.1016/0014-5793(94)00883-3
98. Proia RL (2003) Glycosphingolipid functions: insights from engineered mouse models. Philos Trans R Soc Lond B Biol Sci 358(1433):879–883
99. Kolter T, Proia RL, Sandhoff K (2002) Combinatorial ganglioside biosynthesis. J Biol Chem 277(29):25859–25862
100. Sandhoff R (2010) Very long chain sphingolipids: tissue expression, function and synthesis. FEBS Lett 584(9):1907–1913. https://doi.org/10.1016/j.febslet.2009.12.032
101. Bieberich E, MacKinnon S, Silva J, Li DD, Tencomnao T, Irwin L, Kapitonov D, Yu RK (2002) Regulation of ganglioside biosynthesis by enzyme complex formation of glycosyltransferases. Biochemistry 41(38):11479–11487
102. Maccioni HJ, Quiroga R, Spessott W (2011) Organization of the synthesis of glycolipid oligosaccharides in the Golgi complex. FEBS Lett 585(11):1691–1698. https://doi.org/10.1016/j.febslet.2011.03.030
103. Wennekes T, van den Berg RJ, Boot RG, van der Marel GA, Overkleeft HS, Aerts JM (2009) Glycosphingolipids—nature, function, and pharmacological modulation. Angew Chem Int Ed Engl 48(47):8848–8869. https://doi.org/10.1002/anie.200902620
104. Merrill AH Jr, Wang E, Gilchrist DG, Riley RT (1993) Fumonisins and other inhibitors of de novo sphingolipid biosynthesis. Adv Lipid Res 26:215–234
105. Kolter T, Sandhoff K (1996) Inhibitors of glycosphingolipid biosynthesis. Chem Soc Rev 25:371–381
106. Giraudo CG, Daniotti JL, Maccioni HJ (2001) Physical and functional association of glycolipid N-acetyl-galactosaminyl and galactosyl transferases in the Golgi apparatus. Proc Natl Acad Sci U S A 98(4):1625–1630. https://doi.org/10.1073/pnas.031458398
107. Giraudo CG, Maccioni HJ (2003) Ganglioside glycosyltransferases organize in distinct multienzyme complexes in CHO-K1 cells. J Biol Chem 278(41):40262–40271. https://doi.org/10.1074/jbc.M305455200
108. Uliana AS, Crespo PM, Martina JA, Daniotti JL, Maccioni HJ (2006) Modulation of GalT1 and SialT1 sub-Golgi localization by SialT2 expression reveals an organellar level of glycolipid synthesis control. J Biol Chem 281(43):32852–32860. https://doi.org/10.1074/jbc.M605805200
109. Uliana AS, Giraudo CG, Maccioni HJ (2006) Cytoplasmic tails of SialT2 and GalNAcT impose their respective proximal and distal Golgi localization. Traffic 7(5):604–612. https://doi.org/10.1111/j.1600-0854.2006.00413.x
110. Scheel G, Acevedo E, Conzelmann E, Nehrkorn H, Sandhoff K (1982) Model for the interaction of membrane-bound substrates and enzymes. Hydrolysis of ganglioside GD1a by sialidase of neuronal membranes isolated from calf brain. Eur J Biochem 127(2):245–253
111. Yamashita T, Wada R, Sasaki T, Deng C, Bierfreund U, Sandhoff K, Proia RL (1999) A vital role for glycosphingolipid synthesis during development and differentiation. Proc Natl Acad Sci U S A 96(16):9142–9147
112. Coetzee T, Fujita N, Dupree J, Shi R, Blight A, Suzuki K, Suzuki K, Popko B (1996) Myelination in the absence of galactocerebroside and sulfatide: normal structure with abnormal function and regional instability. Cell 86(2):209–219
113. Bosio A, Binczek E, Stoffel W (1996) Functional breakdown of the lipid bilayer of the myelin membrane in central and peripheral nervous system by disrupted galactocerebroside synthesis. Proc Natl Acad Sci U S A 93(23):13280–13285
114. Takamiya K, Yamamoto A, Furukawa K, Yamashiro S, Shin M, Okada M, Fukumoto S, Haraguchi M, Takeda N, Fujimura K, Sakae M, Kishikawa M, Shiku H, Furukawa K, Aizawa S (1996) Mice with disrupted GM2/GD2 synthase gene lack complex gangliosides but exhibit only subtle defects in their nervous system. Proc Natl Acad Sci U S A 93(20):10662–10667
115. Sheikh KA, Sun J, Liu Y, Kawai H, Crawford TA, Proia RL, Griffin JW, Schnaar RL (1999) Mice lacking complex gangliosides develop Wallerian degeneration and myelination defects. Proc Natl Acad Sci U S A 96(13):7532–7537
116. Takamiya K, Yamamoto A, Furukawa K, Zhao J, Fukumoto S, Yamashiro S, Okada M,

Haraguchi M, Shin M, Kishikawa M, Shiku H, Aizawa S, Furukawa K (1998) Complex gangliosides are essential in spermatogenesis of mice: possible roles in the transport of testosterone. Proc Natl Acad Sci U S A 95(21):12147–12152

117. Kawai H, Allende ML, Wada R, Kono M, Sango K, Deng C, Miyakawa T, Crawley JN, Werth N, Bierfreund U, Sandhoff K, Proia RL (2001) Mice expressing only monosialoganglioside GM3 exhibit lethal audiogenic seizures. J Biol Chem 276(10):6885–6888

118. Chiavegatto S, Sun J, Nelson RJ, Schnaar RL (2000) A functional role for complex gangliosides: motor deficits in GM2/GD2 synthase knockout mice. Exp Neurol 166(2):227–234. https://doi.org/10.1006/exnr.2000.7504

119. Yamashita T, Hashiramoto A, Haluzik M, Mizukami H, Beck S, Norton A, Kono M, Tsuji S, Daniotti JL, Werth N, Sandhoff R, Sandhoff K, Proia RL (2003) Enhanced insulin sensitivity in mice lacking ganglioside GM3. Proc Natl Acad Sci U S A 100(6):3445–3449

120. Tagami S, Inokuchi Ji J, Kabayama K, Yoshimura H, Kitamura F, Uemura S, Ogawa C, Ishii A, Saito M, Ohtsuka Y, Sakaue S, Igarashi Y (2002) Ganglioside GM3 participates in the pathological conditions of insulin resistance. J Biol Chem 277(5):3085–3092. https://doi.org/10.1074/jbc.M103705200

121. Yoon SJ, Nakayama K, Hikita T, Handa K, Hakomori SI (2006) Epidermal growth factor receptor tyrosine kinase is modulated by GM3 interaction with N-linked GlcNAc termini of the receptor. Proc Natl Acad Sci U S A 103(50):18987–18991. https://doi.org/10.1073/pnas.0609281103

122. Nordström V, Willershauser M, Herzer S, Rozman J, von Bohlen und Halbach O, Meldner S, Rothermel U, Kaden S, Roth FC, Waldeck C, Gretz N, de Angelis MH, Draguhn A, Klingenspor M, Gröne HJ, Jennemann R (2013) Neuronal expression of glucosylceramide synthase in central nervous system regulates body weight and energy homeostasis. PLoS Biol 11(3):e1001506. https://doi.org/10.1371/journal.pbio.1001506

123. Proia RL (2004) Gangliosides help stabilize the brain. Nat Genet 36(11):1147–1148. https://doi.org/10.1038/ng1104-1147

124. Simpson MA, Cross H, Proukakis C, Priestman DA, Neville DC, Reinkensmeier G, Wang H, Wiznitzer M, Gurtz K, Verganelaki A, Pryde A, Patton MA, Dwek RA, Butters TD, Platt FM, Crosby AH (2004) Infantile-onset symptomatic epilepsy syndrome caused by a homozygous loss-of-function mutation of GM3 synthase. Nat Genet 36(11):1225–1229

125. Boccuto L, Aoki K, Flanagan-Steet H, Chen CF, Fan X, Bartel F, Petukh M, Pittman A, Saul R, Chaubey A, Alexov E, Tiemeyer M, Steet R, Schwartz CE (2014) A mutation in a ganglioside biosynthetic enzyme, ST3GAL5, results in salt & pepper syndrome, a neurocutaneous disorder with altered glycolipid and glycoprotein glycosylation. Hum Mol Genet 23(2):418–433. https://doi.org/10.1093/hmg/ddt434

126. Harlalka GV, Lehman A, Chioza B, Baple EL, Maroofian R, Cross H, Sreekantan-Nair A, Priestman DA, Al-Turki S, McEntagart ME, Proukakis C, Royle L, Kozak RP, Bastaki L, Patton M, Wagner K, Coblentz R, Price J, Mezei M, Schlade-Bartusiak K, Platt FM, Hurles ME, Crosby AH (2013) Mutations in B4GALNT1 (GM2 synthase) underlie a new disorder of ganglioside biosynthesis. Brain 136(Pt 12):3618–3624. https://doi.org/10.1093/brain/awt270

127. Boukhris A, Schule R, Loureiro JL, Lourenco CM, Mundwiller E, Gonzalez MA, Charles P, Gauthier J, Rekik I, Acosta Lebrigio RF, Gaussen M, Speziani F, Ferbert A, Feki I, Caballero-Oteyza A, Dionne-Laporte A, Amri M, Noreau A, Forlani S, Cruz VT, Mochel F, Coutinho P, Dion P, Mhiri C, Schols L, Pouget J, Darios F, Rouleau GA, Marques W Jr, Brice A, Durr A, Zuchner S, Stevanin G (2013) Alteration of ganglioside biosynthesis responsible for complex hereditary spastic paraplegia. Am J Hum Genet 93(1):118–123. https://doi.org/10.1016/j.ajhg.2013.05.006

128. Schwarzmann G, Hoffmann-Bleihauer P, Schubert J, Sandhoff K, Marsh D (1983) Incorporation of ganglioside analogues into fibroblast cell membranes. A spin-label study. Biochemistry 22(21):5041–5048

129. Sonderfeld S, Conzelmann E, Schwarzmann G, Burg J, Hinrichs U, Sandhoff K (1985) Incorporation and metabolism of ganglioside GM2 in skin fibroblasts from normal and GM2 gangliosidosis subjects. Eur J Biochem 149(2):247–255

130. Lucas M, Gershlick DC, Vidaurrazaga A, Rojas AL, Bonifacino JS, Hierro A (2016) Structural mechanism for cargo recognition by the retromer complex. Cell 167(6):1623–1635. e1614. https://doi.org/10.1016/j.cell.2016.10.056

131. Ghidoni R, Trinchera M, Venerando B, Fiorilli A, Sonnino S, Tettamanti G (1986) Incorporation and metabolism of exogenous GM1 ganglioside in rat liver. Biochem J 237(1):147–155

132. Ghidoni R, Trinchera M, Sonnino S, Chigorno V, Tettamanti G (1987) The sialic

acid residue of exogenous GM1 ganglioside is recycled for biosynthesis of sialoglycoconjugates in rat liver. Biochem J 247(1):157–164
133. Chinnapen DJ, Hsieh WT, te Welscher YM, Saslowsky DE, Kaoutzani L, Brandsma E, D'Auria L, Park H, Wagner JS, Drake KR, Kang M, Benjamin T, Ullman MD, Costello CE, Kenworthy AK, Baumgart T, Massol RH, Lencer WI (2012) Lipid sorting by ceramide structure from plasma membrane to ER for the cholera toxin receptor ganglioside GM1. Dev Cell 23(3):573–586. https://doi.org/10.1016/j.devcel.2012.08.002
134. Saslowsky DE, te Welscher YM, Chinnapen DJ, Wagner JS, Wan J, Kern E, Lencer WI (2013) Ganglioside GM1-mediated transcytosis of cholera toxin bypasses the retrograde pathway and depends on the structure of the ceramide domain. J Biol Chem 288(36):25804–25809. https://doi.org/10.1074/jbc.M113.474957
135. Kok JW, Eskelinen S, Hoekstra K, Hoekstra D (1989) Salvage of glucosylceramide by recycling after internalization along the pathway of receptor-mediated endocytosis. Proc Natl Acad Sci U S A 86(24):9896–9900
136. Schwarzmann G, Hofmann P, Pütz U, Albrecht B (1995) Demonstration of direct glycosylation of nondegradable glucosylceramide analogs in cultured cells. J Biol Chem 270(36):21271–21276
137. Gillard BK, Harrell RG, Marcus DM (1996) Pathways of glycosphingolipid biosynthesis in SW13 cells in the presence and absence of vimentin intermediate filaments. Glycobiology 6(1):33–42
138. Tettamanti G (2004) Ganglioside/glycosphingolipid turnover: new concepts. Glycoconj J 20(5):301–317
139. Tettamanti G, Bassi R, Viani P, Riboni L (2003) Salvage pathways in glycosphingolipid metabolism. Biochimie 85(3–4):423–437
140. Wiegandt H (1985) Gangliosides. In: Wiegandt H (ed) Glycolipids. Elsevier Science Publishers B.V., New York, pp 199–247
141. Krengel U, Bousquet PA (2014) Molecular recognition of gangliosides and their potential for cancer immunotherapies. Front Immunol 5:325. https://doi.org/10.3389/fimmu.2014.00325
142. Naito-Matsui Y, Davies LRL, Takematsu H, Chou H-H, Tangvoranuntakul P, Carlin AF, Verhagen A, Heyser CJ, Yoo S-W, Choudhury B, Paton JC, Paton AW, Varki NM, Schnaar RL, Varki A (2017) Physiological exploration of the long term evolutionary selection against expression of N-glycolylneuraminic acid in the brain. J Biol Chem 292(7):2557–2570. https://doi.org/10.1074/jbc.M116.768531
143. Mlinac K, Fabris D, Vukelic Z, Rozman M, Heffer M, Bognar SK (2013) Structural analysis of brain ganglioside acetylation patterns in mice with altered ganglioside biosynthesis. Carbohydr Res 382:1–8. https://doi.org/10.1016/j.carres.2013.09.007
144. Manzi AE, Sjoberg ER, Diaz S, Varki A (1990) Biosynthesis and turnover of O-acetyl and N-acetyl groups in the gangliosides of human melanoma cells. J Biol Chem 265(22):13091–13103
145. Kohla G, Stockfleth E, Schauer R (2002) Gangliosides with O-acetylated sialic acids in tumors of neuroectodermal origin. Neurochem Res 27(7–8):583–592
146. Watanabe K, Powell M, Hakomori S (1978) Isolation and characterization of a novel fucoganglioside of human erythrocyte membranes. J Biol Chem 253(24):8962–8967
147. Tokuda N, Zhang Q, Yoshida S, Kusunoki S, Urano T, Furukawa K, Furukawa K (2006) Genetic mechanisms for the synthesis of fucosyl GM1 in small cell lung cancer cell lines. Glycobiology 16(10):916–925. https://doi.org/10.1093/glycob/cwl022
148. Hansson HA, Holmgren J, Svennerholm L (1977) Ultrastructural localization of cell membrane GM1 ganglioside by cholera toxin. Proc Natl Acad Sci U S A 74(9):3782–3786
149. Sonnino S, Prinetti A (2016) The role of sphingolipids in neuronal plasticity of the brain. J Neurochem 137(4):485–488. https://doi.org/10.1111/jnc.13589
150. Gulbins E, Walter S, Becker KA, Halmer R, Liu Y, Reichel M, Edwards MJ, Muller CP, Fassbender K, Kornhuber J (2015) A central role for the acid sphingomyelinase/ceramide system in neurogenesis and major depression. J Neurochem 134(2):183–192. https://doi.org/10.1111/jnc.13145
151. Sha S, Zhou L, Yin J, Takamiya K, Furukawa K, Furukawa K, Sokabe M, Chen L (2014) Deficits in cognitive function and hippocampal plasticity in GM2/GD2 synthase knockout mice. Hippocampus 24(4):369–382
152. Ikarashi K, Fujiwara H, Yamazaki Y, Goto J, Kaneko K, Kato H, Fujii S, Sasaki H, Fukumoto S, Furukawa K, Waki H, Furukawa K (2011) Impaired hippocampal long-term potentiation and failure of learning in beta1,4-N-acetylgalactosaminyltransferase gene transgenic mice. Glycobiology 21(10):1373–1381. https://doi.org/10.1093/glycob/cwr090
153. Svennerholm L, Gottfries CG (1994) Membrane lipids, selectively diminished in Alzheimer brains, suggest synapse loss as a

primary event in early-onset form (type I) and demyelination in late-onset form (type II). J Neurochem 62(3):1039–1047

154. Shiozaki K, Takahashi K, Hosono M, Yamaguchi K, Hata K, Shiozaki M, Bassi R, Prinetti A, Sonnino S, Nitta K, Miyagi T (2015) Phosphatidic acid-mediated activation and translocation to the cell surface of sialidase NEU3, promoting signaling for cell migration. FASEB J 29(5):2099–2111. https://doi.org/10.1096/fj.14-262543
155. Shiga K, Takahashi K, Sato I, Kato K, Saijo S, Moriya S, Hosono M, Miyagi T (2015) Upregulation of sialidase NEU3 in head and neck squamous cell carcinoma associated with lymph node metastasis. Cancer science 106(11):1544–1553. https://doi.org/10.1111/cas.12810
156. Takahashi K, Hosono M, Sato I, Hata K, Wada T, Yamaguchi K, Nitta K, Shima H, Miyagi T (2015) Sialidase NEU3 contributes neoplastic potential on colon cancer cells as a key modulator of gangliosides by regulating Wnt signaling. Int J Cancer 137(7):1560–1573. https://doi.org/10.1002/ijc.29527
157. Kopitz J, von Reitzenstein C, Sinz K, Cantz M (1996) Selective ganglioside desialylation in the plasma membrane of human neuroblastoma cells. Glycobiology 6(3):367–376
158. Papini N, Anastasia L, Tringali C, Croci G, Bresciani R, Yamaguchi K, Miyagi T, Preti A, Prinetti A, Prioni S, Sonnino S, Tettamanti G, Venerando B, Monti E (2004) The plasma membrane-associated sialidase MmNEU3 modifies the ganglioside pattern of adjacent cells supporting its involvement in cell-to-cell interactions. J Biol Chem 279(17):16989–16995. https://doi.org/10.1074/jbc.M400881200
159. Sonnino S, Chigorno V, Aureli M, Masilamani AP, Valsecchi M, Loberto N, Prioni S, Mauri L, Prinetti A (2011) Role of gangliosides and plasma membrane-associated sialidase in the process of cell membrane organization. Adv Exp Med Biol 705:297–316. https://doi.org/10.1007/978-1-4419-7877-6_14
160. Sandhoff K, Pallmann B (1978) Membrane-bound neuraminidase from calf brain: regulation of oligosialoganglioside degradation by membrane fluidity and membrane components. Proc Natl Acad Sci U S A 75(1):122–126
161. Pellkofer R, Sandhoff K (1980) Halothane increases membrane fluidity and stimulates sphingomyelin degradation by membrane-bound neutral sphingomyelinase of synaptosomal plasma membranes from calf brain already at clinical concentrations. J Neurochem 34(4):988–992
162. Anheuser S, Breiden B, Schwarzmann G, Sandhoff K (2015) Membrane lipids regulate ganglioside GM2 catabolism and GM2 activator protein activity. J Lipid Res 56(9):1747–1761. https://doi.org/10.1194/jlr.M061036
163. Bierfreund U, Lemm T, Hoffmann A, Uhlhorn-Dierks G, Childs RA, Yuen CT, Feizi T, Sandhoff K (1999) Recombinant GM2-activator protein stimulates in vivo degradation of GA2 in GM2 gangliosidosis AB variant fibroblasts but exhibits no detectable binding of GA2 in an in vitro assay. Neurochem Res 24(2):295–300
164. Riboni L, Sonnino S, Acquotti D, Malesci A, Ghidoni R, Egge H, Mingrino S, Tettamanti G (1986) Natural occurrence of ganglioside lactones. Isolation and characterization of GD1b inner ester from adult human brain. J Biol Chem 261(18):8514–8519
165. Ledeen R, Wu G (2011) New findings on nuclear gangliosides: overview on metabolism and function. J Neurochem 116(5):714–720. https://doi.org/10.1111/j.1471-4159.2010.07115.x
166. Dyatlovitskaya EV, Bergelson LD (1987) Glycosphingolipids and antitumor immunity. Biochim Biophys Acta 907(2):125–143
167. Fredman P (1994) Gangliosides associated with primary brain tumors and their expression in cell lines established from these tumors. Prog Brain Res 101:225–240
168. Ledeen RW, Wu G (2008) Nuclear sphingolipids: metabolism and signaling. J Lipid Res 49(6):1176–1186. https://doi.org/10.1194/jlr.R800009-JLR200
169. Henning R, Stoffel W (1973) Glycosphingolipids in lysosomal membranes. Hoppe Seylers Z Physiol Chem 354(7):760–770
170. Fürst W, Sandhoff K (1992) Activator proteins and topology of lysosomal sphingolipid catabolism. Biochim Biophys Acta 1126(1):1–16
171. Möbius W, Herzog V, Sandhoff K, Schwarzmann G (1999) Gangliosides are transported from the plasma membrane to intralysosomal membranes as revealed by immuno-electron microscopy. Biosci Rep 19(4):307–316
172. Burkhardt JK, Hüttler S, Klein A, Möbius W, Habermann A, Griffiths G, Sandhoff K (1997) Accumulation of sphingolipids in SAP-precursor (prosaposin)-deficient fibroblasts occurs as intralysosomal membrane structures and can be completely reversed by treatment with human SAP-precursor. Eur J Cell Biol 73(1):10–18

173. Kolter T, Sandhoff K (2005) Principles of lysosomal membrane digestion: stimulation of sphingolipid degradation by sphingolipid activator proteins and anionic lysosomal lipids. Annu Rev Cell Dev Biol 21:81–103
174. Wollert T, Hurley JH (2010) Molecular mechanism of multivesicular body biogenesis by ESCRT complexes. Nature 464(7290): 864–869
175. Florey O, Overholtzer M (2012) Autophagy proteins in macroendocytic engulfment. Trends Cell Biol 22(7):374–380
176. Gallala HD, Breiden B, Sandhoff K (2011) Regulation of the NPC2 protein-mediated cholesterol trafficking by membrane lipids. J Neurochem 116(5):702–707
177. Quintern LE, Weitz G, Nehrkorn H, Tager JM, Schram AW, Sandhoff K (1987) Acid sphingomyelinase from human urine: purification and characterization. Biochim Biophys Acta 922(3):323–336
178. Abdul-Hammed M, Breiden B, Adebayo MA, Babalola JO, Schwarzmann G, Sandhoff K (2010) Role of endosomal membrane lipids and NPC2 in cholesterol transfer and membrane fusion. J Lipid Res 51(7):1747–1760. https://doi.org/10.1194/jlr.M003822
179. Gallala H, Sandhoff K (2011) Biological function of the cellular lipid BMP—BMP as a key activator for cholesterol sorting and membrane digestion. Neurochem Res 36(9):1594–1600
180. Kolter T, Winau F, Schaible UE, Leippe M, Sandhoff K (2005) Lipid-binding proteins in membrane digestion, antigen presentation, and antimicrobial defense. J Biol Chem 280(50):41125–41128
181. Appelqvist H, Sandin L, Bjornstrom K, Saftig P, Garner B, Ollinger K, Kagedal K (2012) Sensitivity to lysosome-dependent cell death is directly regulated by lysosomal cholesterol content. PLoS ONE 7(11):e50262. https://doi.org/10.1371/journal.pone.0050262
182. Eskelinen EL, Tanaka Y, Saftig P (2003) At the acidic edge: emerging functions for lysosomal membrane proteins. Trends Cell Biol 13(3):137–145
183. Kolter T, Sandhoff K (2010) Lysosomal degradation of membrane lipids. FEBS Lett 584(9):1700–1712
184. Schuette CG, Pierstorff B, Huettler S, Sandhoff K (2001) Sphingolipid activator proteins: proteins with complex functions in lipid degradation and skin biogenesis. Glycobiology 11(6):81R–90R
185. Kolter T, Sandhoff K (2006) Sphingolipid metabolism diseases. Biochim Biophys Acta 1758(12):2057–2079
186. Schulze H, Sandhoff K (2014) Sphingolipids and lysosomal pathologies. Biochim Biophys Acta 1841(5):799–810. https://doi.org/10.1016/j.bbalip.2013.10.015
187. Sandhoff K, Harzer K (2013) Gangliosides and gangliosidoses: principles of molecular and metabolic pathogenesis. J Neurosci 33(25):10195–10208. https://doi.org/10.1523/JNEUROSCI.0822-13.2013
188. Doering T, Holleran WM, Potratz A, Vielhaber G, Elias PM, Suzuki K, Sandhoff K (1999) Sphingolipid activator proteins are required for epidermal permeability barrier formation. J Biol Chem 274(16):11038–11045
189. Hulkova H, Cervenkova M, Ledvinova J, Tochackova M, Hrebicek M, Poupetova H, Befekadu A, Berna L, Paton BC, Harzer K, Boor A, Smid F, Elleder M (2001) A novel mutation in the coding region of the prosaposin gene leads to a complete deficiency of prosaposin and saposins, and is associated with a complex sphingolipidosis dominated by lactosylceramide accumulation. Hum Mol Genet 10(9):927–940
190. Monti E, Bonten E, D'Azzo A, Bresciani R, Venerando B, Borsani G, Schauer R, Tettamanti G (2010) Sialidases in vertebrates: a family of enzymes tailored for several cell functions. Adv Carbohydr Chem Biochem 64:403–479. https://doi.org/10.1016/S0065-2318(10)64007-3
191. Smutova V, Albohy A, Pan X, Korchagina E, Miyagi T, Bovin N, Cairo CW, Pshezhetsky AV (2014) Structural basis for substrate specificity of mammalian neuraminidases. PLoS ONE 9(9):e106320. https://doi.org/10.1371/journal.pone.0106320
192. Timur ZK, Akyildiz Demir S, Marsching C, Sandhoff R, Seyrantepe V (2015) Neuraminidase-1 contributes significantly to the degradation of neuronal B-series gangliosides but not to the bypass of the catabolic block in Tay-Sachs mouse models. Mol Genet Metab Rep 4:72–82. https://doi.org/10.1016/j.ymgmr.2015.07.004
193. Bonten E, van der Spoel A, Fornerod M, Grosveld G, d'Azzo A (1996) Characterization of human lysosomal neuraminidase defines the molecular basis of the metabolic storage disorder sialidosis. Genes Dev 10(24):3156–3169
194. d'Azzo A, Bonten E (2010) Molecular mechanisms of pathogenesis in a glycosphingolipid and a glycoprotein storage disease. Biochem Soc Trans 38(6):1453–1457. https://doi.org/10.1042/BST0381453
195. Möbius W, Herzog V, Sandhoff K, Schwarzmann G (1999) Intracellular distribution of a biotin-labeled ganglioside, GM1,

by immunoelectron microscopy after endocytosis in fibroblasts. J Histochem Cytochem 47(8):1005–1014
196. Albrecht B, Pohlentz G, Sandhoff K, Schwarzmann G (1997) Synthesis and mass spectrometric characterization of digoxigenin and biotin labeled ganglioside GM1 and their uptake by and metabolism in cultured cells. Chem Phys Lipids 86(1):37–50
197. Bradova V, Smid F, Ulrich-Bott B, Roggendorf W, Paton BC, Harzer K (1993) Prosaposin deficiency: further characterization of the sphingolipid activator protein-deficient sibs. Multiple glycolipid elevations (including lactosylceramidosis), partial enzyme deficiencies and ultrastructure of the skin in this generalized sphingolipid storage disease. Hum Genet 92(2):143–152
198. Harzer K, Paton BC, Poulos A, Kustermann-Kuhn B, Roggendorf W, Grisar T, Popp M (1989) Sphingolipid activator protein deficiency in a 16-week-old atypical Gaucher disease patient and his fetal sibling: biochemical signs of combined sphingolipidoses. Eur J Pediatr 149(1):31–39
199. Schnabel D, Schröder M, Fürst W, Klein A, Hurwitz R, Zenk T, Weber J, Harzer K, Paton BC, Poulos A et al (1992) Simultaneous deficiency of sphingolipid activator proteins 1 and 2 is caused by a mutation in the initiation codon of their common gene. J Biol Chem 267(5):3312–3315
200. Möbius W, van Donselaar E, Ohno-Iwashita Y, Shimada Y, Heijnen HF, Slot JW, Geuze HJ (2003) Recycling compartments and the internal vesicles of multivesicular bodies harbor most of the cholesterol found in the endocytic pathway. Traffic 4(4):222–231
201. Kobayashi T, Beuchat MH, Lindsay M, Frias S, Palmiter RD, Sakuraba H, Parton RG, Gruenberg J (1999) Late endosomal membranes rich in lysobisphosphatidic acid regulate cholesterol transport. Nat Cell Biol 1(2):113–118
202. Oninla VO, Breiden B, Babalola JO, Sandhoff K (2014) Acid sphingomyelinase activity is regulated by membrane lipids and facilitates cholesterol transfer by NPC2. J Lipid Res 55(12):2606–2619. https://doi.org/10.1194/jlr.M054528
203. Kölzer M, Werth N, Sandhoff K (2004) Interactions of acid sphingomyelinase and lipid bilayers in the presence of the tricyclic antidepressant desipramine. FEBS Lett 559(1–3):96–98
204. Hurwitz R, Ferlinz K, Sandhoff K (1994) The tricyclic antidepressant desipramine causes proteolytic degradation of lysosomal sphingomyelinase in human fibroblasts. Biol Chem Hoppe Seyler 375(7):447–450
205. Elojeimy S, Holman DH, Liu X, El-Zawahry A, Villani M, Cheng JC, Mahdy A, Zeidan Y, Bielwaska A, Hannun YA, Norris JS (2006) New insights on the use of desipramine as an inhibitor for acid ceramidase. FEBS Lett 580(19):4751–4756
206. Lüllmann H, Lüllmann-Rauch R, Wassermann O (1978) Lipidosis induced by amphiphilic cationic drugs. Biochem Pharmacol 27(8):1103–1108
207. Suzuki K, Chen GC (1968) GM1-gangliosidosis (generalized gangliosidosis). Morphology and chemical pathology. Pathol Eur 3(2):389–408
208. Fyrst H, Saba JD (2010) An update on sphingosine-1-phosphate and other sphingolipid mediators. Nat Chem Biol 6(7):489–497
209. Blom T, Li Z, Bittman R, Somerharju P, Ikonen E (2012) Tracking sphingosine metabolism and transport in sphingolipidoses: NPC1 deficiency as a test case. Traffic 13(9):1234–1243. https://doi.org/10.1111/j.1600-0854.2012.01379.x
210. van Meer G, Voelker DR, Feigenson GW (2008) Membrane lipids: where they are and how they behave. Nat Rev Mol Cell Biol 9(2):112–124
211. Mouritsen OG, Zuckermann MJ (2004) What's so special about cholesterol? Lipids 39(11):1101–1113
212. Vanier MT (2015) Complex lipid trafficking in Niemann-Pick disease type C. J Inherit Metab Dis 38(1):187–199. https://doi.org/10.1007/s10545-014-9794-4
213. Vanier MT, Millat G (2003) Niemann-Pick disease type C. Clin Genet 64(4):269–281
214. Abdul-Hammed M, Breiden B, Schwarzmann G, Sandhoff K (2017) Lipids regulate the hydrolysis of membrane bound glucosylceramide by lysosomal β-glucocerebrosidase. J Lipid Res 58:563–577
215. Sandhoff K (2013) Metabolic and cellular bases of sphingolipidoses. Biochem Soc Trans 41(6):1562–1568. https://doi.org/10.1042/BST20130083
216. Ballabio A, Gieselmann V (2009) Lysosomal disorders: from storage to cellular damage. Biochim Biophys Acta 1793(4):684–696. https://doi.org/10.1016/j.bbamcr.2008.12.001
217. Mizukami H, Mi Y, Wada R, Kono M, Yamashita T, Liu Y, Werth N, Sandhoff R, Sandhoff K, Proia RL (2002) Systemic inflammation in glucocerebrosidase-deficient mice

with minimal glucosylceramide storage. J Clin Invest 109(9):1215–1221

218. Jeyakumar M, Thomas R, Elliot-Smith E, Smith DA, van der Spoel AC, d'Azzo A, Perry VH, Butters TD, Dwek RA, Platt FM (2003) Central nervous system inflammation is a hallmark of pathogenesis in mouse models of GM1 and GM2 gangliosidosis. Brain 126(Pt 4):974–987
219. Conzelmann E, Sandhoff K (1983) Partial enzyme deficiencies: residual activities and the development of neurological disorders. Dev Neurosci 6(1):58–71
220. Leinekugel P, Michel S, Conzelmann E, Sandhoff K (1992) Quantitative correlation between the residual activity of beta-hexosaminidase A and arylsulfatase A and the severity of the resulting lysosomal storage disease. Hum Genet 88(5):513–523
221. Graber D, Salvayre R, Levade T (1994) Accurate differentiation of neuronopathic and nonneuronopathic forms of Niemann-Pick disease by evaluation of the effective residual lysosomal sphingomyelinase activity in intact cells. J Neurochem 63(3):1060–1068
222. Meivar-Levy I, Horowitz M, Futerman AH (1994) Analysis of glucocerebrosidase activity using N-(1-[14C]hexanoyl)-D-erythroglucosylsphingosine demonstrates a correlation between levels of residual enzyme activity and the type of Gaucher disease. Biochem J 303(Pt 2):377–382
223. Sandhoff K (2016) Neuronal sphingolipidoses: membrane lipids and sphingolipid activator proteins regulate lysosomal sphingolipid catabolism. Biochimie 130:146–151. https://doi.org/10.1016/j.biochi.2016.05.004
224. Vanier M (1983) Biochemical studies in Niemann-Pick disease: I. Major sphingolipids of liver and spleen. Biochim Biophys Acta 750:178–184
225. Gondre-Lewis MC, McGlynn R, Walkley SU (2003) Cholesterol accumulation in NPC1-deficient neurons is ganglioside dependent. Curr Biol 13(15):1324–1329
226. Cheruku SR, Xu Z, Dutia R, Lobel P, Storch J (2006) Mechanism of cholesterol transfer from the Niemann-Pick type C2 protein to model membranes supports a role in lysosomal cholesterol transport. J Biol Chem 281(42):31594–31604. https://doi.org/10.1074/jbc.M602765200
227. Naureckiene S, Sleat DE, Lackland H, Fensom A, Vanier MT, Wattiaux R, Jadot M, Lobel P (2000) Identification of HE1 as the second gene of Niemann-Pick C disease. Science 290(5500):2298–2301
228. Kwon HJ, Abi-Mosleh L, Wang ML, Deisenhofer J, Goldstein JL, Brown MS, Infante RE (2009) Structure of N-terminal domain of NPC1 reveals distinct subdomains for binding and transfer of cholesterol. Cell 137(7):1213–1224
229. Locatelli-Hoops S, Remmel N, Klingenstein R, Breiden B, Rossocha M, Schoeniger M, Koenigs C, Saenger W, Sandhoff K (2006) Saposin A mobilizes lipids from low cholesterol and high bis(monoacylglycerol) phosphate-containing membranes: patient variant saposin A lacks lipid extraction capacity. J Biol Chem 281(43):32451–32460
230. Remmel N, Locatelli-Hoops S, Breiden B, Schwarzmann G, Sandhoff K (2007) Saposin B mobilizes lipids from cholesterol-poor and bis(monoacylglycero)phosphate-rich membranes at acidic pH. Unglycosylated patient variant saposin B lacks lipid-extraction capacity. FEBS J 274(13):3405–3420
231. Bowman EA, Walterfang M, Abel L, Desmond P, Fahey M, Velakoulis D (2015) Longitudinal changes in cerebellar and subcortical volumes in adult-onset Niemann-Pick disease type C patients treated with miglustat. J Neurol 262(9):2106–2114. https://doi.org/10.1007/s00415-015-7819-z
232. Patterson MC, Mengel E, Vanier MT, Schwierin B, Muller A, Cornelisse P, Pineda M, investigators NPCR, Amado-Fondo A, Amraoui Y, Andria G, Arellano M, Audoin B, Azcona C, Barr C, Baruteau J, Baumgartner C, Bell L, Bembi B, Benneddif K, Bernard G, Bobocea N, Bodzioch M, Boettcher T, Bonnan M, Broue P, Bruni A, Caceres M, Camino R, Campbell E, Cances C, Cannell P, Cesar J, Chabrol B, Chakrapani A, Colao R, Collet A, Corsetti T, Cousins A, Covanis A, Cox T, Cuisset JM, Dardis A, Das A, Deegan P, Dengler T, Deodato F, Derralynn H, Di Donato I, Di Rocco M, Dinopoulos A, DomanskaPakiela ES, Engelen M, Eyer D, Fecarotta S, Federico A, Filla A, Fiumara A, Fonseca MJ, Gabrielli O, Garcia T, Garrote J, Gissen P, Giugliani L, Greenberg C, Heron B, Hertzberg C, Higgins F, Hill A, Hiwot T, Hlavata A, Hörbe-Blindt A, Howley E, Hussain N, Illsinger S, Imrie J, Jacklin E, Jones S, Jovanovic A, Kaczmarek V, Kaphan E, Kibaek M, Kleinhans P, Klünemann KH, Koch SM, Koegl-Wallner W, Kolnikova M, Korenke GC, Korinthenberg R, Kumari S, Lachmann R, Lee Ann L, Likopoulou L, Lilienthal E, Link B, Lippold M, Lopez-Laso E, Luecke T, Lundgren J, Mackrell M, Madruga M, Maletta R, Malinova V, Manners J, Marinei R, Marquardt T, Martins E, Martins AM, Martins N, McAlister L,

McCabe A, McKie M, McMahon S, Meehan M, Meldgaard Lund A, Mendozah C, Mengel E, Meyer A, Mielke S, Milligan A, Mir P, Moisa M, Mombelli C, Morris L, Müller vom Hagen J, Munoz B, Murphy E, Nagarajan L, Neto PB, Nevsimalova S, Nia S, Nicolai J, Niemann D, Niktari G, O'Callaghan MDM, Paucar-Arce M, Peers K, Peintinger L, Peralta M, Pérez J, Perez-Poyato M, Petrariu A, Pineda PA, Raiman J, Rask O, Rataj J, Raymond-Gaynor C, Reichelt G, Ribeiro E, Riches V, Roberts A, Roelants J, Rohrbach M, Rokicki D, Rolfs A, Russo C, Rutsch F, Saleem R, Santos M, Schmutz P, Schwahn B, Sedel F, Semotok J, Sharma R, Silska S, Silva A, Simmons L, Sivera R, Skorpen J, Sole G, Souza C, Stadlober-Degwerth M, Starling J, Temudo T, Tomas M, Tranchant C, Uziel G, Valayannopoulous V, Van den Hout H, Van der Tol L, Van Spronsen F, Vellodi A, Verdu A, Vilchez JJ, Vinaixa A, Visser G, Voelker J, Waldek S, Walter A, Walterfang M, Wein U, Widner H, Wilcke C, Wildish L, Wraith E, Wright N, Xaidara A, Yamamoto M, Zallocco F, Zielke S (2015) Stable or improved neurological manifestations during miglustat therapy in patients from the international disease registry for Niemann-Pick disease type C: an observational cohort study. Orphanet J Rare Dis 10:65

233. Schwarzmann G, Breiden B, Sandhoff K (2015) Membrane-spanning lipids for an uncompromised monitoring of membrane fusion and intermembrane lipid transfer. J Lipid Res 56(10):1861–1879. https://doi.org/10.1194/jlr.M056929

234. Vaccaro AM, Tatti M, Ciaffoni F, Salvioli R, Serafino A, Barca A (1994) Saposin-C induces pH-dependent destabilization and fusion of phosphatidylserine-containing vesicles. FEBS Lett 349(2):181–186

235. Conzelmann E, Burg J, Stephan G, Sandhoff K (1982) Complexing of glycolipids and their transfer between membranes by the activator protein for degradation of lysosomal ganglioside GM2. Eur J Biochem 123(2):455–464

236. Wilkening G, Linke T, Sandhoff K (1998) Lysosomal degradation on vesicular membrane surfaces. Enhanced glucosylceramide degradation by lysosomal anionic lipids and activators. J Biol Chem 273(46):30271–30278

237. Sarmientos F, Schwarzmann G, Sandhoff K (1986) Specificity of human glucosylceramide beta-glucosidase towards synthetic glucosylsphingolipids inserted into liposomes. Kinetic studies in a detergent-free assay system. Eur J Biochem 160(3):527–535

238. Murugesan V, Chuang WL, Liu J, Lischuk A, Kacena K, Lin H, Pastores GM, Yang R, Keutzer J, Zhang K, Mistry PK (2016) Glucosylsphingosine is a key biomarker of Gaucher disease. Am J Hematol 91(11):1082–1089. https://doi.org/10.1002/ajh.24491

239. Linke T, Wilkening G, Lansmann S, Moczall H, Bartelsen O, Weisgerber J, Sandhoff K (2001) Stimulation of acid sphingomyelinase activity by lysosomal lipids and sphingolipid activator proteins. Biol Chem 382(2):283–290

240. Linke T, Wilkening G, Sadeghlar F, Mozcall H, Bernardo K, Schuchman E, Sandhoff K (2001) Interfacial regulation of acid ceramidase activity. Stimulation of ceramide degradation by lysosomal lipids and sphingolipid activator proteins. J Biol Chem 276(8):5760–5768

241. Graf CG, Schulz C, Schmälzlein M, Heinlein C, Mönnich M, Perkams L, Püttner M, Boos I, Hessefort M, Lombana Sanchez JN, Weyand M, Steegborn C, Breiden B, Ross K, Schwarzmann G, Sandhoff K, Unverzagt C (2017) Synthetic glycoforms reveal carbohydrate-dependent bioactivity of human saposin D. Angew Chem Int Ed Engl 56:5252–5257. https://doi.org/10.1002/anie.201701362

242. Breiden B, Sandhoff K (2014) The role of sphingolipid metabolism in cutaneous permeability barrier formation. Biochim Biophys Acta 1841(3):441–452. https://doi.org/10.1016/j.bbalip.2013.08.010

243. Henseler M, Klein A, Glombitza GJ, Suzuki K, Sandhoff K (1996) Expression of the three alternative forms of the sphingolipid activator protein precursor in baby hamster kidney cells and functional assays in a cell culture system. J Biol Chem 271(14):8416–8423

244. Tamargo RJ, Velayati A, Goldin E, Sidransky E (2012) The role of saposin C in Gaucher disease. Mol Genet Metab 106(3):257–263

245. Geiger B, Arnon R, Sandhoff K (1977) Immunochemical and biochemical investigation of hexosaminidase S. Am J Hum Genet 29(5):508–522

246. Svennerholm L (1962) The chemical structure of normal human brain and Tay-Sachs gangliosides. Biochem Biophys Res Commun 9:436–441

247. van Echten G, Sandhoff K (1989) Modulation of ganglioside biosynthesis in primary cultured neurons. J Neurochem 52(1):207–214

248. Wilkening G, Linke T, Uhlhorn-Dierks G, Sandhoff K (2000) Degradation of membrane-bound ganglioside GM1. Stimulation by bis(monoacylglycero)phosphate and the activator proteins SAP-B and GM2-AP. J Biol Chem 275(46):35814–35819

Chapter 6

Gangliosides and Tumors

Reiji Kannagi, Bi-He Cai, Hsiang-Chi Huang, Chia-Chun Chao, and Keiichiro Sakuma

Abstract

Tumor-associated gangliosides play important roles in regulation of signal transduction induced by growth-factor receptors including EGFR, FGFR, HGF and PDGFR in a specific microdomain called glycosynapse in the cancer cell membranes, and in interaction with glycan recognition molecules involved in cell adhesion and immune regulation including selectins and siglecs. As the genes involved in the synthesis and degradation of tumor-associated gangliosides were identified, biological functions became clearer from the experimental results employing forced overexpression and/or knockdown/knockout of the genes. Studies on the regulatory mechanisms for their expression also achieved great advancements. Epigenetic silencing of glycan-related genes is a dominant mechanism in glycan alteration at early stages of carcinogenesis. Development of hypoxia resistance involving activation of a transcription factor HIF, and acquisition of cancer stem cell-like characteristics through epithelial–mesenchymal transition are important mechanisms for glycan modulations in the later stages of cancer progression. In the initial stages of studies, the gangliosides which specifically appear in cancers attracted attention under the name of tumor-associated gangliosides. However, it became apparent that not only the cancer-associated gangliosides but also the normal gangliosides present in nonmalignant cells and tissues perform important biological functions, and some of them tend to disappear in cancer cells resulting in the loss of the physiological functions, and this sometimes facilitates progression of cancers.

Key words Cancer stem cells, De novo synthesis, DNA hypermethylation, Embryonic stem cells, Epigenetic silencing, Epithelial–mesenchymal transition (EMT), Glycolipid enriched microdomain (GEM), Glycosynapse, Hematogenous metastasis, Histone modification, Hypoxia inducible factor (HIF), Incomplete synthesis, MicroRNA, Scavenger pathway

Abbreviations

EMT	Epithelial–mesenchymal transition
GEM	Glycolipid enriched microdomain
HIF	Hypoxia inducible factor
iPS cells	Inducible pluripotent stem cells
SSEA	Stage-specific embryonic antigen

Sandro Sonnino and Alessandro Prinetti (eds.), *Gangliosides: Methods and Protocols*, Methods in Molecular Biology, vol. 1804, https://doi.org/10.1007/978-1-4939-8552-4_6, © Springer Science+Business Media, LLC, part of Springer Nature 2018

1 Introduction: Gangliosides as Tumor Markers

Gangliosides are the glycolipids having neuraminic acid residue(s) in their structure. It has long been known that gangliosides and other cell surface glycans undergo drastic changes upon malignant transformation. Tumor-associated gangliosides have been investigated for their potential as diagnostic and therapeutic targets, including GM3, GM2, GD3, GD2, GD1α, SSEA-4, disialyl Gb5 (DSGG), sialyl Lewis X, sialyl Lewis A, and gangliosides containing N-glycolylneuraminic acid (NeuGc). The most recent therapeutic trials include GD2 as a target for the treatment of pediatric neuroblastoma, and GM3 containing NeuGc for the treatment of NSCLC [1–3]. This chapter will introduce the cancer-associated gangliosides, their possible biological and pathological significance, and genes involved in their expression.

2 GM3 and Related Ganglio-Series Glycolipids in Regulation of Growth Factor Signals

2.1 GM3

GM3 is a ganglio-series glycolipid having a simple structure (Table 1), and is a major component of gangliosides in nonmalignant cells. GM3 is known to be reduced upon malignant transformation by oncogenic viruses for a long time [4]. Soon GM3 was found to be enriched in a specific membrane microdomain called glycolipid-enriched microdomain (GEM) or glycosynapse, and to colocalize with growth factor receptors, integrins, Src family kinases, and some tetraspanins such as CD9, CD81, and CD82. GM3 is proposed to have an inhibitory effect on activation of growth factor receptors including EGFR, FGFR, and PDGFR in such membrane microdomains [5, 6]. Suppressive effects of GM3 are strictly dependent on the structure of its sialic acid moiety; for instance, de-*N*-acetyl GM3 does not inhibit but enhance EGFR phosphorylation [7, 8], and mono-chloro-GM3 has a stronger inhibitory effect than GM3 on EGFR activation [9].

The expression level of GM3 is primarily regulated by its synthetic enzyme encoded by the ST3GAL5 gene, and the mRNA levels of ST3GAL5 and GM3 expression are known to correlate with sensitivity of lung cancer cells toward epidermal growth factor receptor (EGFR)-tyrosine kinase inhibitors, gefitinib and AG1478 [10]. Valproic acid, known as an upregulator of ST3GAL5, reportedly increased the level of GM3 and reduced EGFR phosphorylation, leading to inhibition of cancer cell proliferation [11]. Overexpression of ST3GAL5 in ovarian cancer cells is shown to exhibit elevated ganglioside levels, reduced in vitro cell motility, and enhanced expression of the membrane adaptor protein caveo-

Table 1
Examples of tumor-associated gangliosides and glycan-related genes involved in their expression

Gangliosides	Key genes involved	Structure	Antibodies[a]
GM3	*ST3GAL5* *NEU3*	NeuAcα2 → 3Galβ1 → 4Glcβ-Cer	M2590, GMR6
GM2 (Gg_4)	*B4GALNT1*	GalNAcβ1 → 4(NeuAcα2 → 3)Galβ1 → 4Glcβ-Cer	MK1–8, MK1–16
Sd[a] (CAD)	*B4GALNT2*	GalNAcβ1 → 4(NeuAcα2 → 3) Galβ1 → 4GlcNAcβ1 → 3Galβ1 → 4Glcβ-Cer	KM694
GD3	*ST8SIA1, ST3GAL5*	NeuAcα2 → 8NeuAcα2 → 3Galβ1 → 4Glcβ-Cer	R24, GMR19
GD2	*B4GALNT1, ST8SIA1, ST3GAL5*	GalNAcβ1 → 4(NeuAcα2 → 8NeuAcα2 → 3)Galβ1 → 4Glcβ-Cer	14.G2a, GMR7
GD1α	*ST6GALNAC6*	NeuAcα2 → 3GalNAcβ1 → 3(NeuAcα2 → 6) GalNAcβ1 → 4Galβ1 → 4Glcβ-Cer	mAb 122
N-glycolyl GM3	*SLC17A5*	NeuGcα2 → 3Galβ1 → 4Glcβ-Cer	14F7
N-glycolyl GM2	*SLC17A5*	GalNAcβ1 → 4(NeuGcα2 → 3)Galβ1 → 4Glcβ-Cer	MK2–34, YHD-07
SSEA-4 (mono-sialyl Gb5, MSGG)	*ST3GAL2 B3GALT5*	NeuAcα2 → 3Galβ1 → 3GalNAcβ1 → 3Galα1 → 4Galβ1 → 4Glcβ-Cer	MC813–70, B2, RM1, PK8–3
DSGG (disialyl Gb5)	*ST6GALNAC6, ST3GAL2, B3GALT5*	NeuAcα2 → 3Galβ1 → 3(NeuAcα2 → 6) GalNAcβ1 → 3Galα1 → 4Galβ1 → 4Glcβ-Cer	5F3
Sialyl Lewis X	*ST3GAL3/4/6, FUT6/3/7/5*	NeuAcα2 → 3Galβ1 → 4(Fucα1 → 3)GlcNAcβ1 → 3Galβ1 → 4Glcβ-Cer	CSLEX-1, 2H5
Sialyl 6-sulfo Lewis X	*SLC26A2*	NeuAcα2 → 3Galβ1 → 4(Fucα1 → 3) GlcNAcβ1(6-SO_3) → 3Galβ1 → 4Glcβ-Cer	G152, G72
Sialyl Lewis A	*ST3GAL3/4, FUT3/5*	NeuAcα2 → 3Galβ1 → 3(Fucα1 → 4)GlcNAcβ1 → 3Galβ1 → 4Glcβ-Cer	N19–9, 1H4, 2D3
Disialyl Lewis A	*ST6GALNAC6*	NeuAcα2 → 3Galβ1 → 3(Fucα1 → 4)(NeuAcα2 → 6) GlcNAcβ1 → 3Galβ1 → 4Glcβ-Cer	FH7

[a]Further information will be found at the "GlycoEpitope" website: http://www.glycoepitope.jp/

lin-1. In turn, silencing of caveolin-1 by siRNA markedly increases in vitro cell motility and regulates c-Src kinase activity [12–14].

Functional significance of ST3GAL5 seems to depend on the steady-state levels of GM3 in different cellular backgrounds. ST3GAL5 is known to be upregulated in patients with chemoresistant breast cancer [15], rectal cancers [16], and also in breast cancer stem cells induced by epithelial–mesenchymal transition (EMT) [17], and is regarded as a pro-oncogenic gene. In line with this, EMT-related transcription factor Zeb1 is reported to promote ST3GAL5 transcription [18]. Runx (Runx1, 2, and 3) are also known to target ST3GAL5 [19]. ST3GAL5 is also reported to be the direct target of microRNAs including miR-26a, miR-548l and miR-34a in hepatocellular carcinoma cells [20].

Not only the synthetic enzyme but also the degrading enzyme, especially NEU3, is known to regulate intracellular local concentration of GM3. NEU3 is a plasma membrane-associated sialidase, and specifically hydrolyzes gangliosides such as GM3, GD3, and GD1a at neutral pH, while GM2 and GM1a are not good substrates. NEU3 is suggested to colocalize with phospholipase D, and its product, phosphatidic acid, enhances translocation of NEU3 to the cell surface [21]. Transcription of NEU3 is known to be increased in various cancers including colon, renal, ovarian, and prostate cancers [22, 23]. Knockdown of NEU3 in cancer cell lines downregulates EGFR signaling and induces apoptosis, while its overexpression enhances EGFR activation [24, 25]. Neu3-deficient mice were less susceptible than wild-type mice to the colitis-associated colon carcinogenesis induced by azoxymethane and dextran sodium sulfate [26]. The NEU3 gene is known to be regulated by transcription factors Sp1 and Sp3 [27], but the mechanisms for its upregulation in cancers are not fully clear. It is interesting to note that transcription of NEU3 is shown to be induced upon hypoxia, and is proposed to have a physiological role in protecting cells from hypoxia-induced apoptosis [28].

2.2 GM2

An increase of GM2 (Table 1) was detected in various human cancers for a long time, and mRNA for the gene encoding its synthetic enzyme B4GALNT1 is also known to be significantly increased in cancers [29]. Several studies have demonstrated the immunomodulatory roles of GM2 in mediating host immune suppression in various tumors [30, 31]. GM2-mediated tumor cell migration was suggested to be dependent on β1-integrin, and to involve modulation of ERK–MAPK pathway [32]. TALEN-mediated GM2-synthase knockout clones of murine renal cancer cells showed reduced anchorage-independent growth, reduction in tumor growth, and higher cellular adhesion as compared to wild type cells [33]. In cells expressing tetraspanin, CD82, GM2 and/or GM2/GM3 complex inhibits integrin-dependent Met kinase activity and cell motility by binding to CD82 within glycosynaptic microdomain [34]. Cancer stem cells induced through EMT in breast cancer cells were found to exhibit an increased amount of GM2, and concomitant fivefold increase in B4GALNT1 mRNA, which well explains the increase in GM2 [17]. Knockdown of B4GALNT1 in a human breast cancer cell line reduced cell motility and mammosphere formation [17].

The Sd^{a}/CAD antigen is another sialylated glycan having the GalNAcβ1–4 terminus similar to GM2, and found in glycoproteins as well as glycolipids. The GalNAcβ1–4 terminus in Sd^{a} glycan is carried by either α2,3-sialylated type 1 (Galβ1,3GlcNAc), type 2 (Galβ1,4GlcNAc) or core 1 O-linked (Galβ1,3GalNAc) glycans, while the GalNAcβ1–4 terminus is carried by the GM3 structure in GM2. The Sd^{a} glycan is synthesized by the enzyme B4GALNT2, having a different acceptor specificity from that of B4GALNT1 [35]. At present it is not entirely clear which enzyme is responsible for the synthesis of GalNAc-GM1b and GalNAc-GD1a, which is sometimes detected in cancer cells together with GM2 [36]. The B4GALNT2 gene and Sd^{a} glycans are expressed preferentially in nonmalignant epithelial cells, but tend to decrease in cancer cells, leaving the acceptor substrates for synthesis of cancer-associated glycans such as sialyl Lewis X and sialyl Lewis A [37, 38]. Reduced expression of Sd^{a} glycans and B4GALNT2 transcription in cancers was shown to be due to DNA hypermethylation of the gene which occurred upon malignant transformation [37].

GM2 has been regarded as a potent target molecule for cancer therapy. Humanized anti-GM2 antibody has been developed [39, 40]. On the other hand, it is reported that the trial for vaccination of adjuvant ganglioside GM2-KLH/QS-21 did not improve patient outcome in the treatment of stage II melanoma [41].

2.3 GD3 and GD2

GD3 and GD2 (Table 1) are known to be highly expressed in melanomas, neuroblastomas, gliomas, medulloblastomas, and small cell lung cancer cells. These gangliosides are localized in microdomains such as lipid rafts or detergent-resistant microdomains, and enhance cell motility and invasiveness by stimulating focal adhesion kinase and Lyn kinase, which phosphorylate paxillin and p130Cas [42]. Increased expression of ST8SIA1, the gene encoding synthetic enzyme for GD3, is frequently noted not only in neuroectodermal tumors but also in a wide range of cancers. ST8SIA1 is markedly upregulated in breast cancer stem cells [17, 43] and in glioblastoma stem cells [44]. Its increase is known to correlate with patient prognosis in breast cancers [45]. ST8SIA1 is known to constitutively activate also c-Met [46, 47].

GD2 is synthesized from GD3 by B4GALNT1, which is the same enzyme responsible for GM2 synthesis, and is known to be increased in cancer stem cells [17]. Consequently, expression of GD2 and GD3 is markedly enhanced in breast cancer stem cells [17, 43]. The majority of GD2+ cells and GD3+ cells were detected in the CD44^{hi}/CD24^{lo} cancer stem cell-like population in breast cancer cells; and knockdown of ST8SIA1 and B4GALNT1 significantly reduced the expression of GD2 and GD3 and caused a phenotype change from cancer stem cells to noncancer stem cells, which was detected by reduced mammosphere formation and cell motility [17].

Inflammatory transcription factor NFκB (nuclear factor κB) was known to be involved in the transcriptional induction of ST8SIA1 [48], Expression of the ST8SIA1 gene was found to be induced by the transcription factor FOXC2 (forkhead box protein C2), a key downstream effector of several EMT pathways, which links NFκB signaling to activation of ST8SIA1 and generates ST8SIA1-positive cancer stem cells [49]. On the other hand, ST8SIA1 is reported to be frequently DNA methylated in colon cancers [50].

The results of experiments investigating the function of GD3/GD2 or GM3/GM2 need to be interpreted with caution. In the experiments employing overexpression of ST8SIA1, for instance, although it will increase the expression of GD3, at the same time it will frequently reduce the amount of GM3, making it difficult to determine whether the observed functional alteration is due to an increase in the amount of GD3, or to the reduction in the amount of GM3. Overexpression of ST3GAL5, on the other hand, may increase the amount of GM3, but in the presence of a sufficient level of endogenous ST8SIA1, it may induce the increase of GD3 expression rather than GM3. It is not easy to manipulate the amount of a ganglioside without affecting the amounts of other metabolically closely related ganglioside components.

The US Food and Drug Administration approved dinutuximab, an IgG antibody targeted against GD2, for the treatment of pediatric neuroblastoma [1, 2], based on a phase III clinical open-label randomized trial conducted by the Children's Oncology Group. The trial included 226 patients who had responded at least partially to initial therapies. Among patients who received dinutuximab, 66% of them had survived without disease recurrence, whereas only 46% of patients receiving standard therapy did so [51]. Dinutuximab was approved for use in combination with isotretinoin, IL2, and GM-CSF [1].

2.4 GD1α: An Example of Ganglioside Responsible for Organ-Specific Cancer Metastasis?

The O-series ganglioside, GD1α (Table 1), was detected in various animal cell lines and tissues, and also in a small amount in the normal human brain, while information on its expression in human cancers had been limited. However, the ST6GALNAC5 gene, which encodes the key synthetic enzyme for GD1α, was reported to be exclusively expressed in the brain-metastatic cell line of human breast cancers, and to be upregulated in clinical brain metastasis samples from gene expression analyses [52, 53]. Functional validation experiments demonstrated a role for the

ST6GALNAC5 gene in promoting brain metastasis by facilitating passage of cancer cells through the blood brain barrier [52]. These findings suggested that GD1α can be a specific marker for breast cancer cells that undergo organ-specific metastasis to the brain. This was formulated to be the first and most typical example of organ-specific metastasis mediated by specific cell surface glycan [52–54].

Introduction of the ST6GALNAC5 gene to human breast cancer cells was shown to induce GD1α expression [55]. Contrary to expectation, however, overexpression of ST6GALNAC5 in MDA-MB-231 cells reportedly displayed a reduced capacity for interaction and transmigration in the human in vitro model of blood-brain barrier [56]. A careful examination of the relationship in cultured human cancer cell lines between mRNA levels of ST6GALNAC5 or ST6GALNAC6 and GD1α expression using newly generated antibody from ganglioside synthase knockout mice, a good correlation between expression of GD1α and ST6GALNAC5 was noted, while ST6GALNAC6 may not be involved in the synthesis of α-series gangliosides [57]. Expression of GD1α was rare in human cancer cell lines, in that its significant expression was detected only in two out of 71 human cancer cell lines [57].

On the other hand, ST6GALNAC5 was shown to be markedly decreased in most glioma tissues compared to the normal brain [58]. Its forced overexpression in glioma cells was found to inhibit in vitro invasive activity, modify cellular adhesion to fibronectin and laminin, and inhibit in vivo intracranial growth of the glioma cells [59]. These results suggest that ST6GALNAC5 is expressed in normal tissues, and tends to be silenced during malignant transformation at least in the brain. Whether the amount of the product, GD1α, parallels with alteration of ST6GALNAC5 is not yet clear, as the specific antibody to GD1α was not introduced until recently.

3 Gangliosides Carrying NeuGc and Enhanced Sialic Acid Synthesis in Cancers

Gangliosides containing N-glycolylneuraminic acid (NeuGc) instead of N-acetylneuraminic acid (NeuAc), are widely expressed in most mammals, but not in humans, due to an inactivating mutation in the human cytidine monophospho-N-acetylneuraminic acid hydroxylase gene (CMAH). However, NeuGc is frequently detected in sialoconjugates of various human cancers and embryonic tissues. NeuGc in human embryos and cancers is suggested to be of dietary origin, and is incorporated through the scavenger pathway into human sialococonjugates [60]. NeuGc in cultured human cancer cells is proposed to be derived from fetal calf serum in the cell culture medium. The sialic acid transporter,

SLC17A5 (sialin), which plays an important role in the sialic acid scavenger pathway (Fig. 1), is found to be markedly induced by hypoxia, and this was shown to facilitate incorporation of exogenous NeuGc as well as NeuAc into cellular ganglioside [61]. Overexpression of SLC17A5 was shown to increase the amount of NeuGc-containing gangliosides [61], and constitutive activation of the transcription factor HIF (hypoxia inducible factor) in cancer cells is considered to be the major mechanism for the increase of NeuGc in cancer cells. The fact that cancers and embryonic cells are under hypoxic conditions well explains the increase of NeuGc-containing gangliosides in human cancer and embryonic tissues.

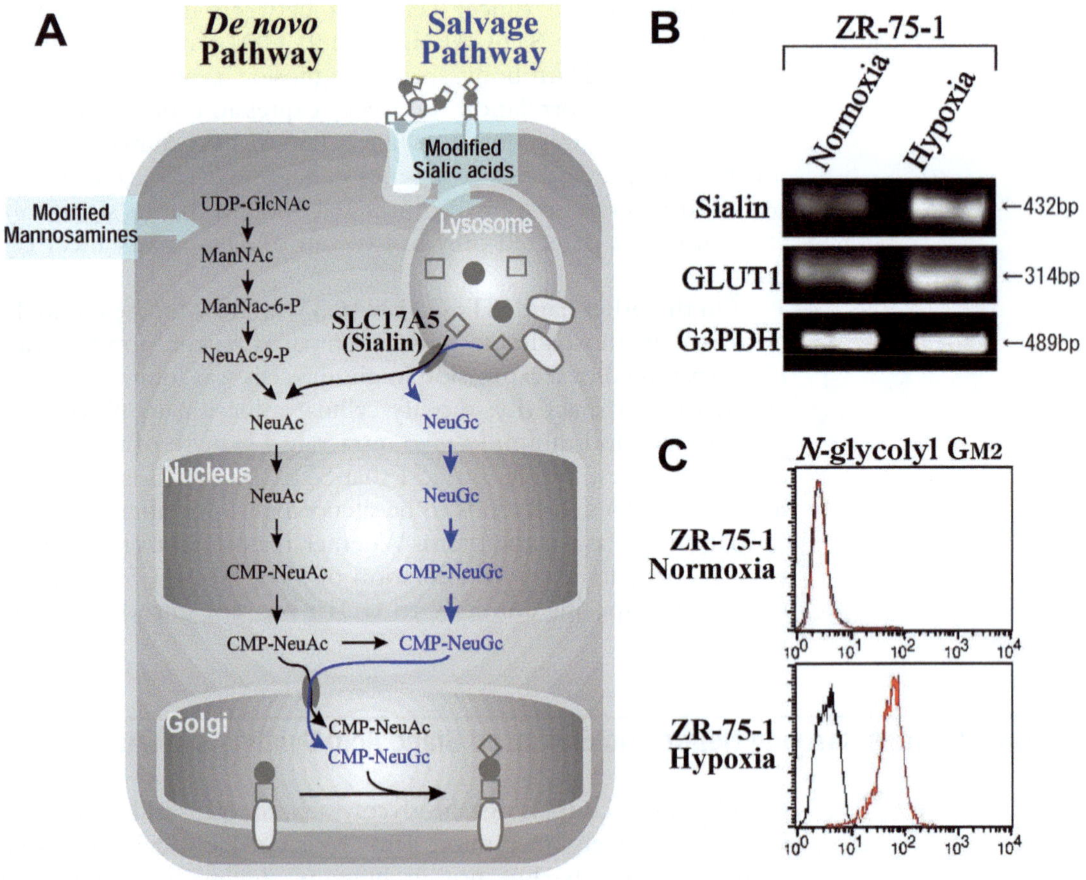

Fig. 1 A scheme showing de novo synthesis pathway and scavenger pathway of sialic acid (**A**). The demand for sialic acid in cancer cells increases under hypoxic conditions at the locally advanced stages, and the scavenger pathway and de novo synthesis pathway of sialic acid are both enhanced in cancers. The scavenger pathway is more preferentially enhanced in cancer cells, and transcription of SLC26A5 (sialin) located in the scavenger pathway is markedly enhanced upon hypoxia (**B**). GLUT1 serves as a positive control which is known to be induced by HIF in panel **B**. This eventually results in the induction and accumulation of gangliosides containing NeuGc, such as N-glycolyl GM2 detected with the MK2–34 antibody, in cancer cells upon hypoxia (**C**). The black lines in panel C indicate staining control. Modified from [61]

To fulfill the increased demand for sialic acid in cancer cells, the scavenger pathway and de novo synthesis pathway of sialic acid are both enhanced in cancers [62], but the enhancement of the scavenger pathway occurs more preferentially in cancer cells. The de novo synthesis pathway is usually employed for vital metabolic labeling of cell surface sialic acids using modified mannosamines [63], but the use of sialic acid mimetics [64] would be more selective in modulating sialic acids in cancer cells.

The enhancement of NeuGc expression in human cancers is so far regarded to be a mere reflection of the increased demand for sialic acid in cancers. The functional roles of gangliosides containing NeuGc in cancer progression, whether glycans containing NeuGc confer any specific growth advantages to cancers compared to those carrying normal sialic acid NeuAc, still remain to be elucidated. Although the exact mechanism is not clear, the inhibitory effect of GM3 on EGFR tyrosine kinase in A431 cells after exposure to NeuGc-GM3 was reported to be lower than that after exposure to GM3 having NeuAc [65].

There may be unknown mechanisms other than the increase of SLC17A5 for the induction of NeuGc in human cancers. The amount of N-glycolyl sialic acid in the cancer cells having the increased levels of NeuGc is usually around a few percent of the total sialic acid, but an ovarian cancer cell line, JHOC-5, was reported to have an extraordinary high level of NeuGc, which reaches around 20% [66]. It is known that endogenous sialidases have preference for NeuAc compared to NeuGc, but it is not clear if this would affect the turnover and accumulation rate of sialoconjugates containing NeuGc. It is noteworthy that gene expression of human CMAH, though mutated and having no enzymatic activity, is known to be significantly upregulated in the human adult stem cell populations of hematopoietic and mesenchymal origin, and CMAH is identified as a novel stem cell marker in these cells [67].

NeuGc-GM3, a ganglioside containing NeuGc (Table 1), has been detected in various human cancers including lung, breast, esophagus, stomach, colon, pancreas, liver cancers, melanoma and neuroblastoma with the aid of specific monoclonal antibody 14F7 [68–72]. Another NeuGc-containing ganglioside, NeuGc-GM2, was also found in a wide variety of human cancers including the cancers in colon, breast and in melanoma, and specific antibodies directed to NeuGc-GM2 were also generated [61, 73, 74].

The gangliosides containing NeuGc are regarded as potential targets in cancer therapy. Co-expression of EGFR and NeuGc-GM3 ganglioside was frequently observed in 63 of 92 patients (68%) with a variety of primary cancers including lung cancers, independent of histological subtype, and therapeutic combinations against these two targets are suggested to be a valid strategy [71, 75, 76]. Racotumomab-alum, an anti-idiotype vaccine targeting

NeuGc-GM3 ganglioside, was submitted to randomized, multicenter clinical trials [77]. After a successful phase II/III study, racotumomab-alum was conditionally approved in Latin American countries as maintenance therapy for advanced non-small cell lung cancer [3]. The Ab3 antibodies induced in the sera in patients after receiving the anti-idiotype vaccine reportedly contain not only anti-NeuGc-GM3 but also anti-NeuGc-GM2 [78–80].

4 SSEA-4, a Globo-Series Ganglioside Found in Human Embryonic Stem Cells

Certain cell surface glycans are known to be good markers for embryonic stem cells. The first example of glycan markers for embryonic stem cells was SSEA-1, which appear in the early stages of murine preimplantation embryos in a stage-specific manner [81, 82]. However, SSEA-1 is not considered to be a good embryonic stem cell marker in humans, because it appears at much later stages in human embryonic development compared to mice. Instead, SSEA-3 and SSEA-4 glycans are regarded as good markers for human embryonic stem cells [83]. SSEA-3 and SSEA-4 glycans appear not only in human embryonic stem cells but also in iPS cells. These glycans were utilized in discriminating the true iPS cell colonies out of many other colonies when human iPS cells were first generated by transfection of OCT3/4, SOX2, KLF4, and c-MYC [84].

SSEA-3/-4 glycans are structurally classified into the globo-series glycolipids (Table 1) [83], which compose a unique series of glycolipids having glycan structures confined to glycolipids, and are not easily detectable in glycoproteins. Nowadays this series of stem cell-associated globoseries glycolipids further contain Globo-H, and disialyl Gb5Cer in addition to SSEA-3/-4, and expression of most of them was shown to be enhanced in embryonic stem cells [85]. Two glycolipids among them, SSEA-4 and disialyl Gb5Cer, contain sialic acid, and hence belong to gangliosides.

Frequent SSEA-4 expression is reported on renal cancer cells [86], testicular germ-cell tumors [87], breast cancer cells [88], basaloid lung cancer [89], epithelial ovarian carcinoma [90], osteosarcoma [91], oral cancer [92], and glioblastoma multiforme [93], in addition to teratocarcinoma [94]. Some of these studies indicate the relationship of SSEA-4 expression with cancer stem cells, and also its implication in EMT [95, 96]. SSEA-4 was found in breast cancer stem cells [88], in a subpopulation of chemotherapy-resistant breast cancer cells and linked to EMT of cancer cells [95]. SSEA-4 is associated with EMT also in prostate cancer cells [96], stem-cell-like cells in embryonal carcinoma of the testis [97] and ovarian cancer stem-cell-like cells [90]. It was also found in small embryonic-like stem cells in ovarian cancer tissues [98] and mesenchymal stem cell-like cells in prostate cancer tissues [99].

The relationship between the level of SSEA-4 expression and tumor malignancy seems to differ in various types of cancers. The expression of SSEA-4 in basaloid lung cancer is associated with a poor prognosis [89]. A trend for higher expression of SSEA-4 associated with a higher grade of astrocytomas was noted in glioblastoma multiforme [93]. On the other hand, there was no correlation between SSEA-4 expression and tumorigenicity in vitro or in vivo in gastric cancers [100]. Expression of SSEA-4 was detected in normal epithelial cells of the ovary, and showed a significant trend to decrease from normal epithelium to benign cystadenoma and to borderline cystadenoma and adenocarcinoma cells, and was lost in 58.7% of epithelial ovarian adenocarcinomas, suggesting that loss or downregulation of SSEA-4 may be involved in the early stage of carcinogenesis of epithelial ovarian cancer [101]. Although SSEA-4 is shown to be expressed in breast cancer stem cells, a recent report demonstrated that the use of SSEA-4 as an isolation marker did not increase the yield of cells capable of forming cell colonies or mammospheres, while the use of SSEA-3 as a marker yielded a population that formed a high percentage of cell colonies and mammospheres from breast cancer cells, suggesting SSEA-3 to be a better cancer stem cell marker than SSEA-4 [102].

SSEA-4 is synthesized through α2,3 sialylation using SSEA-3 as the precursor, and a sialyltransferase encoded by ST3GAL2 is proposed to be the responsible enzyme (Fig. 2) [103]. ST3GAL2 is known to be increased in renal and breast cancers, and its high expression was found to be associated with poor clinical outcome in breast and ovarian cancer patients treated with chemotherapy [95, 103]. ST3GAL2 is also known to be increased in breast cancer stem cells [17]. DNA hypermethylation of ST3GAL2 is reported in healthy individuals [104], while it is markedly hypomethylated in lung cancers [105].

The synthetic precursor SSEA-3 is produced through galactosylation of the substrate Gb4 by B3GALT5 (Fig. 2). Transcription of B3GALT5 is regulated by a retroviral long terminal repeat (LTR) promoter partly driven by HNF1α and HNF1β, and also through DNA methylation and histone modification [106–109]. Knockdown of B3GALT5 caused a reduced proliferation rate and increased apoptosis in breast cancer cells [102]. B3GNT5 is unique in that the encoded enzyme participates not only in SSEA-3/-4 synthesis, but is also involved in the synthetic pathway for another cancer-associated ganglioside, sialyl Lewis A.

The biological function of the SSEA-4 ganglioside is yet unclear. Similar to GM3, the SSEA-4 ganglioside is closely associated with c-Src and FAK in the glycosphingolipid-enriched microdomain (GEM, glycosynapse) in cultured breast cancer cells, and treatment with specific monoclonal antibody greatly enhances their invasiveness and motility [110]. SSEA-4 was reported to bind to FK-506-binding protein 4 (FKBP4), which may be related to

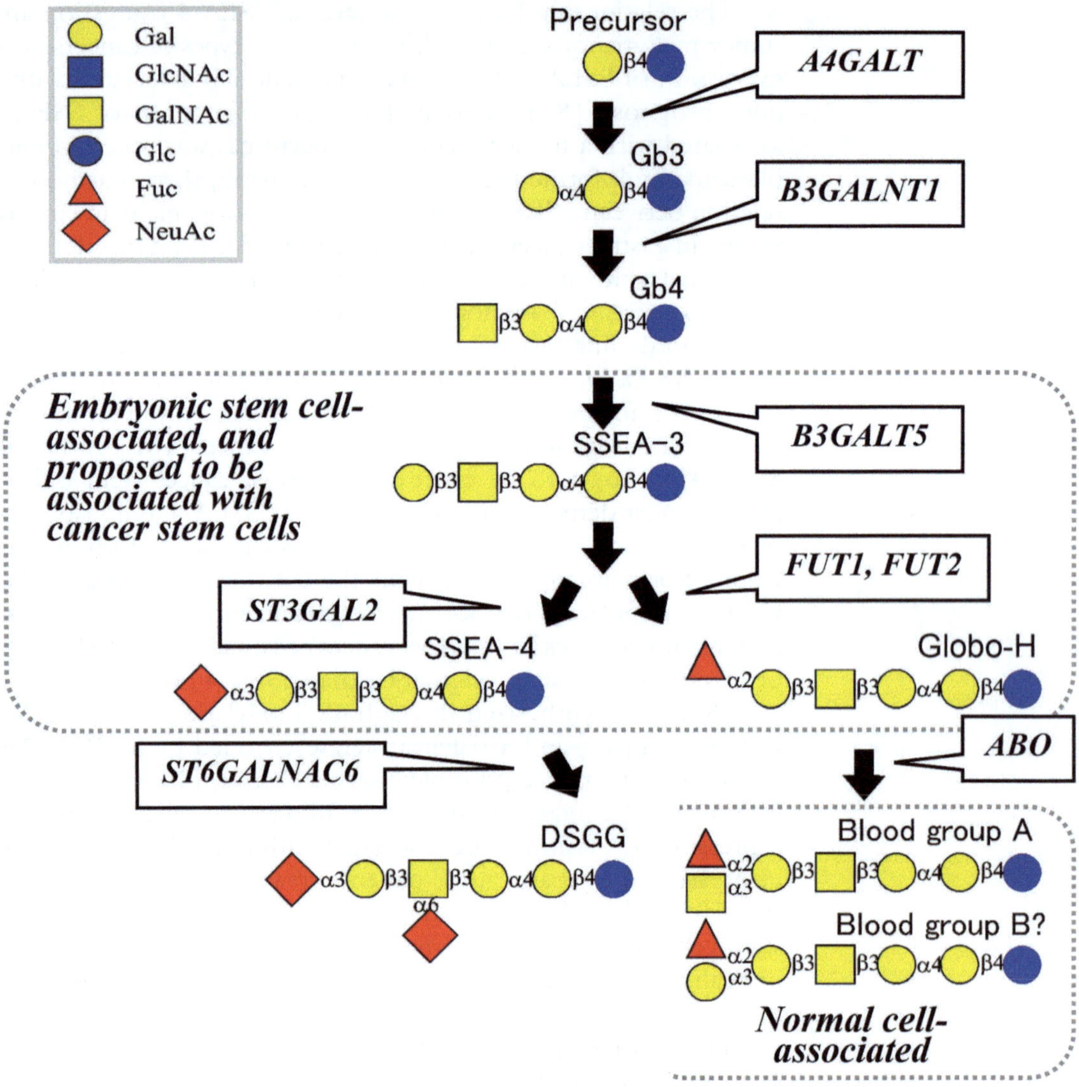

Fig. 2 Synthetic pathway of SSEA-4 ganglioside and related globo-series glycolipids. The genes encoding the glycosyltransferase in each step of synthetic pathway are also shown. Note that the pathway provides not only embryonic/cancer stem cell-associated glycans, but also normal glycans which are not strongly associated with cancers. DSGG is reported to be expressed in renal cancers and correlates with metastatic potential [113], while it is strongly expressed in benign prostate glands and proposed to decrease in the early stage of transformation of prostate cancer from benign glands [114]. Combination of synthetic and degradation enzymes, and expression of the genes encoding them would determine the final product of the pathway in the given tissues

transportation of SSEA-4 to the cell surface and may affect the downstream signaling pathway [111]. ST3GAL2 knockdown in cancer cells reduced SSEA-4 expression, laminin binding, and tumorigenicity [96].

The ganglioside, disialosyl globopentaosylceramide (DSGb5, DSGG), having an extra sialic acid residue linked through α2,6 linkage to penultimate GalNAcβ (Table 1), was first characterized in normal human erythrocytes and chicken skeletal muscle [112]. DSGb5 is found to be expressed in renal cancers and to enhance the migration of renal cell carcinoma cells [113], while in the prostate it is strongly expressed in benign prostate glands, and its expression varied in prostate cancers [114]. Transfer of the second sialic acid in the disialoganglioside is catalyzed by sialyltransferase encoded by ST6GALNAC6 in the kidney, and it was downregulated in renal cancer cell lines and cancer tissues together with the decrease in expression of the disialoganglioside [115]. ST6GALNAC6 is expressed in a variety of nonmalignant tissues and is downregulated in cancers (*see* below in Subheading 5.3). It is not known whether another disialosyl globopentaosylceramide carrying a tandem repeat of sialic acid residues, the NeuAcα2,8NeuAcα2,3-R terminus, are present in cancers, or whether ST8SIA1 can utilize SSEA-4 as a substrate.

A globo-series glycolipid Globo-H was first described as a breast cancer-associated glycolipid. Globo-H is modified at the terminus by fucose through α1–2 linkage, forming O(H)-type structure in the ABO blood group system [116, 117]. Its expression was found in breast cancer stem cells, and is regarded as a suitable candidate target for cancer therapy [88, 118, 119].

Further structural studies on the globoseries glycolipidsindicated that Globo-H is eventually modified further by N-acetylgalactosamine through α 1–3 linkage, forming blood group A-type structure in human erythrocytes (Fig. 2) [120]. Even the glycolipid having the SSEA-1 structure linked to the terminus of SSEA-3 glycolipid was detected in murine kidney [121, 122]. These glycolipids are described in nonmalignant cells and tissues, and their association with cancers is not reported so far.

5 Lacto- and Neolacto-Series Gangliosides and Selectin Family Cell Adhesion Molecules

5.1 Sialyl Lewis X

Sialyl Lewis X (sLex) glycan has long been known to be frequently expressed in human cancers including cancers in the lung, breast, ovary, and digestive organs. This glycan was found to serve as a ligand for E-selectin, a cell adhesion molecule expressed in activated vascular endothelial cells [123–125], and proposed to mediate adhesion of cancer cells to vascular beds during the course of hematogenous metastasis [126]. Expression of sialyl Lewis X on cancer cells is known to correlate with occurrence of hematogenous metastasis or patient prognosis in clinical studies [126]. As this interaction of glycans and glycan-recognition molecule mediates cell adhesion between cancer cells and endothelial cells, it was proposed to also affect angiogenesis in cancer tissues [127].

The gangliosides carrying sialyl Lewis X terminal structure were mainly employed in the earlier studies for elucidation of the glycan structure, distribution among cancers, characterization of antigenic epitope recognized by specific anti-sialyl Lewis X antibodies, and studies on binding to selectins. However, the same terminal structure is carried also by glycoproteins, and nowadays most of the antigenicity and selectin ligand activity at the cell surface are assumed to be exerted by the sialyl Lewis X glycans carried by, rather than glycolipids, cell surface glycoproteins, especially by O-glycans, due to their high quantities and topological factors in the cell membranes.

The ganglioside carrying sialyl Lewis X terminus belongs to neolacto-series glycolipids (Table 1), and is a type 2 chain polylactosamine modified by sialic acid at the terminal galactose through α2,3 linkage and by fucose at the penultimate GlcNAc through α1,3 linkage. The precursor type 2 chain polylactosamine is first modified with sialic acid by α2,3 sialyltransferases, followed by fucosylation by α1,3-fucosyltraseferases. There are several candidates for responsible sialyltransferases, including ST3GAL3 [128], ST3GAL4 [129, 130] or ST3GAL6 [131]. Candidates for fucosyltransferases include FUT6, FUT3, and even FUT7 [132] or FUT5 [133]. Sialyl Lewis X is reported to be degraded by a sialidase, NEU4 [134]. It is not clear if any endogenous fucosidase degrade sialyl Lewis X. Recently, overexpression of FUCA1, the gene encoding the major endogenous fucosidase, was shown to affect selectin binding of human thyroid cancer cells [135], and its knockdown enhanced metastatic activity in breast cancer cells [136]. FUCA1 was found to be a target gene of p53 [137, 138]. Transcription of the genes for candidate α2,3-sialyltransferases is moderately increased in cancers compared to nonmalignant tissues, while that of the genes for candidate fucosyltransferases is variable excluding a few exception [139, 140]. The mechanisms for the cancer-associated expression of sialyl Lewis X were not easy to elucidate, because the correlation of mRNA levels of any genes encoding the candidate α2,3-sialyltransferases or α1,3-fucosyltransferase with the surface expression of sialyl Lewis X was quite variable in cancers.

5.2 Sialyl 6-Sulfo Lewis X

For instance, sialyl 6-sulfo Lewis X (Table 1), which has one extra sulfate residue attached to the C6 position of the penultimate GlcNAc [141], is preferentially expressed in nonmalignant epithelial cells, and its expression is reduced in cancers [142]. This glycan is also a product of the same set of fucosyltransferases and sialyltransferases involved in sialyl Lewis X synthesis. Preferential expression of this glycan in nonmalignant epithelial cells is an evidence against the proposal that induction of some fucosyltransferases or sialyltransferases is the major mechanism for cancer-associated induction of sialyl Lewis X.

Sometime ago we proposed that the loss of 6-sulfation upon malignant transformation could be the major mechanism for induction of sialyl Lewis X expression in cancer cells. Following studies on the expression levels of molecules involved in the glycan 6-sulfation indicated that the gene encoding a sulfate transporter SLC26A2 (DTDST) is markedly downregulated in cancers compared to nonmalignant epithelial cells [143]. Inducible expression of SLC26A2 in cultured cancer cells resulted in induction of sialyl 6-sulfo Lewis X, and reduction of sialyl Lewis X expression as well as suppression of cell proliferation [143]. These findings led us to propose that the reduction of 6-sulfation upon malignant transformation induced accumulation of nonsulfated sialyl Lewis X in cancer cells (Fig. 3).

The pattern of cancer-associated alteration of cell surface glycans had been classified into two major categories in the earlier days; "incomplete synthesis" and "de novo synthesis" [144]. The concept of incomplete synthesis denotes a cancer-associated decrease of activity of an enzyme responsible for a particular step in the glycan synthetic pathway in normal cells and tissues, which results in the loss of mature glycans and specific accumulation of the synthetic precursor for this step in the synthetic cascade, making it appear as a cancer associated glycan. On the other hand, de novo synthesis refers to the synthesis of new glycans which were not present in normal cells and tissues, due to the appearance of an unexpected glycosyltransferase in the given tissues upon malignant transformation. Appearance of sialyl Lewis X had been classified as de novo synthesis, but our results described above indicated that it was in fact the result of incomplete synthesis due to the cancer-associated loss of the 6-sulfation machinery [145, 146].

The disappearance of sialyl 6-sulfo Lewis X and appearance of sialyl Lewis X upon malignant transformation may have biological relevance, because sialyl 6-sulfo Lewis X has been shown to serve as a ligand for siglec-7, while sialyl Lewis X does not [147]. Siglecs are known to be immunosuppressive molecules [148], and this suggests that the normal epithelial cells will lose their glycan ligand for immunosuppressive molecules during the course of carcinogenesis.

The replacement of sialyl 6-sulfo Lewis X with cancer-associated sialyl Lewis X by reduction of SLC26A2 seems to occur through epigenetic silencing. The gene for SLC26A2 exhibits marked histone modification, such as H3K27me3, H4K20me3 and H3K9me3, in cancer cells [143, 146]. Participation of DNA methylation in cancer-associated suppression of this gene was also reported [149, 150].

Although organochemically synthesized gangliosides were used in the initial works proving sialyl 6-sulfo Lewis X to be a specific ligand for L-selectin [141, 151], the gangliosides having such glycan structure had not been described in cells and tissues, and

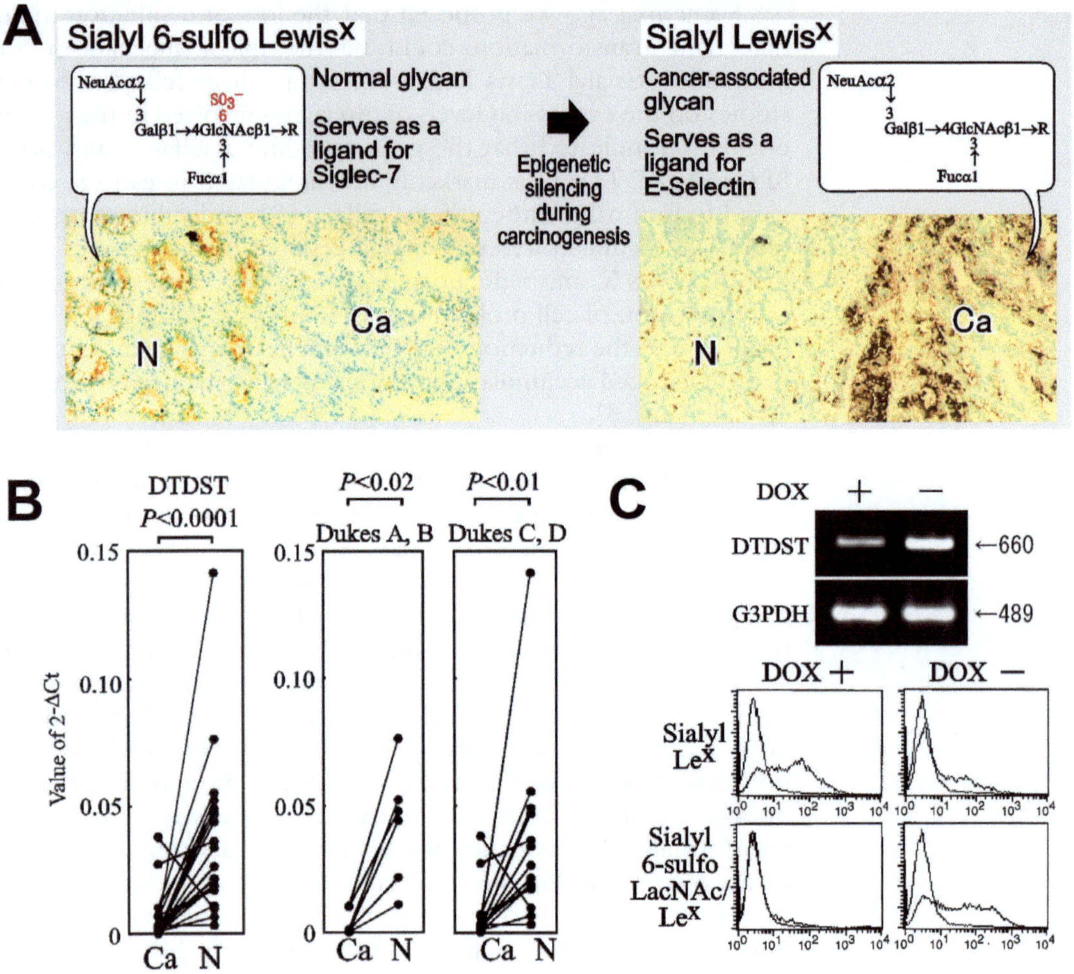

Fig. 3 Sialyl 6-sulfo Lewis X/sialyl Lewis X glycan transition upon malignant transformation. Non-malignant colonic epithelial cells preferentially express sialyl 6-sulfo Lewis X, while colon cancer cells frequently express cancer-associated glycan, sialyl Lewis X, suggesting an impairment of sulfation at C6 position of βGlcNAc by epigenetic silencing occurring during malignant transformation (**A**). R indicates -3Galβ1 → 4Glcβ-Ceramide in the case of gangliosides. Adapted from [142]. The mRNA levels of the gene encoding a sulfate transporter SLC26A2 (DTDST) are reduced at the relatively early stages of cancers (**B**, Ca: Cancer cells; N: Nonmalignant epithelial cells), and the results of studies using Tet-off inducible vector for SLC26A2 (DTDST) indicate its downregulation markedly induce sialyl Lewis X expression (**C**) [143]

most cellular ligands for L-selectin were regarded as glycoproteins carrying sialyl 6-sulfo Lewis X glycan. Quite recently, gangliosides having similar structures were identified in the colon using systematic aminopyridine-labeling of oligosaccharides released by ceramidase treatments [152]. This is a promising method for detection and identification of new cancer-associated ganglioside markers.

Epigenetic silencing or SLC26A2 and disappearance of sialyl 6-sulfo Lewis X can be regarded as relatively early events during the course of carcinogenesis, as it is already evident in sessile serrated adenoma/polyp in the colon [150]. Expression of sialyl Lewis X, thus appeared, seems to be further enhanced during the course of cancer progression. Acquisition of hypoxia-resistant characteristics is important for cancer cells to survive the stages of local tumor through constitutively activating a transcription factor HIF, and expression of sialyl Lewis X is known to be further enhanced by hypoxia from the induction of several genes involved in its synthesis [139, 140]. Generation of cancer stem cell-like cells through EMT is another important change of cancer cells to enhance migration, invasiveness and develop distant metastasis. Expression of sialyl Lewis X is further enhanced in EMT through the activation of c-Myc, while decreasing the activities of competitive synthetic pathways for other glycans by reduction of differentiation-associated transcription factor CDX2 [153].

5.3 Sialyl Lewis A

Sialyl Lewis A (sLea) is the glycan recognized by a monoclonal antibody which was generated against cultured colon cancer cells, and later known to be useful in detecting cancers of the pancreas and bile duct as well as of other digestive organs [154, 155]. This glycan was also found to serve as a ligand for vascular E-selectin similar to sialyl Lewis X [156–158], and proposed to mediate adhesion of cancer cells to endothelial cells in the course of hematogenous metastasis [126, 159]. Expression of sialyl Lewis A on cancer cells of the digestive organs is known to correlate with the risk for hematogenous metastasis and/or patient prognosis in clinical studies [126, 159].

Similar to the case of sialyl Lewis X, the gangliosides carrying sialyl Lewis A terminal structures were employed for elucidation of the glycan structure, distribution among cancers, characterization of antigenic epitope recognized by specific antibodies, and study of binding to selectins in the earlier studies. The same terminal structure, however, is carried also by glycoproteins, and nowadays most of the antigenicity and selectin ligand activity at the cell surface is ascribed to the sialyl Lewis A glycans carried by cell surface glycoproteins, especially by O-glycans, because of their high quantity, density and location at the outmost layers on cell membranes.

The ganglioside carrying sialyl Lewis A terminus belongs to the lacto-series glycolipids (Table 1), and is a type 1 chain polylactosamine modified by sialic acid at the terminal galactose through α2,3 linkage and by fucose at the penultimate GlcNAc through α1,4 linkage. The precursor type 1 chain polylactosamine is first modified with sialic acid by α2,3 sialyltransferases, followed by fucosylation by α1,4 fucosyltraseferases. There are several candidates for responsible sialyltransferases, such as ST3GAL3 [160–162] and less prominently ST3GAL4 [163]. In the case of sialyl Lewis A, ST3GAL6 does not seem to be involved in the synthesis [164]. Candidates for fucosyltransferases include FUT3 and FUT5 [133]. FUT3 is usually assumed to be the enzyme mainly responsible for sialyl Lewis A synthesis, while FUT5 is found to be only sparsely expressed in normal cells and tissues. However, FUT5 was suggested to be mainly responsible for sialyl Lewis x/a synthesis in a cultured human gastric cancer cell line, MKN45, from its mRNA level and *K*m value for acceptor substrates [133]. This finding was later confirmed by using the shRNA approach using the same cell line [165] in which shRNA for FUT5 was shown to affect sialyl Lewis x/a expression and suppress the binding of the gastric cancer cells to E-selectin. Suppression of cell migrating activity was also noted by FUT5 knockdown.

5.4 Disialyl Lewis A

The classical immunohistological findings confirmed cancer-associated expression of sialyl Lewis A, but nonsialylated Lewis A was expressed almost equally in cancer cells and nonmalignant epithelial cells. Moreover, α2,3 α2,6-disialyl Lewis A (Table 1) was found to be preferentially expressed in nonmalignant epithelial cells compared to cancer cells, excluding the cancer-associated induction of α1,4-fucosyltransferases and α2,3-sialyltransferases from the driving force behind the initiation of sialyl Lewis A expression in cancers [166, 167]. The mechanisms for the cancer-associated expression of sialyl Lewis A have long remained unclear.

We have focused on the finding that α2,3 α2,6 disialyl Lewis A (Table 1) is preferentially expressed in nonmalignant epithelial cells in the pancreas and other digestive organs, and tends to disappear in cancer cells [166, 167]. Synthesis of this normal glycan requires an α2,6-sialyltransferase which adds the second sialic acid residue to penultimate GlcNAc through α2,6 linkage, which was identified to be ST6GALNAC6 in the digestive organ including the colon [168]. Expression levels of the ST6GALNAC6 gene were found to be markedly reduced in cancers compared to nonmalignant epithelial cells [169]. Overexpression of ST6GALNAC6 induced strong expression of α2,3 α2,6-disialyl Lewis A in cancer cells with a concomitant decrease in sialyl Lewis A expression [169]. In view of these findings, we proposed that the loss of GlcNAc 6-sialylation upon malignant transformation could be the major mechanism for the induction of sialyl Lewis A in cancer cells, and that this is another example of cancer-associated incomplete synthesis of glycans (Fig. 4).

The disappearance of α2,3 α2,6-disialyl Lewis A and appearance of sialyl Lewis A upon malignant transformation also have biological relevance, because α2,3 α2,6-disialyl Lewis A has been shown to serve as a ligand for siglec-7 and weakly for siglec-9, while sialyl Lewis A has not [147]. Conversely, sialyl Lewis A serves as a ligand for vascular selectins, while α2,3 α2,6-disialyl Lewis A does not bind to selectins.

The normal glycan, α2,3 α2,6-disialyl Lewis A, is again related to immunosuppressive glycan-recognition molecules. Similar to sialyl 6-sulfo Lewis X/sialyl Lewis X transition, here again normal

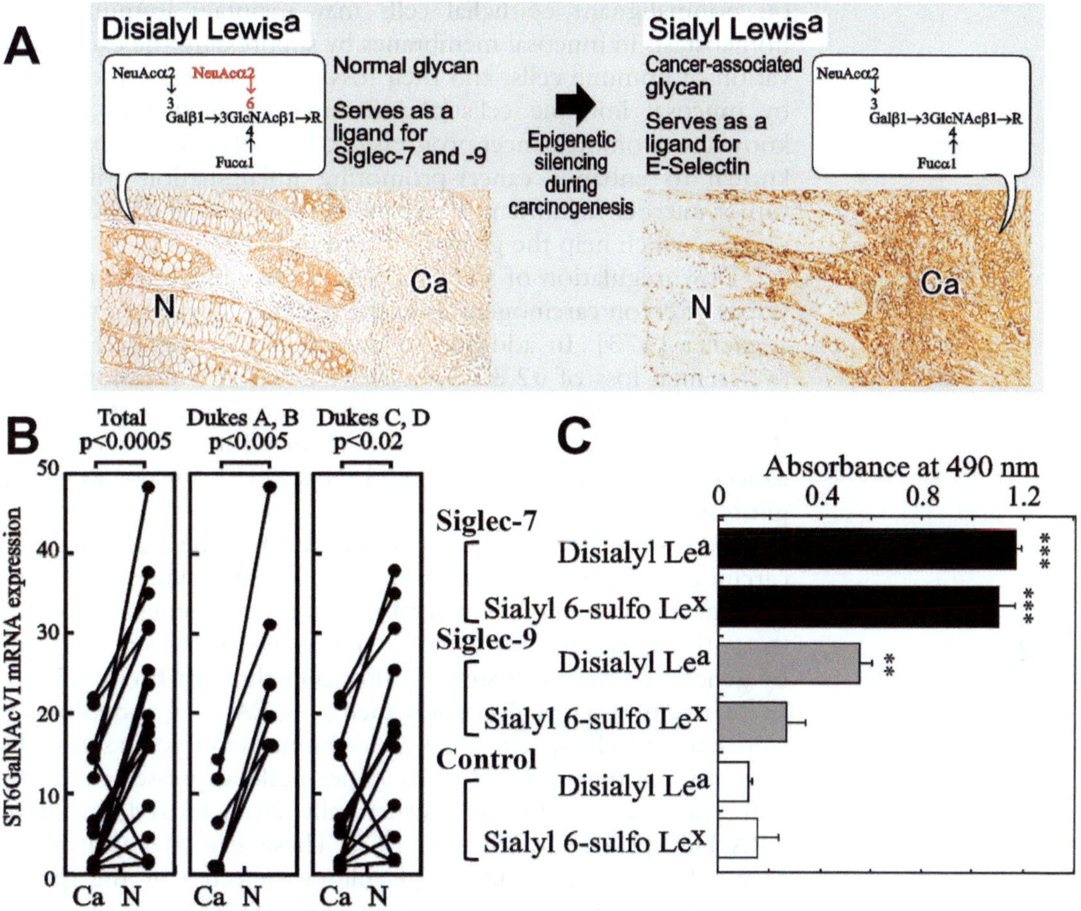

Fig. 4 α2,3 2,6 Disialyl Lewis A/sialyl Lewis A glycan transition upon malignant transformation. Nonmalignant colonic epithelial cells preferentially express disialyl Lewis A, while colon cancer cells frequently express cancer-associated glycan, sialyl Lewis A, suggesting an impairment of sialylation at C6 position of βGlcNAc by epigenetic silencing occurred during malignant transformation (**A**). R indicates -3Galβ1 → 4Glcβ-Ceramide in the case of gangliosides. Adapted from [169]. The mRNA levels of the gene encoding an α2,6 sialyltransferase, ST6GALNAC6, are reduced at the relatively early stages of cancers (**B**, Ca: Cancer cells; N: Nonmalignant epithelial cells), and the results of binding experiments indicate that normal glycans disialyl Lewis A as well as sialyl 6-sulfo Lewis X serve as ligands for recombinant siglecs (**C**) [147]

epithelial cells lose their glycan ligand for immunosuppressive molecules during the course of carcinogenesis, and instead, cancer cells acquire glycan ligands for vascular selectins which enhance tumor angiogenesis and hematogenous metastasis. Inflammatory mediators produced by COX2 have been proposed to facilitate progression of cancers [170, 171]. We previously found that a significant number of tissue macrophage-like cells expressing siglec-7 or siglec-9 is present in normal colonic mucosal membranes, and ligation of either siglec had suppressive effects on the production of COX2 and IL-12 by the macrophage-like cells [147]. Siglec ligands on nonmalignant epithelial cells may maintain immunological homeostasis in mucosal membranes by suppressing the excess activation of immune cells, and their loss would facilitate production by mucosal immune cells of inflammatory mediators that are known to promote cancer progression. Loss of siglecs is sometimes known to enhance cancer-promoting inflammation within the tumor microenvironment by expanding the number of M2 macrophages, which help the growth of tumors [172].

Downregulation of ST6GALNAC6 was observed at the early stages of colon carcinogenesis in the normal–adenoma–carcinoma sequence [173]. In addition to the cancers of digestive organs, preferential loss of α2,3 α2,6 disialyl Lewis A expression is also noted in prostate cancers [174], and cancer-associated decrease of the ST6GALNAC6 mRNA level is reported in a wide variety of cancers including breast [175] and renal [115] cancers, as well as glioblastoma [58].

Expression of sialyl Lewis A, appeared at the early stages of carcinogenesis, is further enhanced in the later stages of cancers through the induction of several glycan-related genes involved in its synthetic pathway mediated by tumor hypoxia [139, 140], and by generation of cancer stem cell-like cells through EMT [153].

The globo-series glycolipids such as SSEA-3, -4 and Globo-H were known to be specific markers for human embryonic stem cells. Recently, another series of glycans specifically expressed in human embryonic stem cells were structurally characterized, including TRA-1-60 [176] and SSEA-5 [177]. These glycans are carried by type 1 lactosamines, and the glycolipids detected by the specific antibodies belong to lacto-series glycolipids. Sialyl Lewis A is carried also by the type 1 chain lactosamine, and is structurally closely related to TRA-1-60 and SSEA-5. These glycans share the same enzymes encoded by the same genes in most steps of their synthetic pathway. Actually, sialyl Lewis A was shown to behave as an embryonic stem cell marker; it is significantly expressed in undifferentiated human embryonic stem cells, and its expression is markedly reduced after 7 days of retinoic acid-induced differentiation [177]. These findings suggest some interrelation between the stemness of embryonic stem cells and that of so-called cancer stem cells.

6 Conclusion

In the initial stages of studies, the gangliosides which specifically appear in cancer cells attracted special attention of researchers under the name of "tumor-associated gangliosides." Recently, their biological functions became clearer than ever; they play important roles in regulation of signal transduction induced by growth-factor receptors in the specific microdomain in the cancer cell membranes, and interact with glycan recognition molecules involved in cell adhesion and immune regulation. Studies on the regulatory mechanisms for their expression have also made great progress, by identifying genes involved in their synthesis and degradation, and transcriptional/post-transcriptional factors implicated in their regulation. Epigenetic silencing of glycan-related genes is an important mechanism in the early stages of carcinogenesis. Development of resistance to hypoxia involving a transcription factor HIF is a major mechanism in the later stages of cancer progression, as well as acquisition of cancer stem cell-like characteristics through EMT. However, as knowledge on biological functions of ganglioside accumulates, it has become apparent that, not only the cancer-associated glycans, but also the normal gangliosides present in nonmalignant cells exert important biological functions, and some of these gangliosides tend to disappear in cancer cells resulting in the loss of physiological functions, and this sometimes facilitates progression of cancers.

References

1. No authors listed (2015) Dinutuximab approved for high-risk neuroblastoma. Cancer Discov 5:OF5
2. Dhillon S (2015) Dinutuximab: first global approval. Drugs 75:923–927
3. Gabri MR, Cacciavillano W, Chantada GL, Alonso DF (2016) Racotumomab for treating lung cancer and pediatric refractory malignancies. Expert Opin Biol Ther 16:573–578
4. Hakomori SI, Murakami WT (1968) Glycolipids of hamster fibroblasts and derived malignant-transformed cell lines. Proc Natl Acad Sci U S A 59:254–261
5. Hakomori SI (2010) Glycosynaptic microdomains controlling tumor cell phenotype through alteration of cell growth, adhesion, and motility. FEBS Lett 584:1901–1906
6. Hakomori SI, Handa K (2015) GM3 and cancer. Glycoconj J 32:1–8
7. Zhou Q, Hakomori S, Kitamura K, Igarashi Y (1994) GM3 directly inhibits tyrosine phosphorylation and De-*N*-acetyl-GM3 directly enhances serine phosphorylation of epidermal growth factor receptor, independently of receptor-receptor interaction. J Biol Chem 269:1959–1965
8. Liu JW, Sun P, Yan Q, Paller AS, Gerami P, Ho N et al (2009) De-N-acetyl GM3 promotes melanoma cell migration and invasion through urokinase plasminogen activator receptor signaling-dependent MMP-2 activation. Cancer Res 69:8662–8669
9. Kawashima N, Qu H, Lobaton M, Zhu Z, Sollogoub M, Cavenee WK et al (2014) Efficient synthesis of chloro-derivatives of sialosyllactosylceramide, and their enhanced inhibitory effect on epidermal growth factor receptor activation. Oncol Lett 7:933–940
10. Noguchi M, Suzuki T, Kabayama K, Takahashi H, Chiba H, Shiratori M et al (2007) GM3 synthase gene is a novel biomarker for histological classification and drug sensitivity against epidermal growth factor receptor tyrosine kinase inhibitors in non-small cell lung cancer. Cancer Sci 98:1625–1632
11. Kawashima N, Nishimiya Y, Takahata S, Nakayama KI (2016) Induction of glycosphingolipid GM3 expression by valproic acid suppresses cancer cell growth. J Biol Chem 291:21424–21433
12. Prinetti A, Aureli M, Illuzzi G, Prioni S, Nocco V, Scandroglio F et al (2010) GM3 synthase overexpression results in reduced cell motility and in caveolin-1 upregulation in human ovarian carcinoma cells. Glycobiology 20:62–77
13. Prinetti A, Prioni S, Loberto N, Aureli M, Nocco V, Illuzzi G et al (2011) Aberrant glycosphingolipid expression and membrane organization in tumor cells: consequences on tumor-host interactions. Adv Exp Med Biol 705:643–667
14. Prinetti A, Cao T, Illuzzi G, Prioni S, Aureli M, Gagliano N et al (2011) A glycosphingolipid/caveolin-1 signaling complex inhibits motility of human ovarian carcinoma cells. J Biol Chem 286:40900–40910
15. Roncati L, Barbolini G, Gatti AM, Pusiol T, Piscioli F, Maiorana A (2016) The uncon-

trolled sialylation is related to chemoresistant metastatic breast cancer. Pathol Oncol Res 22:869–873

16. Choi SY, Jang JH, Kim KR (2011) Analysis of differentially expressed genes in human rectal carcinoma using suppression subtractive hybridization. Clin Exp Med 11:219–226
17. Liang YJ, Ding Y, Levery SB, Lobaton M, Handa K, Hakomori SI (2013) Differential expression profiles of glycosphingolipids in human breast cancer stem cells vs. cancer non-stem cells. Proc Natl Acad Sci U S A 110:4968–4973
18. Mathow D, Chessa F, Rabionet M, Kaden S, Jennemann R, Sandhoff R et al (2015) Zeb1 affects epithelial cell adhesion by diverting glycosphingolipid metabolism. EMBO Rep 16:321–331
19. Kilbey A, Terry A, Jenkins A, Borland G, Zhang Q, Wakelam MJ et al (2010) Runx regulation of sphingolipid metabolism and survival signaling. Cancer Res 70:5860–5869
20. Cai H, Zhou H, Miao Y, Li N, Zhao L, Jia L (2017) MiRNA expression profiles reveal the involvement of miR-26a, miR-548l and miR-34a in hepatocellular carcinoma progression through regulation of ST3GAL5. Lab Investig 97(5):530–542
21. Shiozaki K, Takahashi K, Hosono M, Yamaguchi K, Hata K, Shiozaki M et al (2015) Phosphatidic acid-mediated activation and translocation to the cell surface of sialidase NEU3, promoting signaling for cell migration. FASEB J 29:2099–2111
22. Kakugawa Y, Wada T, Yamaguchi K, Yamanami H, Ouchi K, Sato I et al (2002) Up-regulation of plasma membrane-associated ganglioside sialidase (Neu3) in human colon cancer and its involvement in apoptosis suppression. Proc Natl Acad Sci U S A 99:10718–10723
23. Miyagi T, Takahashi K, Hata K, Shiozaki K, Yamaguchi K (2012) Sialidase significance for cancer progression. Glycoconj J 29:567–577
24. Wada T, Hata K, Yamaguchi K, Shiozaki K, Koseki K, Moriya S et al (2007) A crucial role of plasma membrane-associated sialidase in the survival of human cancer cells. Oncogene 26:2483–2490
25. Mozzi A, Forcella M, Riva A, Difrancesco C, Molinari F, Martin V et al (2015) NEU3 activity enhances EGFR activation without affecting EGFR expression and acts on its sialylation levels. Glycobiology 25:855–868
26. Yamaguchi K, Shiozaki K, Moriya S, Koseki K, Wada T, Tateno H et al (2012) Reduced susceptibility to colitis-associated colon carcinogenesis in mice lacking plasma membrane-associated sialidase. PLoS One 7:e41132
27. Yamaguchi K, Koseki K, Shiozaki M, Shimada Y, Wada T, Miyagi T (2010) Regulation of plasma-membrane-associated sialidase NEU3 gene by Sp1/Sp3 transcription factors. Biochem J 430:107–117
28. Scaringi R, Piccoli M, Papini N, Cirillo F, Conforti E, Bergante S et al (2013) NEU3 sialidase is activated under hypoxia and protects skeletal muscle cells from apoptosis through the activation of the epidermal growth factor receptor signaling pathway and the hypoxia-inducible factor (HIF)-1α. J Biol Chem 288:3153–3162
29. Yuyama Y, Dohi T, Morita H, Furukawa K, Oshima M (1995) Enhanced expression of GM2/GD2 synthase mRNA in human gastrointestinal cancer. Cancer 75:1273–1280
30. Biswas K, Richmond A, Rayman P, Biswas S, Thornton M, Sa G et al (2006) GM2 expression in renal cell carcinoma: potential role in tumor-induced T-cell dysfunction. Cancer Res 66:6816–6825
31. Biswas S, Biswas K, Richmond A, Ko J, Ghosh S, Simmons M et al (2009) Elevated levels of select gangliosides in T cells from renal cell carcinoma patients is associated with T cell dysfunction. J Immunol 183:5050–5058
32. Kundu M, Mahata B, Banerjee A, Chakraborty S, Debnath S, Ray SS et al (2016) Ganglioside GM2 mediates migration of tumor cells by interacting with integrin and modulating the downstream signaling pathway. Biochim Biophys Acta 1863:1472–1489
33. Mahata B, Banerjee A, Kundu M, Bandyopadhyay U, Biswas K (2015) TALEN mediated targeted editing of GM2/GD2-synthase gene modulates anchorage independent growth by reducing anoikis resistance in mouse tumor cells. Sci Rep 5:9048
34. Todeschini AR, Dos Santos JN, Handa K, Hakomori SI (2008) Ganglioside GM2/GM3 complex affixed on silica nanospheres strongly inhibits cell motility through CD82/cMet-mediated pathway. Proc Natl Acad Sci U S A 105:1925–1930
35. Dall'Olio F, Malagolini N, Chiricolo M, Trinchera M, Harduin-Lepers A (2014) The expanding roles of the Sd(a)/Cad carbohydrate antigen and its cognate glycosyltransferase B4GALNT2. Biochim Biophys Acta 1840:443–453
36. Tanaka K, Miyazawa M, Mikami M, Aoki D, Kiguchi K, Iwamori M (2016) Enhanced expression of unique gangliosides with GM2-

determinant in human uterine cervical carcinoma-derived cell lines. Glycoconj J 33:745–754
37. Kawamura YI, Toyota M, Kawashima R, Hagiwara T, Suzuki H, Imai K et al (2008) DNA hypermethylation contributes to incomplete synthesis of carbohydrate determinants in gastrointestinal cancer. Gastroenterology 135:142–151
38. Malagolini N, Santini D, Chiricolo M, Dall'Olio F (2007) Biosynthesis and expression of the Sda and sialyl Lewis x antigens in normal and cancer colon. Glycobiology 17:688–697
39. Yamada T, Bando H, Takeuchi S, Kita K, Li Q, Wang W et al (2011) Genetically engineered humanized anti-ganglioside GM2 antibody against multiple organ metastasis produced by GM2-expressing small-cell lung cancer cells. Cancer Sci 102:2157–2163
40. Li Q, Wang W, Machino Y, Yamada T, Kita K, Oshima M et al (2015) Therapeutic activity of glycoengineered anti-GM2 antibodies against malignant pleural mesothelioma. Cancer Sci 106:102–107
41. Eggermont AM, Suciu S, Rutkowski P, Marsden J, Santinami M, Corrie P et al (2013) Adjuvant ganglioside GM2-KLH/QS-21 vaccination versus observation after resection of primary tumor > 1.5 mm in patients with stage II melanoma: results of the EORTC 18961 randomized phase III trial. J Clin Oncol 31:3831–3837
42. Furukawa K, Hamamura K, Ohkawa Y, Ohmi Y (2012) Disialyl gangliosides enhance tumor phenotypes with differential modalities. Glycoconj J 29:579–584
43. Battula VL, Shi Y, Evans KW, Wang RY, Spaeth EL, Jacamo RO et al (2012) Ganglioside GD2 identifies breast cancer stem cells and promotes tumorigenesis. J Clin Invest 122:2066–2078
44. Yeh SC, Wang PY, Lou YW, Khoo KH, Hsiao M, Hsu TL et al (2016) Glycolipid GD3 and GD3 synthase are key drivers for glioblastoma stem cells and tumorigenicity. Proc Natl Acad Sci U S A 113:5592–5597
45. Ruckhaberle E, Karn T, Rody A, Hanker L, Gatje R, Metzler D et al (2009) Gene expression of ceramide kinase, galactosyl ceramide synthase and ganglioside GD3 synthase is associated with prognosis in breast cancer. J Cancer Res Clin Oncol 135:1005–1013
46. Cazet A, Lefebvre J, Adriaenssens E, Julien S, Bobowski M, Grigoriadis A et al (2010) GD synthase expression enhances proliferation and tumor growth of MDA-MB-231 breast cancer cells through c-Met activation. Mol Cancer Res 8:1526–1535
47. Furukawa K, Kambe M, Miyata M, Ohkawa Y, Tajima O (2014) Ganglioside GD3 induces convergence and synergism of adhesion and hepatocyte growth factor/Met signals in melanomas. Cancer Sci 105:52–63
48. Bobowski M, Vincent A, Steenackers A, Colomb F, Van S, Julien S et al (2013) Estradiol represses the G(D3) synthase gene ST8SIA1 expression in human breast cancer cells by preventing NFκB binding to ST8SIA1 promoter. PLoS One 8:e62559
49. Sarkar TR, Battula VL, Werden SJ, Vijay GV, Ramirez-Pena EQ, Taube JH et al (2015) GD3 synthase regulates epithelial-mesenchymal transition and metastasis in breast cancer. Oncogene 34:2958–2967
50. Mitchell SM, Ross JP, Drew HR, Ho T, Brown GS, Saunders NF et al (2014) A panel of genes methylated with high frequency in colorectal cancer. BMC Cancer 14:54–14
51. Yu AL, Gilman AL, Ozkaynak MF, London WB, Kreissman SG, Chen HX et al (2010) Anti-GD2 antibody with GM-CSF, interleukin-2, and isotretinoin for neuroblastoma. N Engl J Med 363:1324–1334
52. Bos PD, Zhang XH, Nadal C, Shu W, Gomis RR, Nguyen DX et al (2009) Genes that mediate breast cancer metastasis to the brain. Nature 459:1005–1009
53. Lorusso G, Ruegg C (2012) New insights into the mechanisms of organ-specific breast cancer metastasis. Semin Cancer Biol 22:226–233
54. Hu G, Kang Y, Wang XF (2009) From breast to the brain: unraveling the puzzle of metastasis organotropism. J Mol Cell Biol 1:3–5
55. Vandermeersch S, Vanbeselaere J, Delannoy CP, Drolez A, Mysiorek C, Guerardel Y et al (2015) Accumulation of GD1α ganglioside in MDA-MB-231 breast cancer cells expressing ST6GalNAc V. Molecules 20:6913–6924
56. Drolez A, Vandenhaute E, Delannoy CP, Dewald JH, Gosselet F, Cecchelli R et al (2016) ST6GALNAC5 expression decreases the interactions between breast cancer cells and the human blood-brain barrier. Int J Mol Sci 17:E1309
57. Bhuiyan RH, Kondo Y, Yamaguchi T, Tokuda N, Ohkawa Y, Hashimoto N et al (2016) Expression analysis of 0-series gangliosides in human cancer cell lines with monoclonal antibodies generated using knockout mice of ganglioside synthase genes. Glycobiology 26:984–998
58. Kroes RA, Dawson G, Moskal JR (2007) Focused microarray analysis of glyco-gene

expression in human glioblastomas. J Neurochem 103(Suppl 1):14–24

59. Kroes RA, He H, Emmett MR, Nilsson CL, Leach FE III, Amster IJ et al (2010) Overexpression of ST6GalNAcV, a ganglioside-specific α2,6-sialyltransferase, inhibits glioma growth in vivo. Proc Natl Acad Sci U S A 107:12646–12651
60. Samraj AN, Laubli H, Varki N, Varki A (2014) Involvement of a non-human sialic acid in human cancer. Front Oncol 4:33
61. Yin J, Hashimoto A, Izawa M, Miyazaki K, Chen G-Y, Takematsu H et al (2006) Hypoxic culture induces expression of sialin, a sialic acid transporter, and cancer-associated gangliosides containing non-human sialic acid on human cancer cells. Cancer Res 66:2937–2945
62. Go S, Sato C, Yin J, Kannagi R, Kitajima K (2007) Hypoxia-enhanced expression of free deaminoneuraminic acid in human cancer cells. Biochem Biophys Res Commun 357:537–542
63. Laughlin ST, Bertozzi CR (2007) Metabolic labeling of glycans with azido sugars and subsequent glycan-profiling and visualization via Staudinger ligation. Nat Protoc 2:2930–2944
64. Bull C, Boltje TJ, Wassink M, de Graaf AM, van Delft FL, den Brok MH et al (2013) Targeting aberrant sialylation in cancer cells using a fluorinated sialic acid analog impairs adhesion, migration, and in vivo tumor growth. Mol Cancer Ther 12:1935–1946
65. Hayashi N, Chiba H, Kuronuma K, Go S, Hasegawa Y, Takahashi M et al (2013) Detection of N-glycolyated gangliosides in non-small-cell lung cancer using GMR8 monoclonal antibody. Cancer Sci 104:43–47
66. Inoue S, Sato C, Kitajima K (2010) Extensive enrichment of N-glycolylneuraminic acid in extracellular sialoglycoproteins abundantly synthesized and secreted by human cancer cells. Glycobiology 20:752–762
67. Nystedt J, Anderson H, Hirvonen T, Impola U, Jaatinen T, Heiskanen A et al (2010) Human CMP-N-acetylneuraminic acid hydroxylase is a novel stem cell marker linked to stem cell-specific mechanisms. Stem Cells 28:258–267
68. Marquina G, Waki H, Fernandez LE, Kon K, Carr A, Valiente O et al (1996) Gangliosides expressed in human breast cancer. Cancer Res 56:5165–5171
69. Carr A, Mullet A, Mazorra Z, Vazquez AM, Alfonso M, Mesa C et al (2000) A mouse IgG1 monoclonal antibody specific for N-glycolyl GM3 ganglioside recognized breast and melanoma tumors. Hybridoma 19:241–247
70. Blanco R, Rengifo E, Cedeno M, Rengifo CE, Alonso DF, Carr A (2011) Immunoreactivity of the 14F7 Mab raised against N-glycolyl GM3 ganglioside in epithelial malignant tumors from digestive system. ISRN Gastroenterol 2011:645641
71. Blanco R, Dominguez E, Morales O, Blanco D, Martinez D, Rengifo CE et al (2015) Prognostic significance of N-glycolyl GM3 ganglioside expression in non-small cell lung carcinoma patients: new evidences. Pathol Res Int 2015:132326
72. Scursoni AM, Galluzzo L, Camarero S, Lopez J, Lubieniecki F, Sampor C et al (2011) Detection of N-glycolyl GM3 ganglioside in neuroectodermal tumors by immunohistochemistry: an attractive vaccine target for aggressive pediatric cancer. Clin Dev Immunol 2011:245181
73. Miyake M, Ito M, Hitomi S, Ikeda S, Taki T, Kurata M et al (1988) Generation of two monoclonal antibodies that can discriminate *N*-glycolyl and *N*-acetyl neuraminic acid residues of GM2 gangliosides and distribution of the antigens in human lung cancer. Cancer Res 48:6154–6160
74. Kawashima I, Ozawa H, Kotani M, Suzuki M, Kawano T, Gomibuchi M et al (1993) Characterization of ganglioside expression in human melanoma cells: immunological and biochemical analysis. J Biochem (Tokyo) 114:186–193
75. Palomo AG, Santana RB, Perez XE, Santana DB, Gabri MR, Monzon KL et al (2016) Frequent co-expression of EGFR and NeuGcGM3 ganglioside in cancer: it's potential therapeutic implications. Clin Exp Metastasis 33:717–725
76. Herrera ZM, Ramos TC (2014) Pilot study of a novel combination of two therapeutic vaccines in advanced non-small-cell lung cancer patients. Cancer Immunol Immunother 63:737–747
77. Alfonso S, Valdes-Zayas A, Santiesteban ER, Flores YI, Areces F, Hernandez M et al (2014) A randomized, multicenter, placebo-controlled clinical trial of racotumomab-alum vaccine as switch maintenance therapy in advanced non-small cell lung cancer patients. Clin Cancer Res 20:3660–3671
78. Alfonso M, Diaz A, Hernandez AM, Perez A, Rodriguez E, Bitton R et al (2002) An anti-idiotype vaccine elicits a specific response to N-glycolyl sialic acid residues of glycoconju-

gates in melanoma patients. J Immunol 168:2523–2529
79. Diaz A, Alfonso M, Alonso R, Saurez G, Troche M, Catala M et al (2003) Immune responses in breast cancer patients immunized with an anti-idiotype antibody mimicking NeuGc-containing gangliosides. Clin Immunol 107:80–89
80. Hernandez AM, Rodriguez M, Lopez-Requena A, Beausoleil I, Perez R, Vazquez AM (2005) Generation of anti-Neu-glycolyl-ganglioside antibodies by immunization with an anti-idiotype monoclonal antibody: a self versus non-self-matter. Immunobiology 210:11–21
81. Solter D, Knowles BB (1978) Monoclonal antibody defining a stage-specific mouse embryonic antigen (SSEA-1). Proc Natl Acad Sci U S A 75:5565–5569
82. Handa K, Hakomori SI (2016) Changes of glycoconjugate expression profiles during early development. Glycoconj J 34(6):693–699
83. Kannagi R, Cochran NA, Ishigami F, Hakomori S, Andrews PW, Knowles BB et al (1983) Stage-specific embryonic antigens (SSEA-3 and -4) are epitopes of a unique globo-series ganglioside isolated from human teratocarcinoma cells. EMBO J 2:2355–2361
84. Takahashi K, Tanabe K, Ohnuki M, Narita M, Ichisaka T, Tomoda K et al (2007) Induction of pluripotent stem cells from adult human fibroblasts by defined factors. Cell 131:1–12
85. Yu AL, Hung JT, Ho MY, Yu J (2016) Alterations of glycosphingolipids in embryonic stem cell differentiation and development of glycan-targeting cancer immunotherapy. Stem Cells Dev 25:1532–1548
86. Saito S, Orikasa S, Satoh M, Ohyama C, Ito A, Takahashi T (1997) Expression of globo-series gangliosides in human renal cell carcinoma. Jpn J Cancer Res 88:652–659
87. Tokuyama S, Saito S, Takahashi T, Ohyama C, Ito A, Kanto S et al (2003) Immunostaining of stage-specific embryonic antigen-4 in intratubular germ cell neoplasia unclassified and in testicular germ-cell tumors. Oncol Rep 10:1097–1104
88. Huang YL, Hung JT, Cheung SK, Lee HY, Chu KC, Li ST et al (2013) Carbohydrate-based vaccines with a glycolipid adjuvant for breast cancer. Proc Natl Acad Sci U S A 110:2517–2522
89. Gottschling S, Jensen K, Warth A, Herth FJ, Thomas M, Schnabel PA et al (2013) Stage-specific embryonic antigen-4 is expressed in basaloid lung cancer and associated with poor prognosis. Eur Respir J 41:656–663
90. Malecki M, Anderson M, Beauchaine M, Seo S, Tambokan X (2012) TRA-1-60(+), SSEA-4(+), Oct4A(+), Nanog(+) clones of pluripotent stem cells in the embryonal carcinomas of the ovaries. J Stem Cell Res Ther 2:130
91. Zhang W, Ding ML, Zhang JN, Qiu JR, Shen YH, Ding XY et al (2015) mTORC1 maintains the tumorigenicity of SSEA-4(+) high-grade osteosarcoma. Sci Rep 5:9604. https://doi.org/10.1038/:9604
92. Noto Z, Yoshida T, Okabe M, Koike C, Fathy M, Tsuno H et al (2013) CD44 and SSEA-4 positive cells in an oral cancer cell line HSC-4 possess cancer stem-like cell characteristics. Oral Oncol 49:787–795
93. Lou YW, Wang PY, Yeh SC, Chuang PK, Li ST, Wu CY et al (2014) Stage-specific embryonic antigen-4 as a potential therapeutic target in glioblastoma multiforme and other cancers. Proc Natl Acad Sci U S A 111:2482–2487
94. Wenk J, Andrews PW, Casper J, Hata J, Pera MF, von Keitz A et al (1994) Glycolipids of germ cell tumors: extended globo-series glycolipids are a hallmark of human embryonal carcinoma cells. Int J Cancer 58:108–115
95. Aloia A, Petrova E, Tomiuk S, Bissels U, Deas O, Saini M et al (2015) The sialyl-glycolipid stage-specific embryonic antigen 4 marks a subpopulation of chemotherapy-resistant breast cancer cells with mesenchymal features. Breast Cancer Res 17:146–652
96. Sivasubramaniyan K, Harichandan A, Schilbach K, Mack AF, Bedke J, Stenzl A et al (2015) Expression of stage-specific embryonic antigen-4 (SSEA-4) defines spontaneous loss of epithelial phenotype in human solid tumor cells. Glycobiology 25:902–917
97. Malecki M, Tombokan X, Anderson M, Malecki R, Beauchaine M (2013) TRA-1-60+, SSEA-4+, POU5F1+, SOX2+, NANOG+ Clones of pluripotent stem cells in the embryonal carcinomas of the testes. J Stem Cell Res Ther 3:1000134
98. Virant-Klun I, Kenda-Suster N, Smrkolj S (2016) Small putative NANOG, SOX2, and SSEA-4-positive stem cells resembling very small embryonic-like stem cells in sections of ovarian tissue in patients with ovarian cancer. J Ovarian Res 9:12–221
99. Cheng J, Yang K, Zhang Q, Yu Y, Meng Q, Mo N et al (2016) The role of mesenchymal stem cells in promoting the transformation of androgen-dependent human prostate cancer

cells into androgen-independent manner. Sci Rep 6:16993
100. Takaishi S, Okumura T, Tu S, Wang SS, Shibata W, Vigneshwaran R et al (2009) Identification of gastric cancer stem cells using the cell surface marker CD44. Stem Cells 27:1006–1020
101. Ye F, Li Y, Hu Y, Zhou C, Chen H (2010) Stage-specific embryonic antigen 4 expression in epithelial ovarian carcinoma. Int J Gynecol Cancer 20:958–964
102. Cheung SK, Chuang PK, Huang HW, Hwang-Verslues WW, Cho CH, Yang WB et al (2016) Stage-specific embryonic antigen-3 (SSEA-3) and β3GalT5 are cancer specific and significant markers for breast cancer stem cells. Proc Natl Acad Sci U S A 113:960–965
103. Saito S, Aoki H, Ito A, Ueno S, Wada T, Mitsuzuka K et al (2003) Human α2,3-sialyltransferase (ST3Gal II) is a stage-specific embryonic antigen-4 synthase. J Biol Chem 278:26474–26479
104. Izzi B, Binder AM, Michels KB (2014) Pyrosequencing evaluation of widely available bisulfite conversion methods: considerations for application. Med Epigenet 2:28–36
105. Magalhaes A, Marcos-Pinto R, Nairn AV, dela RM, Ferreira RM, Junqueira-Neto S et al (2015) Helicobacter pylori chronic infection and mucosal inflammation switches the human gastric glycosylation pathways. Biochim Biophys Acta 1852:1928–1939
106. Dunn CA, Medstrand P, Mager DL (2003) An endogenous retroviral long terminal repeat is the dominant promoter for human β1,3-galactosyltransferase 5 in the colon. Proc Natl Acad Sci U S A 100:12841–12846
107. Dunn CA, van de Lagemaat LN, Baillie GJ, Mager DL (2005) Endogenous retrovirus long terminal repeats as ready-to-use mobile promoters: the case of primate β3GAL-T5. Gene 364:2–12
108. Trinchera M, Zulueta A, Caretti A, Dall'Olio F (2014) Control of glycosylation-related genes by DNA methylation: the intriguing case of the B3GALT5 gene and its distinct promoters. Biology (Basel) 3:484–497
109. Zulueta A, Caretti A, Signorelli P, Dall'Olio F, Trinchera M (2014) Transcriptional control of the B3GALT5 gene by a retroviral promoter and methylation of distant regulatory elements. FASEB J 28:946–955
110. Steelant WF, Kawakami Y, Ito A, Handa K, Bruyneel EA, Mareel M et al (2002) Monosialyl-Gb5 organized with cSrc and FAK in GEM of human breast carcinoma MCF-7 cells defines their invasive properties. FEBS Lett 531:93–98
111. Hung TC, Lin CW, Hsu TL, Wu CY, Wong CH (2013) Investigation of SSEA-4 binding protein in breast cancer cells. J Am Chem Soc 135:5934–5937
112. Levery SB, Salyan MEK, Steele SJ, Kannagi R, Dasgupta S, Chien JL et al (1994) A revised structure for the disialosyl globo-series gangliosides of human erythrocytes and chicken skeletal muscle. Arch Biochem Biophys 312:125–134
113. Kawasaki Y, Ito A, Kakoi N, Shimada S, Itoh J, Mitsuzuka K et al (2015) Ganglioside, disialosyl globopentaosylceramide (DSGb5), enhances the migration of renal cell carcinoma cells. Tohoku J Exp Med 236:1–7
114. Shimada S, Ito A, Kawasaki Y, Kakoi N, Taima T, Mitsuzuka K et al (2014) Ganglioside disialosyl globopentaosylceramide is an independent predictor of PSA recurrence-free survival following radical prostatectomy. Prostate Cancer Prostatic Dis 17:199–205
115. Senda M, Ito A, Tsuchida A, Hagiwara T, Kaneda T, Nakamura Y et al (2007) Identification and expression of a sialyltransferase responsible for the synthesis of disialylgalactosylgloboside in normal and malignant kidney cells: downregulation of ST6GalNAc VI in renal cancers. Biochem J 402:459–470
116. Kannagi R, Levery SB, Hakomori S (1984) Blood group H antigen with globo-series structure: Isolation and characterization from human blood group O erythrocytes. FEBS Lett 175:397–401
117. Bremer EG, Levery SB, Sonnio S, Ghidoni R, Canevari S, Kannagi R et al (1984) Characterization of a glycosphingolipid antigen defined by the monoclonal antibody MBr1 expressed in normal and neo-plastic epithelial cells of human mammary gland. J Biol Chem 259:14773–14777
118. Chang WW, Lee CH, Lee P, Lin J, Hsu CW, Hung JT et al (2008) Expression of Globo H and SSEA3 in breast cancer stem cells and the involvement of fucosyl transferases 1 and 2 in Globo H synthesis. Proc Natl Acad Sci U S A 105:11667–11672
119. Cheng JY, Wang SH, Lin J, Tsai YC, Yu J, Wu JC et al (2014) Globo-H ceramide shed from cancer cells triggers translin-associated factor X-dependent angiogenesis. Cancer Res 74:6856–6866
120. Clausen H, Watanabe K, Kannagi R, Levery SB, Nudelman E, Arao-Tomono Y et al (1984) Blood group A glycolipid (Ax) with

globo-series structure which is specific for blood group A1 erythrocytes: one of the chemical bases for A1 and A2 distinction. Biochem Biophys Res Commun 124:523–529

121. Sekine M, Taya C, Kikkawa Y, Yonekawa H, Takenaka M, Matsuoka Y et al (2001) Regulation of mouse kidney tubular epithelial cell-specific expression of core 2 GlcNAc transferase. Eur J Biochem 268:1129–1135
122. Suzuki A, Yoshioka S, Sekine M, Yonekawa H, Takenaka M, Kannagi R (2004) Core 2 GlcNAc transferase and kidney tubular cell-specific expression. Glycoconj J 20:151–156
123. Lowe JB, Stoolman LM, Nair RP, Larsen RD, Berhend TL, Marks RM (1990) ELAM-1-dependent cell adhesion to vascular endothelium determined by a transfected human fucosyltransferase cDNA. Cell 63:475–484
124. Phillips ML, Nudelman E, Gaeta FC, Perez M, Singhal AK, Hakomori S et al (1990) ELAM-1 mediates cell adhesion by recognition of a carbohydrate ligand, sialyl-Le^x. Science 250:1130–1132
125. Walz G, Aruffo A, Kolanus W, Bevilacqua M, Seed B (1990) Recognition by ELAM-1 of the sialyl-Le^x determinant on myeloid and tumor cells. Science 250:1132–1135
126. Kannagi R (1997) Carbohydrate-mediated cell adhesion involved in hematogenous metastasis of cancer. Glycoconj J 14:577–584
127. Tei K, Kawakami-Kimura N, Taguchi O, Kumamoto K, Higashiyama S, Taniguchi N et al (2002) Roles of cell adhesion molecules in tumor angiogenesis induced by co-transplantation of cancer and endothelial cells to nude rats. Cancer Res 62:6289–6296
128. Cui HX, Wang H, Wang Y, Song J, Tian H, Xia C et al (2016) ST3Gal III modulates breast cancer cell adhesion and invasion by altering the expression of invasion-related molecules. Oncol Rep 36:3317–3324
129. Higai K, Miyazaki N, Azuma Y, Matsumoto K (2006) Interleukin-1β induces sialyl Lewis X on hepatocellular carcinoma HuH-7 cells via enhanced expression of ST3Gal IV and FUT VI gene. FEBS Lett 580:6069–6075
130. Yoshihama N, Yamaguchi K, Chigita S, Mine M, Abe M, Ishii K et al (2015) A novel function of CD82/KAI1 in sialyl Lewis antigen-mediated adhesion of cancer cells: Evidence for an anti-metastasis effect by down-regulation of sialyl Lewis antigens. PLoS One 10:e0124743
131. Julien S, Ivetic A, Grigoriadis A, QiZe D, Burford B, Sproviero D et al (2011) Selectin ligand sialyl-Lewis x antigen drives metastasis of hormone-dependent breast cancers. Cancer Res 71:7683–7693
132. Martin-Satue M, de Castellarnau C, Blanco J (1999) Overexpression of α(1,3)-fucosyltransferase VII is sufficient for the acquisition of lung colonization phenotype in human lung adenocarcinoma HAL-24Luc cells. Br J Cancer 80:1169–1174
133. Carvalho AS, Harduin-Lepers A, Magalhaes A, Machado E, Mendes N, Costa LT et al (2010) Differential expression of α-2,3-sialyltransferases and α-1,3/4-fucosyltransferases regulates the levels of sialyl Lewis a and sialyl Lewis x in gastrointestinal carcinoma cells. Int J Biochem Cell Biol 42:80–89
134. Shiozaki K, Yamaguchi K, Takahashi K, Moriya S, Miyagi T (2011) Regulation of sialyl Lewis antigen expression in colon cancer cells by sialidase NEU4. J Biol Chem 286:21052–21061
135. Vecchio G, Parascandolo A, Allocca C, Ugolini C, Basolo F, Moracci M et al (2017) Human a-L-fucosidase-1 attenuates the invasive properties of thyroid cancer. Oncotarget 8(16):27075–27092
136. Cheng TC, Tu SH, Chen LC, Chen MY, Chen WY, Lin YK et al (2015) Down-regulation of α-L-fucosidase 1 expression confers inferior survival for triple-negative breast cancer patients by modulating the glycosylation status of the tumor cell surface. Oncotarget 6:21283–21300
137. Ezawa I, Sawai Y, Kawase T, Okabe A, Tsutsumi S, Ichikawa H et al (2016) Novel p53 target gene FUCA1 encodes a fucosidase and regulates growth and survival of cancer cells. Cancer Sci 107:734–745
138. Baudot AD, Crighton D, O'Prey J, Somers J, Sierra GP, Ryan KM (2016) p53 directly regulates the glycosidase FUCA1 to promote chemotherapy-induced cell death. Cell Cycle 15:2299–2308
139. Kannagi R (2004) Molecular mechanism for cancer-associated induction of sialyl Lewis X and sialyl Lewis A expression – the Warburg effect revisited. Glycoconj J 20:353–364
140. Koike T, Kimura N, Miyazaki K, Yabuta T, Kumamoto K, Takenoshita S et al (2004) Hypoxia induces adhesion molecules on cancer cells—a missing link between Warburg effect and induction of selectin ligand carbohydrates. Proc Natl Acad Sci U S A 101:8132–8137
141. Mitsuoka C, Sawada-Kasugai M, Ando-Furui K, Izawa M, Nakanishi H, Nakamura S et al

(1998) Identification of a major carbohydrate capping group of the L-selectin ligand on high endothelial venules in human lymph nodes as 6-sulfo sialyl Lewis x. J Biol Chem 273:11225–11233

142. Izawa M, Kumamoto K, Mitsuoka C, Kanamori A, Ohmori K, Ishida H et al (2000) Expression of sialyl 6-sulfo Lewis x is inversely correlated with conventional sialyl Lewis x expression in human colorectal cancer. Cancer Res 60:1410–1416

143. Yusa A, Miyazaki K, Kimura N, Izawa M, Kannagi R (2010) Epigenetic silencing of the sulfate transporter gene *DTDST* induces sialyl Lewis^x expression and accelerates proliferation of colon cancer cells. Cancer Res 70:4064–4073

144. Hakomori S, Kannagi R (1983) Glycosphingolipids as tumor-associated and differentiation markers. J Natl Cancer Inst 71:231–251

145. Kannagi R, Yin J, Miyazaki K, Izawa M (2008) Current relevance of incomplete synthesis and neo-synthesis for cancer-associated alteration of carbohydrate determinants-Hakomori's concepts revisited. Biochim Biophys Acta 1780:525–531

146. Kannagi R, Sakuma K, Miyazaki K, Lim K-T, Yusa A, Yin J et al (2010) Altered expression of glycan genes in cancers induced by epigenetic silencing and tumor hypoxia: Clues in the ongoing search for new tumor markers. Cancer Sci 101:586–593

147. Miyazaki K, Sakuma K, Kawamura YI, Izawa M, Ohmori K, Mitsuki M et al (2012) Colonic epithelial cells express specific ligands for mucosal macrophage immunosuppressive receptors, siglec-7 and -9. J Immunol 188:4690–4700

148. Varki A, Angata T (2006) Siglecs—the major subfamily of I-type lectins. Glycobiology 16:1R–27R

149. Zhang L, Shi J, Ji M, Liu W, Wang N, Guan H et al (2012) Methylation analysis of drug metabolism and transport genes in papillary thyroid cancer. Adv Sci Lett 17:243–250

150. Inoue A, Okamoto K, Fujino Y, Nakagawa T, Muguruma N, Sannomiya K et al (2015) B-RAF mutation and accumulated gene methylation in aberrant crypt foci (ACF), sessile serrated adenoma/polyp (SSA/P) and cancer in SSA/P. Br J Cancer 112:403–412

151. Kimura N, Mitsuoka C, Kanamori A, Hiraiwa N, Uchimura K, Muramatsu T et al (1999) Reconstitution of functional L-selectin ligands on a cultured human endothelial cell line by co-transfection of α1→3 fucosyltransferase VII and newly cloned GlcNAcβ: 6-sulfotransferase cDNA. Proc Natl Acad Sci U S A 96:4530–4535

152. Shida K, Misonou Y, Korekane H, Seki Y, Noura S, Ohue M et al (2009) Unusual accumulation of sulfated glycosphingolipids in colon cancer cells. Glycobiology 19:1018–1033

153. Sakuma K, Aoki M, Kannagi R (2012) Transcription factors c-Myc and CDX2 mediate E-selectin ligand expression in colon cancer cells undergoing EGF/bFGF-induced epithelial-mesenchymal transition. Proc Natl Acad Sci U S A 109:7776–7781

154. Kannagi R (2007) Carbohydrate antigen sialyl Lewis a -its pathophysiological significance and induction mechanism in cancer progression. Chang Gung Med J 30:189–209

155. Magnani JL, Brockhaus M, Smith DF, Ginsburg V, Blaszczyk M, Mitchell KF et al (1981) A monosialoganglioside is a monoclonal antibody-defined antigen of colon carcinoma. Science 212:55–56

156. Takada A, Ohmori K, Takahashi N, Tsuyuoka K, Yago K, Zenita K et al (1991) Adhesion of human cancer cells to vascular endothelium mediated by a carbohydrate antigen, sialyl Lewis A. Biochem Biophys Res Commun 179:713–719

157. Berg EL, Robinson MK, Mansson O, Butcher EC, Magnani JL (1991) A carbohydrate domain common to both sialyl Le^a and sialyl Le^x is recognized by the endothelial cell leukocyte adhesion molecule ELAM-1. J Biol Chem 266:14869–14872

158. Takada A, Ohmori K, Yoneda T, Tsuyuoka K, Hasegawa A, Kiso M et al (1993) Contribution of carbohydrate antigens sialyl Lewis A and sialyl Lewis X to adhesion of human cancer cells to vascular endothelium. Cancer Res 53:354–361

159. Kannagi R, Izawa M, Koike T, Miyazaki K, Kimura N (2004) Carbohydrate-mediated cell adhesion in cancer metastasis and angiogenesis. Cancer Sci 95:377–384

160. Chung TW, Kim SJ, Choi HJ, Song KH, Jin UH, Yu DY et al (2014) Hepatitis B virus X protein specially regulates the sialyl lewis a synthesis among glycosylation events for metastasis. Mol Cancer 13:222

161. Hu H, Eggers K, Chen W, Garshasbi M, Motazacker MM, Wrogemann K et al (2011) ST3GAL3 mutations impair the development of higher cognitive functions. Am J Hum Genet 89:407–414

162. Grahn A, Barkhordar GS, Larson G (2002) Cloning and sequencing of nineteen tran-

script isoforms of the human α2,3-sialyltransferase gene, ST3Gal III; its genomic organisation and expression in human tissues. Glycoconj J 19:197–210

163. Schnaar JL (2014) ST3 β-galactoside α-2,3-sialyltransferase 3 (ST3GAL3). In: Taniguchi N, Honke K, Fukuda M, Narimatsu H, Yamaguchi Y, Angata T (eds) Handbook of glycosyltransferases and related genes. Springer, Tokyo, pp 657–664
164. Okajima T, Fukumoto S, Miyazaki H, Ishida H, Kiso M, Furukawa K et al (1999) Molecular cloning of a novel α2,3-sialyltransferase (ST3Gal VI) that sialylates type II lactosamine structures on glycoproteins and glycolipids. J Biol Chem 274:11479–11486
165. Padro M, Cobler L, Garrido M, de Bolos C (2011) Down-regulation of FUT3 and FUT5 by shRNA alters Lewis antigens expression and reduces the adhesion capacities of gastric cancer cells. Biochim Biophys Acta 1810:1141–1149
166. Itai S, Arii S, Tobe R, Kitahara A, Kim YC, Yamabe H et al (1988) Significance of 2-3 and 2-6 sialylation of Lewis A antigen in pancreas cancer. Cancer 61:775–787
167. Itai S, Nishikata J, Yoneda T, Ohmori K, Tsunekawa S, Hiraiwa N et al (1991) Tissue distribution of sialyl 2-3 and 2-6 Lewis a antigens and the significance of serum 2-3/2-6 sialyl Lewis a antigen ratio for the differential diagnosis of malignant and benign disorders of the digestive tract. Cancer 67:1576–1587
168. Tsuchida A, Okajima T, Furukawa K, Ando T, Ishida H, Yoshida A et al (2003) Synthesis of disialyl Lewis a structure in colon cancer cell lines by a sialyltransferase ST6GalNAc VI responsible for the synthesis of α-series gangliosides. J Biol Chem 278:22787–22794
169. Miyazaki K, Ohmori K, Izawa M, Koike T, Kumamoto K, Furukawa K et al (2004) Loss of disialyl Lewis[a], the ligand for lymphocyte inhibitory receptor Siglec-7, associated with increased sialyl Lewis[a] expression on human colon cancers. Cancer Res 64:4498–4505
170. Taketo MM (1998) Cyclooxygenase-2 inhibitors in tumorigenesis (part I). J Natl Cancer Inst 90:1529–1536
171. Taketo MM (2012) Roles of stromal microenvironment in colon cancer progression. J Biochem 151:477–481
172. Laubli H, Pearce OM, Schwarz F, Siddiqui SS, Deng L, Stanczak MA et al (2014) Engagement of myelomonocytic Siglecs by tumor-associated ligands modulates the innate immune response to cancer. Proc Natl Acad Sci U S A 111:14211–14216
173. Bowden NA, Croft A, Scott RJ (2007) Gene expression profiling in familial adenomatous polyposis adenomas and desmoid disease. Hered Cancer Clin Pract 5:79–96
174. Young WW Jr, Mills SE, Lippert MC, Ahmed P, Lau SK (1988) Deletion of antigens of the Lewis a/b blood group family in human prostatic carcinoma. Am J Pathol 131:578–586
175. Potapenko IO, Haakensen VD, Luders T, Helland A, Bukholm I, Sorlie T et al (2010) Glycan gene expression signatures in normal and malignant breast tissue; possible role in diagnosis and progression. Mol Oncol 4:98–118
176. Natunen S, Satomaa T, Pitkanen V, Salo H, Mikkola M, Natunen J et al (2011) The binding specificity of the marker antibodies Tra-1-60 and Tra-1-81 reveals a novel pluripotency-associated type 1 lactosamine epitope. Glycobiology 21:1125–1130
177. Tang C, Lee AS, Volkmer JP, Sahoo D, Nag D, Mosley AR et al (2011) An antibody against SSEA-5 glycan on human pluripotent stem cells enables removal of teratoma-forming cells. Nat Biotechnol 29:829–834

Chapter 7

Anti-ganglioside Antibodies in Peripheral Nerve Pathology

Hugh J. Willison

Abstract

Anti-ganglioside antibodies are principally associated with autoimmune peripheral neuropathies. In these disorders, immune attack is inadvertently directed at peripheral nerve by autoantibodies that target glycan structures borne by glycolipids, particularly gangliosides concentrated in nerve myelin and axons. The most thoroughly studied disorder is the acute paralytic disease, Guillain–Barré syndrome (GBS) in which IgG autoantibodies against gangliosides arise following acute infections, notably *Campylobacter jejuni* enteritis. Additionally, chronic autoimmune neuropathies are associated with IgM antibodies directed against many glycolipids including gangliosides. This introductory chapter briefly summarizes the immunological and pathological features of these disorders, focusing on the methodological development of antibody measurement and of animal models.

Key words Gangliosides, Glycolipids, Autoantibody, Peripheral nerve, Autoimmunity, Neuropathy, Animal models

1 The Functional Anatomy of Peripheral Nerve

The peripheral nervous system (PNS) comprises a wide range of neuronal cell types within specialized domains including synapses, axon bundles, and ganglia. Neurons are supported by myelin-forming and nonmyelinating Schwann cells, alongside nonneuronal cells. In functional terms, the PNS subserves motor, sensory, and autonomic functions to perform a wide range of physiological processes; thus the diversity of clinical syndromes arising from even autoimmune peripheral nerve diseases is extensive [1]. The PNS is an immunologically active organ exhibiting both innate and adaptive immune functions [2, 3]. Most of the PNS is sheltered within an immunologically privileged site behind the blood–nerve barrier (the PNS equivalent of the blood–brain barrier). PNS sites more freely exposed to circulating factors include the dorsal and ventral spinal roots and the sensory and motor nerve terminals that are more vulnerable to the effects of anti-ganglioside antibodies [4]. With regard to autoimmunity to gangliosides, the PNS contains many glycosylated molecules potentially capable of acting as

Sandro Sonnino and Alessandro Prinetti (eds.), *Gangliosides: Methods and Protocols*, Methods in Molecular Biology, vol. 1804, https://doi.org/10.1007/978-1-4939-8552-4_7, © Springer Science+Business Media, LLC, part of Springer Nature 2018

antigens, notably myelin and axonal glycolipids, myelin glycoproteins P0, peripheral myelin protein 22 (PMP-22) and myelin associated glycoprotein (MAG), and glycoyslated and/or glycan-binding molecules such as laminins, dystroglycans, glycosaminoglycans, and integrins in the basal lamina [5]. The structural complexity of the PNS, combined with the variations in autoantibody access to specialized compartments such as compacted myelin, makes considerations about the relevance of glycan antigens as pathological targets more complex than might seem at first sight. One general principle for antibody targets is that they should be expressed on the extracellular surface or present in the extracellular milieu, and this applies to many glycolipids and other glycans that act as extracellular recognition molecules. Myelin is thought to have a broadly common molecular composition in all fiber types although subtle differences in the content of particular gangliosides may be present that affect the clinical manifestations of anti-ganglioside antibody mediated disorders [6]. The Schwann cell elaborates many membrane specializations associated with noncompact myelin at paranodal regions where certain glycolipids may be more readily exposed to autoantibodies than they are in compact myelin or axolemmal membranes in the intermodal region of myelin [7].

2 Glycolipid Antigens in Autoimmune Neuropathy

Galactocerebroside (GalC) was the first glycolipid to be identified as an antigen in experimental autoimmune neuritis (EAN), a rodent model of autoimmune neuropathy [8]. Sulfatide is also a prominent lipid antigen in PNS myelin [9, 10]. GalC and sulfatide reside within the exofacial leaflet of the Schwann cell plasma membrane lipid bilayer, where the oligosaccharide is exposed to the extracellular environment and thus available to autoantibody binding. Functionally GalC and sulfatide play important roles in the development and maintenance of myelin, as demonstrated by deficiency states created by targeted disruption of the UDP-galactose ceramide galactosyltransferase (CGT), the key enzyme in galactolipid biosynthesis [9]. Schwann cells also synthesize neolacto-series gangliosides that are localized mainly in myelin, including LM1, also termed sialosylneolactotetraosylceramide, or sialosylparagloboside, and its higher homologue sialosyllactosaminylparagloboside, Hex-LM1 [11]. LM1 and Hex-LM1 share common glycan structures and act as autoantigens in both acute and chronic autoimmune neuropathies [12]. Substitution of sialic acid for a 3-sulfated glucuronic acid (SO_4-3GlcAβ1–3Galβ1-) on the terminal saccharide chain of LM1 and Hex-LM1 generates sulfated glucuronyl paragloboside (SGPG) and its higher lactosaminyl homolog, sulfated glucuronyl lactosaminyl paragloboside (SGLPG). These unusual acidic glycolipids were discovered in autoimmune neuropathy patients with IgM paraproteins reactive to MAG ("anti-MAG" neuropathy). SG(L)PG, MAG, and other peripheral nerve glycoproteins all share this immunologically cross-reactive sulfated glucuronic acid structure as a terminal disaccharide residue [13, 14]. SG(L)PG is present in myelin- and axon-enriched fractions of motor and sensory nerves from human subjects but is not significantly present in rodent nerve [15].

Gangliosides are principally enriched in neuronal membranes, including those in peripheral nerve, but are also minor constituents of myelin [16]. GM1, GD1b, GD1a, and GQ1b are important antigens, but many other gangliosides have also been identified as antibody targets [1, 17]. When considered as antigen targets, ganglioside localization and distribution in the neural plasma membrane is important. Gangliosides are concentrated in association with cholesterol in clusters, referred to as lipid rafts [18], where in the plane of the plasma membrane their close lateral interactions may enhance their ability to bind antibody with high avidity or affinity. Gangliosides in specialized membrane domains may also be binding partners for antibody clearance by endocytosis [19–21]. Activated complement components and regulators such as CD59 that reside in rafts may in turn affect the normal functioning of gangliosides [22] and the pathological effects of anti-ganglioside antibodies [23]. The relative contribution of these factors may vary from site to site within peripheral nerve membranes, and between anti-ganglioside antibodies of differing immunoglobulin class and reactivity. There is little precise knowledge on the distribution of gangliosides in different membrane and raft domains in the PNS; it is clearly important to have more detail on the glycosphingolipid composition and distribution within the PNS in both humans and in experimental animals used for modelling these disorders. These issues are not straightforward, and a complete map of the ganglioside composition of human nerves, their accessibility to autoantibodies, and a comparison between species used for experimental modelling would be a valuable resource.

3 Clinical Considerations

The first human autoimmune neuropathy in which the antigen specificity was identified to be a peripheral nerve glycoconjugate was the "anti-MAG" neuropathy syndrome referred to above. Structural analysis revealed that the critical antigenic epitope was a 3-sulfate-glucuronic acid residue present on both glycoproteins and glycolipids, thereby accounting for the antibody reactivity with multiple molecules, all bearing the same antigenic determinant [24]. Some anti-MAG antibodies additionally cross-react with the glycolipid sulfatide. Antibodies to sulfatide that do not react with sulfated glucuronic acid are also associated with predominantly sensory neuropathies [25, 26]. The clinical phenotype of anti-MAG IgM paraproteinemic neuropathy is usually that of a late onset, chronic, and slowly progressive sensory dominant demyelinating neuropathy [27]. Subsequent to the discovery of "anti-MAG" antibodies, accounting for ~50% of patients with IgM paraproteinemic neuropathy, other patients were found to have IgM paraproteins that react with the NeuAcα2-8NeuAcα2-3-Gal disialosyl epitope in b-series gangliosides GD1b, GD3, GT1b, and GQ1b. The clinical phenotype in this syndrome is characterized by chronic sensory ataxic neuropathy, described under the acronym CANOMAD: *c*hronic *at*axic *n*europathy, *o*phthalmoplegia, Ig*M* paraprotein, cold *a*gglutinins, and *d*isialosyl antibodies [28]. Multifocal motor neuropathy is associated with anti-GM1 ganglioside antibodies in 50% of cases [29]. These antibodies are also of the IgM class but do not normally occur as IgM paraproteins. Three patterns of anti-GM1 antibody specificity are recognized based on epitope mapping [30, 31]: those that react with the terminal Gal beta1–3 GalNAc structure common to GM1, GD1b and asialo-GM1; those that react with the internal sialylated epitope common to GM1 and GM2, and those that react with GM1 monospecifically [32]. The clinical picture of multifocal motor neuropathy is typically of slowly progressive, asymmetrical distal upper limb weakness with minimal or no sensory impairment, although variants on this presentation exist.

Another important group of acute peripheral neuropathies associated with anti-glycolipid antibodies are the Guillain–Barré syndromes (GBS) [33, 34]. The clinical onset of GBS is rapid, and in around 20% of cases leads to total paralysis requiring mechanical ventilation. While recovery is usual, around 20% of affected individuals are still unable to walk 1 year after onset. The commonest form of GBS arises from segmental demyelination of peripheral nerve (acute inflammatory demyelinating polyneuropathy, AIDP) [35]. Although the precise targets for immune attack in AIDP still remain unknown, some evidence suggests that the antibodies may bind to glycolipids and gangliosides enriched in Schwann cell membranes [36]. These include LM1, SGPG, and GM1 ganglioside, each identified in a small proportion of cases. Two GBS variants, acute motor axonal neuropathy (AMAN) and Miller Fisher syndrome (MFS) are consistently and highly associated anti-glycolipid autoantibodies in most cases. In AMAN, rather than the myelin sheath, the primary target is the motor axolemmal membrane [37, 38]. AMAN is very strongly associated with anti-ganglioside antibodies directed to sialylated epitopes on GM1a, GM1b, GD1a, and GalNAc-GD1a [39–42]. The second widely studied variant is MFS [43]. Clinical features in MFS are restricted to limb ataxia, tendon reflex loss, and extraocular nerve paralysis (causing paralysis of eye movements, ophthalmoplegia). This disorder is very strongly associated with antibodies to GQ1b and GT1a gangliosides in the majority of cases [44–46].

4 The Origin of Anti-ganglioside Antibodies

Carbohydrates, including ganglioside glycans, are distinct from protein antigens in that glycans cannot normally be presented by MHC Class II molecules to T cell receptors, whose function is to help the immune system generate adaptive immune responses: Thus they are referred to as T cell independent (TI) antigens [47]. The exception is the CD1 system of antigen presentation in which TCRs can recognize glycans when they are presented by CD1 molecules [48, 49]. Antibody responses to TI antigens provide important components to both innate and adaptive antimicrobial immunity, including protection against lipopolysaccharide (LPS) encapsulated bacteria and are very widely expressed in the natural and adaptive antibody repertoire [50–52]. In GBS, this immunological control over B cell responses to microbial glycans is dysregulated, in many cases associated with molecular mimicry between microbial glycans and host glycolipids [53]. The commonest bacterial infection exhibiting mimicry that triggers GBS is the food and water borne pathogen *Campylobacter jejuni* (*C. jejuni*) [54]. The glycans present on the LOS of GBS-associated *C. jejuni* isolates contain ganglioside-mimicking structures including GM1, GM2, GD1a, GT1a and GD3 [55]. These are immunologically recognizable by monoclonal antibodies and by the anti-ganglioside antibodies in the serum of GBS patients triggered by *C. jejuni* infection [56, 57].

5 Measurement of Anti-ganglioside Antibodies in Human Serum

Anti-ganglioside antibodies are measured in serum, although they can also be estimated in plasma [58]. Cerebrospinal fluid may contain very small amounts of anti-ganglioside antibody derived from the systemic circulation, but in contrast to certain other neuroimmunological disorders there is no evidence that local intrathecal synthesis of antibody occurs in autoimmune neuropathy. Like many other antibodies, serum anti-ganglioside antibodies are stable for short periods at standard ambient temperature and can be temporarily stored at 4 °C without adverse effects. Repeated freeze-thawing of sera should be avoided. In chronic peripheral neuropathy syndromes, anti-ganglioside antibodies are usually IgM class antibodies and measurement can be performed at any stage of the clinical illness. In the acute-onset GBS and clinical variants, serum anti-ganglioside antibodies are generally IgG class but may include IgM and IgA classes as well. They are at their highest levels at clinical onset, then decay over the ensuing weeks, and will also be affected by treatments; therefore samples for antibody measurement should be collected promptly.

Anti-ganglioside antibodies can be detected by several techniques, including enzyme-linked immunosorbent assay (ELISA), immunodot-assay, flow-cytometry and cell surface binding, and glyco-array [59–61]. In most clinical diagnostic situations, assay surfaces are coated with purified gangliosides obtained by in house purification or from commercial sources, and comprise the whole ganglioside molecule—both lipid and glycan. Some investigators add accessory lipids including cholesterol to the coating mixture, or formulate liposomes [62]. A wide range of glycan arrays that contain immobilized ganglio-series glycan headgroups are also in use and under development [63, 64]. Many measurement methods are described and widely available. Assay performance varies widely, both within a single assay and between assays. As such, different techniques should be standardized for consistency between different laboratories [58]. Adaptations of methods may preferentially detect particular types of antibody, believed to be due to variations in the density and the orientation of the glycan headgroup on the immobilized surface. Importantly, different techniques should not necessarily be expected to be fully concordant with each other, or superior to each other, as there is no recognized optimal assay for detecting these antibodies. Despite limitations, ELISA still remains a commonly used method for intact gangliosides as all laboratories are widely conversant with this standard and simple technology.

There is an extensive body of historical evidence that *cis* interactions (lateral interactions in the plane of the plasma membrane) between structurally different glycolipids can substantially enhance or inhibit their ability to bind to antibodies and other lectins [65, 66]. The significance of these interactions in living nerve membranes is of major interest to the anti-glycolipid antibody and neuropathy field. The presence of "cryptic" glycolipid antigens in plasma membranes and the importance of accessory lipids and glycolipids in modulating antibody binding are long recognized, especially when studying antibody binding in immunohistological analyses [67, 68]. One study that very clearly illustrated this point demonstrated that an anti-GM3 monoclonal antibody that bound well to melanoma cell lines expressing GM3 alone was no longer able to bind when GM2, GD2 or GD3 gangliosides were coexpressed with GM3, even when the latter formed 50% of the total ganglioside composition of the cells [69]. This indicates that the antibody-binding GM3 epitope is capable of being masked by adjacent gangliosides. A more recent study has shown masking of GM1 by GD1a on the axolemmal membranes of motor axons that is of relevance in anti-GM1 antibody-associated motor neuropathies [70]. It is equally recognized that combinations of oligosaccharide groups from different glycans can form a distinct selectin epitope termed a" clustered saccharide patch", without detectable binding to the individual oligosaccharides themselves [66]. This concept was prompted by the observation that the selectins had unusually high affinity for their target glycans which was unexpected for apparently monovalent carbohydrate ligands. These patterns of lectin– and antibody–carbohydrate complex interaction have been recently referred to as complex enhanced, complex attenuated or complex independent binding [71].

A series of studies on sera from patients with GBS and chronic motor neuropathies has aroused clinical diagnostic interest in this area [72, 73]. An unusual observation showed that serum IgG antibodies from certain GBS patients showed strong reactivity to the heteromeric mixture of GD1a and GD1b in an equimolar ratio, while failing to bind to either ganglioside alone. In a larger study that comprised screening of GBS sera, complexes of GD1a:GD1b, GM1:GD1a, GM1:GD1b, GM1:GT1b, GD1a:GT1b and GD1b:GT1b were all found to be antibody targets, indicating that clustering of different glyco-epitopes within ganglioside complexes in the plasma membrane act as novel ligands for pathogenic autoantibodies. This phenomenon also appears to occur with IgM antibodies: for example, anti-GM1 antibodies in MMN preferentially bind GM1 when it is in complex with GalC [74, 75]. What remains to be established is how these findings observed in solid-phase immunoassays correlate with glycolipid behavior in the plasma membrane. One study looking at the interaction between GM2 and GM3 has used an anti-GM2:GM3 monoclonal antibody to identify the complex on cell surfaces and shown that it mediates an inhibitory effect on cell motility via a CD82/cMe pathway [12, 76]. Further studies in this area are needed. It is likely that further antibodies to complexes will be identified provided that analytical screening methodology is designed with this concept in mind.

6 The Pathological Effects of Anti-glycolipid Antibodies in Experimental Neuropathy Models

Anti-ganglioside antibodies have been shown to directly injure nerves in experimental animals, although demonstrating this is complex. Anti-ganglioside antibodies might be expected to bind any ganglioside-containing glial, neuronal, or axonal membranes that they can gain access to through the blood–nerve barrier. However steric inhibition locally within the plasma membrane may prevent antibody binding from occurring [69, 70, 77]. Since gangliosides rapidly traffic to and from the plasma membrane through endosomal sorting, membrane-bound antibody must remain on the cell surface for long enough to initiate injury through inflammatory cell recruitment or complement activation, and this rate of clearance may differ between membranes sites and microdomains [19, 20]. The axolemmal membrane is encased by myelin in the internodes, but at the nodes of Ranvier and motor nerve terminals it is very exposed to antibody binding [78]. In the dorsal root ganglion, the plasma membrane of the neuronal cell body and spinal roots is also relatively accessible to autoantibodies [79]. Paranodal, noncompacted myelin membranes, as well as abaxonal Schwann cell plasma membranes will be similarly exposed. One likely explanation for the lack of CNS involvement in anti-glycolipid antibody associated neuropathies, despite the wide distribution of gangliosides in the CNS, is limitation of access through the blood–brain barrier.

Having determined that a target membrane is accessible to an autoantibody, another consideration is whether the density of the glycolipid within the membrane is sufficient to mediate a pathological injury upon antibody binding [80]. Biochemical and immunohistological approaches to establishing this have merits and limitations. The overall glycolipid composition of different nerves as defined biochemically provides no information about microanatomical distribution of gangliosides that could be important when considering disease pathogenesis. Immunohistology, using affinity purified antisera, monoclonal antibodies or ganglioside-binding toxins can reveal fine detail about ganglioside distribution at the cellular and subcellular level. However, many anti-ganglioside antibodies may cross-react with gangliosides and other glycoconjugate antigens bearing similar or often identical structures or be masked from binding by cis-interacting gangliosides [70], making extrapolation of immunohistology data an imprecise indicator of ganglioside localization.

Active and passive immunization models have both been widely used to demonstrate causality and investigate effects of anti-glycolipid antibodies in animals. In active immunization, antigen is administered by injection in an immunologically active form, often combined with adjuvant such that the recipient generates an antibody response to the antigen, which then proceeds to damage the target tissue. In passive immunization, also termed passive transfer, the recipient animal is injected with the relevant antibody or antiserum directed toward the target antigen under study.

Animal models of anti-glycolipid antibodies were first examined in the 1970s by active immunization of rabbits with gangliosides and GalC [8, 81]. Following GalC immunization, rabbits developed flaccid weakness, sensory loss and respiratory paralysis. Pathological findings comprised vesicular demyelination in spinal roots and macrophage infiltration with myelin phagocytosis. Demyelinating lesions corresponded with areas known to have a defective blood–nerve barrier. Subsequently, intraneural injection of rabbit anti-GalC serum produced focal demyelinating lesions in rat sciatic nerves with similar pathology [82]. Local application of anti-galactocerebroside antibody to rat ventral roots was also shown to induce acute conduction block, and activity was lost after complement inactivation, suggesting that the pathological activity of anti-galactocerebroside serum is complement-mediated [83]. Thus, even 40 years ago, the basic principles of anti-glycolipid antibody-induced injury were well established.

The approach of combining both active and passive immunization has now been widely applied to many other anti-glycolipid antibodies in many species. Active immunization of rabbits with SGPG results in an immune response with anti-SGPG antibody induction, hind limb weakness, and slowed nerve conduction velocity in the sciatic nerve [84]. Local microinjection of anti-MAG IgM antibodies into feline nerves induces complement-dependent demyelination with macrophage mediated myelin stripping [85, 86]. In passive immunization experiments in chickens, anti-MAG IgM induces nerve demyelination with the widening of myelin lamellae characteristic of human biopsies [87]. A model of human sensory ataxic neuropathy induced in rabbits by active immunization with GD1b ganglioside well reflects the human counterpart [88]. Immunohistochemical studies showed that GD1b is localized to large DRG neurons in human and rabbit. About 50% of rabbits immunized with GD1b developed ataxia with splayed limbs, but retained normal muscle power. IgG and IgM antibody titers monospecific for GD1b were more prominently raised in affected than unaffected rabbits, indicating the importance of the fine specificity of the immune response exclusively to GD1b. Pathological examination of affected rabbits showed axonal degeneration in the dorsal roots and some neuronal cell bodies in the DRG had either degenerated or disappeared, involving neurotrophin-3-mediated trkC signaling [89]. The ventral root was entirely spared. This active immunization experiment was further supported by demonstrating that degeneration of rabbit sensory neurons could be induced by passive transfer of anti-GD1b antiserum. More recently, application of anti-GD1b antibody to myelinating cell culture has shown both complement dependent and independent effects.

In an anti-GM1 antibody-mediated rabbit model of AMAN, animals become severely paralyzed [90]. Pathological features comprise motor axonal degeneration with node of Ranvier injury in the ventral roots [91]. In murine studies, AMAN-associated antisera and anti-GM1 and anti-GD1a monoclonal antibodies derived from both humans and mice immunized with *Campylobacter jejuni* LOS have been used to induce disease [80]. Axons within the motor nerve terminal and distal motor nerve lying outside the blood–nerve barrier are coated with antibody and severely affected. Deposition of terminal complement pores allowed uncontrolled calcium ingress triggering calpain activation and disruption of nodal and nerve terminal architecture, with concomitant paralysis [92]. An important experimental modification used in mice is to study animals expressing abnormally high amounts of GM1 or GD1a, made possible through the targeted deletion of the biosynthetic enzyme GD3-synthase [93]. This resulted in an absence of b series gangliosides and a compensatory increase in the levels of a series gangliosides. GD3 synthase deficient mice exposed to anti-GM1 or anti-GD1a antibodies develop a highly destructive injury. Inhibition of complement component C5 (that results in an inability to form the terminal membrane attack complex) or earlier classical pathway complement components has a major neuroprotective effect, completely preventing any structural and functional changes [19, 94]. Thus a major pathway of acute pathological injury mediated by anti-glycolipid antibodies in GBS and variants converges at the terminal component of the complement cascade, membrane attack complex.

Experimental animal studies have ALSO been conducted on the anti-GQ1b antibodies that characterize MFS [95]. Toxin, lectin, and antibody based immunohistological studies have shown that the presynaptic NMJ is rich in gangliosides [96]. At the presynaptic neuromuscular junction (NMJ), complex gangliosides including GQ1b are known to act as receptors for botulinum toxins; it therefore seems logical that anti-GQ1b antibodies might also bind at the NMJ where they could induce nerve terminal injury [97]. Using in vitro mouse hemidiaphragm preparations, it was first demonstrated that anti-GQ1b antibodies associated with MFS bind the motor nerve terminal where they locally activate complement and cause destructive injury [98]. In addition to inducing injury to the axon terminal, anti-ganglioside antibodies with different specificities for GQ1b, GT1a and GD3 also target perisynaptic Schwann cells (pSCs) for immune attack [99].

Overall the above animal studies, supported by human autopsy and biopsy studies, clearly indicate that anti-ganglioside antibodies have major pathological effects in animal models and point toward some of the pathological pathways that might be involved, and amenable to therapeutic intervention. However, before embarking on such studies, conceptual and methodological issues outlined above need careful consideration.

7 Conclusions

Recent years have witnessed abundant progress in our understanding of the glycobiology and glycoimmunology of inflammatory neuropathies, and in doing so has both informed basic and clinical science. Clinical and serological data clearly shows a disease-specific correlation between peripheral neuropathies and particular anti-glycolipid antibodies. Many interesting patterns have emerged that support the view that anti-glycolipid antibodies play an active role in pathogenesis. Experimental evidence obtained from human and animal studies continues to support the model of post-infectious neuropathy as a disease involving molecular mimicry between bacterial and neural oligosaccharides. One notable emerging area concerns the interacting relationship between glycolipids and other lipids in the plane of the plasma membrane and how this might either positively or negatively influence autoantibody binding. Methodologies for antibody measurement and animal models are still undergoing refinements that need active consideration when researching this area.

References

1. Willison HJ, Yuki N (2002) Peripheral neuropathies and anti-glycolipid antibodies. Brain J Neurol 125(Pt 12):2591–2625
2. Kieseier BC, Hartung HP, Wiendl H (2006) Immune circuitry in the peripheral nervous system. Curr Opin Neurol 19(5):437–445. https://doi.org/10.1097/01.wco.0000245365.51823.72. 00019052-200610000-00003 [pii]
3. Meyer Zu HG, Hu W, Hartung HP, Lehmann HC, Kieseier BC (2008) The immunocompetence of Schwann cells. Muscle Nerve 37(1):3–13. https://doi.org/10.1002/mus.20893
4. Kanda T (2013) Biology of the blood-nerve barrier and its alteration in immune mediated neuropathies. J Neurol Neurosurg Psychiatry, 84 (2):208–212. https://doi.org/10.1136/jnnp-2012-302312. [pii]
5. Quarles RH, Weiss MD (1999) Autoantibodies associated with peripheral neuropathy. Muscle Nerve 22(7):800–822. https://doi.org/10.1002/(SICI)1097-4598(199907)22:7<800::AID-MUS2>3.0.CO;2-F
6. Ogawa-Goto K, Abe T (1998) Gangliosides and glycosphingolipids of peripheral nervous system myelins-a minireview. Neurochem Res 23(3):305–310
7. Salzer JL, Brophy PJ, Peles E (2008) Molecular domains of myelinated axons in the peripheral nervous system. Glia 56(14):1532–1540. https://doi.org/10.1002/glia.20750
8. Saida T, Saida K, Dorfman SH, Silberberg DH, Sumner AJ, Manning MC, Lisak RP, Brown MJ (1979) Experimental allergic neuritis induced by sensitization with galactocerebroside. Science 204(4397):1103–1106
9. Honke K (2013) Biosynthesis and biological function of sulfoglycolipids. Proc Jpn Acad Ser B Phys Biol Sci 89(4):129–138. DN/JST.JSTAGE/pjab/89.129 [pii]
10. Popko B, Dupree JL, Coetzee T, Suzuki K, Suzuki K (1999) Genetic analysis of myelin galactolipid function. Adv Exp Med Biol 468:237–244
11. Svennerholm L, Bostrom K, Fredman P, Jungbjer B, Lekman A, Mansson JE, Rynmark BM (1994) Gangliosides and allied glycosphingolipids in human peripheral nerve and spinal cord. Biochim Biophys Acta 1214(2):115–123. 0005-2760(94)90034-5 [pii]
12. Gu Y, Chen ZW, Siegel A, Koshy R, Ramirez C, Raabe TD, Devries GH, Ilyas AA (2012) Analysis of humoral immune responses to LM1 ganglioside in guinea pigs. J Neuroimmunol 246(1–2):58–64. https://doi.org/10.1016/j.jneuroim.2012.03.001. S0165-5728(12)00065-3 [pii]
13. Ariga T, Kohriyama T, Freddo L, Latov N, Saito M, Kon K, Ando S, Suzuki M, Hemling ME, Rinehart KL Jr (1987) Characterization of sulfated glucuronic acid containing glycolipids reacting with IgM M-proteins in patients with neuropathy. J Biol Chem 262(2):848–853

14. Chou DK, Ilyas AA, Evans JE, Costello C, Quarles RH, Jungalwala FB (1986) Structure of sulfated glucuronyl glycolipids in the nervous system reacting with HNK-1 antibody and some IgM paraproteins in neuropathy. J Biol Chem 261(25):11717–11725
15. Ilyas AA, Dalakas MC, Brady RO, Quarles RH (1986) Sulfated glucuronyl glycolipids reacting with anti-myelin-associated glycoprotein monoclonal antibodies including IgM paraproteins in neuropathy: species distribution and partial characterization of epitopes. Brain Res 385(1):1–9. 0006-8993(86)91540-4 [pii]
16. Ledeen RW, Yu RK (1982) Gangliosides: structure, isolation, and analysis. Methods Enzymol 83:139–191
17. Kusunoki S (2003) Anti-ganglioside antibodies in Guillain-Barre syndrome; useful diagnostic markers as well as possible pathogenetic factors. Intern Med 42(6):457–458
18. Simons K, Ikonen E (1997) Functional rafts in cell membranes. Nature 387(6633):569–572
19. Cunningham ME, McGonigal R, Meehan GR, Barrie JA, Yao D, Halstead SK, Willison HJ (2016) Anti-ganglioside antibodies are removed from circulation in mice by neuronal endocytosis. Brain 139(Pt 6):1657–1665. https://doi.org/10.1093/brain/aww056
20. Fewou SN, Plomp JJ, Willison HJ (2014) The pre-synaptic motor nerve terminal as a site for antibody-mediated neurotoxicity in autoimmune neuropathies and synaptopathies. J Anat 224(1):36–44. https://doi.org/10.1111/joa.12088
21. Fewou SN, Rupp A, Nickolay LE, Carrick K, Greenshields KN, Pediani J, Plomp JJ, Willison HJ (2012) Anti-ganglioside antibody internalization attenuates motor nerve terminal injury in a mouse model of acute motor axonal neuropathy. J Clin Invest 122(3):1037–1051. https://doi.org/10.1172/JCI59110
22. Ohmi Y, Ohkawa Y, Tajima O, Sugiura Y, Furukawa K, Furukawa K (2014) Ganglioside deficiency causes inflammation and neurodegeneration via the activation of complement system in the spinal cord. J Neuroinflammation 11:61. https://doi.org/10.1186/1742-2094-11-61
23. Morgan BP (2015) The role of complement in neurological and neuropsychiatric diseases. Expert Rev Clin Immunol 11(10):1109–1119. https://doi.org/10.1586/1744666x.2015.1074039
24. Burger D, Simon M, Peruisseau G, Steck AJ (1990) The epitope(s) recognised by HNK-1 antibody and IgM paraprotein in neuropathy is present on several N-linked oligosaccharide structures on human P0 and myelin-associated glycoprotein. J Neurochem 54:1569–1575
25. Lopate G, Kornberg AJ, Yue J, Choksi R, Pestronk A (2001) Anti-myelin associated glycoprotein antibodies: variability in patterns of IgM binding to peripheral nerve. J Neurol Sci 188:67–72
26. Lopate G, Parks BJ, Goldstein JM, Yee WC, Friesenhahn GM, Pestronk A (1997) Polyneuropathies associated with high titre antisulphatide antibodies: characteristics of patients with and without serum monoclonal proteins. J Neurol Neurosurg Psychiatry 62:581–585
27. Nobile-Orazio E, Manfredini E, Carpo M, Meucci N, Monaco S, Ferrari S, Bonetti B, Cavaletti G, Gemignani F, Durelli L, Barbieri S, Allaria S, Sgarzi M, Scarlato G (1994) Frequency and clinical correlates of anti-neural IgM antibodies in neuropathy associated with IgM monoclonal gammopathy. Ann Neurol 36:416–424
28. Willison HJ, O'Leary CP, Veitch J, Blumhardt LD, Busby M, Donaghy M, Fuhr P, Ford H, Hahn A, Renaud S, Katifi HA, Ponsford S, Reuber M, Steck A, Sutton I, Schady W, Thomas PK, Thompson AJ, Vallat JM, Winer J (2001) The clinical and laboratory features of chronic sensory ataxic neuropathy with anti-disialosyl IgM antibodies. Brain J Neurol 124(Pt 10):1968–1977
29. Kornberg AJ, Pestronk A (1995) Chronic motor neuropathies: diagnosis, therapy, and pathogenesis. Ann Neurol 37(Suppl 1):S43–S50
30. Paterson G, Wilson G, Kennedy PGE, Willison HJ (1995) Analysis of anti-GM1 ganglioside IgM antibodies cloned from motor neuropathy patients demonstrates diverse v-region gene usage with extensive somatic mutation. J Immunol 155:3049–3059
31. Townson K, Boffey J, Nicholl D, Veitch J, Bundle D, Zhang P, Samain E, Antoine T, Bernardi A, Arosio D, Sonnino S, Isaacs N, Willison HJ (2007) Solid phase immunoadsorption for therapeutic and analytical studies on neuropathy-associated anti-GM1 antibodies. Glycobiology 17(3):294–303
32. Willison HJ, Paterson G, Kennedy PG, Veitch J (1994) Cloning of human anti-GM1 antibodies from motor neuropathy patients. Ann Neurol 35(4):471–478
33. Goodfellow JA, Willison HJ (2016) Guillain-Barre syndrome: a century of progress. Nat Rev Neurol 12(12):723–731. https://doi.org/10.1038/nrneurol.2016.172
34. Willison HJ, Jacobs BC, van Doorn PA (2016) Guillain-Barre syndrome. Lancet

388(10045):717–727. https://doi.org/10.1016/s0140-6736(16)00339-1

35. Hafer-Macko CE, Sheikh KA, Li CY, Ho TW, Cornblath DR, McKhann GM, Asbury AK, Griffin JW (1996) Immune attack on the Schwann cell surface in acute inflammatory demyelinating polyneuropathy. Ann Neurol 39(5):625–635. https://doi.org/10.1002/ana.410390512
36. Kusunoki S, Kaida K (2011) Antibodies against ganglioside complexes in Guillain-Barre syndrome and related disorders. J Neurochem 116(5):828–832. https://doi.org/10.1111/j.1471-4159.2010.07029.x
37. Hafer-Macko C, Hsieh ST, Li CY, Ho TW, Sheikh K, Cornblath DR, McKhann GM, Asbury AK, Griffin JW (1996) Acute motor axonal neuropathy: an antibody-mediated attack on axolemma. Ann Neurol 40(4):635–644. https://doi.org/10.1002/ana.410400414
38. Ho TW, Hsieh ST, Nachamkin I, Willison HJ, Sheikh K, Kiehlbauch J, Flanigan K, McArthur JC, Cornblath DR, McKhann GM, Griffin JW (1997) Motor nerve terminal degeneration provides a potential mechanism for rapid recovery in acute motor axonal neuropathy after Campylobacter infection. Neurology 48(3):717–724
39. Halstead SK, Kalna G, Islam MB, Jahan I, Mohammad QD, Jacobs BC, Endtz HP, Islam Z, Willison HJ (2016) Microarray screening of Guillain-Barre syndrome sera for antibodies to glycolipid complexes. Neurol Neuroimmunol Neuroinflamm 3(6):e284. https://doi.org/10.1212/nxi.0000000000000284
40. Ho TW, Willison HJ, Nachamkin I, Li CY, Veitch J, Ung H, Wang GR, Liu RC, Cornblath DR, Asbury AK, Griffin JW, McKhann GM (1999) Anti-GD1a antibody is associated with axonal but not demyelinating forms of Guillain-Barre syndrome. Ann Neurol 45(2):168–173
41. Kaida K, Kusunoki S, Kamakura K, Motoyoshi K, Kanazawa I (2003) GalNAc-GD1a in human peripheral nerve: target sites of anti-ganglioside antibody. Neurology 61(4):465–470
42. Yuki N (2001) Infectious origins of, and molecular mimicry in, Guillain-Barre and Fisher syndromes. Lancet Infect Dis 1(1):29–37
43. Fisher M (1956) An unusual variant of acute idiopathic polyneuritis (syndrome of ophthalmoplegia, ataxia and areflexia). N Engl J Med 255(2):57–65
44. Chiba A, Kusunoki S, Obata H, Machinami R, Kanazawa I (1993) Serum anti-GQ1b IgG antibody is associated with ophthalmoplegia in Miller Fisher syndrome and Guillain-Barre syndrome: clinical and immunohistochemical studies. Neurology 43(10):1911–1917
45. Chiba A, Kusunoki S, Shimizu T, Kanazawa I (1992) Serum IgG antibody to ganglioside GQ1b is a possible marker of Miller Fisher syndrome. Ann Neurol 31(6):677–679. https://doi.org/10.1002/ana.410310619
46. Willison HJ, Veitch J, Paterson G, Kennedy PG (1993) Miller Fisher syndrome is associated with serum antibodies to GQ1b ganglioside. J Neurol Neurosurg Psychiatry 56(2):204–206
47. Defrance T, Taillardet M, Genestier L (2011) T cell-independent B cell memory. Curr Opin Immunol 23(3):330–336. https://doi.org/10.1016/j.coi.2011.03.004. S0952-7915(11)00026-4 [pii]
48. Desmazieres A, Zonta B, Zhang A, Wu LM, Sherman DL, Brophy PJ (2014) Differential stability of PNS and CNS nodal complexes when neuronal neurofascin is lost. J Neurosci 34(15):5083–5088. https://doi.org/10.1523/jneurosci.4662-13.2014
49. Mori L, De LG (2012) T cells specific for lipid antigens. Immunol Res 53(1–3):191–199. https://doi.org/10.1007/s12026-012-8294-6
50. Galili U (2016) Natural anti-carbohydrate antibodies contributing to evolutionary survival of primates in viral epidemics? Glycobiology 26(11):1140–1150. https://doi.org/10.1093/glycob/cww088
51. Schneider C, Smith DF, Cummings RD, Boligan KF, Hamilton RG, Bochner BS, Miescher S, Simon HU, Pashov A, Vassilev T, von Gunten S (2015) The human IgG anti-carbohydrate repertoire exhibits a universal architecture and contains specificity for microbial attachment sites. Sci Transl Med 7(269):269ra261. https://doi.org/10.1126/scitranslmed.3010524
52. Zimmermann S, Lepenies B (2015) Glycans as vaccine antigens and adjuvants: immunological considerations. Methods Mol Biol (Clifton, NJ) 1331:11–26. https://doi.org/10.1007/978-1-4939-2874-3_2
53. Willison HJ, Goodyear CS (2013) Glycolipid antigens and autoantibodies in autoimmune neuropathies. Trends Immunol 34(9):453–459. https://doi.org/10.1016/j.it.2013.05.001
54. Yuki N, Taki T, Inagaki F, Kasama T, Takahashi M, Saito K, Handa S, Miyatake T (1993) A bacterium lipopolysaccharide that elicits Guillain-Barre syndrome has a GM1 ganglioside-like structure. J Exp Med 178(5):1771–1775
55. Yuki N (2007) Ganglioside mimicry and peripheral nerve disease. Muscle Nerve

35(6):691–711. https://doi.org/10.1002/mus.20762

56. Boffey J, Nicholl D, Wagner ER, Townson K, Goodyear C, Furukawa K, Furukawa K, Conner J, Willison HJ (2004) Innate murine B cells produce anti-disialosyl antibodies reactive with Campylobacter jejuni LPS and gangliosides that are polyreactive and encoded by a restricted set of unmutated V genes. J Neuroimmunol 152(1–2):98–111. https://doi.org/10.1016/j.jneuroim.2004.04.002
57. Bowes T, Wagner ER, Boffey J, Nicholl D, Cochrane L, Benboubetra M, Conner J, Furukawa K, Furukawa K, Willison HJ (2002) Tolerance to self gangliosides is the major factor restricting the antibody response to lipopolysaccharide core oligosaccharides in Campylobacter jejuni strains associated with Guillain-Barre syndrome. Infect Immun 70(9):5008–5018
58. Willison HJ, Veitch J, Swan AV, Baumann N, Comi G, Gregson NA, Llla I, Jacobs BC, Zielasek J, Hughes RAC (1999) Inter-laboratory validation of an ELISA for the determination of serum anti-ganglioside antibodies. Eur J Neurol 6:71–77
59. Alaedini A, Briani C, Wirguin I, Siciliano G, D'Avino C, Latov N (2002) Detection of anti-ganglioside antibodies in Guillain-Barre syndrome and its variants by the agglutination assay. J Neurol Sci 196(1–2):41–44
60. Ravindranath MH, Muthugounder S, Saravanan TS, Presser N, Morton DL (2005) Human antiganglioside autoantibodies: validation of ELISA. Ann N Y Acad Sci 1050:229–242. https://doi.org/10.1196/annals.1313.024
61. Rinaldi S, Brennan KM, Willison HJ (2012) Combinatorial glycoarray. Methods Mol Biol (Clifton, NJ) 808:413–423. https://doi.org/10.1007/978-1-61779-373-8_28
62. Rongen HA, Bult A, van Bennekom WP (1997) Liposomes and immunoassays. J Immunol Methods 204(2):105–133
63. Hahm HS, Schlegel MK, Hurevich M, Eller S, Schuhmacher F, Hofmann J, Pagel K, Seeberger PH (2017) Automated glycan assembly using the Glyconeer 2.1 synthesizer. Proc Natl Acad Sci U S A 114(17):E3385–e3389. https://doi.org/10.1073/pnas.1700141114
64. Liu Y, McBride R, Stoll M, Palma AS, Silva L, Agravat S, Aoki-Kinoshita KF, Campbell MP, Costello CE, Dell A, Haslam SM, Karlsson NG, Khoo KH, Kolarich D, Novotny MV, Packer NH, Ranzinger R, Rapp E, Rudd PM, Struwe WB, Tiemeyer M, Wells L, York WS, Zaia J, Kettner C, Paulson JC, Feizi T, Smith DF (2016) The minimum information required for a glycomics experiment (MIRAGE) project: improving the standards for reporting glycan microarray-based data. Glycobiology. https://doi.org/10.1093/glycob/cww118
65. Regina TA, Hakomori SI (2008) Functional role of glycosphingolipids and gangliosides in control of cell adhesion, motility, and growth, through glycosynaptic microdomains. Biochim Biophys Acta 1780(3):421–433. https://doi.org/10.1016/j.bbagen.2007.10.008. S0304-4165(07)00240-1 [pii]
66. Varki A (1994) Selectin ligands. Proc Natl Acad Sci U S A 91(16):7390–7397
67. Ritz MF, Lechner-Scott J, Scott RJ, Fuhr P, Malik N, Erne B, Taylor V, Suter U, Schaeren-Wiemers N, Steck AJ (2000) Characterisation of autoantibodies to peripheral myelin protein 22 in patients with hereditary and acquired neuropathies. J Neuroimmunol 104(2):155–163
68. Schwarz A, Futerman AH (1996) The localization of gangliosides in neurons of the central nervous system: the use of anti-ganglioside antibodies. Biochim Biophys Acta 1286(3):247–267. S0304-4157(96)00011-1 [pii]
69. Lloyd KO, Gordon CM, Thampoe IJ, DiBenedetto C (1992) Cell surface accessibility of individual gangliosides in malignant melanoma cells to antibodies is influenced by the total ganglioside composition of the cells. Cancer Res 52(18):4948–4953
70. Greenshields KN, Halstead SK, Zitman FM, Rinaldi S, Brennan KM, O'Leary C, Chamberlain LH, Easton A, Roxburgh J, Pediani J, Furukawa K, Furukawa K, Goodyear CS, Plomp JJ, Willison HJ (2009) The neuropathic potential of anti-GM1 autoantibodies is regulated by the local glycolipid environment in mice. J Clin Invest 119(3):595–610. https://doi.org/10.1172/jci37338
71. Rinaldi S, Brennan KM, Willison HJ (2010) Heteromeric glycolipid complexes as modulators of autoantibody and lectin binding. Prog Lipid Res 49(1):87–95. https://doi.org/10.1016/j.plipres.2009.09.001
72. Kaida K, Kusunoki S (2010) Antibodies to gangliosides and ganglioside complexes in Guillain-Barre syndrome and Fisher syndrome: mini-review. J Neuroimmunol 223(1–2):5–12. https://doi.org/10.1016/j.jneuroim.2010.02.001
73. Kaida K, Morita D, Kanzaki M, Kamakura K, Motoyoshi K, Hirakawa M, Kusunoki S (2004) Ganglioside complexes as new target antigens in Guillain-Barre syndrome. Ann Neurol 56(4):567–571. https://doi.org/10.1002/ana.20222
74. Galban-Horcajo F, Vlam L, Delmont E, Halstead SK, van den Berg L, van der Pol WL,

Willison HJ (2015) The diagnostic utility of determining anti-GM1: GalC complex antibodies in multifocal motor neuropathy: a validation study. J Neuromuscular Dis 2(2):157–165. https://doi.org/10.3233/jnd-150080

75. Pestronk A, Choksi R, Blume G, Lopate G (1997) Multifocal motor neuropathy: serum IgM binding to a GM1 ganglioside-containing lipid mixture but not to GM1 alone. Neurology 48(4):1104–1106

76. Todeschini AR, Dos Santos JN, Handa K, Hakomori SI (2008) Ganglioside GM2/GM3 complex affixed on silica nanospheres strongly inhibits cell motility through CD82/cMet-mediated pathway. Proc Natl Acad Sci U S A 105(6):1925–1930. doi:https://doi.org/10.1073/pnas.0709619104

77. Kusunoki S, Hitoshi S, Kaida K, Arita M, Kanazawa I (1999) Monospecific anti-GD1b IgG is required to induce rabbit ataxic neuropathy. Ann Neurol 45(3):400–403

78. McGonigal R, Rowan EG, Greenshields KN, Halstead SK, Humphreys PD, Rother RP, Furukawa K, Willison HJ (2010) Anti-GD1a antibodies activate complement and calpain to injure distal motor nodes of Ranvier in mice. Brain 133(Pt 7):1944–1960. https://doi.org/10.1093/brain/awq119

79. Kusunoki S, Hitoshi S, Kaida K, Murayama S, Kanazawa I (1999) Degeneration of rabbit sensory neurons induced by passive transfer of anti-GD1b antiserum. Neurosci Lett 273(1):33–36. S0304394099006217 [pii]

80. Goodfellow JA, Bowes T, Sheikh K, Odaka M, Halstead SK, Humphreys PD, Wagner ER, Yuki N, Furukawa K, Furukawa K, Plomp JJ, Willison HJ (2005) Overexpression of GD1a ganglioside sensitizes motor nerve terminals to anti-GD1a antibody-mediated injury in a model of acute motor axonal neuropathy. J Neurosci 25(7):1620–1628

81. Nagai Y (1976) Ganglioside syndrome. Neurosci Lett 2:19604825

82. Saida T, Saida K, Silberberg DH (1979) Demyelination produced by experimental allergic neuritis serum and anti-galactocerebroside antiserum in central nervous system cultures. An ultrastructural study. Acta Neuropathol 48:19–25

83. Sumner A (1982) Electrophysiological and morphological effects of the injection of Guillain-Barré sera in the sciatic nerve of the rat. Rev Neurol (Paris) 138:17–24

84. Yu RK, Ariga T, Kohriyama T, Kusunoki S, Maeda Y, Miyatani N (1990) Autoimmune mechanisms in peripheral neuropathies. Ann Neurol 27(Suppl):S30–S35

85. Hays AP, Latov N, Takatsu M, Sherman WH (1987) Experimental demyelination of nerve induced by serum of patients with neuropathy and an anti MAG IgM M protein. Neurology 37:242–256

86. Willison HJ, Trapp BD, Bacher JD, Dalakas MC, Griffin JW, Quarles RH (1988) Demyelination induced by intraneural injection of human antimyelin-associated glycoprotein antibodies. Muscle Nerve 11(11):1169–1176. https://doi.org/10.1002/mus.880111111

87. Tatum AH (1993) Experimental paraprotein neuropathy, demyelination by passive transfer of human IgM anti-myelin-associated glycoprotein. Ann Neurol 33:502–506

88. Kusunoki S, Shimizu J, Chiba R, Ugawa Y, Hitoshi S, Kanazawa I (1996) Experimental sensory neuropathy induced by sensitisation with ganglioside GD1b. Ann Neurol 39:324–331

89. Hitoshi S, Kusunoki S, Murayama S, Tsuji S, Kanazawa I (1999) Rabbit experimental sensory ataxic neuropathy: anti-GD1b antibody-mediated trkC downregulation of dorsal root ganglia neurons. Neurosci Lett 260(3):157–160

90. Yuki N, Yamada M, Koga M, Odaka M, Susuki K, Tagawa Y, Ueda S, Kasama T, Ohnishi A, Hayashi S, Takahashi H, Kamijo M, Hirata K (2001) Animal model of axonal Guillain-Barre syndrome induced by sensitization with GM1 ganglioside. Ann Neurol 49(6):712–720

91. Susuki K, Nishimoto Y, Yamada M, Baba M, Ueda S, Hirata K, Yuki N (2003) Acute motor axonal neuropathy rabbit model: immune attack on nerve root axons. Ann Neurol 54(3):383–388

92. O'Hanlon GM, Humphreys PD, Goldman RS, Halstead SK, Bullens RW, Plomp JJ, Ushkaryov Y, Willison HJ (2003) Calpain inhibitors protect against axonal degeneration in a model of anti-ganglioside antibody-mediated motor nerve terminal injury. Brain 126(Pt 11):2497–2509. https://doi.org/10.1093/brain/awg254

93. Okada M, Itoh MM, Haraguchi M, Okajima T, Inoue M, Oishi H, Matsuda Y, Iwamoto T, Kawano T, Fukumoto S, Miyazaki H, Furukawa K, Aizawa S, Furukawa K (2002) b-Series ganglioside deficiency exhibits no definite changes in the neurogenesis and the sensitivity to Fas-mediated apoptosis but impairs regeneration of the lesioned hypoglossal nerve. J Biol Chem 277(3):1633–1636

94. Halstead SK, Zitman FM, Humphreys PD, Greenshields K, Verschuuren JJ, Jacobs BC, Rother RP, Plomp JJ, Willison HJ (2008) Eculizumab prevents anti-ganglioside antibody-mediated neuropathy in a murine

model. Brain 131(Pt 5):1197–1208. https://doi.org/10.1093/brain/awm316

95. Plomp JJ, Willison HJ (2009) Pathophysiological actions of neuropathy-related anti-ganglioside antibodies at the neuromuscular junction. J Physiol 587(Pt 16):3979–3999. https://doi.org/10.1113/jphysiol.2009.171702
96. Martin PT (2003) Glycobiology of the neuromuscular junction. J Neurocytol 32(5–8):915–929
97. Bullens RW, O'Hanlon GM, Wagner E, Molenaar PC, Furukawa K, Furukawa K, Plomp JJ, Willison HJ (2002) Complex gangliosides at the neuromuscular junction are membrane receptors for autoantibodies and botulinum neurotoxin but redundant for normal synaptic function. J Neurosci 22(16):6876–6884
98. Plomp JJ, Molenaar PC, O'Hanlon GM, Jacobs BC, Veitch J, Daha MR, Van Doorn PA, Van der Meche FGA, Vincent A, Morgan BP, Willison HJ (1999) Miller Fisher anti-GQ1b antibodies: a-latrotoxin-like effects on motor end plates. Ann Neurol 45:189–199
99. Halstead SK, O'Hanlon GM, Humphreys PD, Morrison DB, Morgan BP, Todd AJ, Plomp JJ, Willison HJ (2004) Anti-disialoside antibodies kill perisynaptic Schwann cells and damage motor nerve terminals via membrane attack complex in a murine model of neuropathy. Brain 127(Pt 9):2109–2123

Chapter 8

Protocols for Glycosyltransferase Assays: Ganglioside Globoside and Lewis-X Intermediate-Lactosylceramide Biosyntheses in Eukaryotic Systems

Subhash Basu, Arun Agarwal, Manju Basu, Rui Ma, and Joseph R. Moskal

Abstract

Protocols for assay of 24 different Glycolipid-Glycosyltransferases (GSL-GLTs) of the eukaryotic systems are described. Problems of quantitating the activities in crude membranes are also described. Different separation methods (for separation of substrate, donors, and the product of the reaction) have been described based on the paper chromatography or high voltage paper electrophoresis in 1.0% $Na_2B_4O_7$. Liquid Scintillation counting system was used for quantitation of the enzymatic product. In the assay of each GSL-GLT it is recommended to compare the selected method to be used with the exact conditions used by the authors published previously. As a test case for these assays the following kinetic parameters for Lactosylceramide Synthase, GalT-2 (UDP-Gal: Glc-Cer β1-4-galactosyltransferase), (Km of glucosylceramide = 1.65×10^{-4} M; Km for UDP-Gal = 0.5×10^{-4} M; V_{max} is determined in the presence of optimum detergent concentrations (2–15 mg/ml of Cutscum–Triton X-100, 2:1); Mn^{++} and Mg^{++}, 10–20 mM) has been reported. The importance of use of GalT-2 assay method (as a model system) in the purified Golgi-rich membranes from 13-day-old embryonic chicken brains (13-ECB) is described.

Key words Glycosyltransferases, Lactosylceramide synthase, GalT-2

1 Introduction

Over the last six decades, elucidation of the structures of numerous Gangliosides of animal origin [1–4] has shown the frequent presence of the terminal structures containing single sialic acid attached to lactosylceramide such as (NeuAc-alpha2-3) Gal-Glc-ceramide (GM3) or disialosyl structure NeuAc (alpha2-8)-NeuAc (alpha2-3) Gal-Glc-ceramide (GD3). In addition to the short chain lactose-c gangliosides longer oligosaccharides-containing gangliosides having GalNAc-in the ganglio family and GlcNAc-in the lactofamily as the third sugar are also widely distributed on the cancer and normal Eukaryotic cell surfaces [5–8]. Over the recent years SA-Le^X and Le^a have also been implicated as cancer markers and are

Sandro Sonnino and Alessandro Prinetti (eds.), *Gangliosides: Methods and Protocols*, Methods in Molecular Biology, vol. 1804, https://doi.org/10.1007/978-1-4939-8552-4_8, © Springer Science+Business Media, LLC, part of Springer Nature 2018

expected to be involved in the metastatic processes [9–14]. The steps of GD1a [15–20], blood-group active GSLs and SA-Le^X [4, 21–31] biosynthesis starting from lactosylceramide ceramide (Fig. 1) have been established in the embryonic tissues (brains) and human carcinoma cells [neuronal (IMR-32) [26], colon (Colo-205) [8, 32, 33], breast (MCF-7, MDA-468, and SKBR-3) [8, 32, 33], and lung [30]. Lactosyl ceramide was first characterized in bovine spleen [34] and in tumor tissues by Rapport and his associates in early sixties. Biosynthesis in vitro of lactosylceramide from glucosylceramide was reported first in 9- to 13-day-old embryonic chicken brain buffy coat membrane mixture rich in mitochondria, synaptosomes [16], and Golgi bodies [17] and then in rat liver Golgi-rich membranes [19]. A few years before the biosynthesis in vitro of GM3 ganglioside was reported in 9-day-old embryonic chicken brain buffy coat as a first step in the GD1a ganglioside biosynthesis from lactosylceramide [15, 35]. Steps of the in vitro biosynthesis of GD1a ganglioside, Globoside, and SA-LeX were established by 1991 [4, 29, 32], and in each case lactosylceramide played a central role (Fig. 1). This chapter is focused on the different in vitro assay methods used in the characterization of each biosynthetic step (ceramide to GD1a and SA-Le^X gangliosides). A detailed overview of the protocols of some of the glycolipid-glycosyltransferase (GSL-GLT) assays has been published earlier [28] except the GlcAT-1 assay [36], which was also

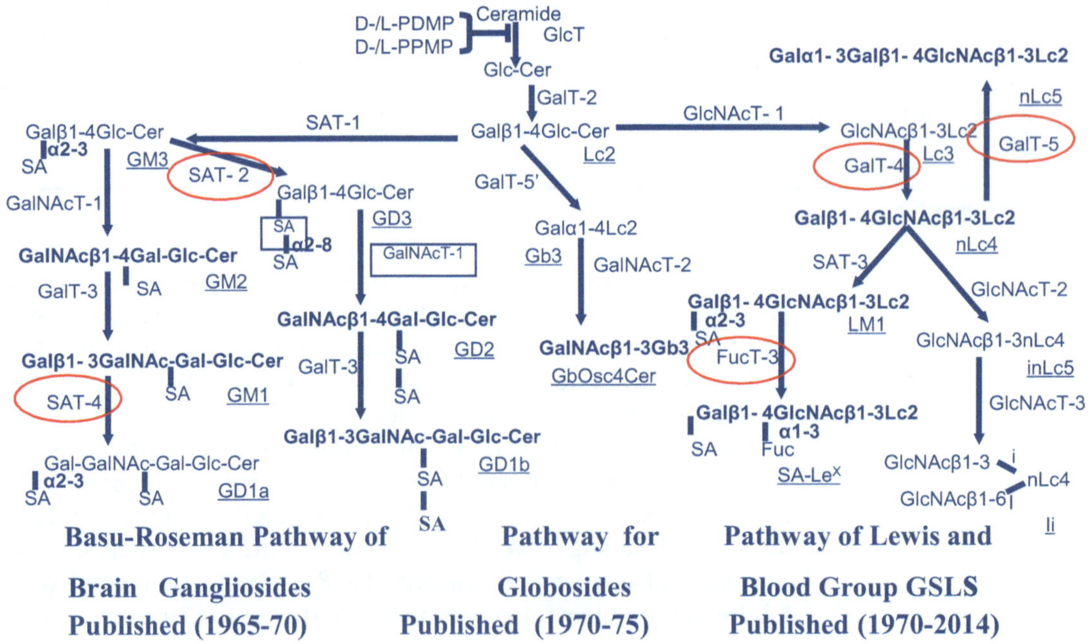

Fig. 1 Biosynthesis of fucosyl-, mono-, and di-sialosyl glycosphingolipids (Basu, S. and collaborators: Glycoconjugate J (2009) 26,647–661)

discovered in ECB. This chapter is designed to discuss the limitations of the different assay methods developed in the Glycolipid field and to recognize the biosynthetic steps well beyond 24 glycolipids as reported here (Fig. 1). These protocols are essential in the study of modulation of Glycolipid biosynthesis in metastatic cancer cells and the development of animal tissues. At the end of this chapter we have reported some kinetic properties of GalT-2 (UDP-Galactose: Glc-ceramide β1-4galactosyltransferase) (Fig. 2a–i), which catalyzes the biosynthesis in vitro of the Lactosyl-ceramide in the presence of Golgi-rich membranes isolated from 13-day-old embryonic chicken brains.

2 Methods

2.1 Protocols for Glycolipid-Glycosyltransferases (GSL-GLT) Assays

The assays for more than 24 GSL-GLTs (Tables 1, 2, and 3) have been developed in this laboratory over the past four decades, based on the theory that formation of each linkage (positional and anomeric) between the sugars is catalyzed by different enzymatic gene products. The specificity of a GSL-GLT depends on the unique structure of the ultimate and penultimate sugars present in the acceptor sphingolipid molecule. In last two to three decades different laboratories have supported the "one enzyme one linkage" theory based on cloning of different GSL-GLT genes and discovering putative sequences of each gene products (Glycosyltransferase enzymes).

Until natural glycolipid-binding proteins are discovered the overall protocol of our assays for GSL-GLTs is based on the following :

$$\underset{\text{(Substrate)}}{\text{Oligosaccharide-ceramide}} + \underset{\text{(Donor)}}{\left[{}^{14}\text{C- or }{}^{3}\text{H-}\right]\text{Sugar-XDP}}$$
$$\rightarrow \left[{}^{14}\text{C / or }{}^{3}\text{H}\right]\text{Sugar-oligosaccharidel-1-ceramide}\left(\left[{}^{14}\text{C- or }{}^{3}\text{H}\right]\text{-GSL-product}\right)$$
$$+\text{XDP or XMP}\left(\text{CMP}\right).$$

2.2 General Procedure for All GSL-GLT Assays

The reactions are carried out in labeled glass baby tubes (6 × 30–50 mm) placed in a plastic baby-tube rack. Both GSL (0.01–0.1 μmol) and a specific neutral detergent (25–100 μg) are dissolved in chloroform–methanol (2:1) and dried completely under nitrogen in a vacuum desiccator before the addition of an appropriate buffer (0.1 M). The plastic rack containing baby tubes is placed in an ice-containing cold water tray (4 °C). All subsequent additions such as metals (0.01 μmol) or other agents are done at 4 °C. After the addition of enzyme preparation (purified membranes suspended in appropriate buffer-containing 50–300 μg protein (measured by Lowry method). In case of a purified Golgi

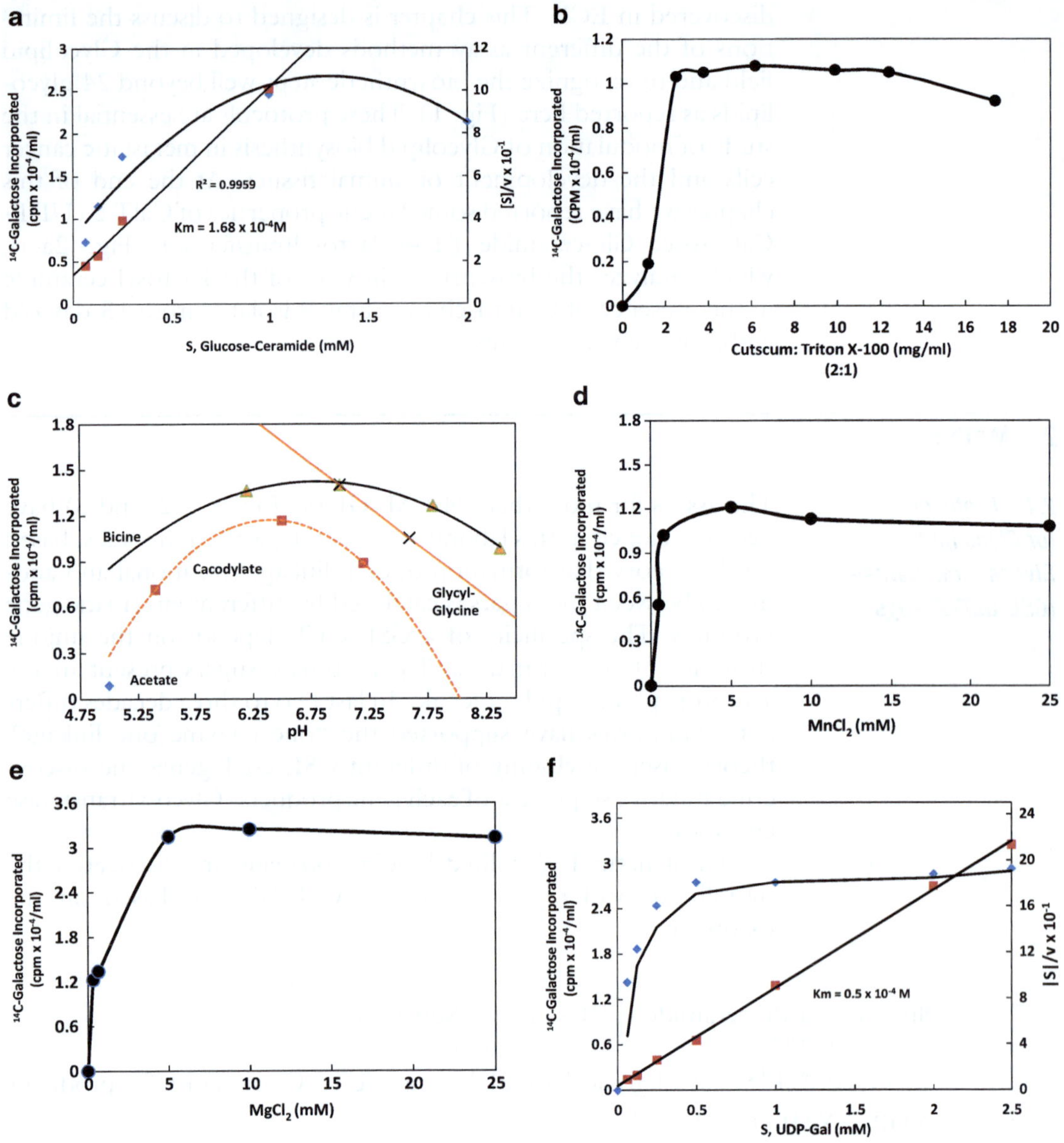

Fig. 2 (**a**) Effect of Glc-CER concentration on ECB GalT-2 activity. (**b**) Effect of detergent concentration on ECB Gal T-2 activity. (**c**) Effect of pH and buffers on ECB GalT-2 activity. (**d**) Effect of $MnCl_2$ concentration on ECB GalT-2 activity. (**e**) Effect of $MgCl_2$ concentration on ECB GalT-2 activity. (**f**) Effect of UDP-Gal concentration on ECB GalT-2 activity. (**g**) Effect of total protein concentration on GalT-2 activity. (**h**) Effect of incubation time on GalT-2 activity. (**i**) Effect of embryonic age on content of GalT-2 activity

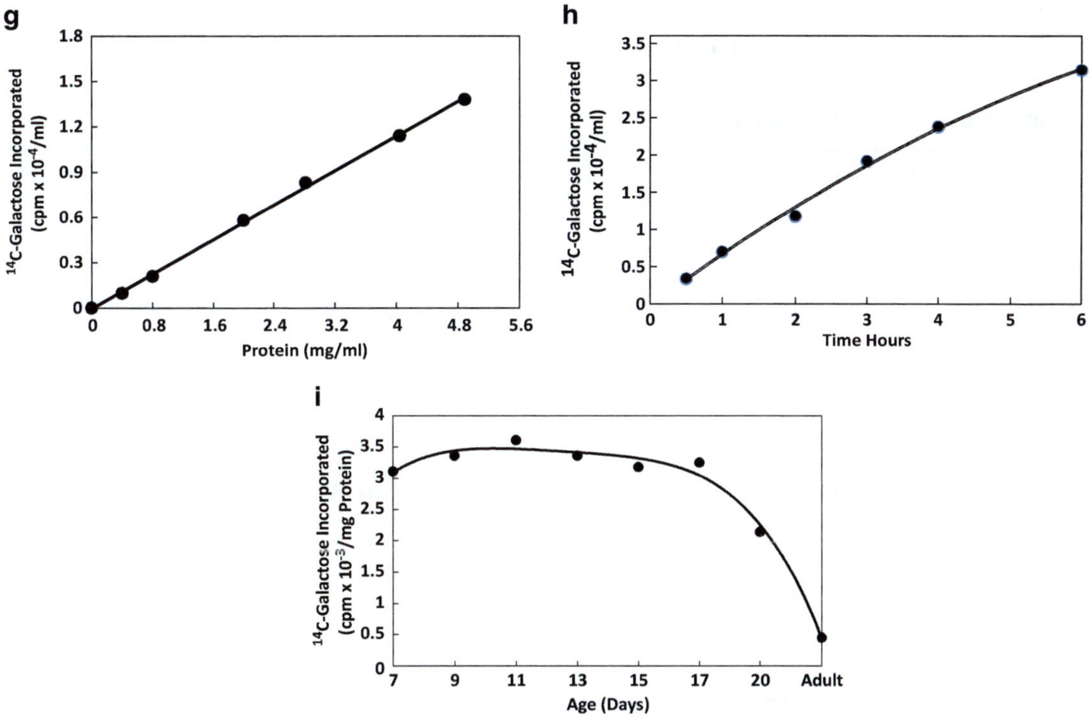

Fig. 2 (continued)

Table 1
Glycosphingolipids: glycosyltransferases (GSL: GLTs) [28]

Abbreviation	Enzyme Name	Linkage	References
Glucosyltransferase			
GlcT	UD-Glc: NFA-Cer	β1-1	[16, 17]
Galactosyltransferases			
GalT-1	UDP-Gal: HFA-Cer	β1-1	[49]
GalT-2	UDP-Gal: Glc-Cer	β1-4	[16, 40]
GalT-3	UDP-Gal: GM2	β1-3	[15]
GalT-4	UDP-Gal: Lc3	β1-4	[21]
GalT-5	UDP-Gal: nLc4	α1-3	[22]
GalT-6	UDP-Gal: Lc2	α1-4	[38]
N-Acetylgalactosaminyltransferases			
GalNAcT-1	UDP-GalNAc: GM3	β1-4	[18]
GalNAcT-2	UDP-GalNAc: Gb3	β1-3	[50]
GalNAcT-3	UDP-GalNAc: Gb4	α1-3	[38, 50]

Table 2
Glycosphingolipids: glycosyltransferases (GSL: GLTs)

Abbreviation	Enzyme name	Linkage	References
N-Acetylglucosaminyltransferases			
GlcNAcT-1	UDP-GlcNAc: Lc2	(β1-3)	[5, 51]
GlcNAcT-2	UDP-GlcNAc: nLc4	(β1-3)	[28, 51]
GlcNAcT-3	UDP-GlcNAc: nLc4	(β1-6)	[28, 51]
Fucosyltransferases			
FucT-2	GDP-Fuc: nLc4 or nLc5	(α1-2)	[52]
FucT-3	GDP-Fuc: Lc3	(α1-3)	[26, 29]
FucT-3	GDP-Fuc: LeX-GSL	(α1-3)	[29, 30]
FucT-4	GDP-Fuc: G_{M1}	(α1-4)	[28]
FucT-3′	GDP-Fuc: IV3NeuAc-LcOse4Cer	(α1-4)	[28]

Table 3
Glycosphingolipids: glycosyltransferases (GSL: GLTs)

Abbreviation	Enzyme name	Linkage	References
Sialyltransferases			
SAT-1 (?)	CMP-NeuAc: GalCer	α2-3	[35, 53]
SAT-1	CMP-NeuAc: Lc2	α2-3	[35, 53]
SAT-2	CMP-NeuAc: G_{M3} or L_{M1}	α2-8	[6, 54]
SAT-3	CMP-NeuAc: nLc4	α2-3	[27, 28, 39]
SAT-4	CMP-NeuAc: G_{M1} or Gg4	α2-3	[35, 53]
Glucuronyltransferase			
GlcAT-1	UDP-GlcA: nLc4Cer	β1-3	[36]

membrane preparation, it is suspended in the incubation buffer (pH 6.5–7.0; 0.1–0.2 M) containing 0.25 M sucrose). In case of a membrane isolated from the 100,000 × *g* pellet (obtained from homogenized cultured cells or cancer cells, a total protein 10–100 μg is used in the tubes. The reaction mixtures are covered tightly with Parafilm to stop the evaporation from the small volume of the reaction mixture during incubation and the rack is placed in the water bath kept at 37 °C. The reaction is terminated by addition of

0.25 μmol of EDTA (pH = 8) plus 25–50 μl of chloroform–methanol (2:1). Before spotting on the chromatographic papers, the tubes are vortexed. Each sample (content) from the small glass baby tube is quantitatively transferred (spotted) on the chromatographic papers with separate glass-capillary tubes (disposable). Capillary tubes are not reused (strictly prohibited for possible contamination of the radioactivity) in between the samples.

The radioactive GSL-products (nanomole or picomole quantities) (stable at −18 °C freezer for at least 7 days) are separated from the radioactive donors by HVWPE (High Voltage Whatman Paper Electrophoresis chromatography; 1500–2000 V; 40 V/cm) for 45 min in the cold room or paper chromatography (descending or ascending at room temperature) with proper solvent systems (1.0% $Na_2B_4O_7$ pH = 9.1) or a mixture of chloroform–methanol–water solution (60–35–8 or other specific proportions (*see* Table 4) as required to separate radioactive products from the substrates) as

Table 4
Recommended assay method for glycosphingolipids: glycosyltransferases (GSL: GLTs)

Abbreviation for methods	Assay method	GLT tested
SGPDC	Whatman SG-81 paper descending chromatography (when ^3H- or ^{14}C-labeled sugar nucleotide is a donor	GalTs; GlcNAcTs; GalNAcTs; FucTs; SATs
WPDC	Whatman 3MM paper descending chromatography (when ^{14}C-labeled sugar nucleotide is a donor	GalTs; GlcNAcTs; GalNAcTs; FucTs; SATs
HVWPE	High voltage Whatman paper electrophoresis	GalTs; FucTs; SATs
SGPAC	Whatman SG-81 paper ascending chromatography	GlcT; GalT-1; GalT-2; GlcNAcT-1
RDAPC	Reverse direction ascending paper chromatography	GalT-4; FucT-3
G25CC	Sephadex G-25 column chromatography	GalT-3; GalNAcT-1; SAT-1; SAT-4
SPKCC	Sep-Pak column chromatography	GalNAcT-1 [16, 17]
MABA	Monoclonal antibody binding assay	GalT-6; GalNAcTs [28]
IGLAC	Iodoglycolipid ascending chromatography	SAT-1 [28]
SCTLC	Short column followed by thin-layer chromatography	SAT-1; SAT-2 [28]
RAGAB	Radioactive glycolipid and monoclonal antibody binding assay	FucT-3 [28]

described below. During spotting or drying under a hood only air at room temperature is applied without an application of hot air. Appropriate areas of the dry chromatographic papers are cut in 1 in. × 1 in. squares and placed in wide mouth scintillation vials. Radioactivity is quantitated in toluene based scintillation system [37].

2.2.1 High Voltage Whatman3MM Paper Electrophoresis (HVWPE)

The Whatman-3MM paper-containing product [^{14}C-Gal-R] and the donor [^{14}C] Gal-UDP (not ^{3}H labeled compounds because ^{3}H-products or reactants quench dpm 90% on Whatman 3MM papers) are spotted 20 cm (= 10 in.) from the top of a 27-in. long strip. A maximum of ten samples can be spotted on a 46 × 57 cm (or 22 in. × 27 in.) sheet of Whatman 3MM paper. The paper is carefully wetted by applying or spraying it with 1.0% $Na_2B_4O_7$ from both sides of the origin [16, 17, 28]. Care is taken not to leave any dry spot at the origin. Any dry regions in the sample spotted area (at the origin) should be soaked with the sodium tetraborate solution using new disposable glass capillary tubes for each lane. Dry spots may cause charring of the samples on the paper. At the end of the electrophoresis, the wet Whatman 3 mm paper is then carefully placed on the bridge of a Savant high-voltage electrophoresis apparatus and lowered into tank containing $Na_2B_4O_7$ at the electrodes (levels of the buffer is maintained same) area with nonflammable organic coolant on the top. The coolant is kept at 15 °C or below by constant circulation of cold water, the electrophoresis is maintained at 2000–3000 V for 50–60 min (40 V/cm). The paper is dried overnight under the hood. The origin, along with 2-in. of paper areas toward the anode side, are cut out and the radioactivity in each 1 in. square is quantitated in toluene-based scintillation systems [37]. For assays of the following enzymes this HVWPE is recommended to obtain faster results with ^{14}C-products of the following GSL-GLT catalyzed reactions only [GlcT-1; GalT-1; GalT-2; GalT-3; GalT-4; GalT-5; SAT-1; SAT-2: SAT-3; SAT-4; Fuct-2; and FucT-3.] (for nomenclature use Fig. 1 and Tables 1, 2, and 3).

2.2.2 Whatman SG-81 Paper Ascending Chromatography (SGPAC)

At the end of a reaction (containing 0.1 ml incubation mixture), 0.05 ml of chloroform–MeOH (2:1) is added and vortexed carefully and spun at a low speed (3000 × g) for 3 min. The total lower layer is quantitatively transferred on the origin of a SG-81 paper (10 × 10 in., containing six lanes (1-in. wide and separated by 0.5 in. blank lane)). Origin-line is almost 1 in. away from the bottom. After spotting, the Whatman Silica Gel-81 paper, which is then clipped on the top arm of a glass U-tube and the whole set up is placed in a wide TLC-plate jar containing freshly poured solvent chloroform–methanol–water (60:30:7 or 60:17:2) [16, 28]. The TLC jar is quickly covered with thin plastic (Saran wrap) and quickly covered with the glass plate. The solvent runs up to the top within less than 2 hours. The solvent front is marked quickly on the SG-81 paper. The SG-81 is removed from the jar and dried under the hood. With lead pencil (ink pen elutes color in toluene based scintillation fluid, which quenches dpm) squares are marked and placed in the clean-dry scintillation vials. Radioactivity is quantitated in a toluene scintillation system-II as described by Moskal et al. in 1974 [37]. Short glycose chains-bound to ceramide (intermediates of Ganglio-, Globo-, and Lacto-family GSLs, such as Glc-ceramide, lactosyl-ceramide, LcOse3-cer, GM3-ganglioside) have been assayed by the SGPAC method. In other words the following GSL-GLTs [GlcT-1; GalT-1; GalT-2; GlcNAcT-1 (Fig. 1 and Tables 1 and 4)] catalyzed reactions are assayed quite rapidly by this method.

2.2.3 Whatman3MM, Whatman-1MM, or SG-81 Paper Descending Chromatographic Method (WSDPC)

When unpurified Golgi preparation or 100,000 × *g* pellets from cultured cells are used as enzyme source the membrane mixture contains broken pieces of membranes from Golgi, plasma, nuclear, lysosomal, and mitochondrial membranes. Incubation mixture kept with these fractions for 1–2 h at 37 °C cleaves some of the radioactive sugar from all these radioactive sugar nucleotides (UDP-[^{14}C] Glc, UDP-^{14}[C]Gal, UDP-[^{14}C]GalNAc, UDP-[^{14}C]GlcNAc, GDP-[^{14}C] Fuc, or CMP-[^{14}C]NeuAc) to free radioactive sugars and sugar phosphates. Contents of the incubation mixtures are spotted by fresh set of capillary tubes on the origin lines on separate lanes (4-in. from the top edge of a chromatography paper (Whatman 3MM, Whatman 1MM, or SG-81) and overnight descending chromatography is performed with a 1.0% $Na_2B_4O_7$ solution. All the radioactive contaminants are removed from the radioactive products (left at the origin) by this method. Dried origin spots containing the radioactive products are quantitated in Toluene Scintillation counting system [37]. This assay method is suitable for assays of all the GSL-GLTs where radioactive GSL-products stays at the origin (Tables 1, 2, 3, and 4) and the unreacted radioactive sugar-nucleotide or its degraded products move away from the origin such as in the assay of GlcT-1, GalT-1, GalT-2, GalT-3, GalT-4, GalT-5, SAT-1, SAT-2, SAT-3, SAT-4, GlcNAcT-1, GlcNAcT-2, GlcNAcT-3, GalNAcT-1, GalNAcT-2, GalNAcT-3, and GlcAT-1 (Fig. 1). This assay method is suitable in a laboratory where a high-voltage apparatus is not available. The detailed assay conditions of individual glycolipid: glycosyltransferases are published previously [28]. When ^{3}H-sugar nucleotides are used as donors instead of Whatman 3MM papers samples are spotted on a full length wider or narrower SG-81 papers to avoid 90% quenching of dpm in the products on a toluene based scintillation counting system [37].

2.2.4 Reverse Direction Ascending Paper Chromatography (RDAPC)

By this method a combination of HVWPE and WSDPC are used. Instead of spotting on the origin line drawn 4-in. away from the top edge of the full length paper, the line is drawn in the middle (almost 8 or 9 in. away of the full length papers (both Whatma-3MM or SG-81). After descending chromatography with 1.0% $Na_2B_40_7$ the papers are dried, are cut one and half inches (from top of the origin) and ascending chromatography is run with 50 mL of suitable solvent system (chloroform–methanol–water = 65:35:8; or 60:17:8; or 60:7:2) in a TLC-jar. Special care is taken for preventing evaporation of the solvent system so that the composition in the solvent system is not altered. Care is also taken from the elution of the product from the origin line to the lower solvent level by maintaining the solvent level at least one half inch below the origin line. After chromatography is done, the papers are dried completely. Chloroform–methanol is completely dried from the papers to avoid quenching during counting beta particles in a liquid scintillation system [37].

The enzyme system which catalyzes mixture of radioactive glycolipid (from added GSL-substrate) and radioactive glycoprotein (from endogenous unknown substrates) products could be separated by this method. All of the GSL-GLTs mentioned under WSDPC could be assayed by this RDPAC assay method. This method is recommend. In the assays of the following GSL-GLTs (GlcT-1, GalT-1, GalT-2, GalT-4, and GlcNAcT-1) where radioactive products move away from the origin (Tables 1, 2, 3, and 4) [28].

2.3 Biosynthesis In Vitro of Lactosylceramide in Embryonic Chicken Brains (GalT-2 Assay in Details)

As soon as buffy coat from embryonic chicken brain homogenate was prepared biosynthesis of GM3 ganglioside was established first by HVWPE assay system in 1965 [15, 16, 28]. After the advent of SGAPC was established [16, 28] biosynthesis of lactosylceramide was published. In this section a detail condition is given for GLT-2 catalyzed

Lactosylceramide in vitro biosynthesis in 13-day-old embryonic chicken brain.

In the mini glass tubes (5 × 40 or 6 × 50 mm) 25 μl of Glc-Cer substrate GSL solution (containing 0.05–0.1 μmol) (*see* Fig. 2a for Km value) in chloroform–methanol (2:1) and 25 μl (containing 0.5–1.0 mg) of detergent solution (Cutscum–Triton X-100; 2:1 dissolved in C-M 2:1) is added (*see* Fig. 2b for detergent saturation curve).

The tubes are placed in a Savant Rotavap for complete removal of chloroform-methanol. The GalT-2 activity is killed in the presence of trace amount of Chloroform-MeOH (2:1). The mini-rack containing the tubes is placed in a glass tray containing ice-cold water. To the mixture of Glc-ceramide and detergent 0.025 mL of glycyl-glycine or Bicin buffer at pH 7.0 is added and mixed on a vortex (*see* buffer effect on Fig. 2c). The GalT-2 activity is active in the presence of cations $Mn^{+2}$and Mg^{+2}. To the incubation tube a 0.01 mL mixture of $MnCl_2$ and $MgCl_2$ (containing 0.5–1.0 μmol) is added (*see* Fig 2d, e for metal saturation curve). After mixing well in 0.005–0.01 ml of UDP-[14C] Gal (specific activity = 2–3 million dpm per micromole) is added while the tubes are immersed in cold water. Before addition a quick checking of the specific activity is always done to record the exact radioactivity added in each tube (*see* Fig. 2f for Km of UDP-Gal). The tube containing 0.04 ml of the incubation mixture now could accommodate 0.06 ml of enzyme preparation from any experimental source containing 0.05–0.3 mg protein (*see* Fig. 2g for protein saturation curve with ECB Golgi-rich purified membrane fraction from 13 day-old ECB preparation at pH = 7.00). The GalT-2 enzyme bound to membranes is stable for at least 40 hours incubation time at 37 °C (*see* Fig. 2g, product formation was proportional up to 0.5 mg/ml pf protein concentration). For routine laboratory work 2–4 h of incubation is recommended. At the end of the incubation, the enzyme activity is stopped by the addition of 10–20 μmol of EDTA to bind metals. The incubation tubes after carefully covered with Parafilm) is stored at −18 °C until these are ready for the second phase of the assay system given above (HVWPE, SGPAC, WSDPC, or RDAPC). For quantitative determination of [^{14}C] lactosylceramide formed by catalytic activity of GalT-2 (using unpurified membrane system or any other biological samplc) RDAPC assay system is recommended. From any tissue homogenate Golgi membranes are purified by our previously published [17, 19] using sucrose gradient ultracentrifugation. For characterization of the ^{14}C-Lactosylceramide product a 100-fold mixture is incubated for overnight and the reaction mixture is mixed with equal volume of chloroform–methanol (2:1) and mixed well at room temperature. The lower layer containing the radioactive product is further purified by silicic acid columns and silica gel-G TLC [16, 28].

2.4 Regulation of Glycolipid Biosynthesis and Degradation during Onset of Apoptosis, Embryogenesis, and Oncogenic Processes

In all these processes the roles of cell surface macromolecules (glycolipids and glycoproteins) are not well understood as yet. Since lactosylceramide is the common precursor of all three families of glycosphingolipids (inner trisaccharide cores of Ganglio family -GalNAc-Gal-Glc-ceramide; Globo family -Gal-Gal-Glc-ceramide; and the Lacto-family -GlcNAc-Gal-Glc-ceramide) the synthesis of lactosyl-ceramide is an intermediate step (*see* Fig. 1). It is expected that the modulation of synthesis of lactosyl-ceramide modulates gangliosides in neuronal cells [37], globosides in kidney cells [38–40], and rat adenocarcinomas [41], or Le^{X}-family glycolipids in cancer embryonic cells (Fig. 3) [7, 8, 25, 26, 29, 33] or normal cells during onset of apoptosis and embryogenesis. Regulation of glycogenes during induction by small organic molecules L-PPMP or D-PDM, betulinic acid, tamoxifen, cisplatin, and melphalan in colon carcinoma cells (Colo-205) [8, 33] and in breast carcinoma cells (SKBR-3, MDA-468, and MCF-3) [7, 8, 33] had been studied in recent years. Both posttranslational expression of glycosyltransferases and transcriptional expression of mRNA of the glycosyltransferase have been quantitated tentatively (Fig. 4) [7, 8, 33]. Biosynthesis of sialyl-oligosaccharides of gangliosides (GD3, GD1a) and SA-Le^{X} (SAT-3, SAT-4, FucT-2, FucT-3)

Name	Structures of Lewis Antigen Epitopes
nLcOse4Cer(nLc4) (Type-2 Chain)	Gaλβ1,4GlcNAcb1,3Galb1,4Glc-Cer
LeX (Type-2 chain)	Galb1,4(Fuca1,3)GlcNAcb1,3Galb1,4Glc-R
SA-LeX (Type-2)	NeuAca2,3Galb1,4(Fuca1,3)GlcNAcb1,3Galb1,4Glc-R
SA-Lea(Type-1 chain)	NeuAca2,3Galb1,3(Fuca1,3)GlcNAcb1,3Galb1,4Glc-R
LeY (Type-2 chain)	Fuca1,2Galb1,4(Fuca1,3)GlcNAcb1,3Galb1,4Glc-R

Mono-, Di-, and Tri –Fucosyl alpha1-3 type-2 structures are also present in Human embryonic cells and cancer cells

Linkages-b=beta/ -a=alpha R-Ceramide or Glycoproteins

Fig. 3 Names and schematic structures of Lewis antigen epitopes

GSL-GLT	Catalyzed Reaction	Enzymatic Activity
GalT-4	Lc3 (GlcNAc-Gal-Glc-Cer)→Galβ-Lc3	Decrease (MCF-7/L-PPMP-2h,6h; SKBR-3/*cis*P, L-PPMP and MDA-468/L-PPMP, 48h)
GalT-5	Lc4) (Gal-GlcNAc-Gal-Glc-Cer)→ Galα-Lc4	Decrease (SKBR-3/L-PPMP-2h,6h; MCF-7,MDA-468/L-PPMP-6h; SKBR-3/*cis*P and MDA-468/L-PPMP-48h)
SAT-2	GM3→GD3	Decrease (MCF-7/*cis*P,L-PPMP-48h)
SAT-4	GM1→GD1a	Decrease (MCF-7/L-PPMP, SKBR-3/*cis*P and MDA-468/L-PPMP-48h)
SAT-4'	Gg4 (Gal-GalNAc-Gal-Glc-Cer)→ GM1b	Decrease (SKBR-3/*cis*P and MDA-468/L-PPMP-48h)
FucT-3	LM1→SA-LeX	Decrease (SKBR-3/L-PPMP-48h)

Fig. 4 Overall conclusion of posttranslational activities of glycosphingolipids: GLTs (cell/agents-time)

were downregulated (Fig. 4) in breast cancer cells after induction of apoptosis (within 2–24 h) produced by L-PPMP, D-PDMP, betulinic acid, tamoxifen, cisplatin, and melphalan [7, 8, 33]. A detailed study is yet to be performed in human normal or carcinoma cells from different organs [42–44].

2.5 Distribution of Le^X-Family Glycosphingolipids on Cancer Cell Surfaces

Several in vivo studies suggested that tumor metastasis depends on the expression of carbohydrate Lewis structures (Table 4). Lewis antigens and their derivatives such as LewisX (Le^X), Sialyl LewisX (SA-Le^X), Sialyl Lewis a (SA-Le^a), Sialyl Leb (SA-Le^b); Lewis Y (Le^Y), were identified as tumor-associated by Koprowski and colleagues [45] almost three decades ago. All these epitopes are expressed as glycosphingolipids as well as glycoproteins. Both Le^X and SA-Le^X are ligands for E- and P-selectin binding, respectively. Both E- and P-selectins are vascular receptors and are expressed on the endothelial cells. L-lectins are expressed on leucocytes. Researchers have proposed that metastatic migration of cancer cells is almost similar to leucocytes migration using interaction between selectin receptor and ligand interaction (Fig. 1). Cancer cell surface Lewis antigens are believed to be involved in the in vivo metastatic propagation in different organs. Polyfucosyl-Type-2 Le^X and Le^Y families of glycolipids have been established by Hakomori and his associates [46, 47]. Research on biosynthesis and regulation of LeX, SA-LeX biosynthesis in cancer cells is the exploding field at the present time. Determination of glycolipid–glycosyltransferase activities as mentioned above and their transcriptional expression as assayed by DNA Microarray assays are needed to conclude any biological functions of the cell surface Glycosphingolipids.

2.6 Assays for Transcriptional Regulations of Glycosyltrans-ferases by SAM (Significance Analysis of Microarrays)

Results of glycobiology microarray assays (Fig. 5) with breast cancer cells treated with L-PPMP have been published [7]. A total of 359 genes [48] with four copies of unique sense 45-mer oligonucleotides were individually synthesized, purified, and immobilized via a 5′ amino linker onto aldehyde-coated microarrays. These DNA microarrays were prepared in the laboratory of Professor Joseph R. Moskal (our collaborator) at Northwestern University. Total RNA from MCF-7, MDA-468, SKBR-3 cells treated with L-PPMP 2 mM for 2 and 24 h were reversely transcribed (Fig. 5) and used as the substrate for RNA amplification and labeling using the Ambion Amino Allyl MessageAmp™ II aRNA Amplification Kit. Each labeled sample with pooled reference human RNA was hybridized with the three arrays. The scanned arrays in acceptable quality were transferred to BlueFuse format and quantified with

1. Drug treatment with 3 cancer cell lines
2. Total RNA isolation with Qiagen Mini RNA kit.
3. Reverse transcription for cDNA synthesis
4. Transcription of cDNA to aRNA
5. aRNA labeling with Cy3 or Cy5
6. Chip hybridization @ 42°C for 24 hours, and wash
7. Imaging scan and data process
8. Data analysis with SAM (Significance Analysis of Microarrays)

Fig. 5 Procedure for DNA-microarray analysis of glyco-related genes (steps)

Cell Line	GLT Gene Name	Linkage Formed	Fold Change
MCF-7	B3GALT5	Galβ1-3GlcNAc-R1	1.19-1.33
	B3GNT3	GlcNAcβ1-3 Gal-R2	1.20
	B3GNT1	GlcNAcβ1-3 Gal-R2	-1.19
MDA-468	UGCGL2	Glcβ1-1 Cer	1.28
	B3GNT4	GlcNAcβ1-3 Gal-R2	1.34
	B3GNT1	GlcNAcβ1-3 Gal-R2	-1.34
SKBR-3	B3GNT4	GlcNAcβ1-3 Gal-R2	1.35
	ST6GAL1	NeuAcα2-6Galβ1-3GlcNAc-R3	-1.55

Core Blood Type: Galβ1-3/4GlcNAcβ1-3Galβ1-4-R1/R2
R1= Glcβ1-1 Cer / R2=OligoN-Protein

Fig. 6 Changes in GLT expressions in apoptotic breast cancer cells (DNA-Microarray/2 μM L-PPMP/2 h)

GeneTraffic (DUO) v3.2-11. Comprehensive statistical analysis with data from GeneTraffic was performed with SAM (Significance Analysis of Microarrays) at 10% false discovery rate (Fig. 6) [7].

3 Discussion

In any biological eukaryotic system if one wants to quantitate the changes occurring in a glycolipid-glycosyl-transferase content or changes occurring in its total activity, it is important to adopt the optimum assay conditions with purified Golgi membranes as described here. Where purified membranes or solubilized membranes are not available at first, optimum assay conditions have to be established. Out of 24 GSL-GLT assays, the GalT-2(UDP-Gal: Glc-Cer β1-4 galactosyltransferase activity in 7- to 17-day-old embryonic chicken brain is reported here (Fig. 2i). In each case analysis of the enzymatic product produced from the increased volume of the micro-incubation mixtures is recommended. Use of four different methods for separation of the radioactive products from the radioactive sugar nucleotides and other compounds are described. Depending on the enzyme source, an investigator is recommended to test all these four methods of separation before large experimental samples are examined. Limitation of each assay system is also mentioned under each method. Activity of each GSL-GLT is determined by the ultimate and penultimate sugar present in the glycolipid acceptor.

The Km values for acceptor GSL or donor sugar nucleotide might change when the enzyme activity is determined in a membrane-bound state or when free enzymes are present in solution solubilized by detergents.

Acknowledgment

We thank Mrs. Dorisanne Nielsen and Mr. Eric Kuehner for their help during preparation of the manuscript. The Jacob Javits Research Award from NIH-NINDS NS-18005, Coleman Cancer Foundation, and NCI grant-CA-14764 to S. Basu; and a grant-in-aid from Siemens Corporation to M. Basu supported this research work.

References

1. Basu S, Basu M (1982) Expression of glycosphingolipid glycosyltransferases in development and transformation. In: Horowitz M (ed) The glycoconjugates V3: glycoproteins, glycolipids and proteoglycans, vol 3. Academic Press, New York, pp 265–285
2. Basu S, Basu M, Dastgheib S, Hawes JW (1999) Biosynthesis and regulation of glycosphingolipids. In: Barton KN D, Meth-Cohen O (eds) Comprehensive natural products chemistry, vol 3. Pergamon Press, New York, pp 107–128
3. Basu S, Das K, Basu M (2000) Glycosyltransferase in glycosphingolipid biosynthesis. In: Ernst B, Sinay P, Hart G (eds) Oligosaccharides in chemistry and biology – a comprehensive handbook. Wiley-VCH Verlag GmbH, Weinheim, pp 329–347
4. Basu S, Ma R, Moskal JR, Basu M (2012) Ganglioside biosynthesis in developing brains and apoptotic cancer cells: X. regulation of glycogenes involved in GD3 and Sialyl-Lex/a syntheses. Neurochem Res 37(6):1245–1255. https://doi.org/10.1007/s11064-012-0762-9
5. Basu S, Basu M, Kyle JW, Chon HC (1984) Biosynthesis in vitro of gangliosides containing Gg- and Lc-cores. Adv Exp Med Biol 174:249–261
6. Higashi H, Basu M, Basu S (1985) Biosynthesis in vitro of disialosylneolactotetraosylceramide by a solubilized sialyltransferase from embryonic chicken brain. J Biol Chem 260(2):824–828
7. Ma R, Matthew Decker N, Anilus V, Moskal JR, Burgdorf J, Johnson JR, Basu M, Banerjee S, Basu S (2009) Post-translational and transcriptional regulation of glycolipid glycosyltransferase genes in apoptotic breast carcinoma cells: VII. Studied by DNA-microarray after treatment with L-PPMP. Glycoconj J 26(6):647–661. https://doi.org/10.1007/s10719-008-9219-4
8. Basu S, Ma R, Boyle PJ, Mikulla B, Bradley M, Smith B, Basu M, Banerjee S (2004) Apoptosis of human carcinoma cells in the presence of potential anti-cancer drugs: III. Treatment of Colo-205 and SKBR3 cells with: cis -platin, tamoxifen, melphalan, betulinic acid, L-PDMP, L-PPMP, and GD3 ganglioside. Glycoconj J 20(9):563–577. https://doi.org/10.1023/B:GLYC.0000043293.46845.07
9. Fukushi Y, Hakomori S, Shepard T (1984) Localization and alteration of mono-, di-, and trifucosyl alpha 1–3 type 2 chain structures during human embryogenesis and in human cancer. J Exp Med 160(2):506–520
10. Mao S, Gao C, Lo CH, Wirsching P, Wong CH, Janda KD (1999) Phage-display library selection of high-affinity human single-chain antibodies to tumor-associated carbohydrate antigens sialyl Lewisx and Lewisx. Proc Natl Acad Sci U S A 96(12):6953–6958
11. Ashizawa T, Aoki T, Yamazaki T, Katayanagi S, Shimizu H, Koyanagi Y (2003) The clinical significance of sialyl Lewis antigen expression in the spread of gastric cancer. Flow cytometric DNA analysis. J Exp Clin Cancer Res 22(1):91–98
12. Silva Z, Tong Z, Cabral MG, Martins C, Castro R, Reis C, Trindade H, Konstantopoulos K, Videira PA (2011) Sialyl Lewisx-dependent binding of human monocyte-derived dendritic cells to selectins. Biochem Biophys Res Commun 409(3):459–464. https://doi.org/10.1016/j.bbrc.2011.05.026
13. Matsumoto S, Imaeda Y, Umemoto S, Kobayashi K, Suzuki H, Okamoto T (2002) Cimetidine increases survival of colorectal cancer patients with high levels of sialyl Lewis-X and sialyl Lewis-A epitope expression on tumour cells. Br J Cancer 86(2):161–167. https://doi.org/10.1038/sj.bjc.6600048
14. Kannagi R (2004) Molecular mechanism for cancer-associated induction of sialyl Lewis X and sialyl Lewis A expression—the Warburg effect revisited. Glycoconj J 20(5):353–364. https://doi.org/10.1023/B:GLYC.0000033631.35357.41

15. Basu S, Kaufman B, Roseman S (1965) Conversion of Tay-Sachs ganglioside to monosialoganglioside by brain uridine diphosphate D-galactose: glycolipid galactosyltransferase. J Biol Chem 240(10):4115–4117
16. Basu S, Kaufman B, Roseman S (1968) Enzymatic synthesis of ceramide-glucose and ceramide-lactose by glycosyltransferases from embryonic chicken brain. J Biol Chem 243(21):5802–5804
17. Basu S, Kaufman B, Roseman S (1973) Enzymatic synthesis of glucocerebroside by a glucosyltransferase from embryonic chicken brain. J Biol Chem 248(4):1388–1394
18. Steigerwald JC, Basu S, Kaufman B, Roseman S (1975) Sialic acids. Enzymatic synthesis of Tay-Sachs ganglioside. J Biol Chem 250(17):6727–6734
19. Keenan TW, Morre DJ, Basu S (1974) Ganglioside biosynthesis. Concentration of glycosphingolipid glycosyltransferases in Golgi apparatus from rat liver. J Biol Chem 249(1):310–315
20. Basu S, Basu M, Kyle JW, De T, Das K, Schaeper RJ (1986) Biosynthesis of gangliosides and blood group glycolipids using solubilized glycosyltransferases. In: Freysz SG (ed) Enzymes of lipid metabolism. Plenum, New York, pp 233–245
21. Basu M, Basu S (1972) Enzymatic synthesis of a tetraglycosylceramide by a galactosyltransferase from rabbit bone marrow. J Biol Chem 247(5):1489–1495
22. Basu M, Basu S (1973) Enzymatic synthesis of a blood group B-related pentaglycosylceramide by an alpha-galactosyltransferase from rabbit bone marrow. J Biol Chem 248(5):1700–1706
23. Basu M, Chien JL, Basu S (1974) Biosynthesis of guinea pig erythrocyte triglycosylceramide by bone marrow beta-N-acetylgalactosaminyltr ansferase. Biochem Biophys Res Commun 60(3):1097–1104
24. Basu M, Basu S, Potter M (1980) Biosynthesis of blood group related glycosphingolipids in T- and B-lymphomas and neuroblastoma cells. In: Sweeley CC (ed) Cell surface glycolipids, vol 128, pp 187–212
25. Presper KA, Basu M, Basu S (1982) Biosynthesis in vitro of a blood group B-active fucose-containing hexaglycosylceramide from neolactopentaosylceramide in bovine spleen. J Biol Chem 257(1):169–173
26. Presper KA, Basu M, Basu S (1978) Biosynthesis in vitro of fucose-containing glycosphingolipids in human neuroblastoma IMR-32 cells. Proc Natl Acad Sci U S A 75(1):289–293
27. Basu M, Basu S, Stoffyn A, Stoffyn P (1982) Biosynthesis in vitro of sialyl(alpha 2-3)neolactotetraosylceramide by a sialyltransferase from embryonic chicken brain. J Biol Chem 257(21):12765–12769
28. Basu M, De T, Das KK, Kyle JW, Chon HC, Schaeper RJ, Basu S (1987) Glycosyltransferases involved in glycolipid biosynthesis. In: Ginsburg V (ed) Methods in enzymology, vol 138. Academic, New York, pp 575–607
29. Basu M, Hawes JW, Li Z, Ghosh S, Khan FA, Zhang BJ, Basu S (1991) Biosynthesis in vitro of SA-Lex and SA-diLex by alpha 1-3 fucosyltransferases from colon carcinoma cells and embryonic brain tissues. Glycobiology 1(5):527–535
30. Holmes EH, Ostrander GK, Hakomori S (1985) Enzymatic basis for the accumulation of glycolipids with X and dimeric X determinants in human lung cancer cells (NCI-H69). J Biol Chem 260(12):7619–7627
31. Radhakrishnan P, Beum PV, Tan S, Cheng PW (2007) Butyrate induces sLex synthesis by stimulation of selective glycosyltransferase genes. Biochem Biophys Res Commun 359(3):457–462. https://doi.org/10.1016/j.bbrc.2007.05.165
32. Basu SC (1991) The serendipity of ganglioside biosynthesis: pathway to CARS and HY-CARS glycosyltransferases. Glycobiology 1(5):469–475
33. Basu S, Ma R, Moskal JR, Basu M, Banerjee S (2012) Apoptosis of breast cancer cells: modulation of genes for glycoconjugate biosynthesis and targeted drug delivery. Adv Exp Med Biol 749:233–255. https://doi.org/10.1007/978-1-4614-3381-1_16
34. Rapport MM, Schneider H, Graf L (1962) Immunochemical studies of organ and tumor lipids. XI. A simplified preparation of cytolipin H. J Biol Chem 237:1056–1059
35. Kaufman B, Basu S, Roseman S (1966) Embryonic chicken brain sialyltransferases. Methods In Enzymol. 8:365-368
36. Das KK, Basu M, Basu S, Chou DK, Jungalwala FB (1991) Biosynthesis in vitro of GlcA beta 1-3nLcOse4Cer by a novel glucuronyltransferase (GlcAT-1) from embryonic chicken brain. J Biol Chem 266(8):5238–5243
37. Moskal JR, Gardner DA, Basu S (1974) Changes in glycolipid glycosyltransferases and glutamate decarboxylase and their relationship to differentiation in neuroblastoma cells. Biochem Biophys Res Commun 61(2):751–758
38. Yeung KK, Moskal JR, Chien JL, Gardner DA, Basu S (1974) Biosynthesis of globoside and

Forssman-related glycosphingolipid in mouse adrenal Y-1 tumor cells. Biochem Biophys Res Commun 59(1):252–260

39. Basu S, Basu M, Higashi H, Evans CH (1982) Biosynthesis and characterization of globoside and Forssman glycosphingolipids in guinea pig tumor cells. In: Makita A, Handa S, Taketomi T, Nagai Y (eds) New vistas in glycolipid research. Plenum, New York, pp 131–137
40. Chatterjee S, Kolmakova A, Rajesh M (2008) Regulation of lactosylceramide synthase (glucosylceramide beta1→4 galactosyltransferase); implication as a drug target. Curr Drug Targets 9(4):272–281
41. Jenis DM, Basu S, Pollard M (1982) Increased activity of a beta-galactosyltransferase in tissues of rats bearing prostate and mammary adenocarcinomas. Cancer Biochem Biophys 6(1):37–45
42. Fukushi Y, Kannagi R, Hakomori S, Shepard T, Kulander BG, Singer JW (1985) Location and distribution of difucoganglioside (VI3NeuAcV3III3Fuc2nLc6) in normal and tumor tissues defined by its monoclonal antibody FH6. Cancer Res 45(8):3711–3717
43. Nishida K, Yamamoto H, Ohtsuki T, Matsuba M, Mukai S, Naito Y, Yoshikawa T, Kondo M (1991) Elevated tissue concentrations of sialyl Lex-i in cancerous tissues compared with those in noncancerous tissues of various organs. Cancer 68(1):111–117
44. Ogiso M, Shogomori H, Hoshi M (1998) Localization of LewisX, sialyl-LewisX and alpha-galactosyl epitopes on glycosphingolipids in lens tissues. Glycobiology 8(1):95–105
45. Koprowski H, Steplewski Z, Mitchell K, Herlyn M, Herlyn D, Fuhrer P (1979) Colorectal carcinoma antigens detected by hybridoma antibodies. Somatic Cell Genet 5(6):957–971
46. Hakomori S, Nudelman E, Levery SB, Kannagi R (1984) Novel fucolipids accumulating in human adenocarcinoma. I. Glycolipids with di- or trifucosylated type 2 chain. J Biol Chem 259(7):4672–4680
47. Hakomori SI, Handa K (2015) GM3 and cancer. Glycoconj J 32(1-2):1–8. https://doi.org/10.1007/s10719-014-9572-4
48. Kroes RA, Panksepp J, Burgdorf J, Otto NJ, Moskal JR (2006) Modeling depression: social dominance-submission gene expression patterns in rat neocortex. Neuroscience 137(1):37–49. https://doi.org/10.1016/j.neuroscience.2005.08.076
49. Basu S, Schultz A, Basu M, Roseman S (1971) Enzymatic synthesis of galactocerebroside by a galactosyltransferase from embryonic chicken brain. J Biol Chem 243:4272–4279
50. Chien JL, Williams T, Basu S (1973) Biosynthesis of a globoside-type glycoshingolipid by an b-N-acetylgalactosaminyltransferase from embryonic chicken brain. J Biol Chem 248:1778-1785
51. Basu M, Basu S (1984) Biosynthesis in vitro of Ii-core glycolipids from neolactotetraosylceramide by b-1 and b1-6 Nacetylglucosaminyltransferases from mouse T-lymphoma. J Biol Chem 259:12557–12562
52. Basu S, Basu M, Chien JL (1975)Enzymatic synthesis of blood group H-related glycosphingolipid by an α-fucosyltransferase from bovine spleen. J Biol Chem 250:2956–2962
53. Basu S (1966) Ph.D. Thesis, University of Michigan. Studies on the biosynthesis of gangliosides.
54. Kaufman B, Basu S, Rosenman S (1968) Enzymatic synthesis of disialogangliosides from monosialogangliosides by sialyltransferases from embryonic chicken brain. J Biol Chem 243:5804-5806

Chapter 9

Mass Spectrometry of Gangliosides

Akemi Suzuki, Minoru Suzuki, Emi Ito, Takahiro Nitta, and Jin-ichi Inokuchi

Abstract

This chapter describes protocols for mass spectrometry (MS) applied to the characterization of ganglioside structures and the determination of ganglioside contents. Matrix-assisted laser desorption/ionization (MALDI) and electrospray ionization (ESI) are often used to ionize biological materials and this chapter covers three protocols for atmospheric pressure MALDI MS (AP-MALDI MS), liquid chromatography-ESI MS (LC-ESI MS), and LC-ESI MS with multiple reaction monitoring (MRM). Purified gangliosides were used in AP-MALDI MS analyses while crude preparations of gangliosides were subjected to LC-ESI MS and LC-ESI MS with MRM. The LC protocol includes conditions for both reversed-phase and normal-phase column chromatography.

Key words Mass spectrometry, MALDI MS, ESI MS, LC-MS, GM3, GM1, GD1a, GD1b, GT1b, GQ1b, Mouse brain, Mouse CD8^{+} T cells, Human serum GM3

1 Introduction

Mass spectrometry (MS) is a fundamental method in the structural characterization of gangliosides and in the determination of changes in ganglioside concentration as a function of physiological and pathological conditions. Characterization of gangliosides has been performed by thin-layer chromatography (TLC) with chemical detection and binding assays using antibodies and carbohydrate recognition reagents. TLC is a simple and low-cost procedure and can provide an overview of ganglioside composition and information on carbohydrate structure with only a small amount of sample. Although TLC can detect mobility shift due to the presence of hydroxylated fatty acids and/or 4-hydroxylated sphinganines as ceramide components, it does not provide further structural information on the ceramide structures. The functions of gangliosides in membrane microdomains have

Sandro Sonnino and Alessandro Prinetti (eds.), *Gangliosides: Methods and Protocols*, Methods in Molecular Biology, vol. 1804,
https://doi.org/10.1007/978-1-4939-8552-4_9, © Springer Science+Business Media, LLC, part of Springer Nature 2018

attracted researchers' attention [1, 2], wherein structural information on the ceramide components of gangliosides is critical. At present, only MS methods can provide structural information on ceramides by consuming practically small amounts of sample. While not described herein, TLC-MS has overcome the aforementioned limitation of TLC and successful applications have been reported [3, 4]. Despite the significant advantages of MS in ganglioside analyses, several challenges remain and are discussed in Subheading 4 of each protocol. This chapter provides three MS protocols and detailed notes for the analysis of gangliosides: atmospheric pressure matrix-assisted laser desorption/ionization MS (AP-MALDI MS) [5], liquid chromatography-electrospray ionization MS (LC-ESI MS) [6–8], and LC-ESI MS with multiple reaction monitoring (MRM) [9]. The ionization of gangliosides, like any other class of molecules subjected to MS, is a key issue. Gangliosides carry single or multiple negative charges due to the presence of sialic acids and can easily provide negatively charged ions. This is a preferable characteristic for sensitive MS detection and gangliosides with carbohydrate chains longer than those of GM1 can provide doubly charged negative ions in the electrospray ionization. However, the glycosidic linkages between sialic acid and Gal/GalNAc, or between sialic acids, are relatively labile and are often broken by the amount of energy required for MALDI ionization. A soft ionization technique such as ESI is required to suppress these breakages. While not discussed herein, this particular drawback of MALDI has been overcome by introducing permethylated derivatives of gangliosides and further permethylated gangliosides have been successfully analyzed by ESI MS with inflow injection [10].

1.1 AP-MALDI MS

MALDI is a widely used ionization method for the analysis of biomolecules because of its efficient ionization and the fact that MALDI-MS instrumentation is relatively simple to use and easy to maintain. MALDI quadrupole ion trap time-of-flight (QIT-TOF) mass spectrometry, which enables MS^n analyses, has been applied to the fine structural characterization of biomolecules including glycoproteins, glycopeptides, and N-glycans. Atmospheric pressure (AP) MALDI has several advantages over MALDI, including easy sample handling and no need for analytes to be kept in vacuo. Therefore, the interpretation of AP-MALDI spectra does not require the consideration of changes in biological samples in a high-vacuum environment. In this protocol, a commercially available AP-MALDI ion source is coupled with a QIT-TOF mass spectrometer for the analysis of monosialo-, disialo-, and trisialo-gangliosides. This analytical system, run in negative ion mode, provides MS, MS^2, and MS^3 spectra that yield fragmentation patterns suitable for the structural characterization of gangliosides [5].

1.2 Structural Characterization of Gangliosides by Liquid Chromatography-Electrospray Ionization Mass Spectrometry (LC-ESI MS)

LC-ESI MS is a preferred method for the structural characterization of gangliosides in mixtures that are extracted from biological materials and have not been extensively purified. However, different protocols requiring different instruments are applied to structural characterization of gangliosides and sensitive detection of their content changes in the crude mixtures. Structural characterization of ceramides requires MS^3 analyses, while high sensitivity can be attained via MS^2 analyses employing MRM, one of the most sensitive detection methods available. In our experiments, the sensitivity requirements for structural characterization and low-level quantitation differ by approximately two orders of magnitude in terms of sample size, although low-level sensitivity depends on the mass spectrometers being used (*see* **Note 1**). One of the key issues for obtaining trustworthy results is the preparation of crude ganglioside mixtures. Therefore, sample preparation is described briefly in the following protocol [6–8].

1.3 Profiling of Ganglioside Composition by LC-ESI MS with Multiple Reaction Monitoring (MRM)

Relative to control samples, ganglioside contents in pathological and experimental samples can be changed. Sensitive detection of these changes is always urgently required. This demand is fulfilled by LC-MS analyses employing MRM [9, 11, 12]. Gangliosides of the nervous system and its derived cells are major components in the acidic fraction of total lipids or in Folch's upper phase. However, gangliosides in tissues and cells originating from other than the nervous system are almost always minor components. This is critical information for LC-MS analysis of crude ganglioside mixtures. Large excesses of nonganglioside molecules dilute the amount of ganglioside in the samples injected into the MS and can suppress ganglioside ionization. MRM selects ions to be analyzed and detects product ions derived from the selected precursor molecules. In this way, large excesses of nonganglioside molecules are eliminated from the analytical system, giving advantage for the analysis of small content molecules. This protocol describes the detection of GM3 in human serum by MRM as an example of such an analysis [9] (*see* **Note 2**).

2 Materials

2.1 Materials for AP-MALDI MS

1. An MS instrument with an ion trap, suitable for MS^n analyses of molecules up to *m/z* 2000, and an AP ionization source.
2. Purified gangliosides such as GM1, GD2, GD1a, GD1b, and GT1b. For this protocol these were purified from bovine brain and GM2 from a brain of a patient with Tay–Sachs' disease (*see* **Note 3**).
3. α-Cyano-4-hydroxycinnamic acid (α-CHCA) as the matrix.
4. HPLC-grade methanol, water, and acetonitrile.
5. A target plate for AP-MALDI MS.

2.2 Materials for LC-ESI MS

1. MS instrument: an LC-ESI MS equipped with an ion trap capable of acquiring MS^n spectra of molecules up to *m/z* 2000.
2. Reagents: water, methanol, isopropanol, acetonitrile, formic acid, and ammonium formate (all LC-MS or MS grade) and stock solutions of 0.4 and 0.1 M ammonium formate (possibly stored at 4 °C for at least 6 months).
3. A C30 reversed-phase column (Develosil C30, 1 × 50 mm, Nomura Chemical, Japan, or equivalent) is required for reversed-phase LC-ESI MS.
4. Solvents for reversed-phase LC-ESI MS: Solvent A: 2.5 mL of 0.4 M ammonium formate stock solution, 0.1 mL of formic acid, 10 mL of water, 87.4 mL of methanol. Solvent B: 2.5 mL of 0.4 M ammonium formate stock solution, 0.1 mL of formic acid, 47.4 mL of methanol, 50 mL of isopropanol.
5. An NH_2 column (Inertsil NH_2 column 1 mm i.d. × 50 mm, GL Science, Japan, or equivalent with silica base) is required for normal-phase LC-ESI MS.
6. Solvents for normal-phase LC-ESI MS: Solvent C: 0.1 mL of 0.1 M ammonium formate stock solution, 83 mL of acetonitrile, 16.9 mL of water. Solvent D: 50 mL of 0.1 M ammonium formate stock solution, 50 mL of acetonitrile.

2.3 Materials for LC-ESI MS with MRM

1. An LC-ESI MS instrument employing MRM.
2. Reagents: ammonium hydroxide (ammonia in water 25–28%), sodium hydroxide, hydrochloric acid, and sodium chloride (all analytical grade); LC-MS-grade acetic acid.
3. Solvents: water, methanol, and isopropanol (all LC-MS grade); HPLC-grade chloroform.
4. Solvent C: 0.1 mL of acetic acid, 0.1 mL of ammonium hydroxide, 20 mL of water, 30 mL of methanol, 50 mL of isopropanol, Solvent D: 0.1 mL of acetic acid, 0.1 mL of ammonium hydroxide, 2 mL of water, 48 mL of methanol, 50 mL of isopropanol.
5. Columns: Develosil C30 column (C30-UG-3 1 × 50 mm, Nomura Chemical Co. Ltd., Japan)
6. The C18 powder was obtained by emptying a Bond Elut C18 cartridge. Disc filters had a mesh size of 0.45 μm.
7. Internal standard GM3: chemically synthesized GM3(d18:1-[^{13}C]16:0).
8. Standard GM3: any purified GM3 for optimizing collision energy.

3 Methods

3.1 AP-MALDI MS

3.1.1 MS analysis

1. Dissolve gangliosides GM1, GM2, GD2, GD1a, GD1b, and GT1b at 1 mg/mL in methanol and dilute tenfold with water.
2. Prepare a matrix solution of 30 mM α-CHCA in acetonitrile–methanol (1:1, v/v).
3. Mix equal volumes of each ganglioside solution and α-CHCA matrix to prepare 50 ng/μL ganglioside solutions.
4. After mixing, apply 1 μL of ganglioside-matrix solution on an AP-MALDI target plate and dry at room temperature.
5. Set up a QIT-TOF MS instrument equipped with an AP-MALDI source in negative ion mode. To maintain mass precision, the temperature inside the instrument must be within 40 ± 0.3 °C. Focus a nitrogen laser at 337 nm (10 Hz) onto a sample zone on a MALDI target plate. The size of the laser spot should be approximately 200 μm. The voltage applied to the MALDI target plate should be ±3.5 kV. The distance between the MALDI target plate and the heated capillary located at the entrance of the ion-introducing block, which is kept at 250 °C, should be 2 mm. The mass spectrum measurement cycle time for obtaining all negative ion MS, MS^2, and MS^3 spectra of each ganglioside, including ion accumulating time, ion cooling time, precursor ion selecting time, and collision induced dissociation time, from *m/z* 500 to 2000 for MS and MS^2 and from *m/z* 100 to 600 for MS^3, should be set to 1.9 s/cycle.

3.1.2 Data Mining

1. The results of GD1a and GD1b analyses are shown in Fig. 1 as a typical example. The MS spectra in Fig. 1a, f show deprotonated molecular ions at *m/z* 1835 and 1863 for GD1a, and at 1835 and 1863 for GD1b, indicating the presence of two major molecular species.
2. During ionization, both GD1a and GD1b eliminate NeuNAc (291 atomic mass units (amu)) from $[M-H]^-$ ions to produce GM1. This indicates that the sialyl linkages of NeuNAc-NeuNAc- in GD1b and the terminal NeuNAc-Hex- in GD1a are weaker than the sialyl linkage of -HexNAc-(NeuNAc-) Hex- (*see* **Note 4**).
3. MS^2 spectra of GM1 structures derived from GD1a and GD1b (Fig. 1b, d, and g, i, respectively) show fragment ions due to the cleavage of glycosidic linkages, 162 and 203 amu for the elimination of Hex and HexNAc, respectively.
4. MS^3 spectra of ceramide ions, shown in Fig. 1c, e, and h, j, respectively, indicate d18:1–18:0 and d20:1–18:0 ceramide structures, based on the annotated ceramide fragmentation pattern in Fig. 1k (*see* **Notes 5** and **6**).

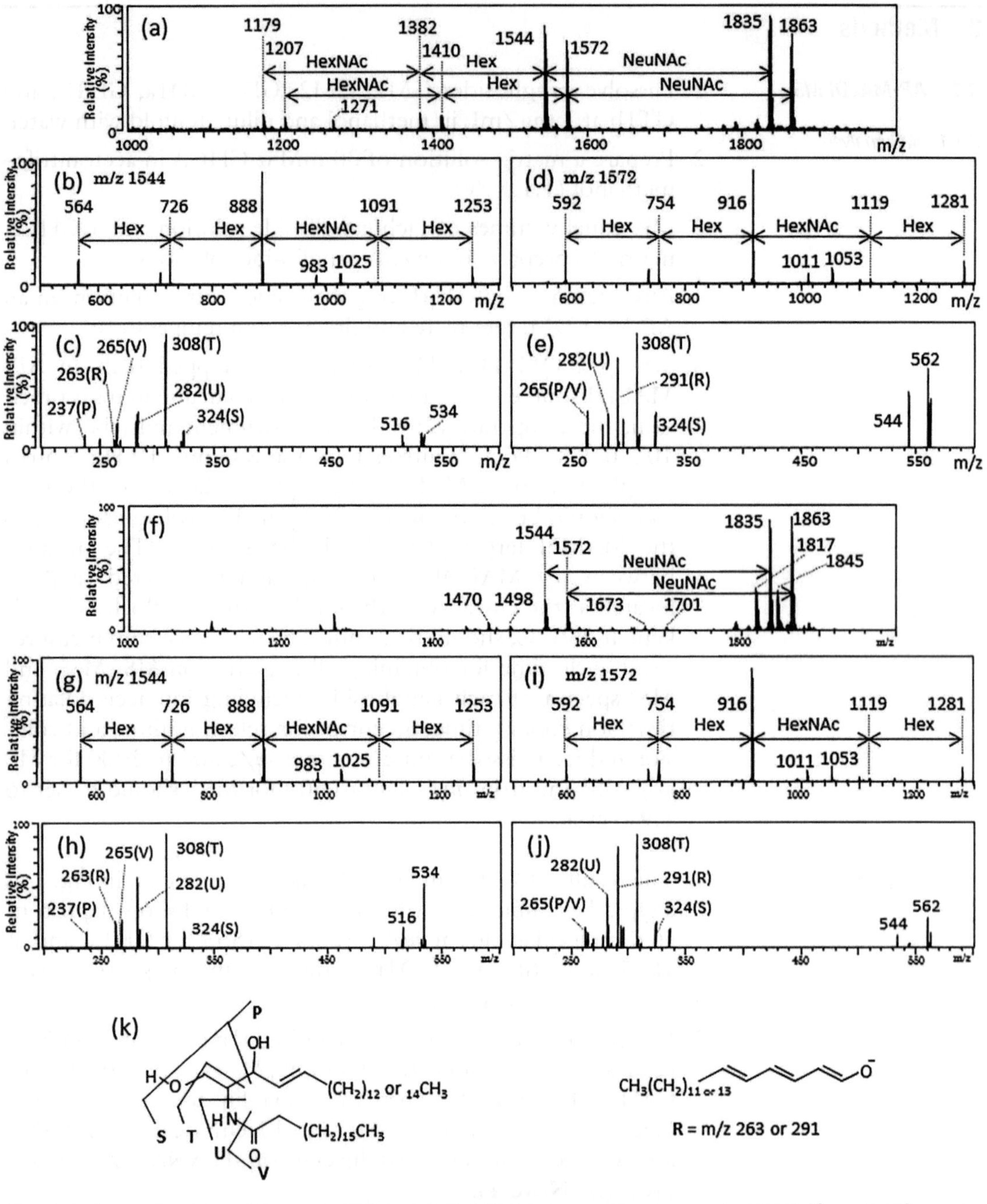

Fig. 1 Analysis of purified GD1a and GD1b by AP-MALDI MS in negative ion mode. MS, MS^2, and MS^3 spectra of GD1a and GD1b by negative ion AP-MALDI-QIT-TOF MS. (**a**) MS spectrum of GD1a, (**b**) MS^2 spectrum selecting the first precursor ion at *m/z* 1544, (**c**) MS^3 spectrum selecting the second precursor ion at *m/z* 564, (**d**) MS^2 spectrum selecting another molecular species of GD1a as the first precursor ion at *m/z* 1572, (**e**) MS^3 spectrum selecting the second precursor ion at *m/z* 592, (**f**) MS spectrum of GD1b, (**g**) MS^2 spectrum selecting the first precursor ion at *m/z* 1544, (**h**) MS^3 spectrum selecting the second precursor ion at *m/z* 564, (**i**) MS^2 spectrum selecting another molecular species of GD1b as the first precursor ion at *m/z* 1572, (**j**) MS^3 spectrum selecting the second precursor ion at *m/z* 592. Annotations for P, R, T, V, U, and S are shown in (**k**)

3.2 LC-ESI MS

Good-quality ganglioside mixtures are critical for obtaining trustworthy results in LC-MS analyses of gangliosides. Example preparation procedures are described briefly in this protocol and several key issues are indicated in Subheading 4.

3.2.1 Preparation of Ganglioside Mixtures

Purified Mouse T-Cell Subsets [6]

Mouse $CD4^+$ and $CD8^+$ cells should be purified using antibodies and subjected to lipid extraction. The total lipids should be separated into neutral and acidic fractions by DEAE-Sephadex chromatography. The acidic fractions should be subjected to a mild alkaline treatment using 0.1 N NaOH. After neutralizing the solutions with acetic acid, salts should be removed on C18 cartridges. Ganglioside mixtures prepared in this way are suitable for the following LC-MS protocol.

Mouse Brain

Total lipids are extracted from mouse brain with 10 mL each of chloroform–methanol (2:1 and 1:2, v/v) and 5 mL of chloroform–methanol–water (30:60:8, v/v). A two-phase separation, known as Folch's partition, is achieved by changing the solution composition to chloroform–methanol–water (8:4:3, v/v). The upper phase should be evaporated to dryness under N_2 in a water bath held at a temperature no higher than 40 °C.

3.2.2 LC Conditions

C30 Reversed-Phase Column LC [6]

1. Programmed elution: 20% Solvent B in Solvent A for 5 min, 20–60% Solvent B for 30 min, 60% Solvent B for 4 min, and 20% Solvent B for 5 min.
2. Flow rate should be 0.05 mL/min.

NH_2 Normal-Phase LC [11]

1. Programmed elution: 0% Solvent D in C for 5 min, from 0% to 76% Solvent D in Solvent C for 15 min, from 76% to 90% Solvent D in C for 5 min, 90% Solvent D in C for 10 min.
2. Flow rate should be 0.05 mL/min.

3.2.3 MS Conditions

1. An MS instrument capable of MS^n analyses should be set up in accordance with the manufacturer's protocol. The MS conditions to be used for C30 reversed-phase and NH_2 normal-phase LC-MS are the same.
2. The MS parameters should be as follows: for MS, acquire three independent scans of each mass spectrum from *m/z* 500 to 2000 with an ion accumulation time of 20 ms. For MS^2, acquire three scans from *m/z* 200 to 2000 with an accumulation time of 50 ms, and for MS^3, acquire three scans from *m/z* 100 to 700 with an accumulation time of 50 ms. Spectra should be acquired in negative ion mode with a detector voltage of 1.9 kV or equivalent. The collision-induced dissociation (CID) energy for MS^2 and MS^3 and the collision gas flow of Ar should be held at their 50% arbitrary values. The temperature of the curved desolvation device (CDL) should be 200 °C.

3. Crude mixtures containing gangliosides dried under N_2 should be dissolved by briefly sonicating in 100 μL of methanol. Aliquots of 1–10 μL should be injected for C30 reversed-phase LC-MS analyses and 1–5 μL aliquots should be injected for NH_2 normal-phase LC-MS (*see* **Note 7**).
4. For the first run, select automatic mode scanning which selects precursor ions for MS^2 and MS^3 analyses according to an algorithm in the MS operating software. If the first run does not give satisfactory results of MS^2 and MS^3 analyses and you know which molecules are to be targeted, precursor ions can be manually selected with *m/z* values of $[M-H]^-$ or $[M-2H]^{2-}$ in the second run. Tables 1 and 2 list theoretical *m/z* values of $[M-H]^-$ or $[M-2H]^{2-}$ for major gangliosides and ceramides, respectively. Solvents used for this protocol give GM3–GM1 as single-charged negative ions. Gangliosides with longer carbohydrate chains, such as GD1, are converted to doubly charged ions.

3.2.4 Data Mining

1. Figure 2 shows negative ion mass chromatograms obtained from the C30 reversed phase LC-ESI MS analyses of mouse $CD8^+$ T-cell gangliosides. These results were obtained using automatic mode scanning as described above in Subheading 3.2.3, **steps 2** and **4**. Accordingly, all mass spectra were recorded, allowing the examination of mass chromatograms of target molecules after the analytical run. The total ion monitoring (TIM) mass chromatogram indicates that the sample contained a variety of molecules. Other chromatograms shown in Fig. 2 trace the theoretical *m/z* values of the $[M-2H]^{2-}$ ion of extended GM1b(NeuNGc), Galβ1-3GalNAcβ1-4(NeuNGcα2-3)Galβ1-3GalNAcβ1-4Galβ1-4Glcβ1-Cer. Each of the assigned peaks was further characterized by MS^2 and MS^3 analyses. MS and MS^2 spectra of *m/z* 1004.5 are shown in Fig. 2b–d. MS^3 spectra do not often provide sufficient information on ceramide structures due to the small amount of precursor ions. A possible solution to this problem is described in Subheading 3.2.3, **step 4** and **Note 8**.
2. Figure 3 shows negative ion mass chromatograms of mouse brain gangliosides obtained from the NH_2 normal-phase LC-MS analyses. These were obtained using automatic mode scanning to trace the major brain gangliosides, as described above. Sufficient peak separation as a function of carbohydrate chain structures was achieved.

3.3 LC-ESI MS with MRM [9]

3.3.1 Preparation of a GM3 Mixture from Human Serum

1. Human serum (50 μL), together with 100 ng of GM3(d18:1-[^{13}C]16:0) as an internal standard, was freeze-dried using Teflon-sealed, screw-capped glass tubes (10 mL in size) and the dried materials were extracted with 10 mL of chloroform–methanol (1:1 v/v) in a water bath at 40 °C for 1 h. The extraction solution was agitated by a total of three brief sonication steps at the onset of, and during, extraction.

Table 1
Theoretical *m/z* values of typical gangliosides

Gangliosides	Formulae[a]	$[M-H]^-$	$[M-2H]^{2-}$
GM3(d18:1–16:0)	$C_{57}H_{104}N_2O_{21}$	1151.7059	
(d18:1–18:0)	$C_{59}H_{108}N_2O_{21}$	1179.7372	
(d18:1–20:0 or d20:1–18:0)	$C_{61}H_{112}N_2O_{21}$	1207.7685	
(d18:1–22:0)	$C_{63}H_{116}N_2O_{21}$	1235.7998	
(d18:1–23:0)	$C_{64}H_{118}N_2O_{21}$	1249.8154	
(d18:1–24:1)	$C_{65}H_{118}N_2O_{21}$	1261.8154	
(d18:1–24:0)	$C_{65}H_{120}N_2O_{21}$	1263.8311	
GM2(d18:1–18:0)	$C_{67}H_{121}N_3O_{26}$	1382.8166	
(d18:1–20:0 or d20:1–18:0)	$C_{69}H_{125}N_3O_{26}$	1410.8479	
GM1(d18:1–18:0)	$C_{73}H_{131}N_3O_{31}$	1544.8694	771.9311[b]
(d20:1–18:0)	$C_{75}H_{135}N_3O_{31}$	1572.9007	785.9467[b]
GD1(d18:1–18:0)	$C_{84}H_{148}N_4O_{39}$	1835.9648	917.4788
(d20:1–18:0)	$C_{86}H_{152}N_4O_{39}$	1863.9961	931.4944
GT1(d18:1–18:0)	$C_{95}H_{165}N_5O_{47}$	2127.0602	1063.0265
(d20:1–18:0)	$C_{97}H_{169}N_5O_{47}$	2155.0915	1077.0421
GQ1(d18:1–18:0)	$C_{106}H_{182}N_6O_{55}$	2418.1556	1208.5742
(d20:1–18:0)	$C_{108}H_{186}N_6O_{55}$	2446.1869	1222.5859

[a]Formulae of gangliosides other than those listed can be obtained as follows: a one-molecule addition for Hex is $C_6H_{10}O_5$, for HexNAc $C_8H_{13}O_5N$, for NeuNAc $C_{11}H_{17}O_8N$ and for NeuNGc $C_{11}H_{17}O_9N$. Neutral losses are 162 amu for Hex, 203 amu for HexNAc, 290 amu for NeuNAc, and 306 amu for NeuNGc
[b]GM1 provides a smaller proportion of $[M-2H]^{2-}$ compared to $[M-H]^-$

Table 2
Theoretical *m/z* values of ceramides

Ceramides	Formulae	$[M-H]^-$
d18:1–16:0	$C_{34}H_{67}NO_3$	536.5048
d18:1–18:0	$C_{36}H_{71}NO_3$	564.5361
d18:1–20:0 or d20:1–18:0	$C_{38}H_{75}NO_3$	592.5674
d18:1–22:0	$C_{40}H_{79}NO_3$	620.5987
d18:1–23:0	$C_{41}H_{81}NO_3$	634.6144
d18:1–24:1	$C_{42}H_{81}NO_3$	646.6144
d18:1–h24:1	$C_{42}H_{81}NO_4$	662.6093
d18:1–24:0	$C_{42}H_{83}NO_3$	648.6300
d18:1–h24:0	$C_{42}H_{83}NO_4$	664.6249

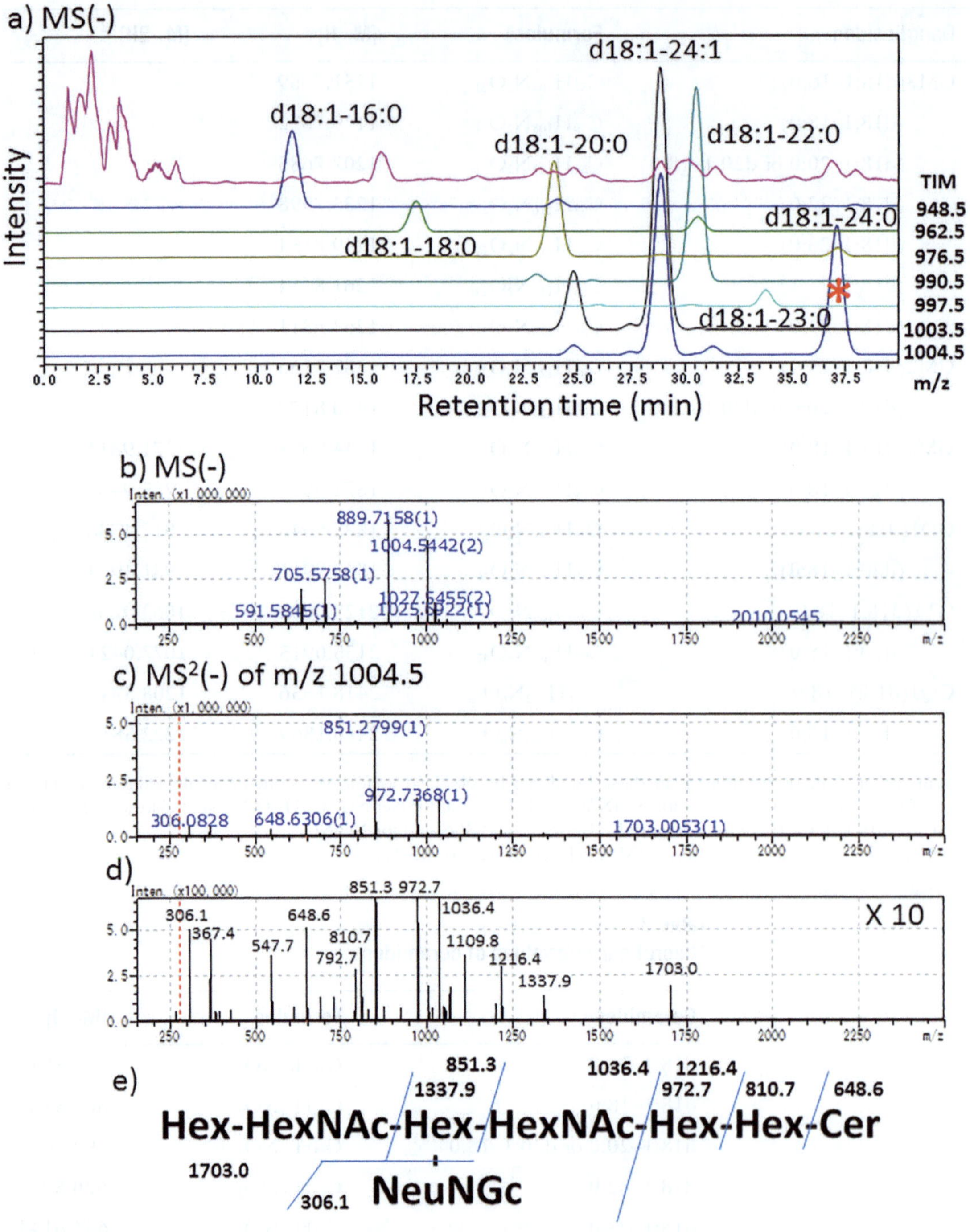

Fig. 2 Mass chromatograms and MS and MS^2 spectra of extended GM1b gangliosides of mouse CD8$^+$ T cells by negative ion mode ESI-LC MS with a C30 reversed-phase column. (**a**) Mass chromatograms of total ions (TIM) and seven molecular species of extended GM1b indicated on the right. (**b**) The MS spectrum acquired at the peak indicated by asterisk in (**a**) contains signals at *m/z* 1004.5, indicating the $[M-2H]^{2-}$ ion. (**c**, **d**) MS^2 spectra acquired from the *m/z* 1004.5 precursor ion. (**e**) The structure of extended GM1b and signal annotations

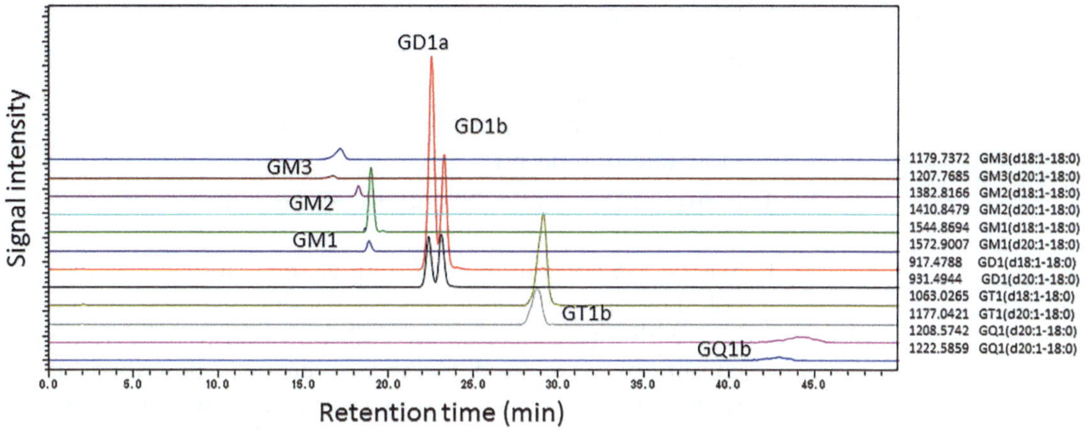

Fig. 3 Mass chromatograms of mouse brain gangliosides acquired by negative ion mode ESI-LC MS with a NH_2 normal-phase column. The monitored *m/z* values are indicated on the right as the theoretical values of $[M-H]^-$ for GM3, GM2, and GM1 and $[M-2H]^{2-}$ for GD1, GT1, and GQ1

2. The supernatant was separated from the precipitate by centrifugation at 720 × *g* for 15 min and the precipitate was extracted once again with 10 mL of chloroform–methanol (1:2, v/v) for 1 h as above.
3. After centrifugation at 720 × *g* for 15 min, the second supernatant was combined with the first and the combined solution was evaporated to dryness under N_2.
4. The residue was vortexed and sonicated in 3 mL of 0.1 N NaOH in methanol until dissolved and then incubated at 40 °C for 2 h with three brief sonication steps.
5. The solution was neutralized with 1 N HCl, and 6 mL of 50 mM NaCl and 200 mg of Bond elute C18 powder were added. The mixture was stirred by a micromixer at 2000 rpm for 30 min. The resulting C18 powder suspension was transferred to a Pasteur pipette with a filter consisting of a glass or quartz wool plug. Any remaining C18 powder in the tube was transferred to the pipette with 3 mL of 50 mM NaCl.
6. The packed powder was washed with 10 mL of water and 10 mL of methanol-water (1: 1, v/v). The eluates with 5 mL of methanol and then 5 mL of methanol–chloroform (1:1, v/v) were combined and dried under N_2.
7. The dried materials were dissolved in 3 mL of methanol and filtered through a disk filter. The tube and filter were then washed with 2 mL of methanol. The combined 5 mL of filter methanol solution was dried and the resulting materials were redissolved in 50 μL of methanol with brief sonication. This solution was injected into the LC-MS.

3.3.2 LC-ESI MS with MRM

LC

1. Prepare Solvents C and D.
2. Program the LC for the following elution: 0% Solvent D in Solvent C for 5 min to 100% Solvent D at 35 min, 100% Solvent D at 39 min, 0% Solvent D at 40 min and until 50 min. Flow rate of 0.05 mL/min.
3. Prepare a Develosil C30 column for analysis.

MS

1. Set the MS conditions according to the manufacturer's protocol. The optimal conditions for routine analyses are applicable, with the exception of collision energy. The following conditions were used for the experiment described above: spray voltage −2.5 kV, vaporizer temperature 370 °C, sheath gas pressure 50 (arbitrary units), auxiliary gas pressure 15 (arbitrary units), capillary temperature 200 °C (Table 3).
2. Select the *m/z* values for the precursor ions of GM3 and the *m/z* values of the product ion as given in Table 2.
3. Define the optimal collision energy (CE) for GM3 in accordance with the manufacturer's protocol, using the purified GM3.
4. Program the MS for Q1 *m/z* and Q3 *m/z* values and optimal CE.
5. Use an injection volume of 1–10 μL.

3.3.3 Data Mining

1. Figure 4 shows the results of one MRM analysis of human serum GM3.
2. The GM3 mixture was detected with d18:1–16:0, 18:1, 20:0, 22:0, 23:0, 24:1, 24:0, and h24:0 as the major molecular species.
3. MS^2 and MS^3 analyses were required to confirm the molecular structures of minor components (*see* **Note 9**).
4. GM3 (d18:1–16:0) was quantifiable relative to the internal standard, GM3(d18:1–[^{13}C]16:0). Accurate quantification of the other GM3 molecular species requires corresponding GM3 internal standards which are labeled with stable isotopes or the determination of their detection efficiency relative to the GM3 internal standard. However, these requirements are not easily fulfilled and in such cases, this protocol can be used to acquire data for comparisons of GM3 molecular species profiles among individuals.

4 Notes

1. Signals detected by MRM may need to be further characterized by MS, MS^2, and/or MS^3 analyses. However, when a large excess of molecules other than gangliosides are introduced

Table 3
Parameters for MRM analysis for GM3

GM3 [M−H]⁻	Q1 *m/z*	Q3 *m/z*	CE (eV)
d18:1–16:0	1152.1	290.1	50
d18:1–[^{13}C]16:0	1168.1	290.1	50
d18:1–18:0	1180.1	290.1	50
d18:1–20:0	1208.1	290.1	50
d18:1–22:0	1236.1	290.1	50
d18:1–23:0	1250.1	290.1	50
d18:1–24:1	1262.2	290.1	50
d18:1–24:0	1264.2	290.1	50
d18:1–h24:0	1280.2	290.1	50

Q quadrupole, *CE* collision energy

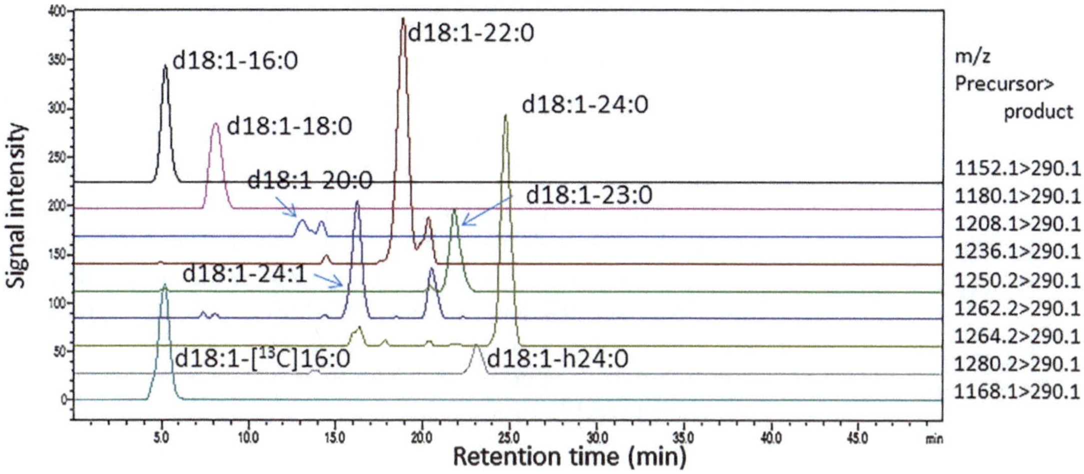

Fig. 4 Mass chromatograms of human serum GM3 by MRM analysis in negative ion mode. Major peaks are assigned as indicated, but minor peaks required further analyses including MS^2 and MS^3. GM3(d18:1–16:0) can be quantified using GM3(d18:1–[^{13}C]16:0) as an ISD. Other molecular species were not quantified but are instead expressed as the peak area ratio of GM3(d18:1–[^{13}C]16:0) and the ISD, assuming identical detection efficiencies for the various GM3 molecular species. Note that this assumption needs to be confirmed with purified or chemically synthesized compounds

into an MS ion trap, they can interfere with the detection of gangliosides due to the suppression of ganglioside ionization.

2. Quantitation of gangliosides in crude preparations of biological samples by LC-ESI MS is not possible without isotope-labeled gangliosides with identical structures. This is due to, among other variables, the molecular structure dependence of detection and ionization efficiencies, hardware differences

between mass spectrometers, and interference of nongangliosides in crude preparations. At present, comparisons between experimental and control samples are valid. However, the comparison should only be performed using MS data obtained using the same MS method on the same mass spectrometer at the same time.

3. With the exception of GM4, gangliosides prepared from mammalian brains contain two major ceramides: d18:1–18:0 and d20:1–18:0. The presence of d20:1 is particularly characteristic of tissues and cells originating from the nervous system. Gangliosides derived from other tissues or body fluids have ceramide components containing variable fatty acids, shown in Fig. 4 as an example in the case of human serum GM3.
4. When gangliosides are spotted onto a MALDI target plate and placed in vacuo, both AP-MALDI and MALDI produce ions due to the elimination of sialic acid residues.
5. GM1 ions derived from GD1a, GD1b, and GT1b, but not their $[M-H]^-$ molecular ions, can be subjected to MS^3 analysis for ceramide characterization. MS^3 analyses of $[M-H]^-$ molecular ions do not yield sufficient amounts of ceramide ions for dependable characterization.
6. The elimination of the sialic acid residues of GD1a, GD1b, and GT1b indicates that this protocol is not suitable for the structural characterization of gangliosides in ganglioside mixtures.
7. In this NH_2 normal-phase LC-MS protocol, injection volumes of methanolic solutions greater than 5 μL can prevent adequate LC separation. Methanol was chosen for sample dissolution because dried ganglioside residues are highly soluble in methanol. If necessary, acetonitrile may be substituted for methanol with special attention to dissolving the residues as much as possible by sonication.
8. The lack of ceramide structural information in MS^3 spectra is due to the low abundance of ceramide precursor ions following MS^2. This effect may be mitigated by running the analysis in manual mode and selecting specific precursor ions for MS^2 and MS^3 analyses.
9. When differences in the composition of minor components between control and experimental or pathological samples are subjected to further studies, such minor peaks are required for additional structural characterization, suggesting enrichment or purification. Samples subjected to MRM analyses as crude mixtures contain a variety of molecules and the selection of precursor and product ions is sometimes insufficient to exclude the influence of nontarget molecules.

References

1. Simons K, Ikonen E (1997) Functional rafts in cell membranes. Nature 387(6633):569–572. https://doi.org/10.1038/42408
2. Aureli M, Grassi S, Sonnino S, Prinetti A (2016) Isolation and analysis of detergent-resistant membrane fractions. Methods Mol Biol 1376:107–131. https://doi.org/10.1007/978-1-4939-3170-5_10
3. Park H, Zhou Y, Costello CE (2014) Direct analysis of sialylated or sulfated glycosphingolipids and other polar and neutral lipids using TLC-MS interfaces. J Lipid Res 55(4):773–781. https://doi.org/10.1194/jlr.D046128
4. Levery SB (2005) Glycosphingolipid structural analysis and glycosphingolipidomics. Methods Enzymol 405:300–369. https://doi.org/10.1016/S0076-6879(05)05012-3
5. Ito E, Tominaga A, Waki H, Miseki K, Tomioka A, Nakajima K, Kakchi K, Suzuki M, Taniguchi N, Suzuki A (2012) Structural characterization of monosialo-, disialo- and trisialo-gangliosides by negative ion AP-MALDI-QIT-TOF mass spectrometry with MS(n) switching. Neurochem Res 37(6):1315–1324. https://doi.org/10.1007/s11064-012-0735-z
6. Nagafuku M, Okuyama K, Onimaru Y, Suzuki A, Odagiri Y, Yamashita T, Iwasaki K, Fujiwara M, Takayanagi M, Ohno I, Inokuchi J (2012) CD4 and CD8 T cells require different membrane gangliosides for activation. Proc Natl Acad Sci U S A 109(6):E336–E342. https://doi.org/10.1073/pnas.1114965109
7. Oikawa N, Hatsuta H, Murayama S, Suzuki A, Yanagisawa K (2014) Influence of APOE genotype and the presence of Alzheimer's pathology on synaptic membrane lipids of human brains. J Neurosci Res 92(5):641–650. https://doi.org/10.1002/jnr.23341
8. Oikawa N, Matsubara T, Fukuda R, Yasumori H, Hatsuta H, Murayama S, Sato T, Suzuki A, Yanagisawa K (2015) Imbalance in fatty-acid-chain length of gangliosides triggers Alzheimer amyloid deposition in the precuneus. PLoS One 10(3):e0121356. https://doi.org/10.1371/journal.pone.0121356
9. Veillon L, Go S, Matsuyama W, Suzuki A, Nagasaki M, Yatomi Y, Inokuchi J (2015) Identification of ganglioside GM3 molecular species in human serum associated with risk factors of metabolic syndrome. PLoS One 10(6):e0129645. https://doi.org/10.1371/journal.pone.0129645
10. Sturgill ER, Aoki K, Lopez PH, Colacurcio D, Vajn K, Lorenzini I, Majic S, Yang WH, Heffer M, Tiemeyer M, Marth JD, Schnaar RL (2012) Biosynthesis of the major brain gangliosides GD1a and GT1b. Glycobiology 22(10):1289–1301. https://doi.org/10.1093/glycob/cws103
11. Ikeda K, Taguchi R (2010) Highly sensitive localization analysis of gangliosides and sulfatides including structural isomers in mouse cerebellum sections by combination of laser microdissection and hydrophilic interaction liquid chromatography/electrospray ionization mass spectrometry with theoretically expanded multiple reaction monitoring. Rapid Commun Mass Spectr 24(20):2957–2965. https://doi.org/10.1002/rcm.4716
12. Ikeda K, Shimizu T, Taguchi R (2008) Targeted analysis of ganglioside and sulfatide molecular species by LC/ESI-MS/MS with theoretically expanded multiple reaction monitoring. J Lipid Res 49(12):2678–2689. https://doi.org/10.1194/jlr.D800038-JLR200

Chapter 10

Visualization of Brain Gangliosides Using MALDI Imaging Mass Spectrometry

Eiji Sugiyama and Mitsutoshi Setou

Abstract

Matrix-assisted laser desorption/ionization imaging mass spectrometry (MALDI-IMS) is a histological method used for various molecules including gangliosides. The method is based on mass spectrometry, and thus target molecules with small structural differences are directly and independently detected. Here we describe a general procedure and related key notes for analyzing major brain gangliosides by MALDI-IMS.

Key words Matrix-assisted laser desorption/ionization, Imaging mass spectrometry, Glycosphingolipids, Ganglioside, Brain

1 Introduction

Matrix-assisted laser desorption/ionization imaging mass spectrometry (MALDI-IMS) is a histological method to examine various molecules including gangliosides [1, 2]. The method is based on mass spectrometry, and there are thus several important advantages compared to other imaging methods such as immunohistochemistry and in situ hybridization: (1) Target molecules are directly detected and no labeling is required. (2) Numerous molecules in a mass range can be analyzed simultaneously. (3) The identification of a detected unknown ion can be achieved. In a study of a given ganglioside, for example, the molecules in a same class containing different lengths of fatty acid chains, such as GM1(d18:1/18:0) and GM1 (d20:1/18:0), can be independently and simultaneously analyzed by MALDI-IMS [3–5], whereas such an analysis is difficult with the use of other methods. Using these advantages, we previously analyzed several gangliosides in human and mouse brain and obtained interesting data [2, 3, 6].

In MALDI-IMS, target molecules in a plane sample are ionized by laser irradiation with the assistance of a matrix, and the

Sandro Sonnino and Alessandro Prinetti (eds.), *Gangliosides: Methods and Protocols*, Methods in Molecular Biology, vol. 1804,
https://doi.org/10.1007/978-1-4939-8552-4_10, © Springer Science+Business Media, LLC, part of Springer Nature 2018

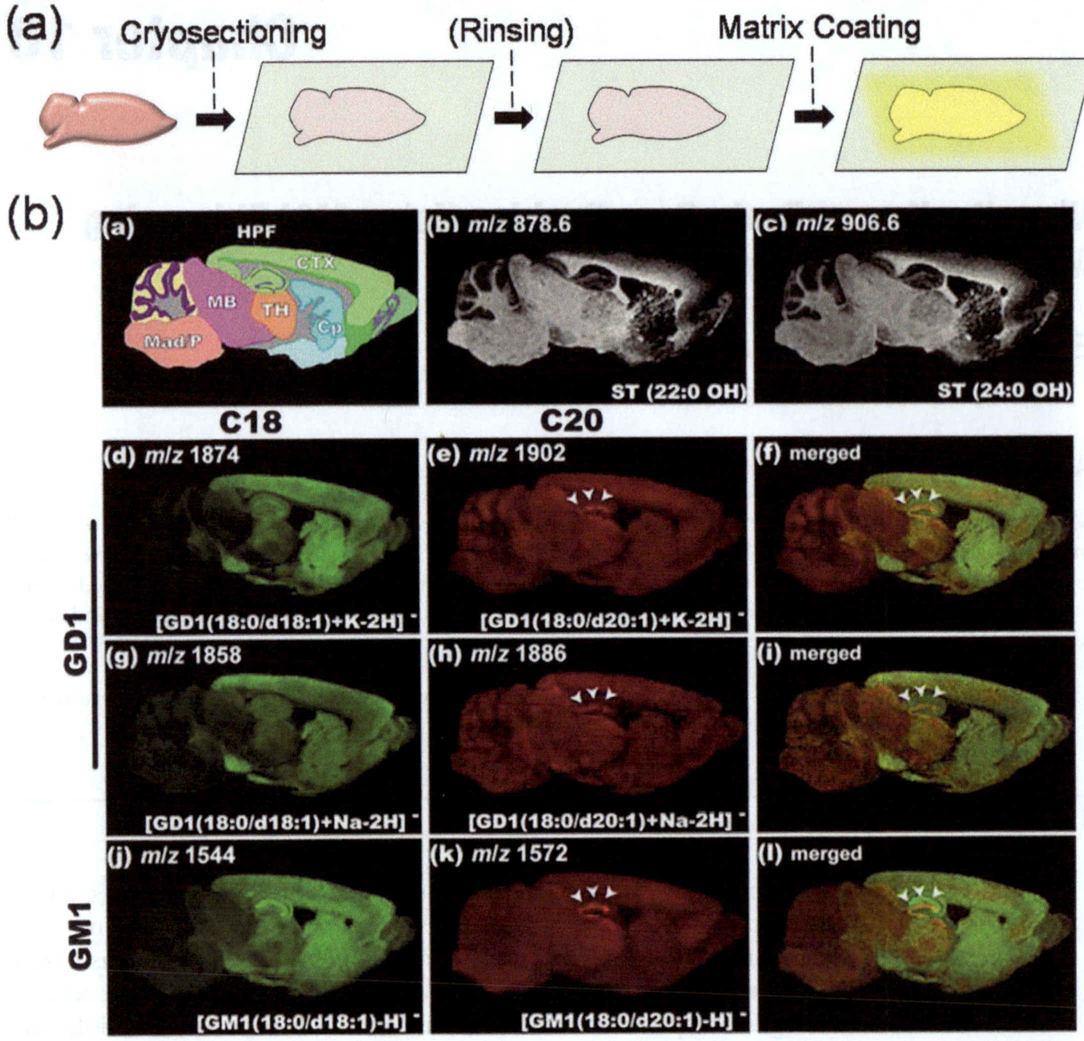

Fig. 1 The sample preparation procedure for analyzing brain gangliosides and a representative result obtained by MALDI-IMS. (**a**) Schematic diagram of the procedure. Details of each step are described in main text. (**b**) A representative result obtained from a mouse brain sample. Panel (**b**) is reprinted from a reference [3] with a license approved by PLoS One

ionized molecules are detected by a mass spectrometer. Here we describe a general procedure and related key notes for analyzing major brain gangliosides by MALDI-IMS (*see* Fig. 1).

2 Materials

1. Optimal cutting temperature (OCT) compound (Sakura Finetek Japan, Tokyo).
2. ITO-coated slide glass (Matsunami Glass Industries, Osaka, Japan) (*see* **Note 1**).

3. Ammonium formate aqueous solution.
4. 2,5-dihydroxyacetophenone (2,5-DHAP).
5. Silica gel particles.

3 Methods

3.1 Collecting the Brain

1. Sacrifice 8-week-old C57BL/6J mice with the protocols approved by the appropriate Animal Care and Use Committee.
2. Collect the brains and rapidly freeze them with powdered dry ice.
3. Stock the frozen brains at −80 °C until use.

3.2 Cryosectioning

1. In a cryostat kept at −16 °C, hold the frozen brain with OCT compound (the embedding media) on a disc. Do not let the target region of the brain become attached to the compound (*see* Fig. 2 and **Note 2**).
2. Cut the brain at 10-μm thickness (*see* **Note 3**).
3. Place the sliced section on the ITO-coated side of the glass slide and thaw-mount the section using a finger. Note that the coated conductive side of the slide must attach to the section to obtain a signal for some instruments.
4. Put the slide into a plastic 50 mL-tube containing silica gel particle covered with a thin paper, and store it at −80 °C until use (*see* **Note 4**).

3.3 Rinsing the Sections (Optional; See Note 5)

1. Put 30 mL of 150 mM ammonium formate aqueous solution in a 10-cm dish on ice.
2. Carefully immerse the prepared slide (after it has returned to room temperature) in the solution for 15 s.
3. Dry the rinsed slide in a desiccator for 15 min.

3.4 Taking Images

1. If needed, spot a mass calibration mixture outside of the section and dry it using a desiccator.
2. Take images including the sections and several coordinates for setting the location of measurement using a scanner or a microscope. Set the appropriate resolution depending on the required spatial resolution for analysis.

3.5 Matrix Coating

Select a matrix and a method for applying the matrix onto the sections (sublimation or spray) depending on the purpose of the study or the available type of ion source equipped with the instruments (*see* **Note 6**). Sublimation of 2,5-DHAP by the following steps is a first choice for an instrument with an atmospheric-pressure MALDI ion source.

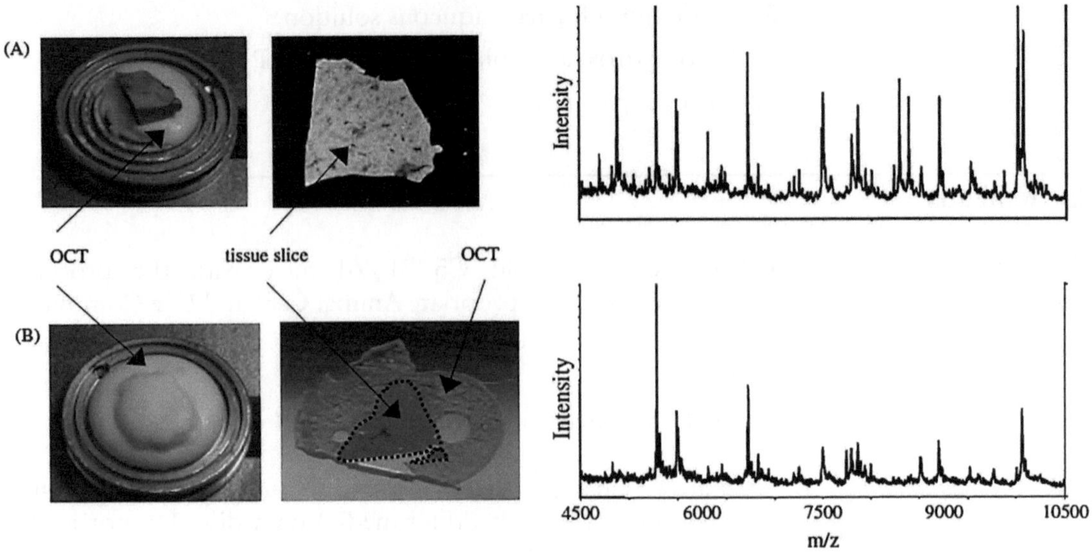

Fig. 2 OCT compound decreases signal intensities in MALDI-IMS. (**A**) Optimal way to adhere the sample to the holder using OCT compound. The resulting spectrum shows intense signals. (**B**) A case in which the tissue is embedded in OCT compound. The section is surrounded by OCT. The resulting spectrum contains about half of the signals compared to that in panel (**A**). Reprinted from [7] with permission from John Wiley & Sons, Ltd

1. Put 600 mg of 2,5-DHAP in a heating box and set it under the prepared sample facing the bottom in a vacuum chamber. Set the distance between the box and the sample at 7.5 cm (*see* **Note 7**).
2. Set a shutter between the box and the sample, close the chamber, and then start the vacuum.
3. When the vacuum pressure reaches 50 Pa, start heating the box, up to 180 °C.
4. Open the shutter for 1.0 min at 20 Pa, and then close the shutter.
5. Cool the box until the temperature reaches 100 °C and then return the pressure inside the box to atmospheric pressure.
6. Take the coated sample slide out of the chamber.

3.6 Measurement of Gangliosides

1. Load the sample slide into an instrument for MALDI-IMS.
2. Find the coordinates captured in the MALDI-IMS instrument and set the positions as the corresponding coordinates in the picture taken before applying the matrix.
3. Select the negative ion mode and a target mass range (*see* **Note 8**).
4. Find appropriate parameters to detect a target signal by test shots outside of the target region (*see* **Note 9**). If needed, perform a mass calibration using the dried standard spot.
5. Select a spatial resolution and run the measurement.

3.7 Data Analysis

1. If necessary, normalize the signal intensity of each pixel by using total ion current (TIC), extracted ion current (XIC), or other methods (*see* **Note 10**).
2. Select the peak(s) of interest with a selecting window. The width of the window is set based on the mass resolution and the precision of the data.

4 Notes

1. A conductive surface is required for some MALDI-IMS instruments to send the ionized molecules from the ion source to the mass analyzer. Note that several instruments do not require such a conductive coating on the slide glass.
2. OCT compound is known to suppress ionization of other molecules (*see* Fig. 2 [7]).
3. Some researchers prepare sections with different thicknesses. The use of thicker sections generally requires a higher amount of matrix [8].
4. The tube prevents the sections from getting wet when they are moved from −80 °C to room temperature.
5. The main purpose of the rinsing described above is to remove potassium (K) and sodium (Na) from the section. The rinsing provides simple mass spectra with higher signal intensity [9], because some gangliosides are detected in multiple ion forms (e.g., $[M-H]^-$, $[M+Na-2H]^-$, $[M+K-2H]^-$) in biological samples [3]. This step needs special caution to avoid breaking the structure of the sections.
6. In previous studies [2, 3, 6, 9], several matrices such as 2,5-DHAP, 2,5-dihydroxybenzoic acid or 9-aminoacridine were used for analyzing gangliosides. Note that 2,5-DHAP is not suitable for some instruments with a vacuum MALDI source because 2,5-DHAP easily sublimates and the sublimation causes a change of ionization efficacy of analytes within a run.

   The method for applying the matrix can be selected in accord with the required spatial resolution and sensitivity. Sublimation provides small crystal sizes and is thus suitable for high spatial resolution (<10 μm). A spray is generally considered to provide higher sensitivity through a liquid extraction step that is not in sublimation, while there is a risk of causing delocalization of the target or producing a large-sized crystal. Some researchers have recently used a combination of the two that involves an initial sublimation step followed by a "recrystallization" step using vapor for other molecule [8, 10].
7. Parameters can be changed depending on the available system. The main parameters are the pressure in the chamber, the temperature of the heating box, the distance between

the heating box and the sample surface, and the sublimation time. The sublimation time is the suitable parameter for controlling the matrix amount. Note again that the optimal matrix amount depends on the thickness of the section or the sample type.

8. Gangliosides are generally detected in the negative ion mode because of a negative charge on their sialic acid moieties. Asialogangliosides, on the other hand, are generally detected in the positive ion mode because they do not have a negative charge on their structure [2].

9. Basic parameters are diameter and power of the laser equipped with the MALDI-IMS instrument. Set the diameter depending on the required spatial resolution and start test shots with a weak power. Increase the power until a target signal comes to be detected. For an adult mouse brain, $[M-H]^-$ ion of a GM1(d18:1/18:0) at m/z 1544.9 should be detected from the cerebral cortex. Note that several molecules such as phosphatidylcholines are known to be fragmented when the laser power is too high. We recommend to check such fragmentation of target molecules before the run. Other parameters could be also critical, but they depend on the type of mass spectrometer used for the analysis.

10. Normalization is performed for the differences in ionization efficacy and the condition of cocrystallization between measuring points. TIC normalization (which is the most frequently selected method of normalization) divides each signal intensity by the sum of all signals. Such normalization works well in some cases [11], but it can cause artifacts depending on the data or the setting. For example, the result of TIC normalization can be easily affected by the setting of the mass range. In contrast, XIC normalization using a signal from an internal standard (which is uniformly applied before matrix) can provide better results [12, 13], although additional optimization is required. Any normalization requires a careful check before and after the normalization to interpret the data correctly.

References

1. Norris J, Caprioli R (2013) Analysis of tissue specimens by matrix-assisted laser desorption/ionization imaging mass spectrometry in biological and clinical research. Chem Rev 113:2309–2342
2. Kitakaze K, Mizutani Y, Sugiyama E et al (2016) Protease-resistant modified human β-hexosaminidase B ameliorates symptoms in GM2 gangliosidosis model. J Clin Invest 126:1691–1703
3. Sugiura Y, Shimma S, Konishi Y et al (2008) Imaging mass spectrometry technology and application on ganglioside study; visualization of age-dependent accumulation of C20-ganglioside molecular species in the mouse hippocampus. PLoS One 3:e3232

4. Colsch B, Jackson SN, Dutta S, Woods AS (2011) Molecular microscopy of brain gangliosides: Illustrating their distribution in hippocampal cell layers. ACS Chem Neurosci 2:213–222
5. Zhang Y, Wang J, Liu J et al (2016) Combination of ESI and MALDI mass spectrometry for qualitative, semi-quantitative and in situ analysis of gangliosides in brain. Sci Rep 6:25289
6. Hirano-Sakamaki W, Sugiyama E, Hayasaka T et al (2015) Alzheimer's disease is associated with disordered localization of ganglioside GM1 molecular species in the human dentate gyrus. FEBS Lett 589:3611–3616
7. Schwartz SA, Reyzer ML, Caprioli RM (2003) Direct tissue analysis using matrix-assisted laser desorption/ionization mass spectrometry: practical aspects of sample preparation. J Mass Spectrom 38:699–708
8. Yang J, Caprioli RM (2011) Matrix sublimation/recrystallization for imaging proteins by mass spectrometry at high spatial resolution. Anal Chem 83:5728–5734
9. Angel PM, Spraggins JM, Baldwin HS, Caprioli RM (2012) Enhanced sensitivity for high spatial resolution lipid analysis by negative ion mode matrix assisted laser desorption ionization imaging mass spectrometry. Anal Chem 84:1557–1564
10. Shimma S, Takashima Y, Hashimoto J et al (2013) Alternative two-step matrix application method for imaging mass spectrometry to avoid tissue shrinkage and improve ionization efficiency. J Mass Spectrom 48:1285–1290
11. Sugiura Y, Konishi Y, Zaima N et al (2009) Visualization of the cell-selective distribution of PUFA-containing phosphatidylcholines in mouse brain by imaging mass spectrometry. J Lipid Res 50:1776–1788
12. Pirman DA, Kiss A, Heeren RMA, Yost RA (2013) Identifying tissue-specific signal variation in MALDI mass spectrometric imaging by use of an internal standard. Anal Chem 85:1090–1096
13. Takai N, Tanaka Y, Saji H (2014) Quantification of small molecule drugs in biological tissue sections by imaging mass spectrometry using surrogate tissue-based calibration standards. Mass Spectrom (Tokyo, Japan) 3:A0025

Chapter 11

Immunoelectron Microscopy of Gangliosides

Takuma Tsuji, Akikazu Fujita, and Toyoshi Fujimoto

Abstract

Because chemical fixatives like aldehydes do not work on most lipid molecules in the membrane, small-scale lipid distribution cannot be identified by immunoelectron microscopy in cells fixed by conventional methods. Here we describe a method for physically stabilizing membranes through quick-freezing and freeze-fracture replica formation and for specifically labeling gangliosides for electron microscopy. This method enables the ultrahigh-resolution mapping of membrane lipids including gangliosides within the two-dimensional plane of membranes.

Key words Ganglioside, GM1, GM3, Quick-freezing, Freeze-fracture, Electron microscopy

1 Introduction

Gangliosides have been regarded as major components of lipid rafts [1], and GM1 has been often used as a marker for rafts in microscopy (e.g., [2]). For this purpose, cholera toxin B subunit tagged with fluorophores is used frequently, but direct application of this ligand to live cells may cause redistribution of GM1 [3]. Moreover, the spatial resolution of fluorescence microscopy may not be high enough to resolve rafts. Electron microscopy, on the other hand, has sufficient spatial resolution, but the molecular distribution in situ needs to be maintained by an appropriate method during sample preparation procedures.

Conventionally, a combination of formaldehyde and glutaraldehyde has been used to fix cells for electron microscopy. The aldehydes stabilize proteins mainly through covalently binding to amino and sulfhydryl groups, but they do not work on most lipids, which lack these reactive groups. In fact, lipids in the membrane have been shown to retain lateral mobility even after aldehyde fixation [4], and in one study, ganglioside GD1a was shown to relocate from one aldehyde-fixed brain slice to another over a millimeter to centimeter distance [5]. These results indicate that conventional

Sandro Sonnino and Alessandro Prinetti (eds.), *Gangliosides: Methods and Protocols*, Methods in Molecular Biology, vol. 1804, https://doi.org/10.1007/978-1-4939-8552-4_11,

© Springer Science+Business Media, LLC, part of Springer Nature 2018

aldehyde fixation is not appropriate for a critical definition of lipid distribution by immunoelectron microscopy.

One possible solution to the above problem would be the development of a new fixative that could react with lipids, but even if such a fixative were developed, the chemical reaction would likely take too much time to be useful in high-resolution studies. This is because lipids move very quickly in the membrane plane. Based on the diffusion coefficient obtained from living cells [6], we can estimate that phospholipids can move approximately 1.9 μm in 1 s.

Accordingly, we propose that physical fixation rather than chemical fixation is the method of choice to capture lipids in situ for observation by electron microscopy. To be specific, we think that a method combining quick-freezing and freeze-fracture will prove most appropriate [7]. In this method, the movement not only of lipids but also of proteins is halted within milliseconds by quick-freezing. Next, each membrane is split into two leaflets by freeze-fracture, and the resultant half-membranes are coated with a thin layer of carbon and platinum grains called the replica. The replica physically holds membrane lipids in the hydrophobic tail portion so that they can be labeled by applying probes to the head group and observed by electron microscopy (Fig. 1) [8]. By this method, we can define the two-dimensional distribution of membrane lipids including gangliosides at the nanometer resolution [9–12].

2 Materials

2.1 Quick Freezing and Freeze Fracturing

1. Gold foil: 20 μm in thickness. Clean it with a mixture of acetone and dilute hydrogen chloride before autoclaving.
2. SDS digestion buffer: Dissolve 2.5% (wt/vol) SDS in 0.1 M Tris–HCl buffer (pH 8.0). To prepare the Tris buffer, dissolve 12.1 g Tris in double-distilled water, adjust pH with HCl and bring the volume to 1000 ml. Store at RT.
3. 50% (wt/vol) glycerol in PBS: Measure 50 g glycerol, add 10 ml 10× phosphate-buffered saline (PBS) and make up the volume to 100 ml with double-distilled water. Stable for more than several months at RT.

2.2 Ganglioside Labeling

1. Tween 20 washing buffer: Dissolve 0.1% (wt/vol) Tween 20 in PBS. Stable for more than several months at RT.
2. Blocking buffer: Dissolve 3% (wt/vol) bovine serum albumin (BSA) in PBS. Add 0.02% sodium azide to prevent bacterial growth. Can be kept at 4 °C for several weeks.
3. Probe dilution buffer: Dissolve 1% (wt/vol) BSA in PBS.
4. Washing buffer: Dissolve 0.1% (wt/vol) BSA in PBS.

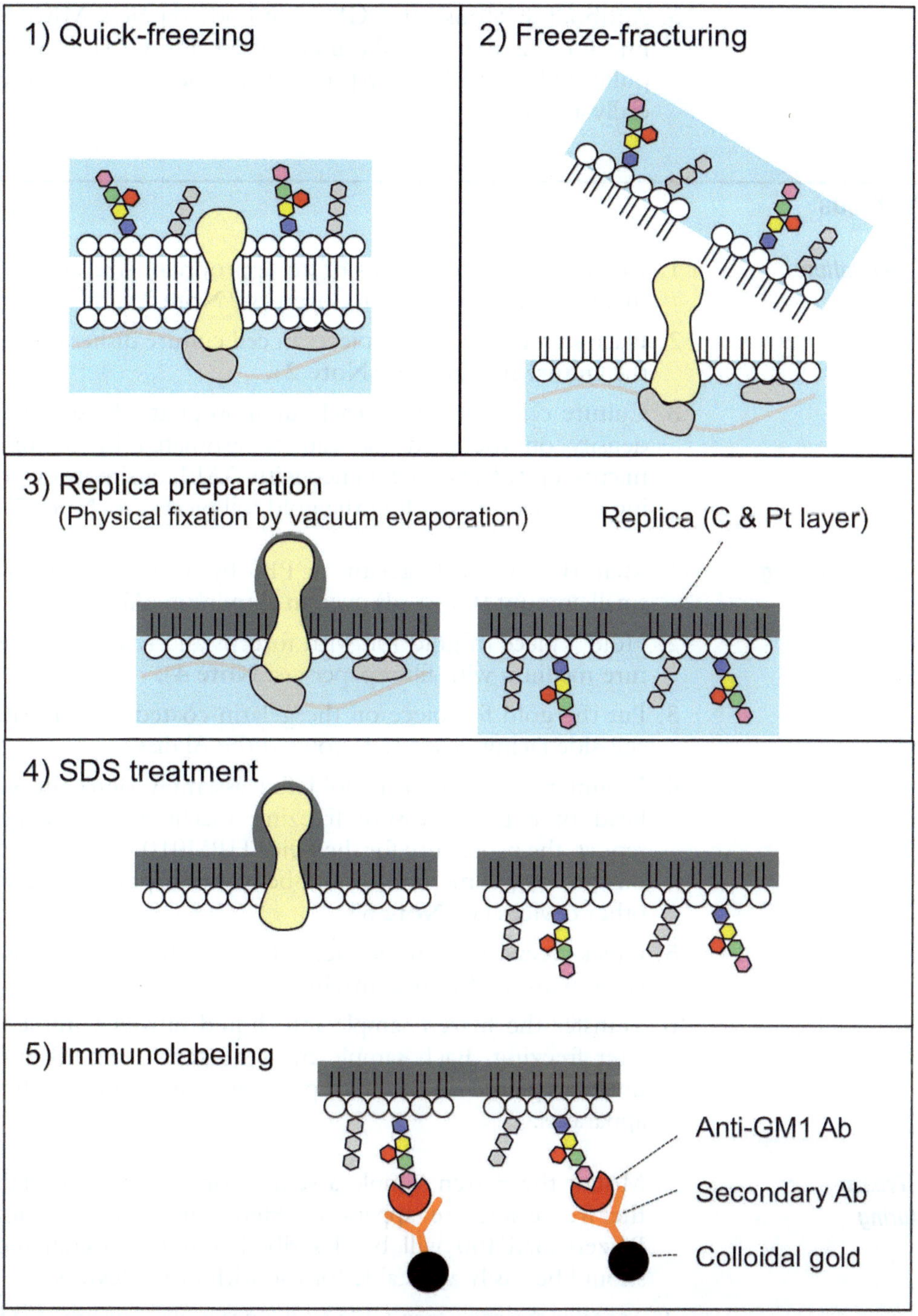

Fig. 1 The outline of the method. (1) Cells cultured on thin gold foil are quick-frozen using a high-pressure freezing device. (2) By freeze-fracturing, membranes are split between two leaflets to expose the hydrophobic interface. (3) Vacuum evaporation of carbon and platinum makes a thin coating, physically stabilizing the membrane monolayer. (4) Treatment of the replica with SDS removes materials that are not covalently bound with the membrane (e.g., cytoskeleton). (5) Anti-GM1 antibody binds to the head group of GM1 molecules held in the replica

5. Antibodies: Mouse anti-GM3 antibody (clone GMR6) works fine for this method. We used rabbit anti-GM1 [13] in our published works [14, 15], but this antibody is not commercially available.

3 Methods

3.1 Cell Culture

1. Cut gold foil into pieces of an appropriate size and sterilize them with autoclave or dry oven (*see* **Note 1**).
2. Place sterilized gold foil pieces in cell culture dishes containing the culture medium (*see* **Note 2**).
3. Culture cells until they reach an appropriate density. The cell density on the gold foil can be estimated by fluorescence microscopy of a sample stained with DAPI, or simply by observing the area surrounding the gold foil pieces (*see* **Note 3**).

3.2 Quick Freezing

1. Melt 10% (wt/vol) gelatin in PBS by warming and spread a small amount (0.5–1 μl) over an aluminum (Al) specimen disk.
2. Hold a piece of gold foil with forceps and remove excess culture medium with filter paper (*see* **Note 4**).
3. Put the gold foil piece on the gelatin-coated Al disk with the cell-side facing down (i.e., toward the Al disk).
4. Mount the Al disk–cell–gold foil assembly onto the sample head of a high-pressure freezing machine. In the present report, the procedure for the Leica HPM010 will be described, but the overall method should be easily adaptable for use with other devices (*see* **Note 5**).
5. Quick-freeze the sample according to the instructions provided with the freezing machine.
6. Transfer the frozen sample into liquid nitrogen immediately after freezing. Each sample must be kept in liquid nitrogen until it is placed on the cold specimen stage of a freeze-fracture apparatus.

3.3 Freeze-Fracturing

1. Mount the frozen sample assembly onto a specimen table of the freeze-fracture apparatus. Here, the procedure for the Balzers BAF400 will be described, but the overall method should be easily adaptable for use with other devices.
2. Transfer the specimen table to the precooled specimen stage of the freeze-fracture apparatus (*see* **Note 6**).
3. Keep the specimen temperature at −120 °C for 5 min, then at −105 °C for 3–5 min, and precool the knife to the lowest possible temperature (e.g., near the temperature of liquid nitrogen).

4. When the vacuum reaches below ~5 × 10^{-6} mmHg, proceed to separate the gold foil from the Al disk with a knife (*see* **Note 7**).
5. Evaporate carbon (C) and platinum-carbon (Pt/C) onto the specimen using an electron beam gun. First, evaporate 2–5 nm of C at an angle of 90° to the specimen surface, then 2 nm of Pt/C at an angle of 45° and finally 10 nm of C at an angle of 90°. The thicknesses of the respective C and Pt/C depositions are controlled by referring to a crystal thickness monitor (*see* **Note 8**).
6. Break the vacuum and transfer the specimen to the SDS digestion buffer (*see* **Note 9**). The replica and underlying gelatin should be released from the Al disk with mild agitation (*see* **Note 10**).
7. Transfer the replicas to a 1.5 ml plastic tube containing the SDS digestion buffer and rotate overnight at 60 °C in an incubator (*see* **Note 11**).
8. Transfer the replicas into 50% glycerol in PBS. They can be kept at −20 °C for several months.

3.4 Ganglioside Labeling and Electron Microscopic Analysis

Carry out all procedures at room temperature unless otherwise specified.

1. Rinse the replicas with Tween 20 washing buffer in a ceramic spot plate by repeated pipetting with a micropipettor bearing a white 10 μl tip.
2. Under a stereomicroscope, aspirate replicas with a minimum volume of solution in a micropipettor and transfer them to a new solution. Typically, 5–10 fragments can be taken up in 1 μl solution. Repeat **steps 1** and **2** five times.
3. Transfer the replicas to the blocking buffer and incubate for at least 30 min.
4. Dilute the primary antibody to an appropriate concentration with the probe dilution buffer (*see* **Notes 12** and **13**). For double labeling of two different antigens, e.g., GM1 and GM3, a mixture of two primary antibodies raised in two different species, e.g., rabbit anti-GM1 and mouse anti-GM3, must be prepared. Centrifuge the probe solution at 21,500 × *g* for 5 min at 4 °C and use the supernatant for incubation. The primary antibody preincubated with an excess amount of antigen, nonimmune IgG, or the dilution buffer alone can be used as controls.
5. Rinse the replicas once with the washing buffer and transfer them into a 0.5 ml tube containing the primary antibody solution. Incubate overnight at 4 °C with continuous rotation.

This process may be shortened by incubating at a higher temperature, e.g., for 60 min at RT or for 30 min at 37 °C.

6. Rinse the replicas by repeated pipetting in the washing buffer. Repeat this process five times by transferring to new solutions.
7. Prepare the colloidal gold-conjugated probe solution in the probe dilution buffer. In cases where an antibody raised in rabbits is used as the primary antibody, incubate replicas with colloidal gold-conjugated anti-rabbit IgG antibody or colloidal gold-conjugated protein A. For double labeling using two primary antibodies, e.g., rabbit anti-GM1 and mouse anti-GM3, use two secondary antibodies conjugated with colloidal gold of different diameters, e.g., goat anti-rabbit IgG conjugated with 5 nm colloidal gold and goat anti-mouse antibody with 10 nm colloidal gold. Centrifuge the colloidal gold solution at 215 × *g* for 5 min at 4 °C and use the supernatant for incubation (*see* **Note 14**). The dilution of colloidal gold conjugates needs to be calibrated for each batch.
8. Transfer the replicas into a 0.5 ml tube containing the secondary probe solution and incubate at 37 °C for 30 min.
9. Rinse the replicas as in **step 6**.
10. Rinse the replicas with distilled water by quickly transferring the solution into two or three wells (*see* **Note 15**).

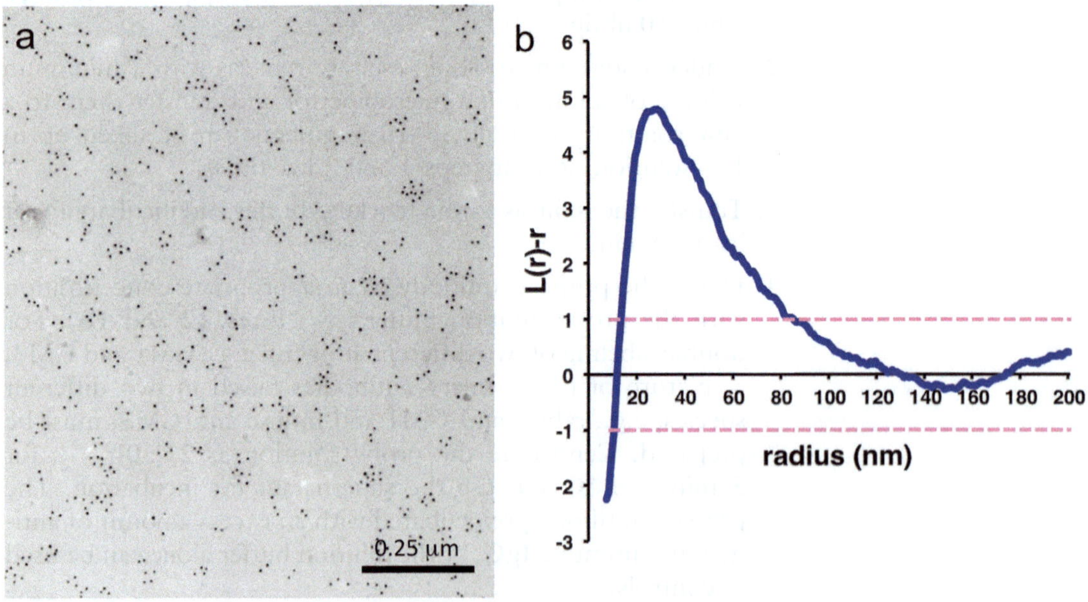

Fig. 2 GM1 in the plasma membrane of mouse fibroblasts in culture. (**a**) Immunoelectron microscopy of GM1. Labeling was done using rabbit anti-GM1 antibody and colloidal gold (5 nm)-conjugated goat anti-rabbit IgG antibody. (**b**) The distribution pattern of gold labels in figure (**a**) was analyzed by Ripley's *K*-function. The mean *L*(*r*)–*r* curve showed maximal deflection from complete spatial randomness (99% confidence interval is shown by a dotted line) at 28 nm, indicating that the gold labeling formed clusters at that radius in this sample

11. Pick up the replicas onto a formvar-coated EM grid.
12. Wick off excess water using a small piece of filter paper and dry the grids.
13. Observe the replicas under a transmission electron microscope operating at 80–100 kV. Gold particles with a 10 nm diameter are easily seen at ×10,000, whereas those with a 5 nm diameter require a higher magnification for observation (Fig. 2a).
14. When necessary, a two-dimensional distribution of gold particles can be analyzed by appropriate point pattern analytical methods (Fig. 2b) (*see* **Note 16**). For this purpose, transform the image to binary data so that only gold particles are visible, and determine the coordinate of each gold particle in the real scale using ImageJ.

4 Notes

1. Several tricks at this step will help to distinguish the two sides of the gold foil after cell seeding: (1) cutting the gold foil into asymmetrical shapes; (2) vacuum evaporation of carbon to one side; (3) scratching one side with sandpaper.
2. Keep the gold foil pieces as flat as possible.
3. Most cells attach and grow well on the clean gold surface. For cells that show poor attachment, coating with poly-L-lysine should help.
4. Take care not to remove so much culture medium that the cells become dry.
5. This step needs to be done as quickly as possible so that the cell is not exposed to unnatural conditions.
6. This step needs to be done quickly to prevent frosting of the specimen surface.
7. Make sure the knife is cooled to the lowest possible temperature before it touches the sample.
8. The first carbon layer is important for the efficient labeling of lipids including gangliosides [15, 16].
9. Avoid hitting the replica on other objects because it is fragile and easily broken into extremely small pieces.
10. A ceramic spot plate is handy for this and the subsequent rinsing steps as the replica is gray in color and can be observed clearly against a white ceramic background.
11. Use at least 0.5 ml of SDS digestion buffer in the 1.5 ml tube to prevent replicas from adhering to the tube wall.
12. The appropriate concentration of the antibody needs to be determined by observing the final result, but 10 μg/ml may be a good starting point.

13. Cholera toxin B subunit also binds to GM1 specifically, but it yields a low level of labeling, probably because it is a pentamer and may require five GM1 molecules in an appropriate disposition.
14. This step is to remove any aggregate of colloidal gold-conjugated reagents. Note that centrifugation is done at a slower speed than for antibodies without colloidal gold conjugation.
15. This rinse and the next pick-up step need to be done for one replica fragment at a time because, once transferred to water, replicas tend to stick to the ceramic wall tightly and become unrecoverable.
16. Ripley's *K*-function is often used for this purpose (Fig. 2b) [15, 17].

Acknowledgment

This work was supported by Grants-in-Aid for Scientific Research from the Ministry of Education, Culture, Sports, Science, and Technology of the Government of Japan to TT (17K15544) and TF (15H02500, 15H05902).

References

1. Simons K, Ikonen E (1997) Functional rafts in cell membranes. Nature 387:569–572
2. Nichols BJ (2003) GM1-containing lipid rafts are depleted within clathrin-coated pits. Curr Biol 13:686–690
3. Fujimoto T (1996) GPI-anchored proteins, glycosphingolipids, and sphingomyelin are sequestered to caveolae only after crosslinking. J Histochem Cytochem 44:929–941
4. Tanaka KA, Suzuki KG, Shirai YM, Shibutani ST, Miyahara MS, Tsuboi H, Yahara M, Yoshimura A, Mayor S, Fujiwara TK, Kusumi A (2011) Membrane molecules mobile even after chemical fixation. Nat Methods 7:865–866
5. Heffer-Lauc M, Lauc G, Nimrichter L, Fromholt SE, Schnaar RL (2005) Membrane redistribution of gangliosides and glycosylphosphatidylinositol-anchored proteins in brain tissue sections under conditions of lipid raft isolation. Biochim Biophys Acta 1686:200–208
6. Schlessinger J, Axelrod D, Koppel DE, Webb WW, Elson EL (1977) Lateral transport of a lipid probe and labeled proteins on a cell membrane. Science 195:307–309
7. Takatori S, Mesman R, Fujimoto T (2014) Microscopic methods to observe the distribution of lipids in the cellular membrane. Biochemistry 53:639–653
8. Fujita A, Cheng J, Fujimoto T (2010) Quantitative electron microscopy for the nanoscale analysis of membrane lipid distribution. Nat Protoc 5:661–669
9. Cheng J, Fujita A, Yamamoto H, Tatematsu T, Kakuta S, Obara K, Ohsumi Y, Fujimoto T (2014) Yeast and mammalian autophagosomes exhibit distinct phosphatidylinositol 3-phosphate asymmetries. Nat Commun 5:3207. https://doi.org/10.1038/ncomms4207
10. Fujimoto K, Umeda M, Fujimoto T (1996) Transmembrane phospholipid distribution revealed by freeze-fracture replica labeling. J Cell Sci 109:2453–2460
11. Fujita A, Cheng J, Tauchi-Sato K, Takenawa T, Fujimoto T (2009) A distinct pool of phosphatidylinositol 4,5-bisphosphate in caveolae revealed by a nanoscale labeling technique. Proc Natl Acad Sci U S A 106:9256–9261
12. Takatori S, Tatematsu T, Cheng J, Matsumoto J, Akano T, Fujimoto T (2016) Phosphatidylinositol

3,5-bisphosphate-rich membrane domains in endosomes and lysosomes. Traffic 17:154–167

13. Kusunoki S, Shimizu J, Chiba A, Ugawa Y, Hitoshi S, Kanazawa I (1996) Experimental sensory neuropathy induced by sensitization with ganglioside GD1b. Ann Neurol 39:424–431
14. Fujita A, Cheng J, Fujimoto T (2009) Segregation of GM1 and GM3 clusters in the cell membrane depends on the intact actin cytoskeleton. Biochim Biophys Acta 1791:388–396
15. Fujita A, Cheng J, Hirakawa M, Furukawa K, Kusunoki S, Fujimoto T (2007) Gangliosides GM1 and GM3 in the living cell membrane form clusters susceptible to cholesterol depletion and chilling. Mol Biol Cell 18:2112–2122
16. Fujita A, Fujimoto T (2007) Quantitative retention of membrane lipids in the freeze-fracture replica. Histochem Cell Biol 128:385–389
17. Prior IA, Muncke C, Parton RG, Hancock JF (2003) Direct visualization of Ras proteins in spatially distinct cell surface microdomains. J Cell Biol 160:165–170

Chapter 12

Nuclear Magnetic Resonance of Gangliosides

Domenico Acquotti, Laura Mauri, and Sandro Sonnino

Abstract

Structure, conformation, and dynamics of sphingolipids can provide substantial help in better understanding sphingolipid–ligand interaction mechanisms. Both the oligosaccharide structure and the ceramide moiety of native glycosphingolipid can be established directly by NMR spectroscopic analysis without the necessity to resort to any other chemical or spectroscopic methods. NMR is a powerful technique to investigate interaction between small ligand, such as ganglioside, and membrane protein.

Key words Nuclear magnetic resonance spectroscopy, Ganglioside, Oligosaccharide, Conformational analysis, Conformational dynamics, Micelles

Abbreviation

1D	One-dimensional
2D	Two-dimensional
Cer	Ceramide
COSY	Correlation spectroscopy
DANTE	Delays alternated by nutation for tailored excitation
Deac-GM1	Neu-GM1
DEPT	Distortionless enhancement by polarization transfer
DMPC	1,2-dimyristoyl-sn-glycero-3-phosphatidylcholine
DMPE	Dimyristoyl phosphatidylethanolamine
DMPG	1,2-Dimyristoyl-sn-glycero-3-phosphoglycerol
DMPS	Dimyristoyl phosphatidylserine
DMSO	Dimethyl-d_6-sulfoxide
DOSY	Diffusion Ordered Spectroscopy
DPC	Dodecylphosphocholine
DPC	Dodecylphosphocholine-d38;
EDTA	Ethylenediaminetetraacetic acid
FA	Fatty acid
Gal	Galactose
GalNAc	N-acetylgalactosamine
Glc	Glucose

Sandro Sonnino and Alessandro Prinetti (eds.), *Gangliosides: Methods and Protocols*, Methods in Molecular Biology, vol. 1804, https://doi.org/10.1007/978-1-4939-8552-4_12, © Springer Science+Business Media, LLC, part of Springer Nature 2018

HMBC	Heteronuclear Multiple Bond Correlation
HOHAHA	Homonuclear Hartman Hahn
HSQC	Heteronuclear Single Quantum Coherence
INEPT	Insensitive nuclei enhanced by polarization transfer
MD	Molecular dynamics
MINSY	Mixing irradiation during NOESY
MLEV	Malcolm LEVitt's decoupling cycle
Neu5Ac	*N*-acetylneuraminic acid
Neu5Gc	Neu5Gc N-glycolylneuraminic acid
NMR	Nuclear Magnetic Resonance
NOE	Nuclear Overhauser effect
NOESY	NOE correlated spectroscopy
PCS	Pseudocontact shift
PFG	Pulsed Field Gradient
PFG-SE	Pulsed Field Gradient Spin Echo
PRE	Paramagnetic Relaxation Enhancement
ROE	Rotating-frame nuclear Overhauser effect
ROESY	Rotating-frame Overhauser Effect Spectroscopy
Sph	Sphingosine
STD	Saturation transfer difference
STEP	Selective TOCSY edited preparation
SUV	Small Unilamellar Vesicles
TOCSY	Total correlated spectroscopy
TPPI	Time proportional phase incrementation

1 Introduction

In this chapter, several aspects of the nuclear magnetic resonance (NMR) spectroscopy of gangliosides largely related to their chemical and physicochemical features are described.

A big advantage of NMR spectroscopy is that the experiments are nondestructive. The major weakness of the method is its poor sensitivity; this disadvantage can be overcome, in part, by new techniques, pulse sequences and instrument improvement (higher magnet fields, cryogenic probes, full digital electronic devices, etc.).

Both the oligosaccharide structure and the ceramide moiety of native glycosphingolipid can be assigned directly by NMR spectroscopic analysis without resorting of any other chemical or spectroscopic methods (Table 1). 1D proton NMR spectrum of a glycosphingolipid dissolved in organic solvent shows only a few well resolved signals. The anomeric, olefinic, and hydroxyl protons in the range of 4.1–6.1 ppm, amide protons over 6.5 ppm, methylene and acetamido groups between 1.6 and 2.5 ppm (Fig. 1). These, in conjunction with vicinal ^{1}H–^{1}H coupling constants ($^{3}J_{HH}$) analysis and with the relative intensities of isolated signals, allow the identification of the number and kind of monosaccharide residues.

Table 1

NMR spectroscopic analysis: structural information may be obtained from NMR data using suitable NMR methods

Structural information	NMR data	NMR methods
Number of sugar residues	^{1}H and ^{13}C chemical shifts Relative abundance	Integrated 1D ^{1}H NMR spectrum 1D ^{13}C NMR spectrum
Monosaccharide residues assignment	^{1}H and ^{13}C chemical shifts ^{1}H vicinal coupling constants Intraresidue NOE	1D ^{1}H NMR spectrum 1D ^{13}C NMR spectrum 1D selective correlation spectroscopy (TOCSY) 1D selective NOE and ROE 2D ^{1}H-^{1}H correlation spectroscopy (COSY, TOCSY) 2D ^{1}H–^{13}C correlation spectroscopy (HMQC, HSQC, HMBC)
Anomeric configuration	^{1}H and ^{13}C chemical shifts ^{1}H vicinal coupling constants ^{1}H–^{13}C coupling constants Interresidue NOE	1D ^{1}H NMR spectrum 1D ^{13}C NMR spectrum 1D selective correlation spectroscopy (TOCSY) 1D ^{1}H–^{1}H selective dipolar spectroscopy (NOESY, ROESY) 2D ^{1}H–^{1}H dipolar spectroscopy (NOESY, ROESY)
Linkage site and sequence	^{1}H and ^{13}C chemical shifts Interresidue NOE Long-range homo- and hetero-nuclear correlation	1D ^{1}H NMR spectrum 1D ^{13}C NMR spectrum 2D ^{1}H–^{1}H correlation spectroscopy (COSY, TOCSY) 2D ^{1}H–^{13}C correlation spectroscopy (HMQC, HSQC, HMBC) 1D ^{1}H–^{1}H selective dipolar spectroscopy (NOESY, ROESY) 2D ^{1}H–^{1}H dipolar spectroscopy (NOESY, ROESY)

Gangliosides in water solution form high molecular weight aggregates. Their hydrodynamic radius vary from 6 to 25 nm, showing such a low mobility so as to prevent obtaining high-resolution NMR spectra [1] (Fig. 2).

It is possible to perform NMR analysis following three different approaches: (1) solubilizing samples into organic solvents where they are present as monomers; (2) investigating gangliosides as mixed micelles with perdeuterated dodecylphosphocholine (DPC) in aqueous solutions, and (3) investigating gangliosides as lipid-modified ganglioside small spherical micelles in aqueous solutions.

Solutions of ganglioside monomers are very useful for ganglioside structural and conformational characterization allowing NMR spectra of the highest resolution. Mixed micelles of gangliosides and DPC mimic the natural membrane environment. They show almost

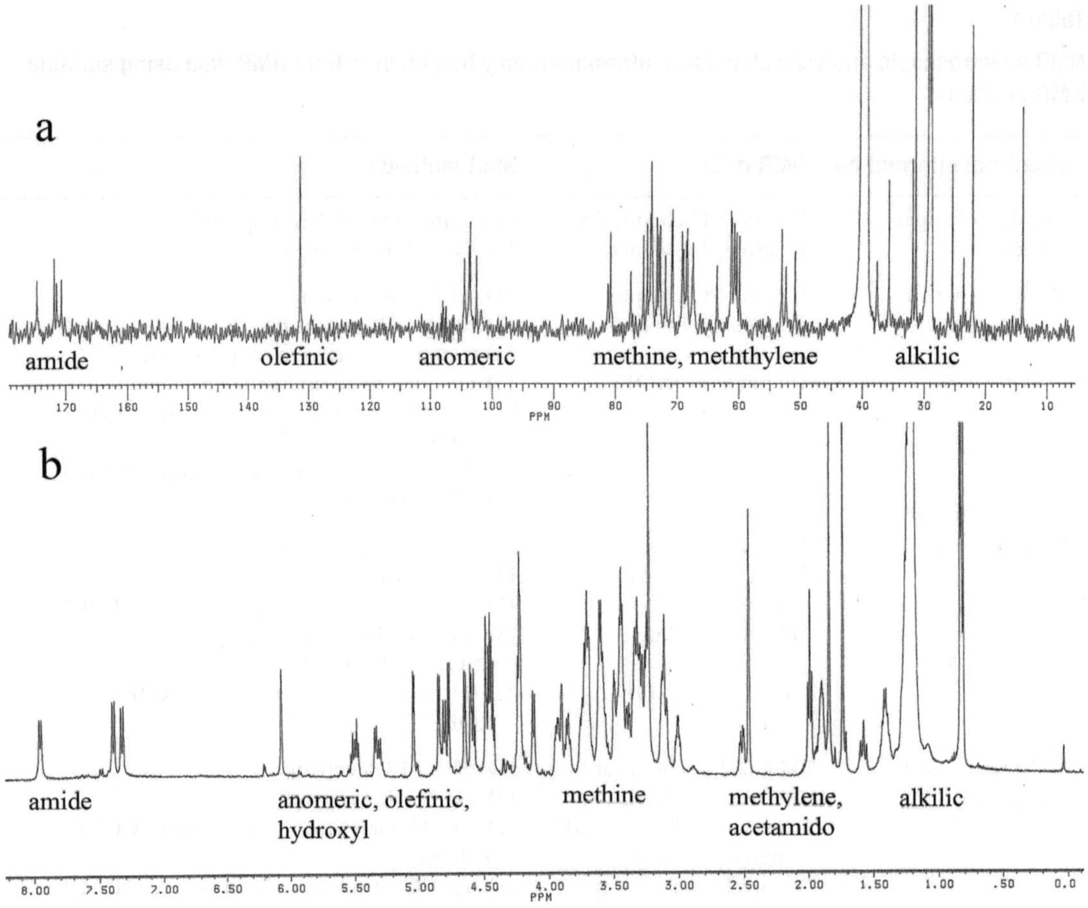

Fig. 1 (**a**) 125 MHz carbon spectrum of GM1(Neu5Gc) ganglioside in DMSO-*d6* solution at 303 K. (**b**) 500 MHz proton spectrum of GM2 ganglioside in DMSO-*d6* solution at 303 K

spherical size and are much smaller in comparison to the pure ganglioside aggregates (12–16 vs. 300–18,000 kDa); consequently they show a much higher rotational mobility which allows to obtain high-resolution NMR spectra [2–4]. Micelles of gangliosides are considered a cluster-like lateral distribution of gangliosides [5].

2 Materials

All solutions are prepared using ultrapure water and analytical grade reagent.

Deuterated water D_2O is a 100% isotopically enriched, NMR solvent.

Dimethylsulfoxide-d_6 (DMSO-d_6) is >99% isotopically enriched and has an HOD peak which is less than 1% of the HOH peak to minimize potential exchange with the analytes.

Chelex 100 resin.

Silica gel 100 resin.

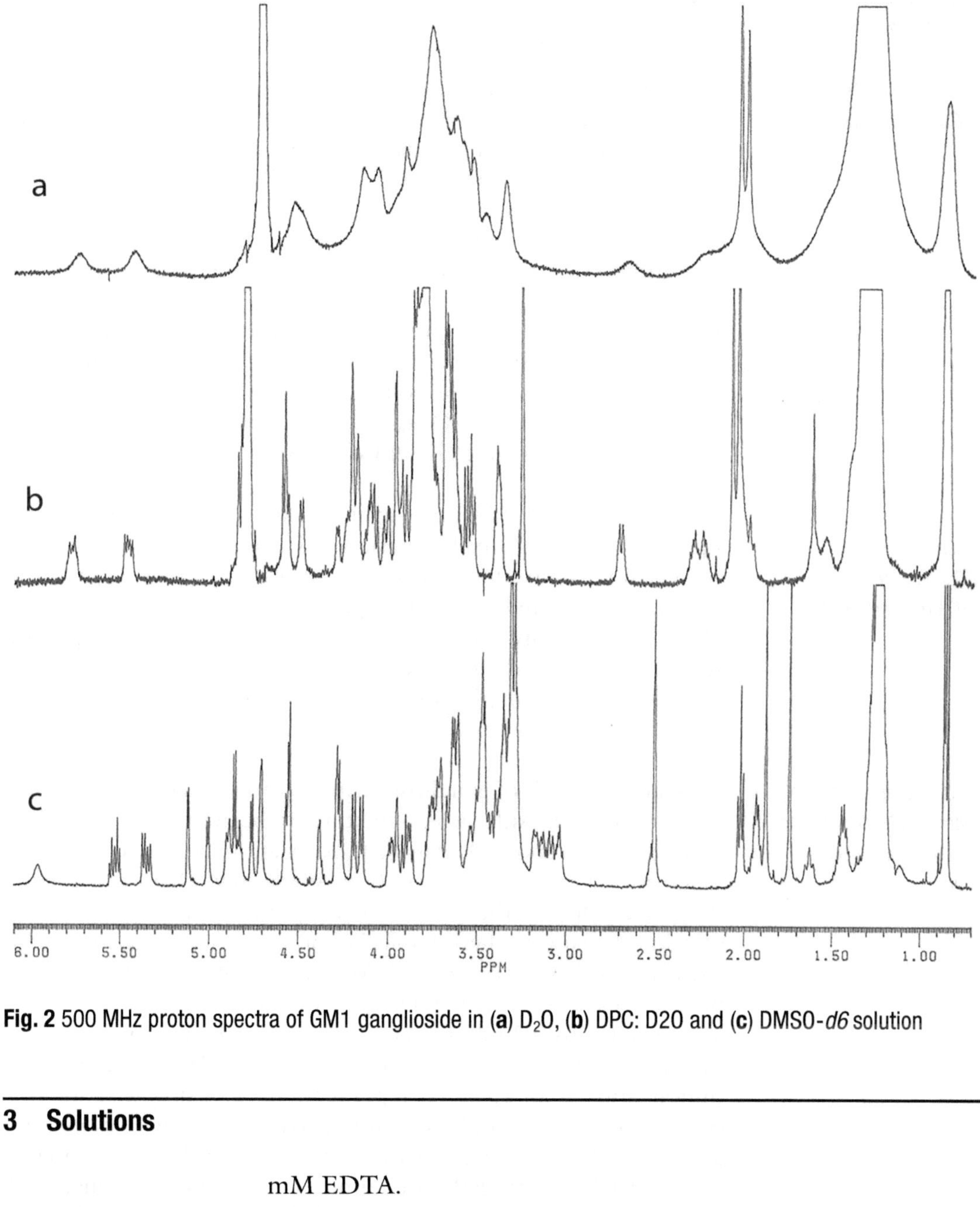

Fig. 2 500 MHz proton spectra of GM1 ganglioside in (**a**) D_2O, (**b**) DPC: D20 and (**c**) DMSO-*d6* solution

3 Solutions

mM EDTA.

DMSO-d_6–D_2O, 98:2.

50 mM, pH 6 deuterated potassium phosphate buffer.

1 M tetramethylammonium hydroxide.

Chloroform–methanol–water, 50:42:11, vol/vol/vol.

Chloroform–methanol–0.3% aqueous $CaCl_2$, 50:42:11, vol/vol/vol.

4 Methods

4.1 Preparation of Gangliosides

A water solution of ganglioside (5–8 mg/ml) is dialyzed with a solution of 2 mM EDTA and then with water. The ganglioside solution is then passed through a 5 × 0.5 cm Chelex 100 column (pH 7) and collected in a small pear-shaped flask. The solution is lyophilized, dried under high vacuum and stored in a nitrogen-fluxed dry box (*see* **Note 1**). Ganglioside structures and their carbon numbering are reported in Chapter 1 of this book.

4.1.1 Samples of Ganglioside Monomers

The small conical flask, prepared as reported above, is opened inside the dry box under nitrogen flux to avoid the inclusion of H_2O from the atmosphere, solubilized in 0.4–0.5 ml of DMSO-d_6 or in 0.4–0.5 ml of DMSO-d_6–D_2O, 98:2 vol/vol, and transferred into a 5 mm NMR tube (*see* **Note 2**, [6]).

4.1.2 Samples of Small Spherical Micelles of Gangliosides and Perdeuterated DPC

Synthetic DPC mixed with a small amount of ganglioside forms in aqueous solutions small spherical micelles of 12–16 kDa [4]. Ganglioside is mixed with DPC at a molar ratio of 1:40, and dissolved in deuterated potassium phosphate buffer (50 mM, 0.4 ml, pD 6). The mixture is exchanged twice with D_2O, with intermediate lyophilization, dried under high vacuum and stored into a dry box. Finally the sample is dissolved in D_2O (0.4 ml) and transferred into a 5 mm NMR tube under a nitrogen flux.

4.1.3 Samples of Small Spherical Micelles of Lipid Modified Gangliosides

Alkaline hydrolysis of gangliosides in butanol–aqueous tetramethylammonium hydroxide yields a mixture of the derivatives lacking the acetyl group(s) and acyl chain [7]. Acetylation of these compounds followed by purification yields a single-lipid-chain ganglioside derivative. Due to the reduced hydrophobic volume, in aqueous solution they form spherical micelles of low molecular mass suitable (*see* Chapter 3 in this book) for NMR measurements [8, 9]. Thirteen micromoles of (GM1) or (GD1a) are dissolved in 2 ml of 1-butanol and mixed with 0.2 ml of aqueous tetramethylammonium hydroxide (1 M). The reaction mixture is refluxed at 100 °C under continuous stirring for 13 h. Then pH is adjusted by dropwise addition of acetic acid to 6–7 and the solution is dried in a rotary evaporator at 50 °C. The residue, dissolved in a few milliliters of water, is applied to 5 × 2 cm RP-18 column. The column is washed with 100 ml of water to remove salt, and then eluted with 200 ml of chloroform–methanol, 2:1, vol/vol, to recover the formed deacetyl- and deacetyldeacyl-ganglioside derivatives.

The chloroform/methanol solution is dried, the residue dissolved in 2 ml of magnesium-dehydrated methanol and mixed with 10 μl of acetic anhydride. After stirring for 30 min at 50 °C, the solution is dried and purified by chromatography on a 50 × 1 cm silica gel column equilibrated and eluted with chloroform–methanol–water,

50:42:11, vol/vol/vol. Column elution is monitored by TLC (silica gel HPTLC plates developed in chloroform–methanol–0.3% aqueous $CaCl_2$, 50:42:11, vol/vol/vol, and revealed by colorimetric detection with anisaldehyde spray reagent). The fractions containing the single-chain ganglioside are collected, pooled, and evaporated until completely dry. Samples are then treated to remove divalent cations and to prepare the NMR tubes as described above.

4.2 Nuclear Magnetic Resonance Experiments

NMR is a powerful tool (*see* **Note 3**) that reveals until 1% of contaminants, including those that often cannot be displayed by colorimetric procedures after chromatography (*see* **Note 4**, [10]). Glycosphingolipids, and in particular gangliosides, are soluble in DMSO-d_6 where they are present in a monomeric form. In this solvent hydroxyl and amido groups do not exchange at room temperature providing a supplementary source of structural information. Moreover, if desired, their resonances may be removed from the spectrum dissolving the sample in *DMSO-d_6:D_2O* (98:2, vol/vol) by chemical exchange with the deuterium nuclei. Thus, high-resolution spectra allow the complete assignment of proton and carbon signals, and then the primary and secondary structural characterization. If determination of chemical shifts of hydroxyl and amido groups is not the aim of the work, they can be exchanged by dissolving samples in DMSO-d_6–D_2O (98:2, vol/vol), so that the exchangeable protons are substituted by deuterium, being their signals disappeared. All spectra are acquired under temperature control (*see* **Note 5**) because hydroxy and amido protons are temperature dependent; their temperature coefficient (dT/dδ) gives information on long-distance interactions such as hydrogen bonds [11]. Moreover, the residual HDO signal frequency can be shifted to a less crowded region of the spectrum by adjusting the temperature. Ganglioside micelles are formed in D_2O, where exchangeable protons disappear, or in H_2O–D_2O, 90:10 vol/vol, where, at a low temperature, labile protons, OH and NH, can be maintained. In the last case, it is mandatory to suppress the large water signal that masks many saccharide frequencies. Due to the variety of experimental situation and spectrometer limitation, there are number of solvent suppression schemes that can be chosen with specific in different situation (*see* **Note 6**). The aim of a solvent suppression scheme is to reduce the solvent signal keeping any possible perturbations of the observable properties in the NMR spectrum as small as possible [6]. *Presaturation*, consisting in the selective saturation of the solvent magnetization with a weak and long pulse at the solvent frequency given before the strong excitation pulse, is the easier method to suppress the water signal. The selective saturation of solvent signal can be obtained with a DANTE sequence [12]: in this case, a train of intense pulses can selectively invert the solvent magnetization. The residual transverse magnetization can be destroyed with a gradient pulse. *NOESY1D* with presaturation can provide very flat baselines. This feature makes it particularly useful for quantitative analysis of very crowded spectra, where the baseline correction by mean of mathematical function can be problematic [13] (*see* **Note 7**). A *low-power presaturation* flavor is the best choice to observe and quantify solute signals close to the solvent peak but significantly reduces exchangeable protons. One-dimensional homonuclear Hartman-Hahn spectroscopy (1D HOHAHA), or total correlation spectroscopy (1D TOCSY), experiments are performed by selective excitation with a DANTE pulse sequence, followed by a MLEV-17 sequence for spin-lock and a *z*-filter, for purging phase or multiplet distortions [12, 14, 15]. The mixing time varies from 50 to 160 ms in 10 ms steps to obtain various degrees of magnetization transfer; 800–1600 scans are accumulated per spectrum. Transient 1D-NOE and -ROE experiments are performed using a composite 90° selective excitation pulse for selective irradiation [16]. The double selective 1D TOCSY-NOESY experiment allows the 1D NOESY experiment to be extended to resonances within overlapped spectral regions [4]. Hu et al. introduced an improved selective TOCSY edited preparation (STEP) function incorporating a novel zero-quantum filter; its application to the selective 1D NOESY experiment permits essentially complete suppression of zero-quantum coherence in a single scan [17, 18]. Residual antiphase distortions due to spin-state mixing are removed using the double difference methodology [19].

A conventional COSY spectrum contains information on scalar spin-coupling networks within each residue of the oligosaccharide chain through the observation of cross-peaks [20]. COSY spectra are run using the instrument sequence library [21] (*see* **Note 8**). The 2D TOCSY experiment for the DPC-water solution of deAc-GM1 is run in the TPPl mode with the MLEV-17 [22] pulse sequence for spin-lock and a saturation DANTE pulse for the residual water signal. The mixing time is 170 ms and the spectral width 3 kHz. Two-dimensional (2D) ROESY spectra are measured using the pulse sequence devised by Rance [23]. To avoid any offset dependence of the spin-lock and any Hartmann-Hahn interference, the radiofrequency (RF) carrier is placed at a distance of 1–1.5 kHz from the center of the spectrum during the spin-lock time, and at the center during evolution and acquisition. The spin-lock is sandwiched between two 170° x pulses whose amplitudes are twice the locking RF power (2.6 kHz) to improve the magnetization transfer efficiency [24–26]. Baseline distortions are minimized by applying a Hahn-echo sequence immediately before the acquisition and adjusting the receiver phase to pure absorption mode (*see* **Note 9**) [27, 28]. Recently, S. Boros and G. Batta introduced a combination of zero-quantum suppression and off-resonance ROESY method to obtain pure phase cross-peaks [29]. The 2D heteronuclear ^{1}H-detected experiments were carried out at 305 K using TPPI for quadrature detection in the F1 dimension. The 2D HSQC (32 scans per t_1 increment) and HSQC-NOESY (96 scans per t_1 increment, 0.25 s mixing time, relaxation delay of 1.1 s) spectra were recorded with t_2 and t_1 acquisition times of 340 and 28.2 ms, respectively. Pulse trains for ^{13}C relaxation and heteronuclear selective NOE measurements were based on the INEPT (for the CH) or DEPT-45° (for the CH_2) magnetization transfer steps [4].

4.2.1 Primary Structure of Gangliosides

The structural analysis of glycosphingolipids is done in three steps: (a) determination of monosaccharide composition and sugar configuration, (b) linkage position and sugar sequence and (c) characterization of ceramide moiety. All the experiments for ganglioside structural characterization are carried out in DMSO-*d6* and DMSO-*d6*:D_2O where gangliosides are in monomeric form and the NMR resolution is at its highest.

4.2.2 Determination of Monosaccharide Composition and Sugar Configuration

In pyranosides, the six-membered rings form a chair of fixed configuration with protons arranged in axial or equatorial position. In mammalian brain glycosphingolipids, monosaccharides are predominantly hexopyranoses having a $^4C_1$ configuration, the same as the D-sugars. Nevertheless, fucose and sialic acid are L-sugars. The ring chair conformation is accompanied by a characteristic vicinal coupling pattern which allows us to recognize the relative configuration of the ring carbons. In fact, the Karplus relation directly correlates $^3J_{HH}$ to the dihedral angles [30, 31]. So if a $^3J_{1,2}$ of 2–4 Hz is observed for an anomeric proton, then a *gauche* configuration of protons 1 and 2 (dihedral angle of ca. 60°) leads to the α-configuration, where glycosidic linkage is oriented below the plane of the ring, as in the case of fucose residue in Fuc-GM1. On the other hand, as in the majority of sugar residues, a larger $^3J_{1,2}$ (6–9 Hz) indicates a *trans* diaxial relationship (dihedral angle of ca. 180°) due to a β-configuration, where glycosidic linkage is oriented above the plane of the ring. Most of the sugar protons crowd within a very narrow frequency range (3.0–4.0 ppm) with severe overlap problems: these derive from the bulk of nonanomeric methine and methylene protons of the different sugar rings having very similar chemical shifts. This leads to a strongly coupled spectrum where many vicinal and geminal coupling constants are very similar. Using several monodimensional (1D) and two-dimensional (2D) different type of experiment at high magnetic field, it is possible to obtain a complete through-bond connectivity analysis and settle all the different spin systems corresponding to the individual sugar rings (Tables 2, 3 and 4). The relative intensities of isolated resonances can be used to establish the homogeneity of the compound. Assignment of a homonuclear 2D spectrum requires a starting signal for the identification of each individual spin system. Usually the anomeric, amide, and hydroxy protons are isolated enough to be a convenient starting point. Within a typical aldohexopyranosyl ring, H1 couples to H2, H2 to H1 and H3, H3 to H2 and H4, and so on. If the coupling constant $^3J_{n,n+1}$ is large enough (5–8 Hz) an intense cross-peak arises but, if no or small couplings (2–3 Hz) are present between vicinal related protons, no or very small cross-peaks occur. For example, for a galactopyranosyl residue, where the $^3J_{45}$ is 2–3 Hz, it is necessary to optimize the magnetization transfer for this *J* to reach protons 5 and 6s. A pure absorption phase sensitive COSY gives cross-peaks containing the entire coupling information concerning the involved protons. However, a single COSY spectrum is not sufficient for an unequivocal assignment of all the resonances of an oligosaccharide chain because of the overlap of many proton signals (*see* **Note 10**). A tool to overcome this problem is the *total correlation spectroscopy TOCSY* or homonuclear Hartmann-Hahn spectroscopy (HOHAHA). It is a very useful experiment, in which the net magnetization is transferred along the chain of scalar coupled spins under spin-locking conditions. In these spectra, a complete spin system can be identified if there is, at least, a well-isolated resonance to be used as a starting point and reasonably large couplings between all the spins of the lattice. In this way, a slice along the diagonal leads to a proton sub-spectrum containing all scalar-coupled protons within a sugar ring. 2D-TOCSY gives poor information about *J* couplings and, even in these spectra, resonance overlapping often does not permit the unambiguous identification of all the protons of the oligosaccharide chain.

Compared to his 2D counterpart, the selective 1D experiment allows considerably simplified spectral interpretation and is irreplaceable for the structure elucidation of small-to-medium sized molecules, although its application is limited to well-resolved resonances only [14, 32]. In the case of oligosaccharides, the selective 1D-TOCSY experiment allows to obtain high-resolution subspectra exciting for each sugar ring some isolated signal, from which the coupling constants are safely measured. In galactosyl residues, i.e., where the $^3J_{45}$ is small, the magnetization transfer is easily recognized from H1 up to H5, not only to H4 as in a 2D-TOCSY spectrum. The hydroxy and amido protons usually show well-separated signals so that their identification is very useful both for the assignment of resonances and for spatial connectivity investigation (Table 4). They can be used as starting point to assign the sialic acid side chains or the methylene protons of the galactosyl residues. For these residues, for instance, the net magnetization can be transferred from OH6 resonance to H6, H6′, and H5. Thus it is possible to assign all the galactose ring resonances combining these spectra with those deriving from the selective excitation of the anomeric protons, which transfer magnetization from H1 to H5. The chemical shifts of the neuraminic glycerol side chain protons can be assigned observing the magnetization transfer from OH8 to OH9, OH7, H7, H8, H9R, and H9S. Nuclear Overhauser Effect (NOE) experiments enable verifying the position of the appended groups, when present, by their dipolar interactions through the space that can occur between protons distant about less than 0.4 nm. NOE

Table 2

NMR chemical shifts (ppm) of nonlabile protons of Neu5Ac and stearic acid containing GM4, GM3, GM2, GM1, GD1a, GalNAc-GD1a, GD1b and *GM3(αHFA), GM3(Neu5Gc), deAc-GM1, GM1(Neu5Gc), GM1-lactone, GalNAc-GD1a (Neu5Ac B, Neu5Gc A) and GalNAc-GD1a (Neu5Gc B, Neu5Ac A), and GD1b-lactone* ganglioside species in DMSO-*d6* solution at 300–324 °C

Saccharide residue	Ganglioside	Proton chemical shift 1	2	3	4	5	6	7	8	9 [sugar] or 4 ↔ (*n* – 1) [FA] or 7 ↔ (*n* – 1) Sph]	11-$COCH_3$ or -CH_3 [Sph and FA]
Glc I	GM3	4.15	3.04	3.33	3.29	3.26	3.62 3.75				
	GM3(αHFA)	*4.20*	*3.04*	*3.35*	*3.32*	*3.30*	*3.63 3.76*				
	GM3(Neu5Gc)	*4.15*	*3.04*	*3.30*	*3.28*	*3.34*	*3.63 3.74*				
	GM2	4.15	3.04	3.33	3.28	3.29	3.61 3.74				
	GM1	4.14	3.03	3.32	3.27	3.39	3.63 3.73				
	deAc-GM1	*4.15*	*3.03*	*3.31*	*3.27*	*3.32*	*3.64 3.71*				
	GM1(Neu5Gc)	*4.15*	*3.04*	*3.30*	*3.28*	*3.34*	*3.63 3.74*				
	GM1-lactone	*4.15*	*3.03*	*nd*	*nd*	*nd*	*nd*				
	GD1a	4.15	3.04	3.33	3.28	3.28	3.64 3.74				
	GalNAc-GD1a	4.14	3.03	3.33	3.28	3.28	3.65 3.72				
	GalNAc-GD1a (Neu5Ac B, Neu5Gc A)	*4.13*	*3.02*	*3.32*	*3.30*	*3.34*	*3.63 3.71*				
	GalNAc-GD1a (Neu5Gc B, Neu5Ac A)										
	GD1b	4.14	3.04	3.34	3.32	3.30	3.63 3.73				
	GD1b-lactone	*4.15*	*3.04*	*3.34*	*3.29*	*3.29*	*3.61 3.75*				

(continued)

Table 2
(continued)

Saccharide residue	Ganglioside	Proton chemical shift 1	2	3	4	5	6	7	8	9 [sugar] or 4 ↔ (*n* – 1) [FA] or 7 ↔ (*n* – 1) Sph]	11-$COCH_3$ or -CH_3 [Sph and FA]
Gal II	GM4	4.06	3.29	3.87	3.70	3.24	3.42 3.47				
	GM3	4.19	3.33	3.93	3.75	3.32	3.46 3.51				
	GM3(αHFA)	*4.20*	*3.32*	*4.00*	*3.74*	*3.33*	*3.47 3.51*				
	GM3(Neu5Gc)	*4.20*	*3.32*	*3.98*	*3.70*	*3.32*	*3.45 3.45*				
	GM2	4.27	3.16	3.74	3.92	3.47	3.46 3.61				
	GM1	4.26	3.13	3.70	3.94	3.47	3.47 3.64				
	deAc-GM1	*4.19*	*3.12*	*3.71*	*3.92*	*3.51*	*3.41 3.63*				
	GM1(Neu5Gc)	*4.27*	*3.13*	*3.74*	*3.94*	*3.74*	*3.38 3.63*				
	GM1-lactone	*4.62*	*4.54*	*4.10*	*nd*	*nd*	*nd*				
	GD1a	4.25	3.18	3.76	3.94	3.44	3.45 3.64				
	GalNAc-GD1a	4.26	3.13	3.70	3.95	3.45	3.62 3.62				
	GalNAc-GD1a (Neu5Ac B, Neu5Gc A)	*4.26*	*3.13*	*3.72*	*3.94*	*3.45*	*3.63 3.38*				
	GalNAc-GD1a (Neu5Gc B, Neu5Ac A)										
	GD1b	4.29	3.24	3.78	3.99	3.46	3.42 3.52				
	GD1b-lactone	*4.33*	*3.26*	*3.92*	*3.93*	*3.51*	*nd*				

GalNAc III	GM2	4.79	3.75	3.39	3.52	3.64	3.39 3.46	1.76
	GM1	4.89	3.90	3.43	3.69	3.64	3.45 3.47	1.73
	deAc-GM1	*4.86*	*3.91*	*3.43*	*3.71*	*3.63*	*3.45 3.47*	
	GM1(Neu5Gc)	*4.90*	*3.92*	*3.50*	*3.73*	*3.64*	*3.47 3.47*	*1.76*
	GM1-lactone	*4.81*	*3.48*	*nd*	*nd*	*nd*	*nd*	*1.87*
	GD1a	4.80	3.96	3.62	3.74	3.62	3.48 3.48	1.75
	GalNAc-GD1a	4.87	3.92	3.38	3.65	3.62	3.43 3.45	1.76
	GalNAc-GD1a(Neu5Ac B, Neu5Gc A)	4.85	*3.92*	*3.46*	*3.66*	*3.59*	*3.47 3.47*	
	GalNAc-GD1a(Neu5Gc B, Neu5Ac A)							
	GD1b	4.93	3.94	3.83	3.85	4.03	3.30 3.49	1.73
	GD1b-lactone	*4.64*	*3.93*	*3.27*	*3.68*	*3.21*	*3.42 3.58*	*1.78*

(continued)

Table 2
(continued)

Saccharide residue	Ganglioside	Proton chemical shift 1	2	3	4	5	6	7	8	9 [sugar] or 4 ↔ (*n* – 1) [FA] or 7 ↔ (*n* – 1) Sph]	11-$COCH_3$ or -CH_3 [Sph and FA]
Gal IV	GM1	4.18	3.32	3.28	3.59	3.36	3.49 3.53				
	deAc-GM1	*4.18*	*3.28*	*3.34*	*3.60*	*3.32*	*3.51 3.53*				
	GM1(Neu5Gc)	*4.22*	*3.30*	*3.27*	*3.60*	*3.35*	*3.52 3.52*				
	GM1-lactone	*4.16*	*3.32*	*nd*	*nd*	*nd*	*nd*				
	GD1a	4.22	3.33	3.95	3.76	3.25	3.53 3.53				
	GalNAc-GD1a	4.18	3.18	3.72	3.91	3.37	3.64 3.64				
	GalNAc-GD1a(Neu5Ac B, Neu5Gc A)	*4.21*	*3.16*	*3.74*	*3.90*	*3.38*	*3.63 3.36*				
	GalNAc-GD1a(Neu5Gc B, Neu5Ac A)										
	GD1b	4.32	3.32	3.27	3.57	3.54	3.58 3.58				
	GD1b-lactone	*4.22*	3.36	*3.27*	*3.68*	*3.38*	*3.52 3.64*				
GalNAc V	GalNAc-GD1a	4.82	3.72	3.37	3.51	3.64	3.47 3.48				1.75
	GalNAc-GD1a(Neu5Ac B, Neu5Gc A)	*4.81*	*3.71*	*3.36*	*3.53*	*3.6*	*3.44 3.44*				
	GalNAc-GD1a(Neu5Gc B, Neu5Ac A)										

Neu5x A	GM4	2.68e 1.45a	3.54	3.42	3.39	3.21	3.56	3.32 3.54	
	GM3	2.74e 1.37*a*	3.61	3.40	3.61	3.21	3.34	3.41 3.61	1.88
	GM3(αHFA)	*2.76e 1.36a*	*3.60*	*3.42*	*3.61*	*3.21*	*3.43*	*3.35 3.63*	*1.88*
	GM3(Neu5Gc)	*2.75e 1.39a*	*3.72*	*3.46*	*3.50*	*3.21*	*3.60*	*3.34 3.63*	1.86
	GM2	2.56e 1.62a	3.70	3.39	3.15	3.16	3.49	3.34 3.61	1.87
	GM1	2.53e 1.67a	3.71	3.34	3.05	3.16	3.47	3.34 3.63	
	deAc-GM1	*2.54e 1.65a*	*3.69*	*2.70*	*3.38*	*3.45*	*3.51*	*3.40 3.70*	*1.88*
	GM1(Neu5Gc)	*2.56e 1.64a*	*3.84*	*3.42*	*3.22*	*3.17*	*3.50*	*3.32R 3.62S*	
	GM1-lactone	*2.37e 1.59a*	*4.11*	*3.75*	*nd*	*nd*	*nd*	*nd*	1.88
	GD1a	2.59e 1.61a	3.72	3.38	3.17	3.21	3.52	3.34R 3.62S	1.87
	GalNAc-GD1a	2.52e 1.62a	3.76	3.39	3.09	3.16	3.48	3.35R 3.60S	
	*GalNAc-GD1a (Neu5Ac B,**Neu5Gc A**)*	*2.53e 1.64a*	*3.86*	*3.43*	*3.18*	*3.17*	*3.47*	*3.32R 3.59S*	*1.86*
	*GalNAc-GD1a(**Neu5Gc B**, Neu5Ac A)*	*2.53e 1.64a*	*3.68*	*3.37*	*3.09*	*3.15*	*3.49*	*3.32R 3.59S*	*3.85*
	GD1b	2.04e 1.74a	4.41	3.52	3.79	3.28	3.83	3.46 3.54	1.87
	GD1b-lactone	*2.38e 1.92a*	*3.98*	*3.54*	*3.48*	*3.49*	*4.08*	*4.71 4.48*	*1.88*

(continued)

Table 2
(continued)

Saccharide residue	Ganglioside	Proton chemical shift 1	2	3	4	5	6	7	8	9 [sugar] or 4 ↔ (*n* – 1) [FA] or 7 ↔ (*n* – 1) Sph]	11-$COCH_3$ or -CH_3 [Sph and FA]
Neu5x B	GD1a			2.72e 1.38a	3.59	3.41	3.40	3.14	3.43	3.33R 3.61	1.88
	GalNAc-GD1a			2.50e 1.64a	3.68	3.39	3.07	3.13	3.49	3.37R 3.59S	1.87
	GalNAc-GD1a (Neu5Gc B, ***Neu5Ac A)***			*2.53e 1.64a*	*3.74*	*3.37*	*3.08*	*3.17*	*3.47*	*3.32R 3.60S*	*1.87*
	*GalNAc-GD1a(****Neu5Gc B****, Neu5Ac A)*			*2.53e 1.64a*	*3.78*	*3.45*	*3.19*	*3.15*	*3.49*	*3.32R 3.60S*	*3.85*
Neu5x C	GD1b			2.91e 1.48a	3.47	3.30	3.06	3.14	3.53	3.32 3.53	1.87
	GD1b-lactone			*2.20e 1.45a*	*4.18*	*3.57*	*3.37*	*3.31*	*3.36*	*3.44 3.60*	*1.89*

Sph	GM3	3.95	3.44	3.79	3.91	5.36	5.54	1.94	1.23	0.85
	GM3(αHFA)	*3.92*	*3.52*	*3.82*	*4.01*	*5.37*	*5.59*	*1.93*		
	GM2	3.99	3.42	3.77	3.88	5.35	5.53	1.93	1.23	0.85
	GM1	3.97	3.43	3.77	3.89	5.35	5.53	1.93	1.23	0.85
	GM1(Neu5Gc)	*3.97*	*3.43*	*3.77*	*3.89*	*5.45*	*5.53*	*1.93*	*1.23*	*0.85*
	GM1-lactone	***nd***		***nd***	***nd***	*5.34*	*5.53*	***nd***	*1.23*	*0.85*
	GD1a	3.96	3.45	3.77	3.82	5.38	5.55	1.93	1.23	0.85
	GalNAc-GD1a	3.98	3.40	3.76	3.87	5.34	5.52	1.92	1.23	0.85
	*GalNAc-GD1a (Neu5Gc B,**Neu5Ac A**)*	*3.42*	*3.96*	*3.75*	*3.86*	*5.34*	*5.52*	*1.92*	*1.23*	*0.85*
	*GalNAc-GD1a(**Neu5Gc B**, Neu5Ac A)*									
	GD1b	3.43	3.97	3.78	3.90	5.35	5.53	1.92	1.23	0.85
	GD1b-lactone	*3.43*	*3.96*	*3.77*	*3.89*	*5.35*	*5.54*	*1.93*	*1.23*	*0.85*

(continued)

Table 2
(continued)

Saccharide residue	Ganglioside	Proton chemical shift 1	2	3	4	5	6	7	8	9 [sugar] or $4 \leftrightarrow (n-1)$ [FA] or $7 \leftrightarrow (n-1)$ Sph]	11-$COCH_3$ or -CH_3 [Sph and FA]
FA	GM3		2.03	1.45						1.23	0.85
	GM2		2.02							1.23	0.85
	GM1		2.02							1.23	0.85
	GM1(Neu5Gc)		*2.03*							*1.23*	*0.85*
	GM1-lactone		*2.01*							*1.23*	*0.85*
	GM1(αHFA)		*3.81*	*1.42a 1.56b*						*1.23*	*0.85*
	GD1b		2.03							1.23	0.85
	GD1b-lactone		*2.03*							*1.23*	*0.85*

Characteristic values for gangliosides containing Neu5Gc, α-Hydroxy fatty acids, or lactonic rings are reported as italic data

Table 3
NMR chemical shifts (ppm) of nonlabile protons of GM1, GM1(*Neu5Gc*), GD1a and GalNAc-GD1a DPC mixed micelles

Residue	Ganglioside	1	2	3	4	5	6	7	8	9	11
Glc I	GM1	4.53	3.44	3.72	3.72	3.64	4.28 3.87				
	GM1-acetyl	*4.49*	*3.39*	*3.68*	*3.66*	*3.60*	*3.86 4.00*				
	GM1(Neu5Gc)	*4.48*	*3.37*	*3.66*	*3.63*	*3.62*	*3.82 4.01*				
	GD1a	4.48	3.35	3.65	3.63	3.61	3.82 4.00				
	GD1a-acetyl	*4.50*	*3.38*	*3.69*	*3.67*	*3.61*	*3.83 4.01*				
	GalNAc-GD1a	4.51	3.38	3.66	3.65	3.63	3.86 4.01				
Gal II	GM1	4.59	3.40	4.22	4.18	3.78					
	GM1-acetyl	*4.56*	*3.38*	*4.18*	*4.17*	*3.78*	*3.82*[a] *3.82*[a]				
	GM1(Neu5Gc)	*4.56*	*3.39*	*4.19*	*4.17*	*3.78*	*3.63 3.80*				
	GD1a	4.55	3.38	3.62	4.15	3.78	3.83 3.83				
	GD1a-acetyl	*4.55*	*4.41*	*4.18*	*4.15*	*3.77*	*3.85*[c] *3.85*[c]				
	GalNAc-GD1a	4.56	3.41	4.18	4.15	3.79	3.86 3.86				
GalNAc III	GM1	4.86	4.11	3.86	4.22	3.77					2.04
	GM1-acetyl	*4.85*	*4.09*	*3.83*	*4.19*	*3.77*	*3.80*[b] *3.80*[b]				*2.04*
	GM1(Neu5Gc)	*4.83*	*4.08*	*3.83*	*4.19*	*3.76*	*3.85 3.85*				
	GD1a	4.82	4.07	3.83	4.19	3.75	3.79 3.79				2.03
	GD1a-acetyl	*4.83*	*4.09*	*3.84*	*4.20*	*3.77*	*3.81*[d] *3.81*[d]				*2.04*
	GalNAc-GD1a	4.84	4.10	3.83	4.16	3.75	3.82 3.82				

(continued)

Table 3
(continued)

Residue	Ganglioside	1	2	3	4	5	6	7	8	9	11
Gal IV	GM1	4.60	3.58	3.69	3.98	3.72					
	GM1-acetyl	*4.58*	*3.56*	*3.67*	*3.94*	*3.72*	*3.79 3.79*				
	GM1(Neu5Gc)	*4.58*	*3.55*	*3.68*	*3.95*	*3.72*	*3.80 3.80*				*2.06*[c]
	GD1a	4.64	3.56	4.11	3.97	3.69	3.75 3.75				
	GD1a-acetyl	*4.64*	*3.57*	*4.11*	*3.98*	*3.70*	*3.76 3.76*				
	GalNAc-GD1a	4.65	3.40	4.16	4.13	3.79	3.78 3.78				
GalNAc V	GalNAc-GD1a	4.83	3.97	3.71	3.98	3.79	3.82 3.82				
Neu5Ac A	GM1			1.98[a] 2.72[e]	3.59	3.86	3.55	3.64	3.79	3.68 3.93	2.09
	GM1-acetyl			*1.95*[a] *2.70*[e]	*3.85*	*3.83*	*3.53*	*3.63*	*3.79*	*3.67R 3.91S*	*2.09*
	GM1(Neu5Gc)			*1.97*[a] *2.71*[e]	*3.92*	*3.93*	*3.65*	*3.63*	*3.80*	*3.66R 3.90S*	*2.04*
	GD1a			1.91[a] 2.72[e]	3.82	3.83	3.53	3.61	3.78	3.66 3.89	4.15
	GD1a-acetyl			*1.94°* *2.70*[e]	*3.83*	*3.83*	*3.53*	*3.63*	*3.80*	*3.68 3.93*	*2.06*
	GalNAc-GD1a			1.95[a] 2.74[e]	3.85	3.85	3.55 3.56	3.62	3.73	3.70 3.92	2.08
Neu5Ac B	GD1a			1.80[a] 2.78[e]	3.72	3.86	3.65	3.61	3.91	3.67 3.90	2.08
	GD1a-acetyl			*1.83*[a] *2.77*[e]	*3.72*	*3.87*	*3.65*	*3.62*	*3.92*	*3.68 3.93*	*2.07*[f]
	GalNAc-GD1a			1.95[a] 2.72[e]	3.85	3.85	3.55 3.56	3.62	3.73	3.70 3.92	

Sph	GM1		4.10	4.13		5.45	5.83		
	GM1-acetyl	*4.12 3.82*	*3.99*	*4.12*		*5.45*	*5.76*		*1.99*
	GM1(Neu5Gc)	*4.20 3.75*	*3.94*	*4.10*		*5.35*	*5.77*		*2.02*
	GD1a	4.18 3.79	3.95	4.10		5.45	5.75		
	GD1a-acetyl	*4.11 3.82*	*4.00*	*4.12*		*5.45*	*5.77*		
	GalNAc-GD1a	3.79 4.19	3.96	4.12		5.46	5.76		

GM1-acetyl and *GD1a-acetyl*, containing an acetyl group as acyl moiety, are pure micelles in D_2O solution at 310 K. All gangliosides are referenced to TMS indirectly by setting Sph 4 at 5.45 ppm

[a, b; c, d; e,f] These couples of values may be interchangeable

Table 4
NMR chemical shifts (ppm) of labile protons of Neu5Ac and stearic acid containing GM3, GM2, GM1, GD1a, GalNAc-GD1a and GD1b and *GM3(αHFA)*, *deAc-GM1*, *GM1(Neu5Gc)*, *GalNAc-GD1a(Neu5Ac B, Neu5Gc A)* and *GalNAc-GD1a(Neu5Gc B, Neu5Ac A)*, ganglioside species in DMSO-*d6* solution at different temperatures

Residue	Ganglioside	2	3	4	6	7	8	9	11
Glc I	GM3	4.98	4.49		4.33				
	GM3(αHFA)	*4.39*		*4.27*	*4.33*				
	GM2	5.06	4.51		4.48				
	GM1	5.13	4.55		4.59				
	deAc-GM1	*5.14*	*4.58*		*4.60*				
	GM1(Neu5Gc)	*5.04*			*4.47*				
	GD1a	4.98	4.51		4.43				
	GalNAc-GD1a	5.12	4.55		4.55				
	GD1b	5.06	4.61		4.55				
	GalNAc-GD1a(Neu5Ac B, Neu5Gc A)	*5.13*	*4.56*		*4.54*				
	GalNAc-GD1a(Neu5Gc B, Neu5Ac A)								

Gal II	GM3	4.39		4.27	4.5
	GM3(αHFA)	*4.89*		*4.26*	*4.5*
	GM2	4.87			4.23
	GM1	5.03			4.30
	deAc-GM1	*5.12*			*4.31*
	GM1(Neu5Gc)	*4.91*			*4.23*
	GD1a	4.72			4.19
	GalNAc-GD1a	4.99			4.38
	GD1b	4.94			4.15
	GalNAc-GD1a(Neu5Ac B, Neu5Gc A)	*4.93*			*4.27*
	GalNAc-GD1a(Neu5Gc B, Neu5Ac A)				
GalNAc III	GM2	7.35	4.48	4.24	4.47
	GM1	7.68		4.34	4.58
	deAc-GM1	*7.57*		*4.33*	*4.56*
	GM1(Neu5Gc)	*7.53*		*4.23*	*4.51*
	GD1a	7.18		4.04	4.43
	GalNAc-GD1a	7.61		4.65	4.52
	GD1b			4.16	5.01
	GalNAc-GD1a(Neu5Ac B, Neu5Gc A)	*7.50*		*4.55*	*4.54*
	GalNAc-GD1a(Neu5Gc B, Neu5Ac A)				

(continued)

Table 4 (continued)

Residue	Ganglioside	2	3	4	6	7	8	9	11
Gal IV	GM1	3.68	4.78	4.40	4.86				
	deAc-GM1	*3.73*	*4.80*	*4.42*	*4.91*				
	GM1(Neu5Gc)	*3.67*	*4.66*	*4.30*	*4.82*				
	GD1a	3.87		4.22	5.03				
	GalNAc-GD1a	4.46	4.71		4.45				
	GD1b		4.71	4.30	5.26				
	GalNAc-GD1a(Neu5Ac B, Neu5Gc A)	*4.71*			*4.59*				
	GalNAc-GD1a(Neu5Gc B, Neu5Ac A)								
GalNAc V	GalNAc-GD1a	7.43	4.30	4.30	4.49				
	GalNAc-GD1a(Neu5Ac B, Neu5Gc A)	*7.38*	*4.27*	*4.32*	*4.50*				
	GalNAc-GD1a(Neu5GcB, Neu5Ac A)	*7.34*	*4.53*						

Neu5x A	GM3	4.77	8.05	4.55	6.15	3.94	
	GM3(αHFA)	*4.65*	*8.05*	*4.51*	*6.18*	*3.93*	
	GM2	4.67	7.98	4.63	6.10	4.54	
	GM1	4.72	8.03	4.71	5.97	4.89	
	deAc-GM1	*5.30*		*4.64*	*6.06*	*5.03*	*5.44*
	GM1(Neu5Gc)	*4.81*	8.08	*4.60*	*5.99*	*4.67*	
	GD1a	4.74	8.02	4.65	5.98	4.59	
	GalNAc-GD1a	4.75	*7.65*	4.69	5.96	4.91	*5.59*
	GalNAc-GD1a(Neu5Ac B, ***Neu5Gc A****)*	*4.83*	*7.65*	*4.64*	*5.91*	*4.80*	
	GalNAc-GD1a(Neu5GcB, ***Neu5Ac A****)*	*4.79*	*8.09*	*4.74*	*5.89*	*4.80*	
	GD1b	4.51		4.18		3.65	
	GD1b-lactone		*8.26*[a]				
Neu5x B	GD1a	4.69	7.90	4.49	6.1	4.36	
	GalNAc-GD1a	4.67	8.01	4.75	6.07	4.72	
	*GalNAc-GD1a(****Neu5Ac B****, Neu5Gc A)*	*4.70*	*8.06*	*4.74*	*6.05*	*4.65*	
	GalNAc-GD1a(Neu5GcB, ***Neu5Ac A****)*	*4.74*	*7.65*	*4.66*	*6.08*	*4.66*	*5.59*
	GD1b-lactone		*8.05*[a]				
Neu5Ac C	GD1b	6.09		4.92	6.19	3.97	

(continued)

Table 4
(continued)

Residue	Ganglioside	2	3	4	6	7	8	9	11
Sph	GM3	7.37	4.74						
	GM2	7.41	4.80						
	GM1	7.42	4.80						
	GM1 (Neu5Gc)	*7.41*	*4.78*						
	GD1a	7.38	4.73						
	GalNAc-GD1a	7.47	4.85						
	GalNAc-GD1a (Neu5Ac B, Neu5Gc A)	*7.51*	*4.82*						
	GalNAc-GD1a (Neu5Gc B, Neu5Ac A)								
	GD1b	7.43							
	GD1b-lactone	*7.50*							

GM3 at 323 K. GM2 at 310 K (II6, A8, and A9 at 315 K). GM1 at 297 K. GD1a at 324 K. GalNAc-GD1a at 300 K, *GalNAc-GD1a* (*Neu5Ac B, Neu5Gc A*) and *GalNAc-GD1a* (Neu5GcB, Neu5Ac A) at 310 K. GD1b at 313 K

[a]May be interchanged

depends on the spin–spin dipolar interaction and, consequently, on the distance between protons, so that the NOE connectivities between the anomeric and the aglycone protons are usually very effective. NOEs depend on resonance frequency ω_o and molecular rotational correlation time τ_c, which is itself dependent on molecular size [33, 34]. In the range of molecular mobility of oligosaccharides, NOE can be positive or negative. Thus, oligosaccharides may show positive or negative NOE levels, as a direct result of their molecular size, so that NOESY become undetectable when $\omega_0\tau_c$, the product of spectrometer angular frequency and molecular rotational correlation time, is approximately equal to unity. In the rotating frame, on the contrary, all molecules behave as if they were in the positive NOE regime and ROE increases from 38.5 to 67.5%, and does not vanish for any $\omega_0\tau_c$ value. Moreover, ROESY experiments allow to avoid the phenomenon of spin diffusion, a mechanism of propagation of NOE along the molecular backbone in the NOESY experiments, which generates cross peaks between protons which have not a direct dipolar interaction. In addition ROESY allows discriminating genuine dipolar interaction from cross-peaks due to chemical exchanging protons [35]. Therefore, NOE experiments in the rotating frame (ROE) are often preferred to the conventional NOE experiments for polysaccharides (from trisaccharides to decasaccharides). The side chain proton set can be assigned to the proper sialic acid ring by determining NOE contacts between OHs and some protons of the closer sugar residue [36]. 2D-NOESY and 2D-ROESY experiments, in principle, give spatial information for all nuclei in a single spectrum but the interpretation may be very difficult in crowded spectra as are that of carbohydrates. Again, 1D selective methods are a useful alternative, especially after the advent of pulsed field gradient (PFGs) which get the suppression of unwanted signal, i.e. solvents, very efficient [37, 38]. In double selective 1D TOCSY–NOESY experiment, the magnetization is first transferred from the selectively excited frequencies to the spin lattice neighbor protons via scalar coupling and then transferred to spatially close protons via dipolar coupling. This allows the 1D NOESY experiment to be extended to resonances within overlapped spectral regions [4, 39]. The combined use of selective TOCSY edited preparation (STEP) and a zero-quantum filter ensures that the final spectra are free of distortions, which is crucial to the reliable detection of weak NOE enhancements [17]. If the proton signal overlapping is so extended that the above-mentioned homonuclear techniques are not sufficient to assign all resonances, ^{13}C NMR spectroscopy is a very powerful tool for glycosphingolipid analysis because of its larger chemical shift range, which minimizes the signal overlapping, and its minor complexity due to the lack of spin–spin coupling (Fig. 1). Conventional 1D spectra require very long acquisition time or large amount of samples because of the natural low abundance of ^{13}C. Inverse-detection techniques provide heteronuclear correlation maps that relate a carbon atom to the resonance of its directly bonded proton splitted by a large $^1J_{CH}$ scalar coupling [40]. The *heteronuclear single quantum coherence* (HSQC) experiment shows the cross-peaks obtained at carbon frequency F1 (vertical axis) and proton frequency F2 (horizontal axis). This simplifies the complete assignment of both the ^1H and ^{13}C resonances [41, 42]. The last step is the location of the appended groups, if present. The substitution of hydroxyl and amido groups with an acyl, a glycolyl, a glycolic, or a glycoside group leads to chemical shift changes of the nuclei at the substitution site and some other nearby resonance in both ^1H and ^{13}C spectra.

4.2.3 Linkage Position and Sugar Sequence

Once each sugar residue has been completely characterized, it is necessary to identify the glycosidic linkages and their sequence to complete the structure determination. The protons across the glycosidic linkages are four bonds apart, so that they do not show any scalar coupling and, consequentially, no cross-correlation can be observed using COSY or TOCSY experiments. Two different approaches can be used: the *glycosylation-induced shift* method and the *ab initio* method that does not require any prior knowledge. The first method, which allows comparing any unknown structure to other structures previously examined by NMR, is because the glycosylation induces chemical shift changes of both protons and carbons involved in the new linkage and their vicinal, whereas the other nuclei are much less affected. For the proton spectrum, this method can be applied only to oligosaccharides for which the assignments are already unambiguously available. Glycosylation shifts in ^{13}C NMR are relatively regular: the glycosylated carbon shifts to lower fields by 4–10 ppm and the resonances of the adjacent carbon atoms shift upfield by a small amount, whereas the other carbon resonances remain virtually unaffected. These effects can be correlated with the configuration at the anomeric center of both pyranose units, because they depend on the spatial disposition of protons that cause polarization of the C-H bonds [43]. The correlation between ^{13}C chemical shifts for both glycone and aglycone carbons with one of the torsional angles (ψ) can be used for the determination of the conformation of the glycosidic linkage [44]. The sequence of an oligosaccharide chain can be assigned without prior knowledge of the type of the structure using long-range heteronuclear ^1H–^{13}C correlation and homonuclear nuclear Overhauser effect (NOE) data. The identification of the aglyconic proton at the linkage site is unequivocal, except when there is an equatorial proton vicinal to it, e.g., H4 in 3-glycosylated galactose residues, which exhibits NOEs of approximately the same magnitude. In these cases, the problem is circumvented with the chemical shift analysis: substitution by a second sugar unit at a sugar ring induces a larger chemical shift change for the proton at the linkage site than to the vicinal one. For sialic acids the diagnostic contacts are between the H3ax of the sugar ring and some protons of the adjacent sugar residue, i.e., H4 of the inner Gal residue of many gangliosides. Measurements of the average spin lattice relaxation time for sugar carbon, which in glycosides increases with the growth of distance from the aglycone, together with the total assignment of the OH resonances can confirm the NOE results. HMBC allows the assignment of the quaternary carbons via two- or three-bond ^{1}H–^{13}C correlations. The sugar sequence can be confirmed by the HMBC spectra that show cross-peaks between the anomeric protons and the carbons at the linkage position [45].

4.3 Characterization of Ceramide Moiety

Some proton resonances of ceramide are easily recognizable in the spectra (Fig. 1a): for GM1, for instance, the 0.85, 1.23, 1.92, 2.03, 5.35, and 5.53 ppm signals belong to methyl, alkyl, allylic, α-carbonyl methylene and olefinic resonances, respectively. The olefinic proton resonances of sphingosine, which are assigned between 5.3 and 5.6 ppm, show vicinal coupling constant J_{45} of 15 Hz, clearly indicating a *trans* configuration. The NH and the OH of sphingosine are approximately at 7.5 and 4.8 ppm, respectively, in the temperature range 297–323 K. The presence of α-hydroxylated fatty acids is indicated by the appearance of a new doublet at 5.5 ppm; moreover, the α-carbonyl methylene protons shift to low field at 3.8 ppm. Sometimes the presence of this α-OH can influence the glucose anomeric resonance, as in GM3. A 1D-TOCSY spectrum by selective excitation of the two different OH groups and/or of the olefinic resonances allows unambiguous assignment of the sphingosine and the fatty acid moieties [46]. The content of sphingosine on the total long-chain base content, sphingosine and sphinganine, can be determined by comparison of the olefinic and anomeric resonance intensities. The evaluation of the length of the lipid chains, as the average total number of carbons, requires a very accurate integration of ^{13}C spectrum, acquired without NOE and respecting a relaxation delay of seven time the T1, while the precise knowledge of the long-chain base and fatty acid distributions requires the use of other analytical methods.

4.4 Secondary Structure of Gangliosides

Detailed information on the spatial disposition of the saccharide units is obtained on gangliosides dissolved in dimethylsulfoxide, the structures being determined for an isolated molecule [3, 47–52]. In view of extending the results to the physiological membrane, the need to study aqueous solution systems becomes significant. Pure ganglioside aggregates present in water media are generally too big for high resolution NMR and do not allow to detect any conformational changes reflecting possible gangliosides–phospholipid interactions.

Thus, both the effects of the aqueous solvent and of the phospholipid surface on the ganglioside secondary structure are investigated using small micelles of dodecylphosphocholine containing one ganglioside molecule per micelle (Fig. 2) [2, 4, 5, 40, 47, 52, 53]. The model system consisting of the small spherical micelles formed by semisynthetic gangliosides, bearing a single tail lipid moiety, like GD1a(acetyl) and GM1(acetyl), also gives information on the conformational, dynamic, and interaction properties of the ganglioside oligosaccharide polar head in an enriched in ganglio-sides environment, thus mimicking a ganglioside-clustered mem-brane microdomain [4]. Combining the NMR results, it is clear that only a few very strong and cooperative intramolecular interre-sidual interactions primarily determine the conformational proper-ties of ganglioside oligosaccharide, independently from the environment and the particular composition of the lipidic portion.

4.4.1 Conformation

NMR investigation of solution conformation of biomolecules of any class is mainly based on the proton–proton distance informa-tion gained from the quantitative evaluation of NOE data. Oligosaccharides show few structurally relevant NOE contacts, very often only one contact per glycosidic linkage, in contrast to proteins, oligopeptides, and nucleic acids, which, usually, display a large number of NOE connectivities useful for the modeling of well-founded three-dimensional (3D) structure.

This situation can be improved using hydroxy and amido groups as long-range sensors to provide additional contact infor-mation (Table 5). Since these groups protrude approximately twice as far as C-linked protons, a number of additional interatomic con-tacts can be observed for each disaccharide fragment of the oligo-saccharide chain, and even for not directly linked rings. In this way, the amount of experimental data available for a verification of theo-retically calculated conformations are greatly extended [36].

The analysis of coupling constants and temperature coefficients (dT/dδ) for amido and hydroxy protons allows the identification of intra- and/or inter-residual hydrogen bonds that gives further conformational information. An oligosaccharide structure can be a linear or branched chain of quite rigid sugar residues with a well-defined 3D structure.

Usually the connected groups also assume a preferred confor-mation. In particular, the sialic acid side chain of gangliosides, which in principle should be very flexible, adopts a well-defined conformation up to atom C8, where H6-C6-C7-H7 (θ_1) and H7-C7-C8-H8 (θ_2) dihedral angles can be derived from vicinal $^3J_{6,7}$ and $^3J_{7,8}$, less than 2 and 8.5 Hz respectively. Although the Karplus equation gives two alternative solutions for each ^3J value, the correct one can be chosen by "NOE labeling," i.e., by locating the vicinal protons firmly localized in the molecule.

Dihedral angles $\theta_1 = -60°$ and $\theta_2 = -160°$ give the best agree-ment for the observed NOE contacts between ring and side chain protons and vicinal coupling constants, where small $^3J_{67}$ and large $^3J_{78}$ may directly be compared to $^3J_{45}$ in galactose ($J < 1.5$ Hz,

Table 5
Interresidue ^{1}H-^{1}H NOE interactions for ganglioside saccharide moiety that can be found for the branched oligosaccharide structure GalNAc-[Neu5Ac-]Gal-GalNAc-[Neu5Ac-Neu5Ac-] Gal-Glc-Cer

α-Neu5Ac-(2-3)-ß-Gal-(1-3)-ß-GalNAc-				
B-3ax : IV-3	IV-1 : III-2	B-8 : III-11		
B-OH8 : IV-3	IV-1 : III-3	B3eq-III1		
B-8 : IV-3	IV-1 : III-4	B3eq-III5		
B-3eq : IV-OH2	IV-1 : III-11			
B-NH : IV-OH6	IV-1 : III-NH			
	IV-6 : II-OH6			
	IV-OH6 : III-NH			
	IV-OH2 : III-11			
ß-GalNAc-(1-4)-[α-Neu5Ac-(2-3)]-ß-Gal-				
V-1 : IV-4	B-3ax : IV-3	B-8 : V-1		
V-NH : IV-2	B-3ax : IV-OH2	B-OH8 : V-1		
V-11 : IV-OH2	B-3eq : IV-OH2	B-OH8 : V-5		
V-11 : IV-2	B-8 : IV-3	B-OH8 : V-NH		
V-OH6 : IV-OH6	B-OH7 : IV-4	B-OH9 : V-OH6		
	B-OH8 : IV-4	B-OH9 : V-5		
		B-9R : V-OH6		
ß-GalNAc-(1–4)-[α-*Neu5Ac*-(2–8)-α-Neu5Ac-(2–3)]-ß-Gal-				
III-1 : II-4	III-1 : A-8	A-3ax : II-3	C-*3eq* : III-1	C-*3ax* : A-6
III-11 : II-2	III-1 : A-OH8	A-8 : II-4	C-*3eq* : III-5	C-*6* : A-9a
III-NH : II-2	III-5 : A-OH8	A-OH8 : II4		C-*8* : A-9a
III-OH6 : II-OH6	III-5 : A-OH9			C-*3eq* : A-8

(continued)

Table 5 (continued)

III-OH6 : A-OH9	III-11 : A-OH2
	III-6 : A-OH9
	III-OH6 : A-9R
	III-OH6 : A-OH9
	III-NH : A-OH8
ß-Gal-(1-4)-ß-Glc-	**ß-Glc-(1-1)-Cer**
II-1 : I-3	I-1 : Cer-1′
II-1 : I-4	I-1 : Cer-2
II-1 : I-6′	I-1 : Cer-3
II-1 : I-OH3	I-OH2 : Cer-NH
II-1 : I-OH6	I-OH2 : Cer-NH
II-6 : I-OH3	
II-OH2 : I-6	
II-OH2 : I-6′	
II-OH2 : I-OH3	
II-OH2 : I-OH6	

H5-C5-C4-H4 = −60°) and glucose (J = 10 Hz, H5-C5-C4-H4 = 180°) rings of a ganglioside molecule. The small $^3J_{8,\mathrm{OH8}}$ coupling constant (<2 Hz vs. 6 Hz for a freely rotating OH group), the small temperature shift coefficient of the OH8 and the strong OH8/H6 NOE contact suggest the presence of a hydrogen bond between OH8 and either the carboxylic O1 or the O6 oxygen. The former of these two hydrogen bonds seems to be more probable in view of the exceptionally large lowfield shift of OH8 (ca. 6 ppm), as compared with the chemical shift of Glc IOH3 (ca. 4.6 ppm) hydrogen bonded to Gal IIO5 ring oxygen. The situation does not change after deletion of the N-acetyl group of the neuraminic acid. In fact, the vicinal coupling constants $^3J_{67}$ and $^3J_{78}$ remain unchanged in deac-GM1, and proton OH8 shows the same behavior. Moreover, it appears that proton OH7 is hydrogen bonded with the carbonyl oxygen. In fact, this proton shows a significantly smaller coupling constant (8.2 vs. 3.7 Hz) and chemical shift temperature coefficient (6.12 vs. 3.27×10^{-3} ppm) ongoing from deac-GM1 to GM1 [3], and the NOE distance to proton H5 changes from 0.29 to 0.23 nm. The existence of the aforementioned hydrogen bonding is supported by the favorable conformation of the acetamido group, characterized by a single torsional angle H-N-C5-H5 close to 150°, a value that is in excellent agreement with measured NOE contacts and observed vicinal coupling constant. The vicinal coupling constants for prochiral methylene protons OH-9R and OH-9S have significantly different values and can be readily assigned by "NOE labeling" to proton H7 [47]. Assuming that the hydroxymethyl group can occupy three staggered conformations, the population ratio for *gauche-gauche*, *gauche-trans*, and *trans-gauche* conformers obtained from extended Karplus equation is equal to 0.5:0.5:0 [31]. Moreover, the conformation of the sialic acid side chain remains almost the same when gangliosides are solubilized in organic solvents or anchored in DPC micelles in aqueous solution, as inferred from the coupling constants and NOE interaction with the GalNAc residue. Therefore, if every ring has a fixed rigid conformation, then the 3D structure of an oligosaccharide chain can be described by a set of torsional angles H1′-C1′-O1-Cx (ϕ) and C1′-O1′-Cx-Hx (ψ) connecting two sugar units through the glycosidic bond. For sialic acids, these are defined by C1′-C2′-O2′-Cx (ϕ) and C2′-O2′-Cx-Hx (ψ) [54]. As already seen, interresidual NOEs can serve as a basis for the determination of glycosylation sites and sequences, and can be converted into interatomic distances. Reliable distance values are calculated from the cross-peak volumes obtained from ROESY spectra, measured in the linear buildup region of NOE. Several proton pairs for the reference distance could be found in the saccharide rings; for instance, all of the H1–H5, H1–H3, and H3–H5 interactions arise from protons separated by 0.25 nm from each other. The interatomic distances are therefore calculated safely by applying the two-spin approximation with equation $r_{ij} = r_{kl} (V_{kl}/V_{ij})^{1/6}$, where V_{ij} and V_{kl} are the cross-peak volumes for unknown (r_{ij}) and calibration (r_{kl}) distances, respectively. This equation relies on the assumption that all dipole–dipole interactions are modulated by a single correlation time, which has been checked by calculating, with the same reference volume and interatomic distance, several r_{kl} concerning proton pairs at a known distance and located in different sugar moieties. The derived interatomic distances can be directly used to determine the torsional angles ϕ,ψ (Table 6) with a simple geometrical iterative calculation, such as the distance-mapping procedure. Unrestrained molecular mechanics (MM) and dynamics (MD) calculations can confirm the results. An unconstrained Monte Carlo/Energy Minimization (MC/EM) conformational search performed on GM1 ganglioside led to results that are fully compatible with the experimental NMR data [55]. Application of ^{1}H NMR to large micelles of a single ganglioside component leads to a detailed study of a surface system with a high carbohydrate density, which might mimic the presence of phase-separated surface domains on cell membranes. High-resolution ^{1}H NMR studies of the micelles show few interresidual dipole–dipole interactions between CH protons of the oligosaccharide chain; thus, hydroxyl protons are assigned (Table 7) and their interactions investigated. Based on NOE results, the detected interactions were

Table 6
Glycosidic torsional angles (ϕ,ψ) pairs in various ganglioside species. Accuracy is within ± (10–15)°

Compound	System	ϕ,ψ (°)						
		V-IV	B-IV	C-A	IV-III	III-II	A-II	II-I[a]
GM4	DMSO-d_6 [47]						−153,-26; −85, 12; 61,-14	
GM3	DMSO-d_6 [51]						−161,-25; −78, 19; 84, 34	55, −5; −15,-45
GM1	DMSO-d_6 [3] DPC [40] ^2H_2O [5] MC/EM [55]				25, 30; 30, −40 42, 11; 36, −30 flexible 51, 2	50, 28 33, 15 36, 8 20, 29	−165, −16 −163, −28 −162,-29 −155,-28	55, 0 47, −1 flexible flexible
GD1b	DMSO-d_6 [49]			89, 10	53, 10	41, 9	175, 5	54, 3
GD1b-L	DMSO-d_6 [49]			−43, 119	49,-16	38, 13	−176, −18	54, 3
GD1a	DMSO-d_6 [52] DPC		−71, 1; −159, −17		31, 1; −26,-20	32, 17	−162,-29	53, −1
GalNAc-GD1a	DMSO-d_6 [52] DPC	36, 15	−163,-25		54, 21; −32, −23	31, 21	−162,-28	49, 0

[a]These data refer to the most populated conformation

Table 7
NMR chemical shifts (ppm) of labile protons of GM1-acetyl micelles and DPC/GD1a mixed micelles in water solution at 285 and 305 K, respectively

		Chemical shift of hydroxyl protons at position							
Residue	**Ganglioside**	**2**	**3**	**4**	**5**	**6**	**7**	**8**	**9**
Glc I	GM1-acetyl	n.d.	5.84			n.d.			
	GD1a	6.95	5.85			5.41			
Gal II	GM1-acetyl	6.51				n.d.			
	GD1a	6.60				n.d.			
GalNAc III	GM1-acetyl	7.53		n.d.		n.d.			
	GD1a	7.60		5.85		n.d.			
Gal IV	GM1-acetyl	n.d.	n.d.	n.d.		n.d.			
	GD1a	5.62	n.d.	5.95		n.d.			
Neu5Ac A	GM1-acetyl			6.34	8.22		5.86	6.24	n.d.
	GD1a			6.45	8.25		5.75	6.25	3.66
Neu5Ac B	GD1a			6.41	8.25		5.75	6.22	n.d.
Sph	GM1-acetyl	8.15	n.d.						
	GD1a	8.22	6.20						

n.d. not detected

in complete agreement with those expected for a ganglioside head group that maintained the same conformational features already observed for a monomer in DMSO and for a single ganglioside molecule per DPC micelle in water [3, 4, 9, 53]. It can be seen that, qualitatively, there are no new and no missing contacts at the level of the saccharide portion of the GM1(acetyl) micelles with respect to the other mentioned systems. Thus, it is possible to regard all of the contacts as belonging to a single monomer. Even if NMR-invisible contact(s) cannot be excluded, the results suggest that the carbohydrate-enriched surface does not influence the main conformational features of the single ganglioside oligosaccharide chain.

Possible contacts between monomers inserted into a surface can be investigated in ganglioside mixed micelles in D_2O formed by Neu5Ac and Neu5Gc containing compounds. The spectral parameters of the sialic acid ring protons for GM1(Neu5Ac) and GM1(Neu5Gc) are different [40]. These differences make possible to check for any NOE originating from interresidual interactions between two sialic acid residues of distinct monomers. Note that the sialic acid belongs to the branched region of the oligosaccharide chain, possibly a favorable location for intermonomer interactions, but no Neu5Ac/Neu5Gc NOE contacts are detected.

A further investigation of ganglioside conformation is carried out by analyzing the water-ganglioside exchange process. By running off resonance ROESY and setting the angle θ between the effective field and the external static field equal to 35.3°, the dipolar cross-relaxation is made to vanish. Only chemical exchange is effective in determining magnetization transfer from water protons to ganglioside OH and NH [56, 57]. Thus, it is possible to establish that at 285–276 K and pH 7.4, the water–ganglioside interaction is completely dominated by exchange in the low mixing time range. The low-temperature, long correlation time and aggregation state of the sample lead to a water–ganglioside exchange in the time scale of 10 ms, an NMR accessible time range. Hydroxyl proton exchange is then followed in the evolution time range of 1–120 ms. The exchange rates are accurately calculated only for Gal IIOH2, Neu5Ac AOH4 and AOH8, their resonances being isolated in the spectrum. By comparing the exchange times of IIOH2, AOH4 and AOH8, similar behavior can be seen for IIOH2 and AOH4, whereas the AOH8 exchange time and its temperature dependence appear quite peculiar. Comparing the exchange times of AOH4 and IIOH2, the latter's constant larger exchange time can be attributed to a reduced solvent accessibility, according to the position depth of the group within the hydrophilic shell of the micelle. On the contrary, despite its relatively external position, AOH8 exchanges less easily than AOH4 and IIOH2 and its exchange time shows a brusque increment at the lowest temperature studied. The fact that the activation energy for the AOH8 proton exchange process is higher than that for the IIOH2 and AOH4 confirms that AOH8 is involved in intramolecular interactions. This agrees with the observation that it belongs to the interaction region between the sialic acid lateral chain and GalNAc, which is probably less accessible in water. Moreover, AOH8 has been shown to have a hydrogen bond with the carboxylic group (note the strong AOH8:A6 NOE contact), thus making the sialic acid structure quite rigid [4, 53].

4.4.2 Dynamics

A critical analysis of the cross peak contacts and, consequently, of the derived distances is necessary. Sometimes two contacts are incompatible with a single conformer; the terminal fragment Gal-GalNAc, i.e., is described by the IV1–III2, IV1–III4, and IV1–IIINH interactions (*see* Table 5), that cannot be satisfied simultaneously by a single rigid structure.

Thus, it is necessary to take into account the possibility of a flexible linkage due to the presence of at least two different conformations. Another interesting example is referred to the terminal sialic acid, i.e., in GD1a, that behaves differently from the internal one. In GD1a, the contact BOH8:IV3 is observed, and the IV3:B3ax distances of 0.27 nm is longer than II3:A3ax (0.21 nm) but significantly shorter than in DMSO solution (0.36 nm) indicating a strong solvent effect. A molecular dynamics simulation revealed the existence of four conformers where the

Table 8

Relaxation parameters ($T_{1\rho}$ [s], NOE) for ^{13}C nuclei in GDla/DPC micelles in D_2O solution at 305 K. The estimated average standard error is 10% for $T_{1\rho}$ and 20% for NOE values. Data for the C7 atoms of internal and terminal Neu5Ac residues are not included in the analysis because of spectral overlap

Residue	Parameter 1	2	3	4	5	6[a]	8	9[a]	
Neu5AcB	$T_{1\rho}$			0.14	0.13	0.14	0.13	0.15	0.17
	NOE			1.6	1.6	1.6	1.6	1.7	1.6
Gal VI	$T_{1\rho}$	0.14	0.15	0.15	0.07	0.14	0.22		
	NOE	1.5	1.7	1.6	1.2	1.5	2.05		
GalNAc III	$T_{1\rho}$	0.07	0.08	0.07	0.06	0.08	0.16		
	NOE	1.1	1.2	1.1	1.2	1.2	1.9		
Neu5Ac A	$T_{1\rho}$			n.d.	0.08	0.08	0.08	0.08	0.13
	NOE			1.1	1.1	1.2	1.3	1.1	1.5
Gal II	$T_{1\rho}$	0.07	0.08	0.08	0.08	0.07	0.16		
	NOE	1.1	1.1	1.0	1.0	1.1	1.7		
Glc I	$T_{1\rho}$	0.08	0.08	0.08	0.08	0.08	0.09		
	NOE	1.2	1.2	1.2	1.3	1.2	1.4		
Sph	$T_{1\rho}$		0.07	0.08					
	NOE		1.2	1.4					

[a]The relaxation parameters for the methylene carbons should be treated with caution since they might contain systematic errors due to cross-correlation effects

$\phi,\psi_{B\text{-}IV}$ flips between two distinct angles, −74,1 and −159,−8, and the two distances BOH8:IV3 and IV3:B3ax changed from 0.42 to 0.23 nm indicating that the experimental values might correspond to an average distance [52]. These results are compatible with those other authors have found in compounds carrying the same terminal fragment, among them GM4, GM3 [46, 51, 58]. The conformational dynamics of GD1a ganglioside anchored into a DPC micelle were investigated by means of multinuclear magnetic resonance spectroscopy. The observed $^1H/^1H$ NOE interactions, once again, revealed conformational averaging of the terminal α-Neu5Ac-(2-3)-β-Gal and β-Gal-(1-3)-β-GalNAc glycosidic linkages. The pronounced flexibility of this terminal trisaccharide moiety was substantiated by probing ^1H–^{13}C dipolar interactions through two-dimensional proton-detected ^{13}C, $T_{1\rho}$, and NOE measurements [4]. With the exception of the IV4 spin, the data in Table 8 show significantly different $T_{1\rho}$ and NOE values for the terminal Gal IV and Neu5Ac B residues as compared to the rest of

Table 9

Average ^{13}C relaxation rates and dynamics parameters, obtained by nonlinear least squares optimization, for GDla/DPC micelles in aqueous solution. The relaxation data for the methylene carbons and Gal IV4 were not included in the analysis

Residue	R_1 [s^{-1}]	R_2 [s^{-1}]	NOE	τ_m [ns]	τ_i [ns]	S^2
Neu5Ac B, Gal IV	2.3 ± 0.2	7.1 ± 1.0	1.6 ± 0.2	2.8 ± 0.1	0.34 ± 0.1	0.55 ± 0.08
GalNAc, Neu5Ac A, Gal II, Glc I	2.2 ± 0.2	12.5 ± 2.0	1.2 ± 0.2	2.8 ± 0.1	0	1

the molecule. The relaxation data are converted into motional parameters assuming two correlation times for the terminal fragment Neu5Ac-Gal and a single correlation time for the GalNAc[Neu5Ac-]-Gal-Glc (Table 9). The presence of the fast internal motions of the Gal IV and Neu5Ac B rings, reflected in the τ_i and S^2 parameters, agrees with the proton NOE data and those previously reported [47].

Recently, the conformational characterization of oligosaccharides in solution was investigated by paramagnetism-assisted NMR techniques [59]. This technique makes use of paramagnetic probes such as lanthanide tags and spin labels attached to the reducing end of the oligosaccharide, which cause paramagnetic effects such as pseudocontact shift (PCS) and paramagnetic relaxation enhancement (PRE) as NOE-independent sources of long-distance information [60–63]. The paramagnetic NMR data were used for experimental validation of a conformational space proposed by theoretical approaches. A PCS-based NMR experiment was applied to conformational characterization of a series of ganglioside oligosaccharides, including the GM1 pentasaccharide, to confirm data derived from molecular dynamics (MD) simulation [64–66]. The role of glycosphingolipids, gangliosides in particular, in modulating interactions of proteins or peptide with membrane lipid bilayer can be evaluated using a number of NMR techniques [67]. Saturation transfer difference (STD) NMR involves excitation and saturation of receptor signals and then the transfer of the magnetization to the altered intensities of the ligand signals give information about their proximity to the receptor. In STD experiment, the focus is on the signals of the ligand, without any need of assignment of receptor signals. This experiment requires only small quantities of unlabeled receptor, i.e., ganglioside micelle ligand [68]. Diffusion NMR experiments such as pulsed field gradient spin echo (PFG-SE) allows one to measure the diffusion coefficients as well as get information about the extent of binding between two species present in the solution. Magnetic field-gradient is used to spatially mark the diffusing molecules. The diffusion coefficients of peptides are reduced depending upon their extent of binding with GM1 micelles [69–71]. Diffusion ordered spectroscopy (DOSY) relies on differences in translation diffusion as a means to separate components in a solution mixture. It is a NMR experiment used widely for the determination of diffusion constants, which are indicative of molecular size, shape and association between individual components in a mixture [67, 72–76].

5 Notes

1. The use of an oil vacuum pump needs a liquid nitrogen trap to condense the oil vapors, to avoid the presence of a high signal due to silicon compound around zero ppm in the NMR spectrum
2. The water signal can pose a major problem because it can obscure different methine signals. Although the solvent signal can be suppressed by different sequences, the less solvent into nmr tube, the better the spectrum [6]. The same reason forces using a very high-quality deuterated solvent.
3. Measurement must be run on 500 MHz or higher field spectrometers.
4. Contamination of the ganglioside solution with other glycosphingolipids, amphiphilic compounds, or lipids must be avoided because they can lead to very large mixed aggregates in water, causing severe signals broadening. For the same reason the sample solution must be free of bivalent cations that strongly bind to the ganglioside oligosaccharide chain and lead to signal broadening and slow proton deuterium exchange [10].
5. All spectra are acquired under temperature control in the range 297 ÷ 333 K with a sweep width of 10 ÷ 12 ppm and – 10 ÷ 200 ppm for ^{1}H and ^{13}C spectra, respectively. The relaxation delay is in the range 1.0 ÷ 1.5 s. Chemical shifts are referenced indirectly by setting the signal of residual DMSO-d_5 at 2.49 ppm and at 39.5 ppm for ^{1}H and ^{13}C spectra, respectively. In water solution ^{1}H and ^{13}C chemical shifts are respectively referenced to separated resonances such as the sphingosine olefinic H-4 resonance at 5.45 ppm and to the anomeric resonance of Gal (IV) at 104.8 ppm, respectively. In DMSO, some chemical shifts may be sensitive to the water content.
6. If important solute peaks are close to the solvent, but you must also observe exchangeable protons, use a *watergate* flavor [5]. If suppression is critical with exchangeables, or if it is crucial to eliminate the solvent peak, use *excitation sculpting* with a double pulsed field gradient spin echo (DPFGSE). This method produces excellent suppression but also a broader notch bandwidth about the solvent peak. Finally, if multiple

solvent peaks must be reduced, use *wet*. *Wet* is a prominent method used with mixed solvents (e.g., if the sample derives from a HPLC separation) but it can reduce intensity of exchangeable protons. Most of these subsequences can be inserted in every kind of monodimensional and bidimensional NMR experiment.

7. NOESY 1D is run with presaturation during the whole length of recycle delay of 5 s, centered at the HDO residual signal (around 4.7 ppm) with a γB2 power of about 25 Hz. Mixing time is 10 ms.
8. All 2D spectra are recorded with 512 t1 increments and 32–64 scans collected for each t1. Whenever possible, the gradient-based 2D experiments, which reduce the experimental time, allow very good solvent suppression and avoid a lot of artifacts, are recommended [21].
9. The mixing time is in the range of 75–300 ms; the spectral width is 4–5 kHz. After zero filling, the time-domain spectra are transformed to give a 4 K × 2 K real points matrix with the resolution of 3.91 and 1.95 Hz/point in the F1 and F2 dimension, respectively.
10. A number of experiments can be run to overcome this problem and simplify the spectrum. The double quantum filtered COSY (DQF-COSY) suppresses the signals of the isolated protons, such as those arising from solvents or isolated methyl and acetyl groups. Moreover, this experiment provides pure absorption cross- and diagonal-peaks showing a thinner diagonal and allowing a better resolution of cross-peaks close to the diagonal. The triple-quantum filtered COSY (TQF-COSY) suppresses all the spin systems that contain less than three or more mutually scalar coupled spins; such a system in hexopyranosides is H5, H6, and H6′, whose protons are often difficult to assign.

References

1. Sonnino S, Cantu L, Corti M, Acquotti D, Venerando B (1994) Aggregative properties of gangliosides in solution. Chem Phys Lipids 71(1):21–45
2. Eaton HL, Hakomori SI (1988) Conformation of micelle-bound gangliosides by 2D proton NMR spectroscopy. 195th ACS Natl Mtg, 91 CARB
3. Acquotti D, Poppe L, Dabrowski J, Von der Lieth CW, Sonnino S, Tettamanti G (1990) Three-dimensional structure of the oligosaccharide chain of GM1 ganglioside revealed by a distance-mapping procedure: a rotating and laboratory frame nuclear overhauser enhancement investigation of native glycolipid in dimethyl-sulfoxide and in water dodecylphosphocholine solutions. J Am Chem Soc 112(21):7772–7778
4. Poppe L, Van Halbeek H, Acquotti D, Sonnino S (1994) Carbohydrate dynamics at a micellar surface: Gd1a headgroup transformations revealed by NMR spectroscopy. Biophys J 66(5):1642–1652
5. Brocca P, Berthault P, Sonnino S (1998) Conformation of the oligosaccharide chain of G(M1) ganglioside in a carbohydrate-enriched surface. Biophys J 74(1):309–318. https://doi.org/10.1016/S0006-3495(98)77788-4
6. Guéron M, Plateau P (2007) Water signal suppression in NMR of biomolecules. In: eMagRes. Wiley. doi:https://doi.org/10.1002/9780470034590.emrstm0590

7. Sonnino S, Kirschner G, Ghidoni R, Acquotti D, Tettamanti G (1985) Preparation of GM1 ganglioside molecular species having homogeneous fatty acid and long chain base moieties. J Lipid Res 26(2):248–257
8. Sonnino S, Cantù L, Corti M, Acquotti D, Kirschner G, Tettamanti G (1990) Aggregation properties of semisynthetic GM1 ganglioside (II3Neu5AcGgOse4Cer) containing an acetyl group as acyl moiety. Chem Phys Lipids 56(1):49–57
9. Brocca P, Cantù L, Sonnino S (1995) Aggregation properties of semisynthetic GD1a ganglioside (IV3Neu5AcII3Neu5AcGgOse4Cer) containing an acetyl group as acyl moiety. Chem Phys Lipids 77(1):41–49
10. Brocca P, Acquotti D, Sonnino S (1993) ^{1}H-NMR study on ganglioside amide protons: evidence that the deuterium exchange kinetics are affected by the preparation of samples. Glycoconj J 10(6):441–446
11. Poppe L, Van Halbeek H (1991) Nuclear-magnetic-resonance of hydroxyl and amido protons of oligosaccharides in aqueous-solution: evidence for a strong intramolecular hydrogen bond in sialic acid residues. J Am Chem Soc 113(1):363–365
12. Morris GA, Freeman R (1978) Selective excitation in Fourier transform nuclear magnetic resonance. J Magn Reson 29(3):433–462. https://doi.org/10.1016/0022-2364(78)90003-3
13. Gallo V, Intini N, Mastrorilli P, Latronico M, Scapicchio P, Triggiani M, Bevilacqua V, Fanizzi P, Acquotti D, Airoldi C, Arnesano F, Assfalg M, Benevelli F, Bertelli D, Cagliani LR, Casadei L, Cesare Marincola F, Colafemmina G, Consonni R, Cosentino C, Davalli S, De Pascali SA, D'Aiuto V, Faccini A, Gobetto R, Lamanna R, Liguori F, Longobardi F, Mallamace D, Mazzei P, Menegazzo I, Milone S, Mucci A, Napoli C, Pertinhez T, Rizzuti A, Rocchigiani L, Schievano E, Sciubba F, Sobolev A, Tenori L, Valerio M (2015) Performance assessment in fingerprinting and multi component quantitative NMR analyses. Anal Chem 87(13):6709–6717. https://doi.org/10.1021/acs.analchem.5b00919
14. Subramanian S, Bax A (1987) Generation of pure phase NMR subspectra for measurement of homonuclear coupling-constants. J Magn Reson 71(2):325–330
15. Sørensen OW, Rance M, Ernst RR (1984) z Filters for purging phase- or multiplet-distorted spectra. J Magn Reson 56(3):527–534. https://doi.org/10.1016/0022-2364(84)90317-2
16. Sklenar V, Feigon J (1990) Simplification of DNA proton nuclear magnetic resonance spectra by homonuclear Hartmann-Hahn edited two-dimensional nuclear Overhauser enhancement spectroscopy. J Am Chem Soc 112(14):5644–5645
17. Hu H, Bradley SA, Krishnamurthy K (2004) Extending the limits of the selective 1D NOESY experiment with an improved selective TOCSY edited preparation function. J Magn Reson 171(2):201–206. https://doi.org/10.1016/j.jmr.2004.08.018
18. Thrippleton MJ, Keeler J (2003) Elimination of zero-quantum interference in two-dimensional NMR spectra. Angew Chem Int Ed Engl 42(33):3938–3941. https://doi.org/10.1002/anie.200351947
19. Cano K, Thrippleton MJ, Keeler J, Shaka AJ (2004) Improved transient NOE buildups in coupled spin systems. 45th Experimental NMR Conference, Pacific Grove, USA
20. Marion D, Wuthrich K (1983) Application of phase sensitive two-dimensional correlated spectroscopy (COSY) for measurements of ^{1}H-^{1}H spin-spin coupling constants in proteins. Biochem Biophys Res Commun 113(3):967–974
21. Hurd RE (1990) Gradient-enhanced spectroscopy. J Magn Reson 87(2):422–428
22. Bax A, Davis DG (1985) Mlev-17-based two-dimensional homonuclear magnetization transfer spectroscopy. J Magn Reson 65(2):355–360
23. Rance M (1987) Improved techniques for homonuclear rotating-frame and isotropic mixing experiments. J Magn Reson 74(3):557–564
24. Zhao DZ, Fujiwara T, Nagayama K (1989) Suppression of off-resonance effects in 1D and 2D ROESY. J Magn Reson 81(3):628–630
25. Farmer BT, Brown LR (1987) The effect of resonance offset on the cross-relaxation rate in the rotating frame. J Magn Reson 72(1):197–202
26. Bax A (1988) Correction of cross-peak intensities in 2D spin-locked NOE spectroscopy for offset and Hartmann-Hahn effects. J Magn Reson 77(1):134–147
27. Davis DG (1989) Elimination of baseline distortions and minimization of artifacts from phased 2D NMR spectra. J Magn Reson 81(3):603–607
28. Marion D, Bax A (1988) Baseline distortion in real-fourier-transform NMR spectra. J Magn Reson 79(2):352–356
29. Boros S, Batta G (2016) Offset-compensated and zero-quantum suppressed ROESY pro-

vides accurate ^{1}H-^{1}H distances in small to medium-sized molecules. Magn Reson Chem 54(12):947–952

30. Altona C, Haasnoot CAG (1980) Prediction of anti and gauche vicinal proton-proton coupling constants in carbohydrates: a simple additivity rule for pyranose rings. Org Magn Reson 13(6):417–429. https://doi.org/10.1002/mrc.1270130606
31. Haasnoot CAG, de FAAM L, Altona C (1980) The relationship between proton-proton NMR coupling constants and substituent electronegativities. An empirical generalization of the Karplus equation. Tetrahedron 36(19):2783–2792 https://doi.org/10.1016/0040-4020(80)80155-4
32. Robinson PT, Pham TN, Uhrin D (2004) In phase selective excitation of overlapping multiplets by gradient-enhanced chemical shift selective filters. J Magn Reson 170(1):97–103
33. Neuhaus D, Williamson MP (1989) The nuclear overhauser effect in structural and conformational analysis. VCH, New York
34. Dabrowski J (1987) Application of two-dimensional NMR methods in the structural analysis of oligosaccharides and other complex carbohydrates. In: Croasmun WR, Carlson RMK (eds) Two-dimensional NMR spectroscopy. VCH, NY, pp 349–386
35. Bothner-By AA, Stephens RL, Lee J, Warren CD, Jeanloz RW (1984) Structure determination of a tetrasaccharide: transient nuclear Overhauser effects in the rotating frame. J Am Chem Soc 106(3):811–813 https://doi.org/10.1021/ja00315a069
36. Dabrowski J, Poppe L (1989) Hydroxyl and amido groups as long-range sensors in conformational analysis by nuclear Overhauser enhancement: a source of experimental evidence for conformational flexibility of oligosaccharides. J Am Chem Soc 111(4):1510–1511 https://doi.org/10.1021/ja00186a063
37. Stott K, Keeler J, Van QN, Shaka AJ (1997) One-dimensional NOE experiments using pulsed field gradients. J Magn Reson 125(2):302–324. https://doi.org/10.1006/jmre.1997.1110
38. Stonehouse J, Adell P, Keeler J, Shaka AJ (1994) Ultrahigh-quality NOE spectra. J Am Chem Soc 116(13):6037–6038. https://doi.org/10.1021/ja00092a092
39. Boudot D, Roumestand C, Toma F, Canet D (1990) Pseudo-3D NMR of proteins with selective excitation by DANTE-Z. One-dimensional TOCSY-NOESY experiment. J Magn Reson 90(1):221–227. https://doi.org/10.1016/0022-2364(90)90382-J
40. Brocca P, Acquotti D, Sonnino S (1996) Nuclear Overhauser effect investigation on GM1 ganglioside containing N-glycolylneuraminic acid (II3Neu5GcGgOse4Cer). Glycoconj J 13(1):57–62
41. Palmer AG, Cavanagh J, Wright PE, Rance M (1991) Sensitivity improvement in proton-detected two-dimensional heteronuclear correlation NMR spectroscopy. J Magn Reson 93(1):151–170. https://doi.org/10.1016/0022-2364(91)90036-S
42. Willker W, Leibfritz D, Kerssebaum R, Bermel W (1993) Gradient selection in inverse heteronuclear correlation spectroscopy. Magn Reson Chem 31(3):287–292. https://doi.org/10.1002/mrc.1260310315
43. Baumann H, Jansson P-E, Kenne L (1991) Synthesis and nuclear magnetic resonance studies of some L-fucosyl-containing disaccharides. J Chem Soc Perkin Trans 1(9):2229–2233. https://doi.org/10.1039/P19910002229
44. Bock K, Brignole A, Sigurskjold BW (1986) Conformational dependence of ^{13}C nuclear magnetic resonance chemical shifts in oligosaccharides. J Chem Soc Perkin Trans 2(11):1711–1713 https://doi.org/10.1039/P29860001711
45. Lundqvist LC, Kaszowska M, Sandstrom C (2015) NMR study of the O-specific polysaccharide and the core oligosaccharide from the lipopolysaccharide produced by Plesiomonas shigelloides O24:H8 (strain CNCTC 92/89). Molecules 20(4):5729–5739. https://doi.org/10.3390/molecules20045729
46. Sonnino S, Acquotti D, Cantù L, Chigorno V, Valsecchi M, Casellato R, Masserini M, Corti M, Allevi P, Tettamanti G (1994) Synthesis and aggregative properties of GM1 ganglioside (IV3Neu5AcGgOse4Cer) containing D-(+)-2-hydroxystearic acid. Chem Phys Lipids 69(2):95–104
47. Poppe L, Dabrowski J, von der Lieth CW, Numata M, Ogawa T (1989) Solution conformation of sialosylcerebroside (GM4) and its NeuAc(α 2-3)gal beta sugar component. Eur J Biochem 180(2):337–342
48. Scarsdale JN, Prestegard JH, Yu RK (1990) NMR and computational studies of interactions between remote residues in gangliosides. Biochemistry 29(42):9843–9855. https://doi.org/10.1021/bi00494a014
49. Acquotti D, Fronza G, Ragg E, Sonnino S (1991) Three dimensional structure of GD1b and GD1b-monolactone gangliosides in dimethylsulphoxide: a nuclear Overhauser effect investigation supported by molecular dynamics calculations. Chem Phys Lipids 59(2):107–125

50. Levery SB (1991) ^{1}H-NMR study of GM2 ganglioside: evidence that an interresidue amide-carboxyl hydrogen bond contributes to stabilization of a preferred conformation. Glycoconj J 8(6):484–492 https://doi.org/10.1007/bf00769848
51. Siebert HC, Reuter G, Schauer R, Von der Lieth CW, Dabrowski J (1992) Solution conformations of GM3 ganglioside containing different sialic acid residues as revealed by NOE-based distance mapping, molecular mechanics, and molecular dynamics calculations. Biochemistry 31(30):6962–6971. https://doi.org/10.1021/bi00145a014
52. Acquotti D, Cantù L, Ragg E, Sonnino S (1994) Geometrical and conformational properties of ganglioside GalNAc-GD1a, IV4GalNAcIV3Neu5AcII3Neu5AcGgOse4 Cer. Eur J Biochem 225(1):271–288
53. Poppe L, Von der Lieth CW, Dabrowski J (1990) Conformation of the glycolipid globoside head group in various solvents and in the micelle-bound state. J Am Chem Soc 112(21):7762–7771. https://doi.org/10.1021/ja00177a042
54. Klyne W, Prelog V (1960) Description of steric relationships across single bonds. Experientia 16(12):521–523. https://doi.org/10.1007/bf02158433
55. Bernardi A, Raimondi L (1995) Conformational analysis of GM1 oligosaccharide in water solution with a new set of parameters for the Neu5Ac moiety. J Org Chem 60(11):3370–3377. https://doi.org/10.1021/jo00116a021
56. Desvaux H, Berthault P, Birlirakis N, Goldman M (1994) Off-resonance ROESY for the study of dynamic processes. J Magn Reson Ser A 108(2):219–229. https://doi.org/10.1006/jmra.1994.1114
57. Desvaux H, Berthault P, Birlirakis N, Goldman M, Piotto M (1995) Improved versions of off-resonance ROESY. J Magn Reson Ser A 113(1):47–52. https://doi.org/10.1006/jmra.1995.1054
58. Breg J, Kroon-Batenburg LMJ, Strecker G, Montreuil J, Vliegenthart JFG (1989) Conformational analysis of the sialyl$\alpha(2\rightarrow3/6)$ N-acetyllactosamine structural element occurring in glycoproteins, by two-dimensional NOE ^{1}H-NMR spectroscopy in combination with energy calculations by hard-sphere exo-anomeric and molecular mechanics force-field with hydrogen-bonding potential. Eur J Biochem 178(3):727–739 https://doi.org/10.1111/j.1432-1033.1989.tb14504.x
59. Yagi-Utsumi M, Kato K (2015) Structural and dynamic views of GM1 ganglioside. Glycoconj J 32(3–4):105–112. https://doi.org/10.1007/s10719-015-9587-5
60. Ying Z, Takumi Y, Koichi K (2013) New NMR tools for characterizing the dynamic conformations and interactions of oligosaccharides. Chem Lett 42(12):1455–1462. https://doi.org/10.1246/cl.130789
61. Yamaguchi Y, Yamaguchi T, Kato K (2014) Structural analysis of oligosaccharides and glycoconjugates using NMR. In: Yu RK, Schengrund C-L (eds) Glycobiology of the nervous system. Springer, New York, NY, pp 165–183. https://doi.org/10.1007/978-1-4939-1154-7_8
62. Erdelyi M, d'Auvergne E, Navarro-Vazquez A, Leonov A, Griesinger C (2011) Dynamics of the glycosidic bond: conformational space of lactose. Chem 17(34):9368–9376 https://doi.org/10.1002/chem.201100854
63. Canales A, Mallagaray A, Perez-Castells J, Boos I, Unverzagt C, Andre S, Gabius HJ, Canada FJ, Jimenez-Barbero J (2013) Breaking pseudo-symmetry in multiantennary complex N-glycans using lanthanide-binding tags and NMR pseudo-contact shifts. Angew Chem Int Ed Engl 52(51):13789–13793. https://doi.org/10.1002/anie.201307845
64. Yamamoto S, Zhang Y, Yamaguchi T, Kameda T, Kato K (2012) Lanthanide-assisted NMR evaluation of a dynamic ensemble of oligosaccharide conformations. Chem Commun (Camb) 48(39):4752–4754. https://doi.org/10.1039/c2cc30353a
65. Zhang Y, Yamamoto S, Yamaguchi T, Kato K (2012) Application of paramagnetic NMR-validated molecular dynamics simulation to the analysis of a conformational ensemble of a branched oligosaccharide. Molecules 17(6):6658–6671. https://doi.org/10.3390/molecules17066658
66. Zhang Y, Yamaguchi T, Satoh T, Yagi-Utsumi M, Kamiya Y, Sakae Y, Okamoto Y, Kato K (2015) Conformational dynamics of oligosaccharides characterized by paramagnetism-assisted NMR spectroscopy in conjunction with molecular dynamics simulation. Adv Exp Med Biol 842:217–230. https://doi.org/10.1007/978-3-319-11280-0_14
67. Khatun UL, Gayen A, Mukhopadhyay C (2014) Capability of ganglioside GM1 in modulating interactions, structure, location and dynamics of peptides/proteins: biophysical approaches. Glycoconj J 31(6–7):435–447 https://doi.org/10.1007/s10719-014-9554-6
68. Viegas A, Manso J, Nobrega FL, Cabrita EJ (2011) Saturation-transfer difference (STD) NMR: a simple and fast method for ligand screening and characterization of protein bind-

ing. J Chem Educ 88(7):990–994. https://doi.org/10.1021/ed101169t

69. Chatterjee C, Majumder B, Mukhopadhyay C (2004) Pulsed-field gradient and saturation transfer difference NMR study of enkephalins in the ganglioside GM1 micelle. J Phys Chem B 108(22):7430–7436. https://doi.org/10.1021/jp037553r
70. Chatterjee C, Mukhopadhyay C (2005) Binding and folding of melittin in the presence of ganglioside GM1 micelles. J Biomol Struct Dyn 23(2):183–192. https://doi.org/10.1080/07391102.2005.10507058
71. Chatterjee C, Mukhopadhyay C (2005) Interaction and structural study of kinin peptide bradykinin and ganglioside monosialylated 1 micelle. Biopolymers 78(4):197–205. https://doi.org/10.1002/bip.20278
72. Sørland GH, Aksnes D, Gjerdåker L (1999) A pulsed field gradient spin-echo method for diffusion measurements in the presence of internal gradients. J Magn Reson 137(2):397–401. https://doi.org/10.1006/jmre.1998.1670
73. Malliavin TE, Louis V, Delsuc MA (1998) The DOSY experiment provides insights into the protegrin-lipid interaction. J Chim Phys 95(2):178–186
74. Morris GA (2007) Diffusion-ordered spectroscopy. In: eMagRes. Wiley. doi:https://doi.org/10.1002/9780470034590.emrstm0119.pub2
75. Johnson CS (1999) Diffusion ordered nuclear magnetic resonance spectroscopy: principles and applications. Prog Nucl Magn Reson Spectrosc 34(3):203–256. https://doi.org/10.1016/S0079-6565(99)00003-5
76. Hounsell EF (2006) NMR of carbohydrates, lipids and membranes. In: Webb GA (ed) nuclear magnetic resonance, vol 35. The Royal Society of Chemistry, p 362–388. https://doi.org/10.1039/9781847555236-00362

Chapter 13

Methods for the Preparation of Anti-ganglioside Monoclonal Antibodies

Noriyuki Yuasa and Ikuo Kawashima

Abstract

The antibody preparation method to the glycolipid is basically same to the method to the protein. However, because the immunogenicity of the carbohydrate moieties of glycolipid is generally low in the mouse, the development of the anti-glycolipid antibody using purified glycolipid as an immunogen was not yet established. Here, we describe a method using a purified ganglioside adsorbed onto *Salmonella minnesota* for the efficient production of an anti-ganglioside mouse monoclonal antibody that recognizes the carbohydrate moieties of the ganglioside.

Key words Monoclonal antibody, *Salmonella Minnesota R595*, Myeloma cell line, Cell fusion, Hybridoma, Glycolipid

1 Introduction

Since a preparation method of the monoclonal antibody was established in 1975 [1], a lot of monoclonal antibodies that recognize the carbohydrate moiety of glycoconjugates, including glycolipids, glycoproteins, and mucins, have been developed. The structure of the carbohydrate moiety has been applied not only in basic research such as functional analysis but has also found application in the field of medicine for the diagnosis and treatment of cancers.

A large number of these antibodies were developed through studies of tumor antigens and differentiation markers. Briefly, a tumor cell line or a differentiated cell line was used as an antigen to prepare a hybridoma, and then screening was performed to isolate an antibody that reacted with the cell line, resulting in a specific antigen being identified as a carbohydrate moiety [2]. The process of developing such antibodies is fundamentally the same as that of developing anti-protein antibodies. However, the establishment of anti-carbohydrate moieties antibodies using purified glycolipids as immunogens is difficult because of the low immunogenicity of carbohydrate moieties [3, 4].

Sandro Sonnino and Alessandro Prinetti (eds.), *Gangliosides: Methods and Protocols*, Methods in Molecular Biology, vol. 1804, https://doi.org/10.1007/978-1-4939-8552-4_13,

© Springer Science+Business Media, LLC, part of Springer Nature 2018

Of course, by using traditional methods, although it was not the purpose to obtain anti-glycolipid antibodies, numerous anti-glycolipid antibodies have been obtained by immunizing mice with whole tumor cells and embryonic cells. Several improved protocols obtaining anti-glycolipids have been established: mice are immunized with the glycosphingolipid conjugate with keyhole limpet hemocyanin, or with glycolipids of liposomes containing lipid-A, or with purified glycosphingolipids adsorbed on *S. minnesota* [5–7].

Here, we introduce a method using a purified ganglioside adsorbed onto *S. minnesota* for the efficient production of an anti-ganglioside mouse monoclonal antibody. Using this method, we succeeded in the establishment of much anti-ganglioside antibodies [8] (Fig. 1).

2 Materials

2.1 Preparation of Acid-Treated S. minnesota R595

1. L-broth: 1% Bacto tryptone, 0.5% yeast extract, and 0.5 g NaOH; dissolve in 900 mL water and adjust pH with 0.1 N NaOH. Fill up to 1 L with water. Sterilize by autoclaving for 20 min. Store at room temperature.
2. 1% acetic acid: 0.1 mL acetic acid dissolved in 10 mL distilled water.
3. Erlenmeyer flask (500 mL, 5 L).
4. Desiccator.

2.2 Immunization of Mice

1. Phosphate-buffered saline (PBS): Dissolve 8 g NaCl, 0.2 g KCl, 1.44 g Na_2HPO_4, 0.24 g KH_2PO_4 in 800 mL of H_2O. Adjust the pH to 7.4 with 1 N HCl, and then add H_2O to 1 L. Sterilize them by autoclaving for 20 min. Store at room temperature.
2. Ethanol.
3. Purified ganglioside.
4. Mice: C_3H/HeN (aged 7 weeks) mice were purchased from CREA (Tokyo, Japan).

2.3 Producing Hybridoma

1. Mouse myeloma cell line: PAI cells (JCRB cell bank, JCRB0113).
2. Medium A: RPMI1640 supplemented with 60 mg/L kanamycin.
3. Medium B: Medium A supplemented with 10% fetal calf serum (*see* **Note 3**), 2 mM L-glutamine, 10 mM HEPES, and 1 mM sodium pyruvate.

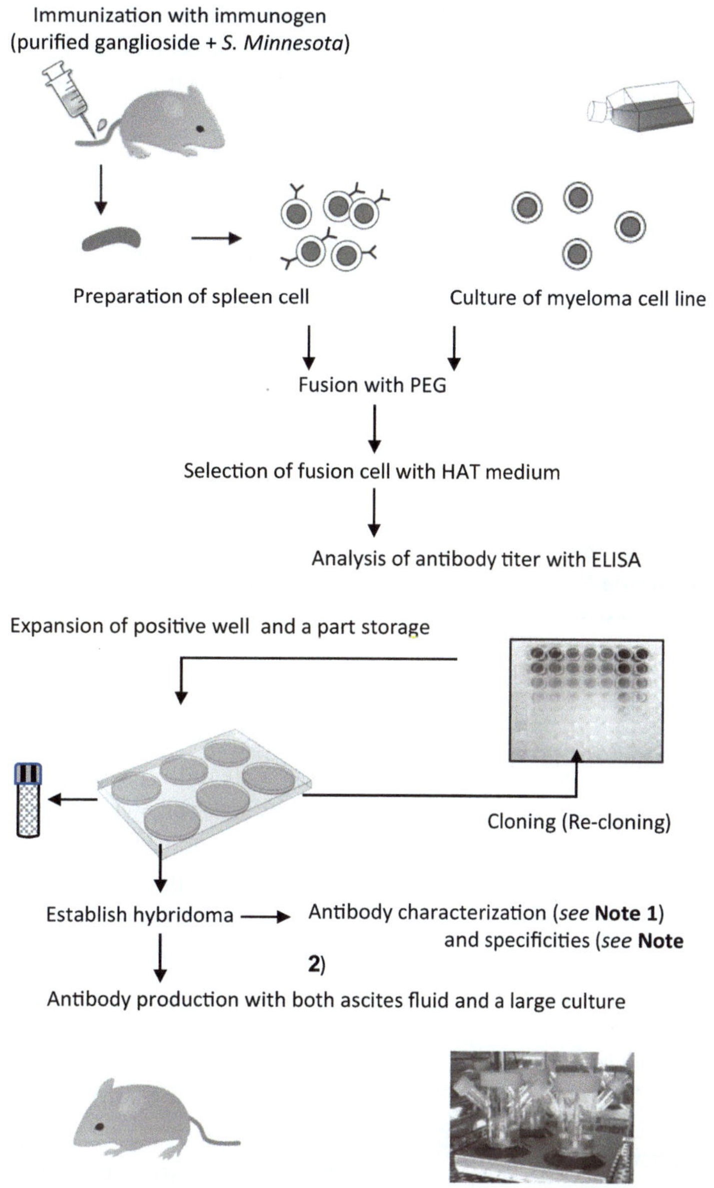

Fig. 1 A summary of the procedures employed to prepare monoclonal antibodies. On the onset, an appropriate screening procedure must be established. First, mice are injected with an immunogen preparation. Once a good humoral immune response appears in the immunized animal, the cell fusion step can be initiated. Polyethylene glycol is the most common agent used for cell fusion [9, 10]. After the fusion, cells are diluted in selective medium and plated in multi-well plates (e.g., 96-well plate). Hybridomas develop and are ready to be screened in 1 week after the fusion. Cells of the positive wells are expanded in 6- or 12-well culture dishes, and in most cases, recloning (subcloning) is required. After characterization of antibodies, monoclonal antibodies can be produced from ascites fluid or culture supernatant. The monoclonal antibodies can be purified with ease using conventional purification methods

4. Tris–NH_4Cl: Make up to 100 mL working solution: mix 90 mL of 0.16 M NH_4Cl and 10 mL of 0.17 M Tris, pH 7.65; Adjust to pH 7.2 with HCl (450 mL NH_4Cl + 50 mL Tris).
5. 50% PEG4000.
6. HAT medium: obtained from GIBCO.
7. HT medium: obtained from GIBCO.
8. Cell culture plate: 96-well, 24-well, and 6-well.
9. Serum tube.
10. 100 mm dish.
11. Mesh filter 10#.

3 Methods

3.1 Preparation of Acid-Treated S. minnesota R595

1. Transfer 100 mL culture medium (L-broth) to a 500 mL Erlenmeyer flask; inoculate *S. minnesota* R595 and preculture overnight at 37 °C.
2. Transfer half of the preculture medium to a 5 L flask; add 950 mL culture medium and shake overnight at 37 °C intensely.
3. Centrifuge the culture medium at 10,000 × *g* for 10 min.
4. Collect the cell pellet; resuspend the pellet for washing and centrifuge by the following solution, water, ethanol, acetone, and diethyl ether sequentially
5. Dry-up the final pellet in a desiccator.
6. Resuspend 200 mg of the bacterial pellet in 10 mL 1% acetic acid and heat at 100 °C for 2 h; wash with distilled water, lyophilize, and store frozen until use.

3.2 Immunization of Mice

3.2.1 Preparation of Immunogen

1. Dissolve 10 μg purified ganglioside in 50 μL ethanol and sonicate. Add 800 μL PBS and incubate at 37 °C.
2. Resuspend the acid-treated *S. minnesota* R595 in PBS and adjust the concentration to 1 mg/mL.
3. Mix the solution from **step 1** and 250 μL of the solution from **step 2**; incubate at 37 °C for 10 min.
4. Store this immunogen frozen until use.

3.2.2 Immunization

When using animals in research, follow the guidelines of the ethics committee of the institution and regional regulations.

5. Use C_3H/HeN mice for immunization (*see* **Note 4**).
6. Administrate 200 μL of the immunogen via the tail vein on days 0, 4, 7, 11, and 21.
7. Collect blood from orbital sinus and mesure the antibody titer of the serum on days 0, 11 and 21.

3.3 Creating a Hybridoma

Cryopreserved stocks of myeloma cells should be thawed at least 2 weeks prior to the fusion. The myeloma cells should be healthy and rapidly growing before the fusion.

3.3.1 Culture of Mouse Myeloma Cells (PAI) for Fusion (see **Note 5**)

1. After confirming the good revival, 1 day before the fusion split the myeloma cells into fresh 250 mL of medium B at a final cell density of 2×10^4 cells/mL.
2. After 3–4 days, count the number of myeloma cells and check the cell viability. Cell viability of PAI cells should be higher than 90% before fusion. Finally dilute $2–3 \times 10^7$ cells in 10 mL of medium A.

3.3.2 Preparation of Splenocytes for Fusion

3. Aseptically remove the spleens from immunized mice (3 days after the final immunization), and place them in a 100 mm tissue culture dish containing 20 mL of medium A (prewarmed to 37 °C).
4. Untie the spleen using a blunt pincette on a mesh (#10) and disrupt any cell clump by pipetting. Transfer the cell suspension into a sterile centrifuge tube.
5. Lyse the red blood cells using Tris–NH4Cl solution (10 mL) by promptly mixing and then centrifuge (200 × *g*) for 5 min. Discard the supernatant and resuspend the cells in appropriate volume of medium A.
6. Count the cells and adjust the cell number to 2.0×10^8 cells in 20 mL of medium A.

3.3.3 Cell Fusion

7. Gently mix the spleen and myeloma (PAI) cells in a sterile 50 mL conical tube. Centrifuge the cells at 200 × *g* for 5 min and discard the supernatant. Tap the cell pellet vigorously to loosen the cells (this will make easier to dissolve the pellet in the next step) (*see* **Note 6**).
8. In a drop-wise manner, slowly add 0.8 mL of 50% PEG 4000 in cell suspension with constant agitation using a pipette tip. In addition, gently swirl the cells for 1 min.
9. Stop the fusion by adding 10 mL of medium A by gentle swirling and centrifuge the cells at 200 × *g* for 5 min.
10. Discard the supernatant and resuspend the cell pellet in 40 mL of HAT medium. Then count the cells and adjust the cell number to 2.5×10^6 cells/mL HAT medium.
11. Plate 200 μL of cell suspension per well in a 96-well plate and incubate the cells at 37 °C in a CO_2 incubator until the formation of HAT-resistant colonies. Colonies are visible by eyes by about 7–10 days after plating.
12. Transfer the supernatant to a 96-well plate in duplicates for ELISA screen. Replace the lost volume (e.g., 100 μL) with fresh HAT medium.

3.3.4 Screening and Cloning (see **Note 7***)*

When the colonies are visible and ready to screen, mark the wells with preferred numbering system. Single-cell ensures that the cells producing the antibody of interest are truly monoclonal and that the secretion of the antibody is stably maintained.

13. Transfer the supernatant to a 96-well plate in duplicates for ELISA screen. Aseptically remove about 50–100 μL of culture supernatant by using micropipette. Replace the lost volume (e.g., 100 μL) with fresh HAT medium.
14. When the positive well is identified, count the cells and adjust the cell number to 1.5 cells/mL in 20 mL of HT medium.
15. Plate 200 μL of cell suspension per well in a 96-well plate and incubate the cells at 37 °C in a CO_2 incubator. Colonies are visible by eyes by about 7–10 days after plating.

4 Notes

Antibodies

A part of the antibody developed this method can be obtained from a reagent company (Tokyo chemical industry Co. Ltd., Tokyo, Japan).

1. Antibody characterization

   The property of the antibody is determined using a mouse immunoglobulin decision kit and electrophoresis, in accordance with conventional methods [11].
2. Binding specificities of the antibody

   Binding specificities of the antibody is determined by the authentic glycolipid by using ELISA and immunostaining with TLC [11].
3. Fetal bovine serum

   The lot of serum used for hybridoma formation is confirmed in advance to be appropriate for myeloma cell line proliferation. Lot checks using different cell lines are not of much use.
4. Mouse selection

   BALB/c and C57BL/6 mice are commonly used for immunization. The authors compared the ganglioside immunogenicity among 10 types of inbred mouse. The results indicated that the immunogenicity of gangliosides in mice is dependent on the mouse lineage and ganglioside molecule type.

   Based on these results, when the production of monoclonal antibodies by immunizing C_3H/HeN mice with gangliosides was attempted, it was possible to produce antibodies extremely efficiently compared to previous methods.

5. Mouse myeloma cell line

   Many mouse myeloma cell lines have been established, with the NS-1 line being representative. The benefits of the PAI line that the authors use are as follows: [1] they are stable cells that are comparatively easy to culture; and [2] they are adherent, so it is easy to determine colony formation. In addition, since a relatively large number of cells (1×10^7 cells) are needed during cell fusion, cells in the logarithmic growth phase are always used for this purpose.

6. Cell fusion conditions

   The ratio of myeloma cells to mouse spleen cells used by the authors is 10:1. A ratio in the range of 1:1 to 10:1 is considered acceptable.

7. Screening

   The authors use ELISA. Regardless of the method employed, using a highly reliable assay is the key to successful antibody production.

Acknowledgment

We would like to thank Dr. Tadashi Tai and colleagues; most of these techniques were developed in collaboration with Dr. Tai's laboratory.

References

1. Köhler G, Milstein C (1975) Continuous cultures of fused cells secreting antibody of predefined specificity. Nature 256:495–497
2. Tai T, Kawashima I, Ogura K (2007) Anti-carbohydrate antibodies. In: Kamerling JP (ed) Comprehensive glycoscience. from chemistry to systems biology, vol 3. Elsevier, Amsterdam, pp 765–785
3. Magnani JL (1987) Mouse and rat monoclonal antibodies directed against carbohydrates. Methods Enzymol 138:484–491
4. Kannagi R (2000) Monoclonal anti-glycosphingolipid antibodies. Methods Enzymol 312:160–179
5. Galanos C, Luderitz O, Westphal O (1971) Preparation and properties of antisera against the lipid-A component of bacterial lipopolysaccharides. Eur J Biochem 24:116–122
6. Zhang S, Graeber LA, Helling F, Ragupathi G, Adluri S, Lloyd KO, Livingston PO (1996) Augmenting the immunogenicity of synthetic MUC1 peptide vaccines in mice. Cancer Res 56:3315–3319
7. Kannagi R (2002) Use of liposomes containing carbohydrates for production of monoclonal antibodies. Methods Mol Biol 199:203–218
8. Tai T, Kawashima I, Ozawa H, Kotani M, Ogura K (1997) Cell type-specific expression of ganglioside antigens in the central nervous system. Pure Appl Chem 69:1903–1910
9. Galfre G, Howe S, Milstein C, Butcher GW, Howard JC (1977) Antibodies to major histocompatibility antigens produced by hybrid cell lines. Nature 266:550–552
10. Gefter ML, Margulies DH, Scharff MD (1977) A simple method for polyethylene glycol-promoted hybridization of mouse myeloma cells. Somatic Cell Genet 3:231–236
11. Kotani M, Ozawa H, Kawashima I, Ando S, Tai T (1992) Generation of one set of monoclonal antibodies specific for a-pathway ganglio-series gangliosides. Biochim Biopys Acta 1117:97–103

Chapter 14

Chemical Synthesis of Gangliosides

Akihiro Imamura and Makoto Kiso

Abstract

Various methods for the chemical synthesis of gangliosides have been investigated to date and numerous natural gangliosides and their structural analogues have been synthesized during the past three decades. Key technologies in the synthesis of gangliosides include α-selective sialylation and introduction of the ceramide moiety into the oligosaccharide chain. This chapter introduces two major strategies for ganglioside synthesis—the most commonly used strategy and the recently developed glucosylceramide cassette approach. Synthetic procedures for selected reactions are also presented.

Key words Ganglioside, Glycolipid, Total synthesis, Sialylation, Glycosylation

1 Introduction

Gangliosides, which are glycosphingolipids containing one or more sialic acid residues, are present in all animal cell membranes and participate in many biological processes [1–5]. Gangliosides are found in especially high abundance in the nervous system, and several neuronal gangliosides have been linked to neurological disorders such as Alzheimer's disease and Parkinson's disease [6]. Gangliosides have received increasing attention from not only biologists but also synthetic chemists due to the abundance of research into their physiological and pathological implications. Natural gangliosides pose a number of synthetic challenges, and many synthetic organic chemists have contributed to the development of various methods for their total synthesis. These challenges include regioselective and stereoselective sialylation as well as introduction of the ceramide moiety into the oligosaccharide chain. Although reliable methods for α-sialylation have been developed and used in the synthesis of natural gangliosides and their analogues [7, 8], linking the flexible ceramide moiety to a large glycan remains a challenging undertaking. The typical procedure for connecting the lipid and glycan units is first to prepare the entire oligosaccharide framework and then to link it either to 2-azide sphingosine, which

Sandro Sonnino and Alessandro Prinetti (eds.), *Gangliosides: Methods and Protocols*, Methods in Molecular Biology, vol. 1804, https://doi.org/10.1007/978-1-4939-8552-4_14,

© Springer Science+Business Media, LLC, part of Springer Nature 2018

serves as a ceramide precursor, or to the ceramide moiety directly. This approach has proved effective for small gangliosides such as GM4 and GM3; for the synthesis of complex gangliosides, however, the oligosaccharide donor generally couples to the lipid acceptor in low yield. To overcome these synthetic challenges, another method was recently developed called as the glucosyl ceramide (GlcCer) cassette approach [9, 10]. This approach involves the coupling of glucose and ceramide (forming GlcCer) early in the synthesis, and has been used for the efficient total synthesis of a series of natural gangliosides in satisfactory overall yields.

1.1 Overview of the Chemical Synthesis of Gangliosides

Figure 1 shows a typical synthetic sequence for natural gangliosides. In nearly all ganglioside syntheses reported to date, α-sialylation is performed at an early stage due to its poor stereoselectivity. Various strategies for α-selective sialylation have been developed, including achieving stereocontrol by the nitrile solvent effect, varying the nature of the leaving group or promoter, and modifying the structure of sialyl donors at the C-1, C-3, and C-5 positions. In particular, it has been found in the past decade that changing the C-5 position to an amide, imide, carbamate,

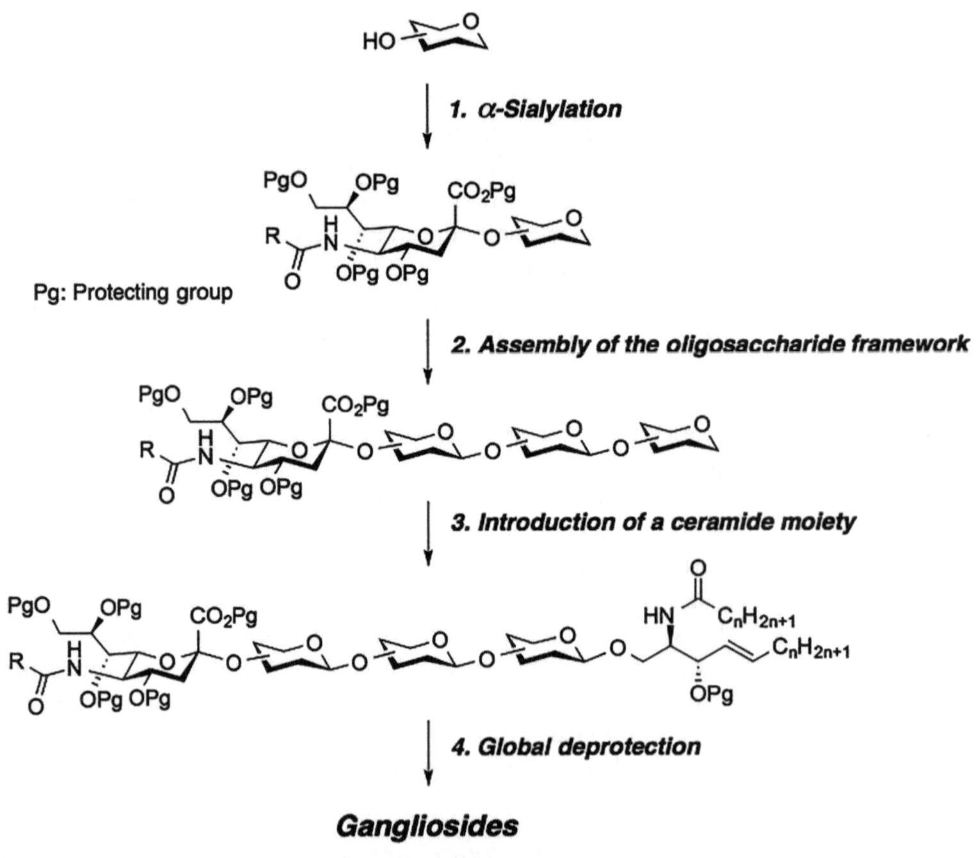

Fig. 1 Typical synthetic sequence for natural gangliosides

azide, isothiocyanate, ureido, or 5-*N*,4-*O*-oxazolidinone is effective in improving α-selectivity [11]. Yet stereoselective sialylation remains a challenge and currently requires laborious column chromatographic separation of glycosylated products, including stereoisomers and other by-products, to obtain the pure α-sialosides. To overcome this difficulty, a unique purification method for α-sialosides has been developed in which 1,5-lactamization is used to separate α-sialosides from mixtures of glycosylated products, allowing specific isolation of the desired α-sialoside [12, 13] (Fig. 2). After formation of the α-sialoside, the glycan units are constructed through a step-by-step process. The sequence of steps and the choice of protecting groups are critical for efficiently assembling the oligosaccharide framework. The final step in the conventional approach to the total synthesis of gangliosides involves the introduction of the ceramide moiety. Derivatives of 2-azide sphingosine are typically used as ceramide precursors to introduce the lipid moiety into the glycan unit, because direct substitution of ceramide generally gives unsatisfactory results with low yields for the glycosylation. However, since 2000, the GlcCer cassette approach has been utilized as a new method for introducing the lipid moiety [9, 10] (Fig. 3). This approach drastically improves the efficiency of the lipid–glycan coupling step and thus may become a widely used technique for ganglioside synthesis in the future.

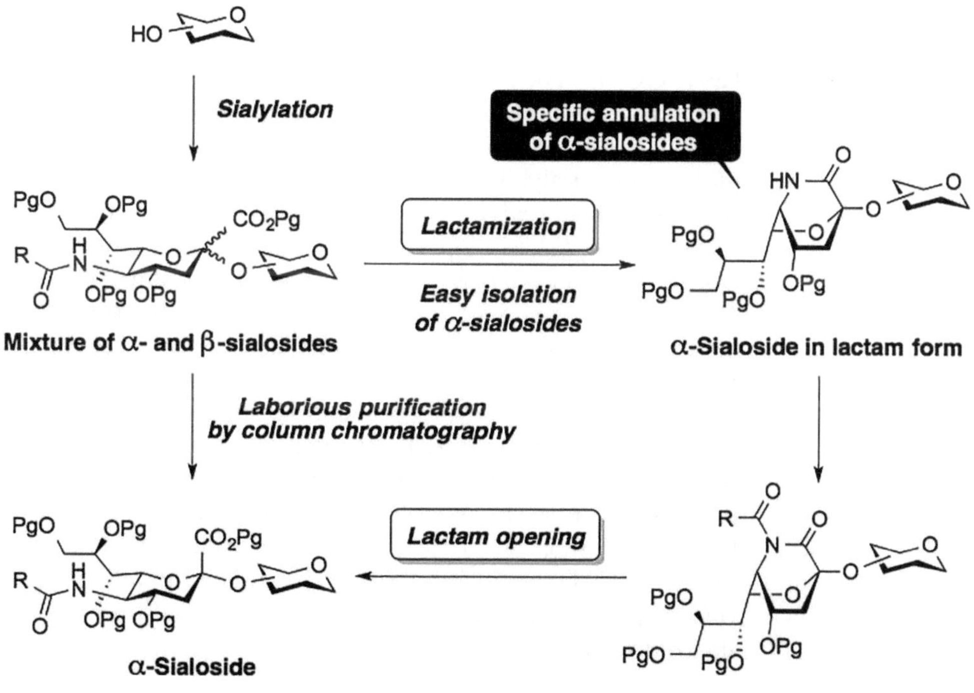

Fig. 2 Isolation of α-sialosides based on specific annulation of α-sialosides by 1,5-lactamization

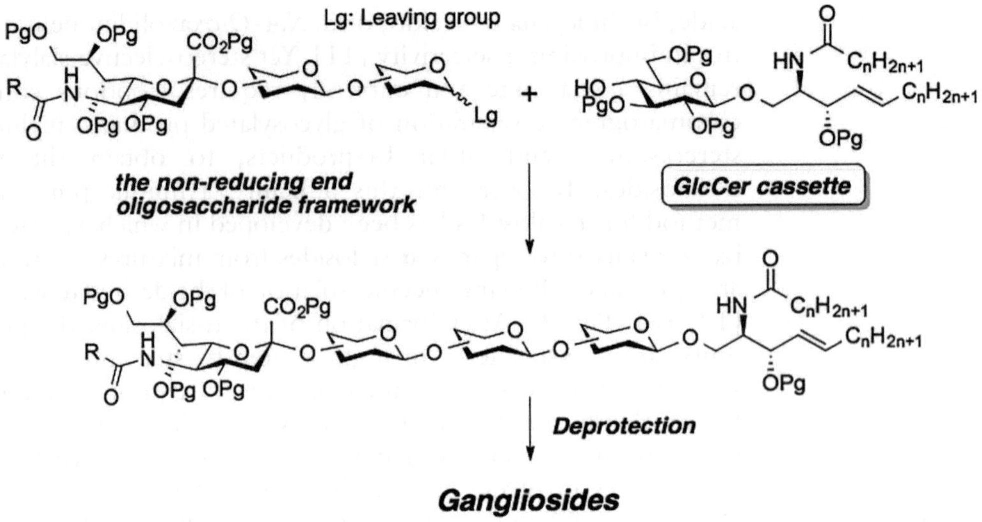

Fig. 3 Outline of the GlcCer cassette approach

1.2 Total Syntheses of Natural Gangliosides

Natural mammalian gangliosides that have been chemically synthesized are as follows: the lacto series (sialyl Lc4 [14], sialyl Lewis A [15], disialyl Lewis A [16]), the neolacto series (sialyl nLc4 [14], sialyl Lewis X [17, 18], VIM-2 [19]), the gala series (GM4 [20, 21], GD4 [to our knowledge, there is no publication], GT4 [22]), the hematoside series (GM3 [23, 24], GD3 [25, 26], GT3 [22]), the ganglio series (GM2 [27, 28], GM1 [27, 29, 30], GM1b [31, 32], GalNAc-GM1b [33, 34], GD1a [29, 30], GalNAc-GD1a [35], GD1c [36], GT1a [36], GD2 [37, 38], GD1b [39], GT1b [39], GQ1b [10, 39, 40], GM1α [41], GD1α [42, 43], GT1aα [44, 45], GQ1bα [46], GQ1β [47]), the lacto-ganglio hybrid (X1 [48], X2 [49]), the globo series (sialyl Gb5 [50, 51]), and the isoglobo series (sialyl iGb5 [52]). The structures of all these gangliosides arc illustrated in Fig. 4. Recently, a trend in the field of ganglioside synthesis appears to be a shift toward gangliosides isolated from echinoderms. Some of these gangliosides have not only unique and complex structures but also greater neuritogenic activity than that of the gangliosides isolated from vertebrates [53]. However, the conventional synthetic approaches are not always applicable to the synthesis of echinodermatous gangliosides possessing unique structures. Fundamental advances in the synthetic methodology required for the synthesis of these structurally complex gangliosides have enabled the total synthesis of echinodermatous gangliosides and their analogues including HLG-2 [54, 55], Hp-s6 [54], DSG-A [56], HPG-7 [57], LLG-3 [58, 59], GAA-7 [60, 61], PNG-2A [62], and GP3 [63], the structures of which are shown in Fig. 5.

a

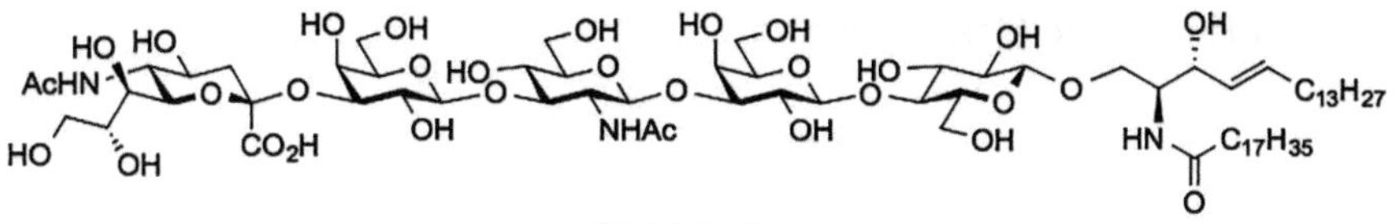

Sialyl Lc4

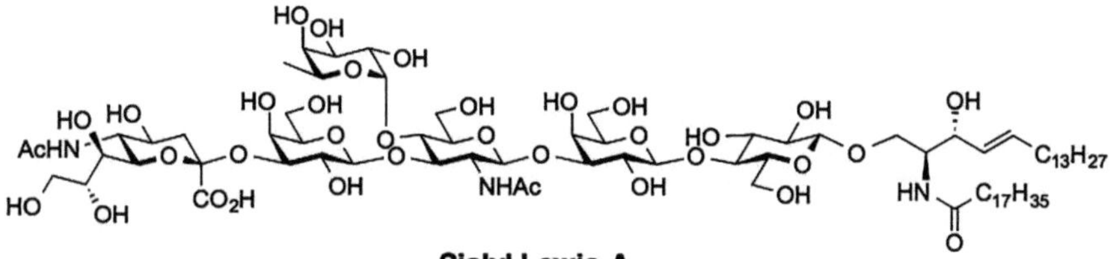

Sialyl Lewis A

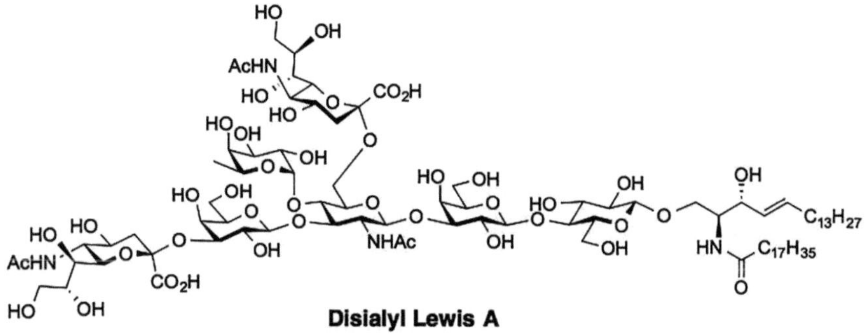

Disialyl Lewis A

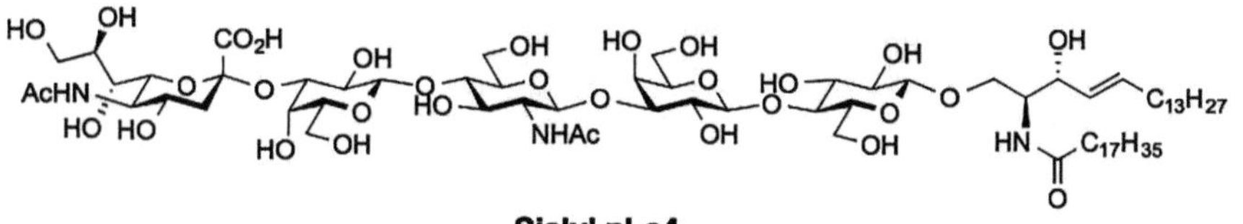

Sialyl nLc4

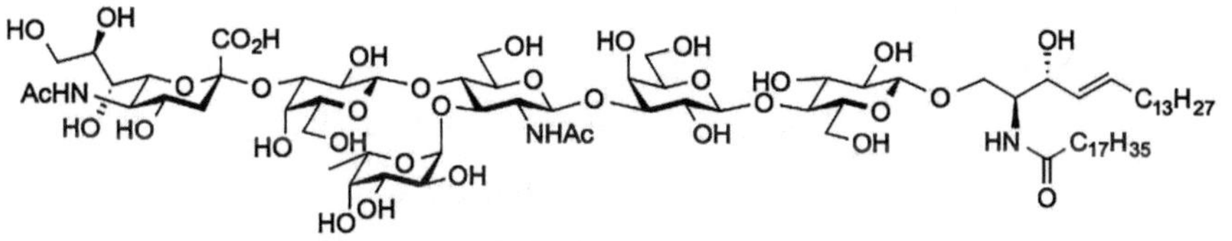

Sialyl Lewis X

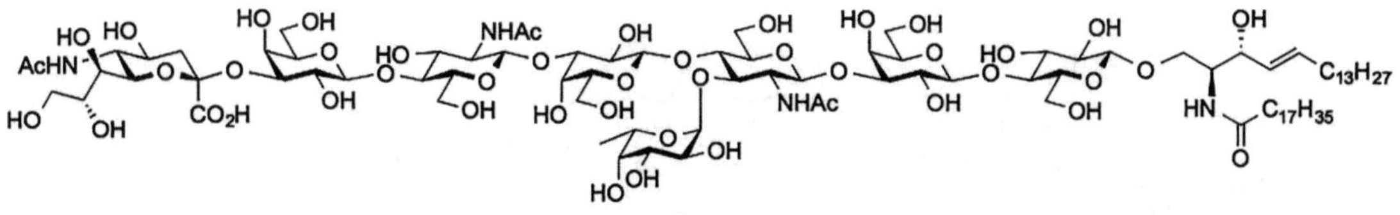

VIM-2

Fig. 4 (**a**) Natural mammalian gangliosides for which the total synthesis has been achieved via a chemical approach. (**b**) Natural mammalian gangliosides for which the total synthesis has been achieved via a chemical approach. (**c**) Natural mammalian gangliosides for which the total synthesis has been achieved via a chemical approach

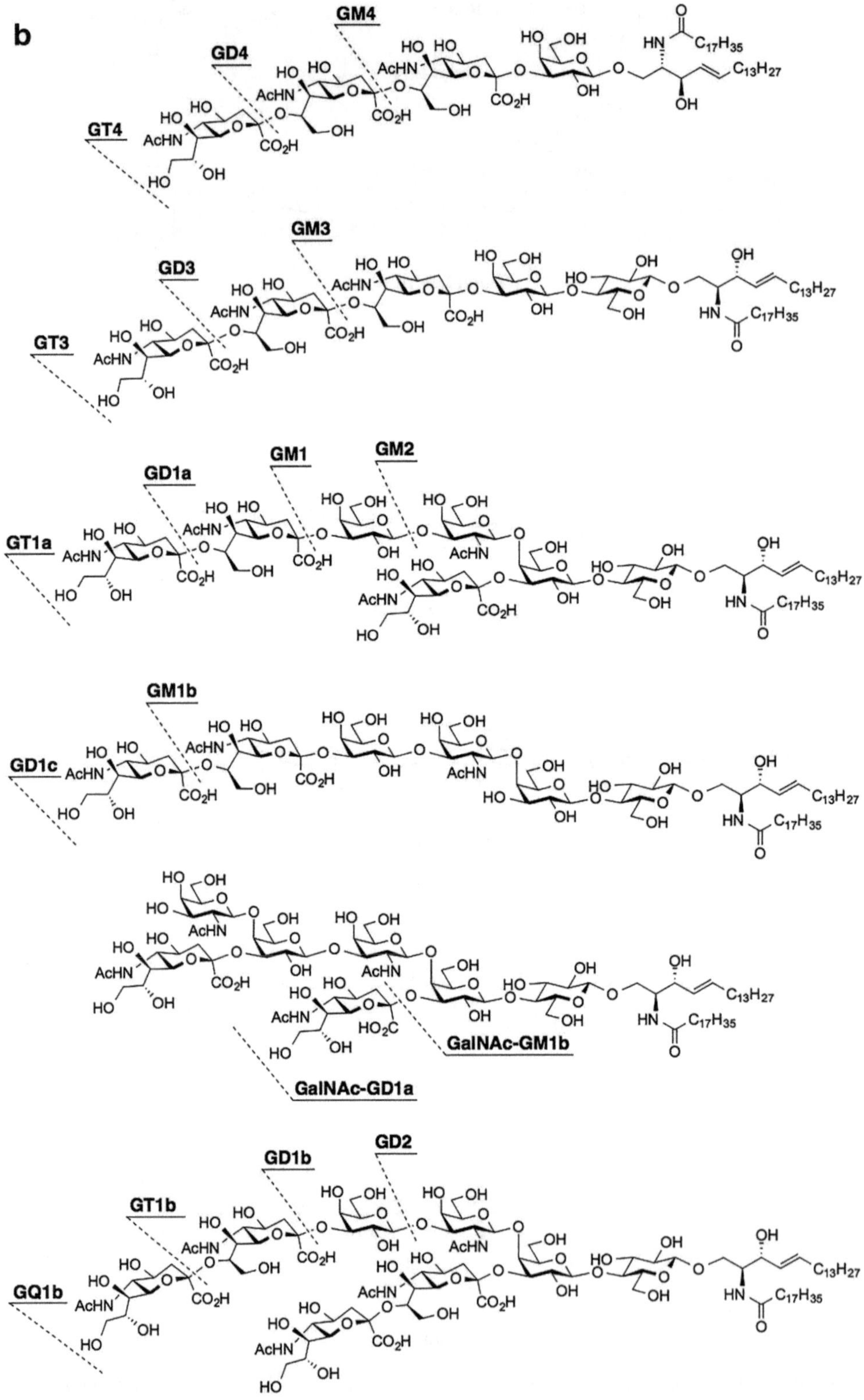

Fig. 4 (continued)

c

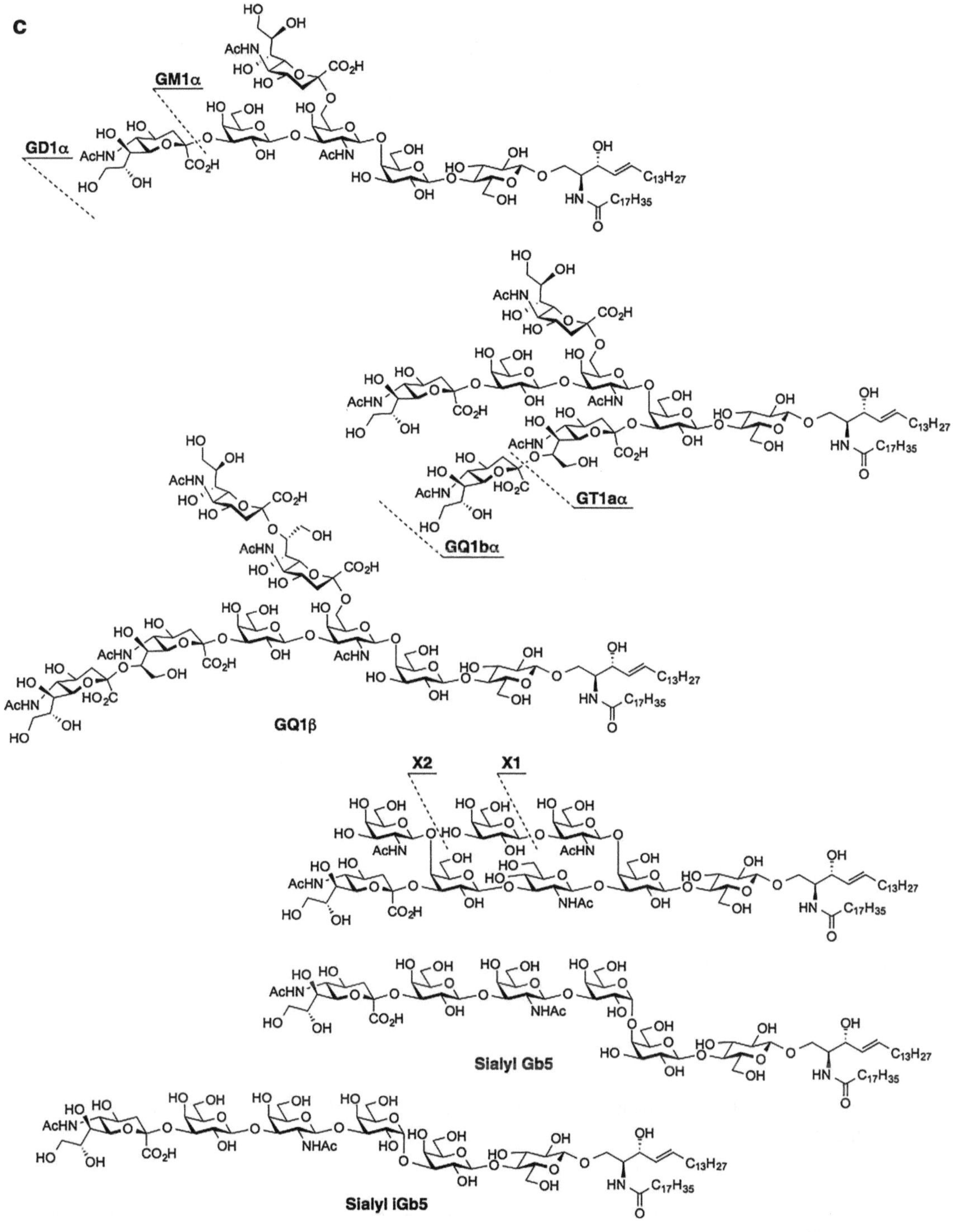

Fig. 4 (continued)

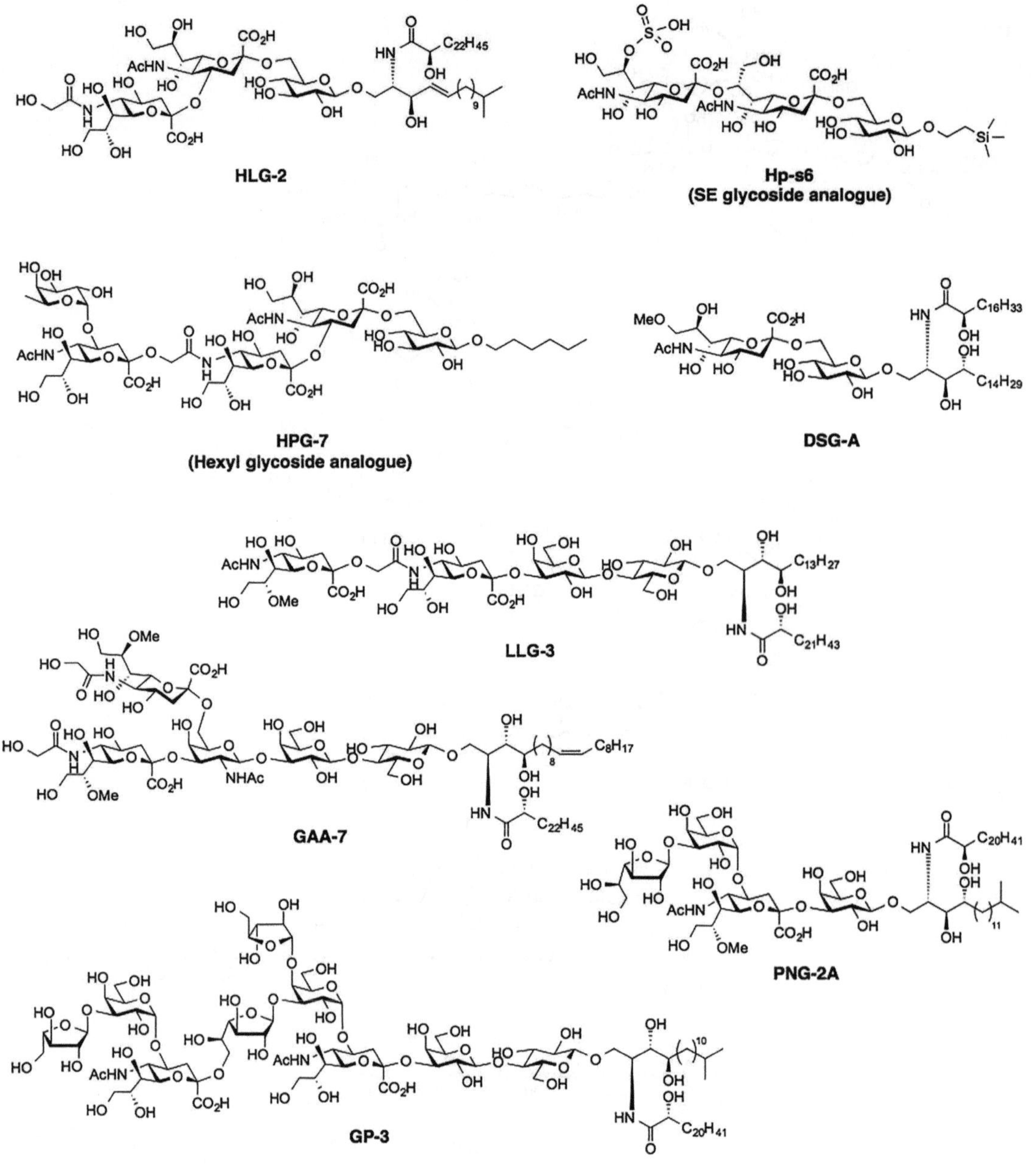

Fig. 5 Natural echinodermatous gangliosides (or their analogs) for which the total synthesis has been achieved via a chemical approach

2 Materials

2.1 General Materials for Chemical Reactions

1. Reagent-grade chemicals were purchased from commercial suppliers and used without further purification.
2. Anhydrous CH_2Cl_2 or CH_3CN (*see* **Note 1**).

3. TLC plates (silica gel $60F_{254}$ on glass plate) for monitoring reactions (*see* **Note 2**). TLC analysis was performed on Merck TLC plates or Merck High Performance TLC plates (HPTLC) and compounds were detected by either exposure to UV light (254 nm) or staining (e.g., 10% H_2SO_4 in ethanol, phosphomolybdic acid in ethanol, *p*-anisaldehyde stain) followed by heating.
4. Silica gel for column chromatography to purify synthesized products (*see* **Note 3**). The quantity of silica gel was usually estimated as 100 to 200 times the weight of sample to be loaded onto the column. Solvent systems in chromatography are specified in as volume ratios.
5. Reactions were carried out in appropriate round-bottomed glass flasks under a positive pressure of argon (*see* **Note 4**), unless otherwise noted.
6. Drying agent (3/4 Å molecular sieves [3/4 Å MS], acid-washed molecular sieves [AW-300], Drierite® [anhydrous $CaSO_4$]). Molecular sieves and Drierite were dried at 300 °C for 2 h (for Molecular sieves), 210 °C for 1 h (for Drierite) (*see* **Note 5)** in a muffle furnace prior to use.
7. Celite®.
8. Structural characterization of products was usually performed by ^1H and ^{13}C NMR spectroscopy and high-resolution mass spectrometric analyses (e.g., ESI-TOF). NMR spectroscopy was performed on a standard 500 MHz or higher spectrometer in $CDCl_3$ (*see* **Note 6**) for protected compounds or mixed solvent system (e.g., $CDCl_3/CD_3OD/D_2O$ = 5:5:1) for natural gangliosides.
9. Evaporation and concentration were carried out under vacuum.

2.2 α-Selective Sialylation of a Galactose Acceptor with an N-Troc-Sialyl Donor

1. Glycosyl donor.
2. Glycosyl acceptor.
3. CH_3CH_2CN (*see* **Note 7**).
4. Molecular sieves: 3 Å MS.
5. Argon (*see* **Note 4**).
6. *N*-iodosuccinimide (NIS).
7. Trifluoromethanesulfonic acid (TfOH).
8. $NaHCO_3$.
9. EtOAc.
10. $Na_2S_2O_3$.
11. Brine (sat. aq. NaCl).
12. Na_2SO_4 (*see* **Note 8**).

13. 4:1 toluene–EtOAc.
14. *n*-hexane–EtOAc.

2.3 Synthesis of a Sialoside Based on the Lactamization Strategy

1. TLC plates for monitoring reactions.
2. Silica gel for column chromatography to purify synthesized products.
3. Glycosyl donor.
4. Glycosyl acceptor.
5. CH_3CH_2CN.
6. Molecular sieves: 3 Å MS.
7. IBr (1 M solution in CH_2Cl_2).
8. AgOTf.
9. HPTLC plates for monitoring reactions.
10. 1:3 *n*-hexane–EtOAc or 30:1 $CHCl_3$–MeOH.
11. $NaHCO_3$.
12. $CHCl_3$.
13. $Na_2S_2O_3$.
14. Na_2SO_4.
15. 1:1 $CHCl_3$–MeOH.
16. EtOH.
17. Argon.
18. 1,4-diazabicyclo[2.2.2]octane (DABCO).
19. 10:1 $CHCl_3$–MeOH for monitoring reactions.
20. $CHCl_3$.
21. 2 M HCl, H_2O, sat. aq. $NaHCO_3$, and brine.
22. 50:1 → 40:1 $CHCl_3$–MeOH
23. DMF.
24. 1,8-Diazabicyclo[5.4.0]undec-7-ene (DBU).
25. EtOAc or 20:1 $CHCl_3$–MeOH).
26. AcOH.
27. Dry toluene.
28. 3:1 → 1:1 n-hexane–EtOAc.
29. Di-*tert*-butyl dicarbonate (Boc_2O).
30. 4-dimethylaminopyridine (DMAP).
31. MeCN.
32. 1:1 *n*-hexane–EtOAc, 1:3 *n*-hexane–EtOAc.
33. 2:1 → 1:1 *n*-hexane–EtOAc).
34. MeOH.

35. NaOMe (28% solution in MeOH).
36. 10:1 $CHCl_3$–MeOH.
37. Muromac (H^+) resin.
38. 100:1 → 10:1 $CHCl_3$–MeOH.

2.4 Assembly of a Ganglioside Framework via the GlcCer Cassette Approach

1. TLC plates for monitoring reactions.
2. AW-300 (*see* **Note 9**).
3. Oligosaccharide donor.
4. GlcCer cassette acceptor.
5. Argon.
6. Trimethylsilyl trifluoromethanesulfonate (TMSOTf).
7. 4:1 toluene–MeOH.
8. $NaHCO_3$.
9. $CHCl_3$.
10. Na_2SO_4.
11. 8:1 → 7:1 toluene–MeOH.

3 Methods

3.1 α-Selective Sialylation of a Galactose Acceptor with an N-Troc-Sialyl Donor (Fig. 6)

1. Glycosyl donor **1** (4.00 g, 5.58 mmol) and glycosyl acceptor **2** (1.74 g, 3.72 mmol) are dissolved in CH_3CH_2CN (47 mL) (*see* **Note 7**) and 3 Å MS (1.80 g) are added to the solution.
2. The mixture is stirred for 1 h at room temperature under an argon atmosphere and then cooled to −50 °C (*see* **Note 10**).
3. *N*-Iodosuccinimide (NIS; 1.88 g, 8.37 mmol) and trifluoromethanesulfonic acid (TfOH; 74 μL, 0.840 mmol) are added to the mixture.
4. After stirring for 30 min, the reaction mixture is quenched by adding sat. aq. $NaHCO_3$ and filtered through a pad of Celite that is then washed with EtOAc.
5. The filtrate and washings are combined, successively washed with sat. aq. $Na_2S_2O_3$ and brine (sat. aq. NaCl).

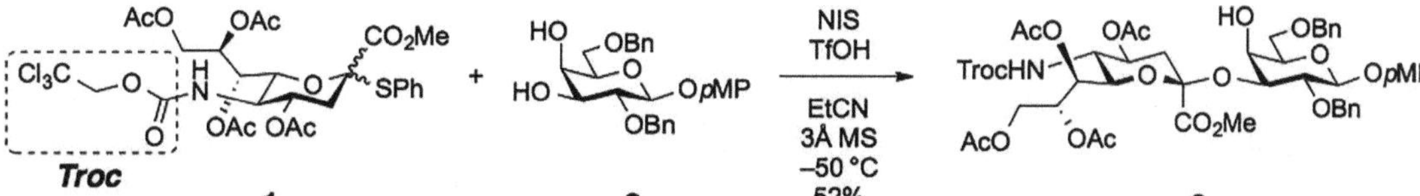

1 **2** **3**

Fig. 6 α-Selective sialylation of a galactose acceptor with an *N*-Troc-sialyl donor

6. The organic layer is dried over Na_2SO_4 (*see* **Note 9**), filtered through cotton that is then washed with EtOAc, and then concentrated.
7. Flash silica gel column chromatography (4:1 toluene–EtOAc) of the resulting residue followed by crystallization from *n*-hexane–EtOAc gave α-sialylated product **3** (2.05 g, 52%) as a white crystal.

3.2 Synthesis of a Sialoside Based on the Lactamization Strategy (Fig. 7)

1. Glycosyl donor **4** (124 mg, 0.177 mmol) and glycosyl acceptor **5** (55 mg, 0.118 mmol) are placed in a pear-shaped flask and dissolved in CH_3CH_2CN (1.5 mL), of which solution is then transferred into a two-neck round bottom flask containing 3 Å MS (150 mg) by using a cannula and the suspension was stirred for 30 min at room temperature.
2. The mixture is cooled to −80 °C and stirred for an additional 15 min.
3. Subsequently, IBr (1 M solution in CH_2Cl_2, 0.26 mL, 0.265 mmol) and AgOTf (20 mg, 79.6 μmol) are added to the mixture.
4. After stirring for 1 h at −80 °C while the reaction is monitored by HPTLC (1:3 *n*-hexane/EtOAc or 30:1 $CHCl_3$–MeOH, developed twice), an additional portion of AgOTf (8 mg, 31.8 μmol) is added to the mixture at −80 °C.
5. After stirring for 2 h in total, the reaction mixture is quenched with sat. aq. $NaHCO_3$ and then filtered through a pad of Celite that is then washed with $CHCl_3$.
6. The combined filtrate and washings are washed with sat. aq. $Na_2S_2O_3$, dried over Na_2SO_4, and concentrated.

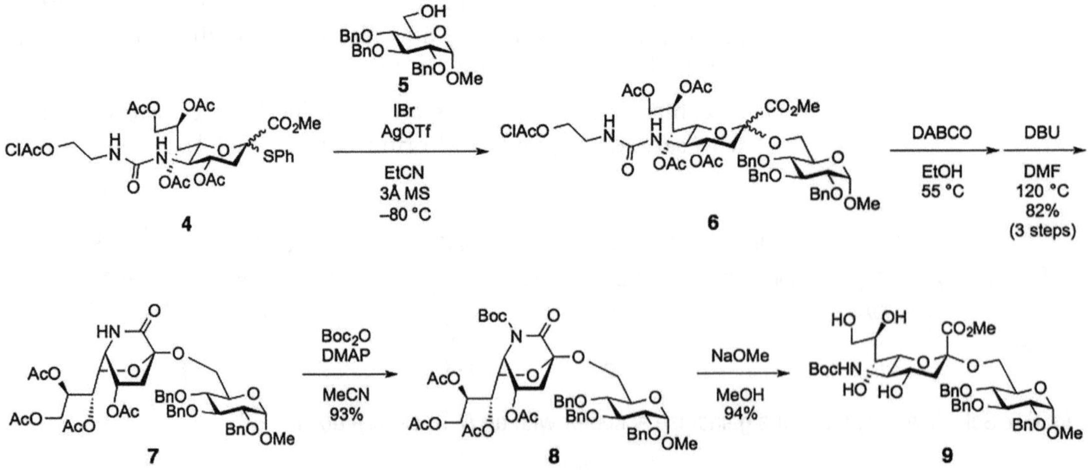

Fig. 7 Synthesis of a sialoside via the lactamization strategy

7. The resulting residue is roughly purified by gel filtration column chromatography (LH-20, 1:1 $CHCl_3$–MeOH) to give sialylated products **6** (α, β-mixture).
8. Compound **6** is dried under high vacuum for 1.5 h and then dissolved in EtOH (5.9 mL) under an argon atmosphere.
9. 1,4-Diazabicyclo[2.2.2]octane (DABCO; 198 mg, 1.77 mmol) is added to the solution at room temperature.
10. After stirring for 2 h at 55 °C while the reaction is monitored by TLC (10:1 $CHCl_3$–MeOH), EtOH is evaporated from the reaction mixture and then the residue was diluted with $CHCl_3$.
11. The organic layer is successively washed with 2 M HCl, H_2O, sat. aq. $NaHCO_3$, and brine; dried over Na_2SO_4; and concentrated.
12. The resulting residue is roughly purified by flash silica gel column chromatography (50:1 → 40:1 $CHCl_3$–MeOH) to give the corresponding alcohol compound.
13. The residue is dissolved in DMF (11.8 mL).
14. 1,8-Diazabicyclo[5.4.0]undec-7-ene (DBU; 26 μL, 0.177 mmol) is added to the solution at room temperature.
15. After stirring for 4.5 h at 120 °C while the reaction is monitored by TLC (EtOAc or 20:1 $CHCl_3$–MeOH), the reaction mixture is cooled to room temperature.
16. The reaction is quenched by the addition of AcOH at room temperature and stirred for a few minutes.
17. The mixture is coevaporated with dry toluene.
18. The resulting residue is purified by flash silica gel column chromatography (3:1 → 1:1 *n*-hexane–EtOAc) to give lactamized α-sialoside **7** (90 mg, 82% over three steps).
19. Di-*tert*-butyl dicarbonate (Boc_2O; 11 μL, 48.5 μmol) and a catalytic amount of 4-dimethylaminopyridine (DMAP) are added to a solution of **7** (28 mg, 32.3 μmol) in MeCN (1.3 mL) at room temperature.
20. After stirring for 3 h at room temperature while the reaction is monitored by TLC (1:1 *n*-hexane–EtOAc, 1:3 *n*-hexane–EtOAc), the solvent is removed by evaporation.
21. The resulting residue is purified by silica gel column chromatography (2:1 → 1:1 *n*-hexane–EtOAc) to give **8** (29 mg, 93%).
22. Finally, to a solution of **8** (47 mg, 48.8 μmol) in MeOH (4.9 mL) is added NaOMe (28% solution in MeOH, 26 mg, 488 μmol) at room temperature.
23. After stirring for 1 h at room temperature while the reaction is monitored by TLC (10:1 $CHCl_3$–MeOH), the reaction is neu-

tralized with Muromac (H^+) resin and filtered through cotton that is then washed with MeOH.

24. The filtrate and washings are concentrated and the resulting residue was purified by silica gel column chromatography (PSQ-60B, 100:1 → 10:1 $CHCl_3$–MeOH) to give **9** (38 mg, 94%).

3.3 Assembly of a Ganglioside Framework via the GlcCer Cassette Approach (Fig. 8)

1. AW-300 (200 mg) (*see* **Note 10**) is added to a solution of oligosaccharide donor **10** (146 mg, 47.4 μmol) and GlcCer cassette acceptor **11** (85 mg, 72.2 μmol) in CH_2Cl_2 (1.2 mL) at room temperature.
2. The mixture is stirred for 2 h at room temperature under an argon atmosphere and then cooled to 0 °C.
3. Trimethylsilyl trifluoromethanesulfonate (TMSOTf; 1.0 μL, 5.52 μmol) is added to the mixture.
4. After stirring for 1 h at 0 °C while the reaction is monitored by TLC (4:1 toluene/MeOH), the reaction mixture is quenched by adding sat. aq. $NaHCO_3$ and filtered through a pad of Celite that is then washed with $CHCl_3$.
5. The combined filtrate and washings are washed with brine, dried over Na_2SO_4, and concentrated.
6. The resulting residue is purified by flash silica gel column chromatography (8:1 → 7:1 toluene–MeOH) to give **12** (177 mg, 91%).

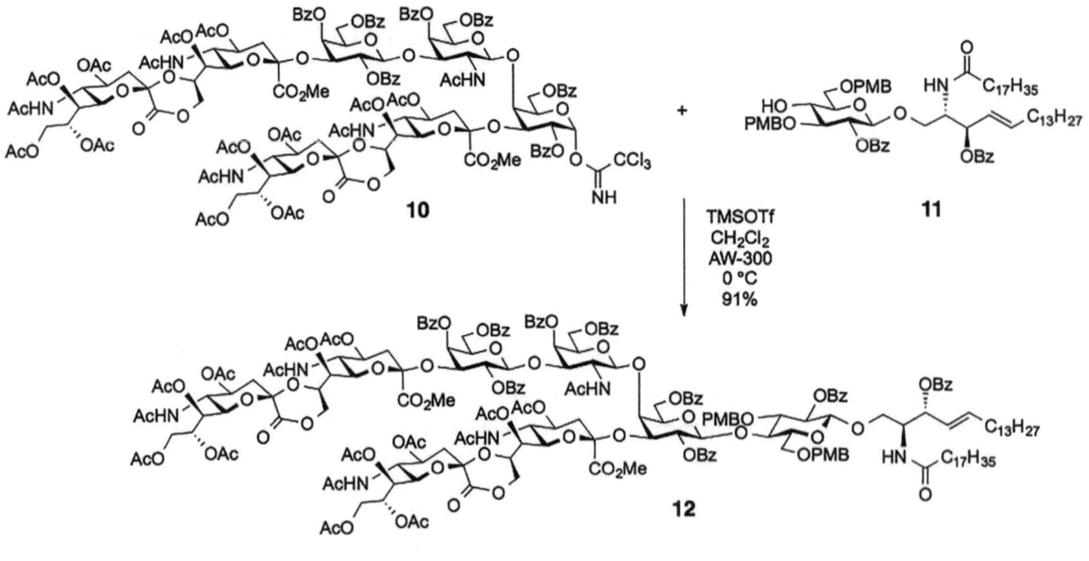

Fig. 8 Assembly of a ganglioside framework via the GlcCer cassette approach

4 Notes

1. Instead of CH_3CN, CH_3CH_2CN is used to low the reaction temperature.
2. Various solvent systems were used for thin layer chromatography (TLC) analysis. For protected carbohydrate molecules, combination of solvents such as ethyl acetate–*n*-hexane, ethyl acetate–toluene, and acetone–chloroform (or dichloromethane) was usually chosen. For compounds including sialic acid residues with higher polarity, the use of methanol–chloroform, methanol–toluene, and methanol–ethyl acetate systems often performs good separations.
3. When higher polar solvent systems like methanol–chloroform (1:3) are used as eluents to purify fully deprotected compounds, a porous fine silica gel, Iatrobeads, is often used.
4. Instead of an argon gas, nitrogen is also available as an inert gas.
5. Drierite should be heated at exactly 210 °C (425 °F) for 1 h.
6. Signal peaks in ^{1}H NMR spectra are sometimes broadened. In such cases, the use of deuterated higher-polar solvents such as CD_3CN, acetone-d_6, and DMSO-d_6 is effective to avoid signal peak broadening. Also, increasing the measurement temperature can often improve the spectral resolution.
7. CH_3CH_2CN should be freshly distilled prior to use because commercially available ones sometimes show a basic property, which could impede the glycosylation reaction.
8. $MgSO_4$ is also used for drying the organic layer.
9. When the trichloroacetimidate group is used as the leaving group where a tiny amount of acid catalyst is added, either acid-washed molecular sieves (AW-300) or 5 Å MS should be chosen because 3/4 Å MS shows a slightly basic property, which could impede the glycosylation reaction.
10. Constant temperature bath with magnetic stirrer is used to maintain the reaction temperature.

References

1. Sonnino S, Mauri L, Chigorno V, Prinetti A (2006) Gangliosides as components of lipid membrane domains. Glycobiology 17:1R–13R
2. Todeschini AR, Hakomori S (2008) Functional role of glycosphingolipids and gangliosides in control of cell adhesion, motility, and growth, through glycosynaptic microdomains. Biochim Biophys Acta 1780:421–433
3. Lopez PHH, Schnaar RL (2009) Gangliosides in cell recognition and membrane protein regulation. Curr Opin Struct Biol 60:537–544
4. Yu RK, Tsai Y-T, Ariga T, Yanagisawa M (2011) Structures, biosynthesis, and functions of gangliosides-an overview. J Oleo Sci 60:537–544
5. Lloyd KO, Furukawa K (1998) Biosynthesis and functions of gangliosides: recent advances. Glycoconj J 15:627–636

6. Schnaar RL, Gerardy-Schahn R, Hildebrandt H (2014) Sialic acids in the brain: gangliosides and polysialic acid in nervous system development, stability, disease, and regeneration. Physiol Rev 94:461–518
7. Ando H, Kiso M (2008) Selective α-Sialylation. In: Fraser-Reid BO, Tatsuta K, Thiem J (eds) Glycoscience, vol 2, 2nd edn. Springer-Verlag, Berlin Heiderberg, pp 1313–1359
8. Hanashima S (2011) Recent strategies for stereoselective sialylation and their application to the synthesis of oligosaccharides. Trends Glycosci Glycotech 23:111–121
9. Sakamoto H, Nakamura S, Tsuda T, Hashimoto S (2000) Chemoselective glycosidation strategy based on glycosyl donors and acceptors carrying phosphorus-containing leaving groups: a convergent synthesis of ganglioside GM3. Tetrahedron Lett 41:7691–7695
10. Imamura A, Ando H, Ishida H, Kiso M (2009) Ganglioside GQ1b: efficient total synthesis and the expansion to synthetic derivatives to elucidate its biological roles. J Org Chem 74:3009–3023
11. De Meo C, Priyadarshani U (2008) C-5 modifications in *N*-acetyl-neuraminic acid: scope and limitations. Carbohydr Res 343:1540–1552
12. Tanaka H-N, Ando H, Ishida H, Kiso M, Ishihara H, Koketsu M (2009) Synthetic study onα(2→8)-linked oligosialic acid employing 1,5-lactamization as a key step. Tetrahedron Lett 50:4478–4481
13. Tanase M, Imamura A, Ando H, Ishida H, Kiso M (2016) A 5-ureido-modified sialyl donor: a tool for the synthesis of α-sialosides. Org Lett 18:1454–1457
14. Kameyama A, Ishida H, Kiso M, Hasegawa A (1990) Stereoselective synthesis of sialyl-lactotetraosylceramide and sialylneolactotetraosylceramide. Carbohydr Res 200:269–285. (for Sialyl Lc4 and Sialyl nLc4)
15. Kameyama A, Ishida H, Kiso M, Hasegawa A (1994) Total synthesis of tumor-associated ganglioside, sialyl Le^a. J Carbohydr Chem 13:641–654. (for Sialyl Lewis A)
16. Ando T, Ishida H, Kiso M (2003) First total synthesis of α-(2→3)/α-(2→6)-disialyl lactotetraosyl ceramide and disialyl Lewis a ganglioside as cancer-associated carbohydrate antigens. Carbohydr Res 338:503–514. (for Disialyl Lewis A)
17. Kameyama A, Ishida H, Kiso M, Hasegawa A (1991) Total synthesis of sialyl Lewis X. Carbohydr Res 209:c1–c4. (for Sialyl Lewis X)
18. Kameyama A, Ishida H, Kiso M, Hasegawa A (1991) Total synthesis of tumor-associated ganglioside, sialyl Lewis X. J Carbohydr Chem 10:549–560. (for Sialyl Lewis X)
19. Ehara T, Kameyama A, Yamada Y, Ishida H, Kiso M, Hasegawa A (1996) Total synthesis of VIM-2 ganglioside isolated from human chronic myelogenous leukemia cells. Carbohydr Res 281:237–252
20. Numata M, Sugimoto M, Koike K, Ogawa T (1987) Total synthesis of sialosylcerebroside, GM_4. Carbohydr Res 163:209–225. (for GM4)
21. Murase T, Kameyama A, Kartha KPR, Ishida H, Kiso M, Hasegawa A (1989) A facile, regio and stereoselective synthesis of ganglioside GM4 and its position isomer. J Carbohydr Chem 8:265–283. (for GM4)
22. Ando H, Ishida H, Kiso M, Hasegawa A (1997) A synthetic approach to the c-series gangliosides containing sialyl-α-(2→8) sialyl-α-(2→8)sialic acid: synthesis of ganglioside GT4, α-(2→6)GT4 and GT3. Carbohydr Res 300:207–217. (for GT4 and GT3)
23. Sugimoto M, Ogawa T (1985) Synthesis of a hematoside (GM3-ganglioside) and a stereoisomer. Glycoconj J 2:5–9. (for GM3)
24. Murase T, Ishida H, Kiso M, Hasegawa A (1989) A facile, regio- and stereo-selective synthesis of ganglioside GM3. Carbohydr Res 188:71–80. (for GM3)
25. Ito Y, Numata M, Sugimoto M, Ogawa T (1989) Highly stereoselective synthesis of ganglioside GD3. J Am Chem Soc 111:8508–8510. (for GD3)
26. Ishida H-K, Ohta Y, Tsukada Y, Kiso M, Hasegawa A (1993) A synthetic approach to polysialogangliosides containing α-sialyl-(2→8)-sialic acid: total synthesis of ganglioside GD3. Carbohydr Res 246:75–88. (for GD3)
27. Sugimoto M, Numata M, Koike K, Nakahara Y, Ogawa T (1986) Total synthesis of gangliosides GM_1 and GM_2. Carbohydr Res 156:c1–c5. (for GM1 and GM2)
28. Hasegawa A, Nagahama T, Ohki H, Kiso M (1989) A facile total synthesis of ganglioside GM2. J Carbohydr Chem 11:699–714. (for GM2)
29. Hasegawa A, Nagahama T, Kiso M (1992) A facile, systematic synthesis of ganglio-series gangliosides: Total synthesis of gangliosides GM1 and GD1a. Carbohydr Res 235:C13–C17. (for GM1 and GD1a)
30. Hasegawa A, Ishida H-K, Nagahama T, Kiso M (1993) Total synthesis of gangliosides GM1

and GD1a. J Carbohydr Chem 12:703–718. (for GM1 and GD1a)

31. Sugimoto M, Fujikura K, Nunomura S, Horisaki T, Ito Y, Ogawa T (1990) A stereocontrolled total synthesis of a *ganglio*-ganglioside GM1b, IV[3]NeuAcαGgOse$_4$Cer. Tetrahedron Lett 31:385–388. (for GM1b)
32. Prabhanjan H, Ishida H, Kiso M, Hasegawa A (1991) A facile total synthesis of ganglioside GM1b and its positional analog. Carbohydr Res 211:c1–c5. 220:127–143 (for GM1b)
33. Sugimoto M, Fujikura K, Nunomura S, Ito Y, Ogawa T (1990) Total synthesis of an extended *ganglio*-ganglioside, IV[4]GalNAcβGM1b. Tetrahedron Lett 31:1435–1438. (for GalNAc-GM1b)
34. Konishi M, Imamura A, Fujikawa K, Ando H, Ishida H, Kiso M (2013) Extending the glucosyl ceramide cassette approach: application in the total synthesis of ganglioside GalNAc-GM1b. Molecules 18:15153–15181. (for GalNAc-GM1b)
35. Fujikawa K, Nakashima S, Konishi M, Fuse T, Komura N, Ando T, Ando H, Yuki N, Ishida H, Kiso M (2011) The first total synthesis of ganglioside GalNAc-GD1a, a target molecule for autoantibodies in Guillain–Barré syndrome. Chem Eur J 17:5641–5651. (for GalNAc-GD1a)
36. Ishida H-K, Ando H, Ito H, Ishida H, Kiso M, Hasegawa A (1997) Total synthesis of gangliosides GD1c and GT1a. J Carbohydr Chem 16:413–428. (for GD1c and GT1a)
37. Matsuzaki Y, Nunomura S, Ito Y, Sugimoto M, Nakahara Y, Ogawa T (1993) Stereocontrolled synthesis of GD$_2$. Carbohydr Res 242:C1–C6. (for GD2)
38. Hasegawa A, Ishida H-K, Isogai Y, Ishida H, Kiso M (1993) Total synthesis of ganglioside GD2. J Carbohydr Chem 12:1217–1222. (for GD2)
39. Ishida H-K, Ishida H, Kiso M, Hasegawa A (1994) Total synthesis of gangliosides GQ1b and the related polysialogangliosides. Tetrahedron Asymmetry 5:2493–2512. (for GD1b, GT1b and GQ1b)
40. Ishida H-K, Ishida H, Kiso M, Hasegawa A (1994) Total synthesis of gangliosides GQ1b. Carbohydr Res 260:C1–C6. (for GQ1b)
41. Hotta K, Komba S, Ishida H, Kiso M, Hasegawa A (1994) Synthesis of α-series ganglioside GM1α. J Carbohydr Chem 13:665–677. (for GM1α)
42. Prabhanjan H, Kiso M, Hasegawa A (1991) Total synthesis of a disialoganglioside GD1α. Carbohydr Res 222:c1–c4. (for GD1α)
43. Prabhanjan H, Aoyama K, Kiso M, Hasegawa A (1992) Synthesis of disialoganglioside GD1α and its positional isomer. Carbohydr Res 233:87–99. (for GD1α)
44. Ito H, Ishida H, Kiso M, Hasegawa A (1997) First total synthesis of ganglioside GT1aα. Carbohydr Res 304:187–189. (1998) 306:581–585 (for GT1aα)
45. Ito H, Ishida H, Waki H, Ando S, Kiso M (1999) Total synthesis of a cholinergic neuron-specific ganglioside GT1aα: a high affinity ligand for myelin-associated glycoprotein (MAG). Glycoconj J 16:585–598. (for GT1aα)
46. Hotta K, Ishida H, Kiso M, Hasegawa A (1995) First total synthesis of a cholinergic neuron-specific ganglioside GQ1bα. J Carbohydr Chem 14:491–506. (for GQ1bα)
47. Hotta K, Kawase T, Ishida H, Kiso M, Hasegawa A (1995) A total synthesis of β-series ganglioside GQ1β. J Carbohydr Chem 14:961–975. (for GQ1β)
48. Nakashima S, Ando H, Imamura A, Yuki N, Ishida H, Kiso M (2011) A first total synthesis of a hybrid-type ganglioside associated with amyotrophic lateral sclerosis-like disorder. Chem Eur J 17:588–597. (for X1)
49. Nakashima S, Ando H, Saito R, Tamai H, Ishida H, Kiso M (2012) Efficiently synthesizing lacto-ganglio-series gangliosides by using a glucosyl ceramide cassette approach: the total synthesis of ganglioside X2. Chem Asian J 7:1041–1051. (for X2)
50. Lassaletta JM, Schmidt RR (1995) Versatile approach to the synthesis of globoside glycosphingolipids synthesis of sialyl-galactosylgloboside. Tetrahedron Lett 36:4209–4212. (for SGG)
51. Ishida H, Miyawaki R, Kiso M, Hasegawa A (1996) First total synthesis of sialyl globopentaosyl ceramide (V[3]Neu5AcGb$_5$Cer) and its positional isomer (V[6]Neu5AcGb$_5$Cer). J Carbohydr Chem 15:163–182. (for SGG)
52. Ishida H, Miyawaki R, Kiso M, Hasegawa A (1996) Systematic synthesis of α-sialyl-(2→3)- and -(2→6)-isoglobopentaosylceramides (V[3]Neu5AciGb$_5$Cer and V[6]Neu5AciGb$_5$Cer). Carbohydr Res 284:179–190. (for Sialy iGb5)
53. Higuchi R, Inagaki M, Yamada K, Miyamoto T (2007) Biologically active gangliosides from echinoderms. J Nat Med 61:367–370
54. Ando H, Koike Y, Koizumi S, Ishida H, Kiso M (2005) 1,5-Lactamized sialyl acceptors for various disialoside syntheses: novel method for the synthesis of glycan portions of Hp-s6 and HLG-2 gangliosides. Angew Chem Int Ed 44:6759–6763

55. Iwayama Y, Ando H, Ishida H, Kiso M (2009) A first total synthesis of ganglioside HLG-2. Chem Eur J 15:4637–4648
56. Wu Y-F, Tsai Y-F, Guo J-R, Yu C-P, Yu H-M, Liao C-C (2014) First total synthesis of ganglioside DSG-A possessing neuritogenic activity. Org Biomol Chem 12:9345–9349
57. Iwayama Y, Ando H, Tanaka H-N, Ishida H, Kiso M (2011) Synthesis of the glycan moiety of ganglioside HPG-7 with an unusual trimer of sialic acid as the inner sugar residue. Chem Commun 47:9726–9728
58. Tamai H, Ando H, Tanaka H-N, Hosoda-Yabe R, Yabe T, Ishida H, Kiso M (2011) The total synthesis of the neurogenic ganglioside LLG-3 isolated from the starfish *Linckia laevigata*. Angew Chem Int Ed 50:2330–2333
59. Yamagishi M, Hosoda-Yabe R, Tamai H, Konishi M, Imamura A, Ishida H, Yabe T, Ando H, Kiso M (2015) Structure-activity relationship study of the neuritogenic potential of the glycan of starfish ganglioside LLG-3. Mar Drugs 13:7250–7274
60. Tamai H, Imamura A, Ogawa J, Ando H, Ishida H, Kiso M (2015) First total synthesis of ganglioside GAA-7 from starfish *Asterias amurensis versicolor*. Eur J Org Chem 2015:5199–5211
61. Tamai H, Ando H, Ishida H, Kiso M (2012) First synthesis of a pentasaccharide moiety of ganglioside GAA-7 containing unusually modified sialic acids through the use of *N*-Troc-sialic acid derivative as a key unit. Org Lett 14:6342–6345
62. Goto K, Suzuki T, Tamai H, Ogawa J, Imamura A, Ando H, Ishida H, Kiso M (2015) Total synthesis and neuritogenic activity evaluation of ganglioside PNG-2A from the starfish *Protoreaster nodosus*. Asian J Org Chem 4:1160–1171
63. Goto K, Sawa M, Tamai H, Imamura A, Ando H, Ishida H, Kiso M (2016) The total synthesis of starfish ganglioside GP3 bearing a unique sialyl glycan architecture. Chem Eur J 22:8323–8331

Chapter 15

Radioactive Gangliosides for Biological Studies

Laura Mauri, Simona Prioni, Maria Grazia Ciampa, and Sandro Sonnino

Abstract

In this chapter, we present the preparation of gangliosides isotopically labelled with ^{3}H or ^{14}C. The methods do not present specific difficulties and can be used in any radiochemical laboratory. Some procedures can be applied to both gangliosides and neutral glycosphingolipids.

Key words Gangliosides, Sphingolipid, Radiolabeling, Tritium, ^{14}C

1 Introduction

Glycosphingolipids, in particular gangliosides, key components of membrane domains enriched in sphingolipids, cholesterol, and a specific subset of proteins involved in the process of signal transduction, are involved in the regulation of several cellular functions.

Radioactive glycosphingolipids are key derivatives in development of research methods aimed to better understand the physiological role played by these molecules at the cell surface. Indeed, many experiments on cultured cells have shown that exogenously administered gangliosides bind to cells, become components of the cell membrane, reside sometime in the plasma membranes, and enter the physiological sphingolipid metabolic and trafficking pathways [1–6].

The new analytical technologies allow for the detection of small amounts of radioactivity. This allow for the addition of a very low quantity of the appropriate exogenous ganglioside to the cell, avoiding any perturbation deriving from a large increase of the ganglioside content at the membrane level.

The simplest way to produce isotopically radioactive gangliosides is to administer ^{3}H or ^{14}C-labeled biosynthetic precursors such as sugars, fatty acids, or sphingosines, to animals or to cells in culture. Unfortunately, the resulting ganglioside specific radioactivity is low. Moreover, the radioactive ganglioside mixture

Sandro Sonnino and Alessandro Prinetti (eds.), *Gangliosides: Methods and Protocols*, Methods in Molecular Biology, vol. 1804, https://doi.org/10.1007/978-1-4939-8552-4_15, © Springer Science+Business Media, LLC, part of Springer Nature 2018

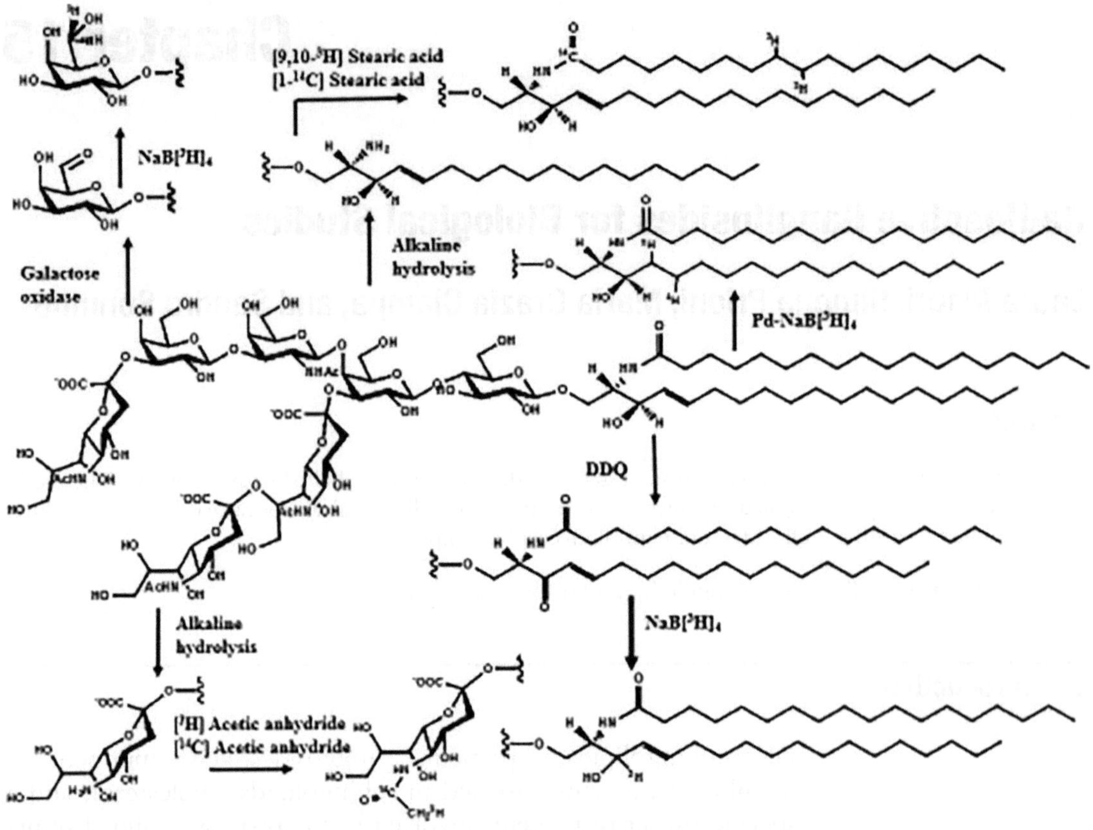

Fig. 1 Method scheme for the preparation of gangliosides containing [^{3}H] and [^{14}C] isotopically inserted into the sugar and lipid moieties

obtained in this way, must be purified and fractionated to give single homogeneous gangliosides.

Chemical and enzymatic procedures were developed for the radiolabeling of highly purified glycosphingolipids.

In this chapter we review the radiolabeling methods, sometimes with further modifications, capable of introducing radioactivity into specific position of both oligosaccharide and the ceramide moieties (Fig. 1) without changing the natural structure because the labeling is isotopic.

All procedures allow to synthesize several gangliosides (Table 1) with high specific radioactivity.

2 Materials

1. Galactose oxidase EC 1.1.3.9.
2. Triton X-100.
3. HPTLC silica gel $60F_{254}$.

Table 1
[^{3}H] and [^{14}C]gangliosides prepared in the laboratory of the Authors (S.S.)

	Treated ganglioside	Purified substrate for radiolabeling	Radioactive reagent	Reagent radioactivity		Ganglioside radioactivity	
Radioactive ganglioside	mg	mg		mCi	Ci/mmol	mCi	Ci/mmol
[6-^{3}H *(Gal-IV)*]GM1	20	15–18	NaB[^{3}H]H_4	100	5–10	8–15	1–2
[6-^{3}H *(Gal-IV)*]AGM1	20	15–18	NaB[^{3}H]H_4	100	5–10	8–15	1–2
[6-^{3}H *(Gal-IV)*]GD1b	20	13–15	NaB[^{3}H]H_4	100	5–10	8–15	1–2
[6-^{3}H *(GalNAc)*]GM2	20	7–8	NaB[^{3}H]H_4	100	5–10	4–8	1–2
[3-^{3}H *(C18sphingosine)*] GM3	50	20–25	NaB[^{3}H]H_4	100	5–10	2–4	1–2
[3-^{3}H *(C20sphingosine)*] GM3	50	20–25	NaB[^{3}H]H_4	100	5–10	4–8	1–2
[3-^{3}H *(C18sphingosine)*] GM2	50	20–25	NaB[^{3}H]H_4	100	5–10	3–6	1–2
[3-^{3}H *(C20sphingosine)*] GM2	50	20–25	NaB[^{3}H]H_4	100	5–10	3–6	1–2
[3-^{3}H *(C18sphingosine)*] GM1	50	20–25	NaB[^{3}H]H_4	100	5–10	3–6	1–2
[3-^{3}H *(C20sphingosine)*] GM1	50	20–25	NaB[^{3}H]H_4	100	5–10	3–6	1–2
[3-^{3}H *(C18sphingosine)*] Fuc-GM1	50	20–25	NaB[^{3}H]H_4	100	5–10	3–6	1–2
[3-^{3}H *(C20sphingosine)*] Fuc-GM1	50	20–25	NaB[^{3}H]H_4	100	5–10	3–6	1–2

(continued)

Table 1
(continued)

Radioactive ganglioside	Treated ganglioside	Purified substrate for radiolabeling	Radioactive reagent	Reagent radioactivity		Ganglioside radioactivity	
	mg	mg		mCi	Ci/mmol	mCi	Ci/mmol
[3-^{3}H *(C18sphingosine)*] GD1a	50	20–25	$NaB[^{3}H]H_4$	100	5–10	3–6	1–2
[3-^{3}H *(C20sphingosine)*] GD1a	50	20–25	$NaB[^{3}H]H_4$	100	5–10	3–6	1–2
[11-^{3}H *(Neu5Ac)*]GM3	15	4–5	$NaB[^{3}H]H_4$	25	4–10	7–10	2–5
[11-^{3}H *(Neu5Ac)*]GM2	15	4–5	$(C[^{3}H]H_3CO)^2O$	25	4–10	7–10	2–5
[11-^{3}H *(Neu5Ac)*]GM1	15	4–5	$(C[^{3}H]H_3CO)^2O$	25	4–10	7–10	2–5
[10-^{14}C *(Neu5Ac)*]GM1	15	4–5	$(CH_3[^{14}C]CO)^2O$	0.5	0.10–0.12	0.1–0.2	0.05–0.06
[1-^{14}C *(stearoyl)*]GM3	15	5–6	$CH_3(CH_2)16[^{14}C]$ COOH	0.25	0.05–0.06	0.2–0.3	0.05–0.06
[1-^{14}C *(stearoyl)*]GM2	15	5–6	$CH_3(CH_2)16[^{14}C]$ COOH	0.25	0.05–0.06	0.2–0.3	0.05–0.06
[1-^{14}C *(stearoyl)*]GM1	15	5–6	$CH_3(CH_2)16[^{14}C]$ COOH	0.25	0.05–0.06	0.2–0.3	0.05–0.06

The position of labeling, the amount of radioactivity used for labeling and the final amount of radioactive ganglioside are reported. All gangliosides were isotopically labelled.

4. Silica gel 60.
5. [^{3}H]$NaBH_4$.
6. $NaBH_4$.
7. Amberlite® IR120 H^+ form.
8. ^{3}H acetic anhydride.
9. Acetic anhydride.
10. PtO_2.
11. ^{3}H stearic acid.
12. Pentafluorophenol.
13. *N,N'*-dicyclohexylcarbodiimide (DCC).

2.1 Solutions

1. Chloroform–methanol, 2:1, by vol.
2. 5 mM EDTA-25 mM sodium phosphate buffer, pH 7.0.
3. Chloroform–methanol–2-propanol–water, 60:25:10:5, by vol.
4. Propanol-0.1 M NaOH, 7:3, by vol.
5. Chloroform–methanol–100 mM aqueous KCl– 5 M formic acid, 30:50:3:3, by vol.
6. 5 M sulfuric acid.
7. 4 M KOH.
8. Chloroform–methanol–2-propanol–formic acid, 30:40:10:6, by vol.
9. Dichloromethane–methanol–water, 60:35:8, by vol.
10. Chloroform–methanol–2-propanol–water, 60:25:10:5, by vol.
11. Chloroform–methanol–0.2% aqueous $CaCl_2$, 50:42:11, by vol.
12. Propanol–water, 7:3, by vol.
13. Acetonitrile–water, 1:1, by vol.
14. Acetonitrile–5 mM phosphate buffer, pH 7.0, 60:40, by vol.
15. Acetonitrile-5 mM phosphate buffer, pH 7.0, 50:50, by vol.
16. 1 M KOH in 1-propanol.
17. Chloroform–methanol–2-propanol–water, 60:30:5:5, by vol.
18. Chloroform–methanol–30 mM aqueous $CaCl_2$–100 mM aqueous KCl, 50:50:4:8, by vol.
19. Dichloromethane–methanol–water, 60:35:5 by vol.
20. Chloroform–methanol–0.3% aqueous $CaCl_2$, 50:42:11, by vol.
21. Chloroform–methanol–2-propanol–water, 60:25:5:5, by vol.

3 Methods

3.1 Preparation of Gangliosides Radio-labeled on the Oligo-saccharide Chain

3.1.1 Labeling at the C6 of Galactose. The Galactose Oxidase-Sodium Boro[^{3}H]Hydride Method [7–9]

Galactose oxidase (E.C.1.1.3.9.) oxidizes the primary alcohol group of galactose, and with less yield that of *N*-acetyl galactosamine, to aldehyde. The following treatment with sodium boro[^{3}H] hydride gives back the original alcohol group, thus causing the tritium to be specifically introduced at the sugar carbon 6.

The enzyme is purely active on the terminal Gal or GalNAc of glycosphingolipids that in solution form aggregates of high molecular mass [10]. The addition of Triton X-100 to the gangliosides, produces mixed micelles allowing the enzyme to interact and oxidize the external primary alcohol. In such a condition, the average oxidation yield is as follows: 90% on GM1, 80% on GD1b, 30% on GM2, and 88% on AGM1. The enzyme has no, or very low, activity on internal Gal. As an example, the oxidation yield on GD1a is 2–3%.

Tritium labelled GM2, GM1, GM1 derivative containing 12-aminododecanoyl chain, GD1b and tetrahexosylceramide Gg4Cer were prepared with the galactose oxidase/sodium boro[^{3}H]hydride method.

Method of preparation. Thirteen micromoles of sphingolipid in 1 mL of chloroform–methanol, 2:1 by vol, are mixed with 180 mg of high-quality nonoxidized Triton X-100 in 10 mL of the same solvent. The solution is dried and the residue dissolved in 8 mL of 5 mM EDTA–25 mM sodium phosphate buffer, pH 7.0, containing 450 Units of galactose oxidase. The mixture is maintained at 37 °C under continuous stirring for 24 h; an additional 450 Units of the enzyme are added after 6 h. Following a further 24 h, the solution is dried, solubilized in chloroform–methanol–2-propanol–water, 60:25:10:5 by vol, and chromatographed on a 25 × 2 cm silica gel 60 column, equilibrated and eluted with the same solvent system.

Column elution is monitored by HPTLC using chloroform–methanol–100 mM aqueous KCl: 5 M formic acid, 30:50:3:3, by vol., as solvent system, and an anisaldehyde spray reagent for revealing. The fractions containing the oxidized ganglioside are collected, pooled, and evaporated to dryness.

Purification of the oxidized ganglioside is carried out quickly to avoid production of byproducts. The product is stored as dry powder, at −20 °C and under argon.

The oxidized sphingolipid is dissolved in propanol–0.1 M NaOH, 7:3, by vol, in a screw-capped tube, and treated at room temperature with solid [^{3}H]$NaBH_4$. After 5 h, under stirring, 10 mg of not-radioactive $NaBH_4$ are added and the reaction is allowed to continue for a further 30 min under stirring. The reaction mixture is evaporated under nitrogen pressure to a very small volume, vapors being carefully condensed in a liquid nitrogen trap.

At the end of the evaporating/condensing apparatus, there is a trap of 5 M sulfuric acid to oxidize tritium and keep back residual tritiated water. Finally, the wet residue is dissolved in 1 mL of water and dialyzed.

The distribution of radioactivity into the glycan structure, determined by chemical hydrolysis of the radioactive compound followed by determination of the specific radioactivity of the formed compounds resulted to be: 83–85% on GalNAc and 15–17% on Gal for GM2; 97–98% on external Gal; and 2–3% on GalNAc for GM1, GD1b, and tetrahexosylceramide.

3.1.2 Labeling at the Sugar Acetyl Groups. The De-Acetylation/Re-N-[^{3}H]/[^{14}C]Acetylation Method [11–16]

Alkaline hydrolysis of monosialogangliosides GM3, GM2, or GM1 yields a mixture of the derivative lacking the acetyl groups and/or the fatty acyl chain, in a ratio and yield depending on the chemical conditions and on the glycolipid structure. Reacetylation of the amino groups with radioactive acetic anhydride is a simple reaction, but acetic anhydride is unstable and easily degrades to acetic acid. The procedure has been applied to the preparation of radioactive GM3, GM2, GM1 and to their species having acetylated sphingosine.

Method of preparation. In a round-bottom flask, 30 μmol of ganglioside are dissolved in 1 mL of water, prewarmed to 60 °C, and mixed with an equal volume of 4 M KOH, prewarmed to 60 °C. The reaction mixture is refluxed at 90 °C under vigorous stirring. After 3.5 h, the solution was chilled and neutralized with strong cation exchange resin (i.e., Amberlite® IR120 H^+ form) and freeze-dried. Deacetylganglioside is purified by flash chromatography using a silica gel 60 column, equilibrated and eluted with the solvent system chloroform–methanol–2-propanol–formic acid 30:40:10:6 by vol. Column elution is monitored by HPTLC developed in chloroform–methanol–100 mM aqueous KCl–5 M formic acid, 30:50:3:3, by vol, and visualized with an anisaldehyde spray reagent). The fractions containing the deacetyl-ganglioside or the deacetyl-deacyl-ganglioside are collected, pooled and evaporated to dryness.

The purified deacetyl-ganglioside is dissolved (2 mg/mL) in magnesium dehydrated methanol and mixed with radioactive acetic anhydride to have a final molar ratio of 1:1. After stirring for 30 min at room temperature, 10 μL of not-radioactive acetic anhydride are added, the mixture is stirred for an additional 30 min and the solution is treated as reported in section A, with traps of NaOH than sulfuric acid being used to keep back radioactive acetic acid. The addition of cold acetic anhydride can be avoided to increase the ganglioside specific radioactivity. In this case, radioactive ganglioside and the unreacted deacetyl-ganglioside are separated by chromatography on a 50 × 1 cm silica gel 100 column equilibrated and eluted with dichloromethane–methanol–water, 60:35:8, by vol.

3.2 Preparation of Gangliosides Radiolabeled on the Ceramide

3.2.1 The DDQ-Sodium Boro[³H]Hydride Method [17–20]

2,3-dichloro-5,6-dicyanobenzoquinone (DDQ) specifically oxidizes allylic hydroxyl groups. By reduction with sodium boro[^{3}H]hydride, the formed carbonyl group is reduced back to the original alcohol group, so that tritium is introduced at the carbon position 3.

To avoid oxidation of sialic acid, the DDQ treatment is carried out on inverted mixed micelles of ganglioside and Triton X-100 dispersed in toluene. The oxidation procedure has been applied to ganglioside GM3, GM2, GM1, Fuc-GM1 and GD1a, with yield of 60–65%. When applied to disialosyl chain containing gangliosides, GD1b and GT1b, the yield was very poor due to the production of several uncharacterized products.

Reduction of the carbonyl group of the 3-keto-sphingolipid is not stereospecific and produces a mixture of *erythro* and *threo* diastereoisomers and the purification of the natural 2R3S structure is carried out by reversed-phase HPLC.

Method of preparation. 13 μmol of ganglioside GM3, GM2, GM1, Fuc-GM1 or GD1a, are dissolved in 10 mL of chloroform–methanol, 2:1 by vol, and mixed with 10 mL of a solution of Triton X-100 in the same solvent (60 mg/mL). The solvent is evaporated and the residue dissolved in 10 mL of a DDQ solution in dehydrated toluene (60 mg/mL). The mixture is maintained at 37 °C for 40 h under constant vigorous stirring in a screw-capped flask. The solvent is removed under vacuum at 37 °C. The residue is suspended in cold acetone, while that from the oxidation of GM3 is suspended in 10 mL of cold acetonitrile. After centrifugation at 12,000 × *g* the supernatant containing Triton X-100 and DDQ, is discarded. The treatment is repeated up to obtain a white pellet. The oxidized gangliosides GM3 and GM2 gangliosides are purified by flash chromatography on a 25 × 2 cm silica gel 60 column, Column is equilibrated and eluted with the solvent system chloroform–methanol–2-propanol–water, 60:25:10:5 by vol. for purification of oxidized GM3 and GM2, or in the ratio of 60:25:5:5 by vol., for oxidized GM1, Fuc-GM1, and GD1a. Column elution is monitored by TLC using chloroform–methanol–0.2% aqueous $CaCl_2$, 50:42:11 by vol, and an anisaldehyde spray reagent for revealing.

The oxidized gangliosides are stored as dry powder, at −20 °C and under argon.

The purified 3-keto-ganglioside is dissolved in 5 mL of propanol–water, 7:3 by vol, in a screw-capped tube, and treated at room temperature with solid [^{3}H]$NaBH_4$. Then 10 mg of cold $NaBH_4$ are added and the reaction continued for further 30 min. The solution containing the radioactive ganglioside is then treated as reported above in section 3.1.1.

3–5 mg (or up to 4–5 mCi) of ganglioside are dissolved in 0.5 mL of acetonitrile–water, 1:1, by vol, in a microtube and introduced into a syringe-loading sample injector equipped with a 1 mL loop. The microtube is washed five times with 100 μL of acetonitrile–water, 1:1, by vol, the washings being added to the previous sample in order to minimize loss of material. The gangliosides species, erythro and threo, C18 sphingosine and C20 sphingosine (when present), are separated on a reversed-phase LiChrosphere RP8 column, 250 × 4 mm I.D., 5 μm average particle diameter (Merck, Darmstadt, F.R.G.) equilibrated and eluted at a flow rate of 10 mL/min with the solvent system acetonitrile–5 mM phosphate buffer, pH 7.0, 60:40, by vol, for GM3, GM2, GM1, and Fuc-GM1; and 50:50, by vol, for GD1a. The separation of GM3 species is carried out at 45 °C and that of all the other gangliosides at room temperature. The elution profile is monitored by flow-through detection of both the UV absorbance at 195 nm and the radioactivity content.

3.2.2 Labeling at C4 and C5 of Sphinganine. The Sphingosine [^{3}H]H_2 Reduction Method [21]

Gangliosides containing radioactive sphinganine are prepared by catalytic tritium reduction of gangliosides which contain sphingosine. The yield of reaction is 100%.

Two μmol of ganglioside dissolved in 3–4 mL of 96% ethanol are treated with tritium gas produced from sodium boro[^{3}H] hydride for 36 h at room temperature in the presence of 1–2 mg of PtO_2 as catalyst, under constant stirring. At the end of the reaction the mixture is centrifuged at 9000 rpm (7600 rcf) for 20 min at 4 °C and the supernatant filtered through paper to remove the unsoluble catalyst. The solution is then evaporated to dryness, solubilized in 1–2 mL, dialyzed at 4 °C and mixed with 2–3 volumes of propanol or methanol.

3.2.3 Labeling at the Fatty Acid. The Specific GM1 Deacylation—Re-[^{3}H]/[^{14}C]Acylation Method [11]

GM1 hydrolysis in KOH-deoxygenated anhydrous propanol produced lyso-GM1, the GM1 derivative lacking only the fatty acid chain. By acylation with radioactive fatty acid the GM1 structure is reconstituted.

Method of preparation. Propan-1-ol is deoxygenated by 30-min bubbling with a stream of argon (20–30 L/min) and maintained in a closed bottle under moderate argon pressure. A solution of GM1 (1.4 mM) in deoxygenated propan-1-ol, warmed at 90 °C, is mixed in a round-bottom flask provided with a screw cap (open top) and a natural rubber septum, with a 90 °C prewarmed deoxygenated propan-1-ol potassium hydroxide solution (1 M) to obtain a final KOH concentration of 0.2 M. Two stainless-steel needles were introduced in the vial septum, one connected to the argon bottle, and the reaction mixture was maintained for 30 min under argon flux of 5 mL/min. Then the needles are removed and the reaction is continued at 90 °C and under continuous stirring for 6 h. The reaction mixture is then dried, and the residue dissolved in water (5 mg starting GM1/0.1 mL) dialyzed for 3 h, and freeze-dried. The purification of lyso-GM1 is carried out by silica gel flash chromatography (25 × 1.5 cm/50 mg of starting GM1) using chloroform–methanol–2-propanol–water, 60:30:5:5, by vol, as the eluting solvent. The elution profile was monitored by TLC using chloroform–methanol–30 mM aqueous $CaCl_2$:100 mM aqueous KCl, 50:50:4:8 by vol.

Radioactive stearic acid, pentafluorophenol, and *N,N'*-dicyclohexylcarbodiimide (DCC) at a molar ratio of 1:1.3:1 are dissolved in anhydrous dichloromethane. The reaction is carried out under vigorous stirring at RT overnight. The reaction mixture is filtered to remove the precipitated dicyclohexylurea. The filtered solution containing the activated radioactive stearic acid is dried and the residue solubilized in anhydrous dimethylformamide (0.04 mmol/mL). The solution is added to lyso-GM1 (final stearic acid–ganglioside derivative molar ratio of 1:1) and the reaction left at room temperature for 24 h under stirring. The solution is then lyophilized and the residue is purified by chromatography on a 50x1 cm silica gel column chromatography equilibrated and eluted with the solvent system dichloromethane–methanol–water, 60:35:5 by vol. Column elution is monitored by TLC (silica gel HPTLC plates developed in chloroform–methanol–0.3% aqueous $CaCl_2$, 50:42:11, by vol, and revealed by autoradiography). The fractions containing the radioactive GM1 are collected, evaporated to dryness and immediately solubilized in methanol.

3.3 Labeling at the Fatty Acid. General Procedure [16, 18, 22]

The preparation of gangliosides containing a radioactive fatty acid chain requires the preparation of the lyso-ganglioside prepared and purified as follow.

In a round-bottom flask, 50 μmol of ganglioside is dissolved in 15 mL of dry methanol, prewarmed to 60 °C, and mixed with an equal volume of 2 M KOH, prewarmed to 60 °C. The reaction mixture is refluxed at 90 °C under vigorous stirring. After 12 h, the solution was chilled and neutralized with strong cation exchange resin (i.e., Amberlite® IR120 H^+ form) and dried under vacuum. The purification of lyso-ganglioside is carried out by silica gel flash chromatography (25 × 1.5 cm/50 mg of starting ganglioside) using chloroform–methanol–2-propanol–water, 60:25:5:5, by vol for GM3, GM2 GM1, Fuc-GM1 and GD1a as the eluting solvent. The elution profile was monitored by TLC using chloroform–methanol–0.2% aqueous $CaCl_2$, 50:42:11, by vol. Reacylation with radioactive fatty acid is carried out as in section 3.2.1.

4 Notes

1. The commercial chemicals must be the purest available.
2. The starting gangliosides must be as pure as possible and lyophilized before use.

3. For the purification of radioactive ganglioside derivatives, $CHCl_3$ has been substituted with CH_2Cl_2, in order to reduce radiolysis reactions yielding radioactive by-products.
4. The use of oil vacuum pump must be avoided and, if absolutely necessary, a liquid nitrogen trap should be used to condense the oil vapors.
5. Silica gel chromatography columns are always thoroughly washed before use with the solvent system subsequently used to elute lipid components adsorbed on the silica gel.
6. Reactions requiring anhydrous conditions must be carried out in a controlled-atmosphere chamber; glass apparatus are dried overnight in an oven at 120 °C, taken out just before use and connected to a tube filled with desiccant $CaCl_2$ granules.
7. Anhydrous solvent can be prepared using 3-Ă and a 4-Ămolecular sieve, depending on the solvent.
8. Anhydrous triethylamine is dried over KOH pellets.

References

1. Chigorno V, Tettamanti G, Sonnino S (1996) Metabolic processing of gangliosides by normal and Salla human fibroblasts in culture. A study performed by administering radioactive GM3 ganglioside. J Biol Chem 271(36):21738–21744
2. Radsak K, Schwarzmann G, Wiegandt H (1982) Studies on the cell association of exogenously added sialo-glycolipids. Hoppe-Seyler's Zeitschrift fur physiologische Chemie 363(3):263–272
3. Schwarzmann G, Hoffmann-Bleihauer P, Schubert J, Sandhoff K, Marsh D (1983) Incorporation of ganglioside analogues into fibroblast cell membranes. A spin-label study. Biochemistry 22(21):5041–5048
4. Chigorno V, Pitto M, Cardace G, Acquotti D, Kirschner G, Sonnino S, Ghidoni R, Tettamanti G (1985) Association of gangliosides to fibroblasts in culture: a study performed with GM1 [14C]-labelled at the sialic acid acetyl group. Glycoconj J V2(3):279–291
5. Orlando P, Cocciante G, Ippolito G, Massari P, Roberti S, Tettamanti G (1979) The fate of tritium labeled GM1 ganglioside injected in mice. Pharmacol Res Commun 11(9):759–773
6. Trinchera M, Ghidoni R, Sonnino S, Tettamanti G (1990) Recycling of glucosylceramide and sphingosine for the biosynthesis of gangliosides and sphingomyelin in rat liver. Biochem J 270(3):815–820
7. Ghidoni R, Tettamanti G, Zambotti V (1977) Labeling of natural substrates for the radiochemical assay of enzymes involved in lipid storage diseases: a general procedure for tritiation of gangliosides. Biochem Exp Biol 13(1):61–69
8. Leskawa KC, Dasgupta S, Chien JL, Hogan EL (1984) A simplified procedure for the preparation of tritiated GM1 ganglioside and other glycosphingolipids. Anal Biochem 140(1):172–177
9. Fra AM, Masserini M, Palestini P, Sonnino S, Simons K (1995) A photo-reactive derivative of ganglioside GM1 specifically cross-links VIP21-caveolin on the cell surface. FEBS Lett 375(1–2):11–14
10. Sonnino S, Cantu L, Corti M, Acquotti D, Venerando B (1994) Aggregative properties of gangliosides in solution. Chem Phys Lipids 71(1):21–45
11. Sonnino S, Acquotti D, Kirschner G, Uguaglianza A, Zecca L, Rubino F, Tettamanti G (1992) Preparation of lyso-GM1 (II3Neu5AcGgOse4-long chain bases) by a one-pot reaction. J Lipid Res 33(8):1221–1226
12. Taketomi T, Kawamura N (1970) Preparation of lysohematoside from hematoside of equine erythrocyte and its chemical and hemolytic properties. J Biochem 68(4):475–485
13. Tayot JL, Tardy M (1980) Isolation of cholera toxin by affinity chromatography on porous

silica beads with covalently coupled ganglioside GM1. Adv Exp Med Biol 125:471–478

14. Higashi H, Basu S (1982) Specific 14C labeling of sialic acid and N-acetylhexosamine residues of glycosphingolipids after hydrazinolysis. Anal Biochem 120(1):159–164
15. Neuenhofer S, Schwarzmann G, Egge H, Sandhoff K (1985) Synthesis of lysogangliosides. Biochemistry 24(2):525–532
16. Sonnino S, Kirschner G, Ghidoni R, Acquotti D, Tettamanti G (1985) Preparation of GM1 ganglioside molecular species having homogeneous fatty acid and long chain base moieties. J Lipid Res 26(2):248–257
17. Ghidoni R, Sonnino S, Masserini M, Orlando P, Tettamanti G (1981) Specific tritium labeling of gangliosides at the 3-position of sphingosines. J Lipid Res 22(8):1286–1295
18. Sonnino S, Ghidoni R, Gazzotti G, Kirschner G, Galli G, Tettamanti G (1984) High performance liquid chromatography preparation of the molecular species of GM1 and GD1a gangliosides with homogeneous long chain base composition. J Lipid Res 25(6): 620–629
19. Gazzotti G, Sonnino S, Ghidoni R, Orlando P, Tettamanti G (1984) Preparation of the tritiated molecular forms of gangliosides with homogeneous long chain base composition. Glycoconj J 1(2):111–121. https://doi.org/10.1007/bf01213725
20. Negroni E, Chigorno V, Tettamanti G, Sonnino S (1996) Evaluation of the efficiency of an assay procedure for gangliosides in human serum. Glycoconj J 13(3):347–352
21. Schwarzmann G (1978) A simple and novel method for tritium labeling of gangliosides and other sphingolipids. Biochim Biophys Acta 529(1):106–114
22. Valiente O, Mauri L, Casellato R, Fernandez LE, Sonnino S (2001) Preparation of deacetyl-, lyso-, and deacetyl-lyso-GM(3) by selective alkaline hydrolysis of GM3 ganglioside. J Lipid Res 42(8):1318–1324

Chapter 16

Synthesis of Fluorescent Gangliosides

Günter Schwarzmann

Abstract

Labeled gangliosides are invaluable tools to study their transport and metabolism within cells as well as to determine their distribution in membranes, and their interaction with membrane lipids and proteins. Here I describe established procedures to synthesize ganglioside derivatives with a fluorescent tag either attached to its sialooligosaccharide or ceramide portion. These procedures are chosen as to minimize the integrity of the ganglioside molecule and hence, to leave their native skeleton formally intact. The α-position of the stearic acid residue is favorable for the attachment both of hydrophilic and of lipophilic dyes. In some other cases, and starting from lyso-gangliosides, procedures are described by which a fluorescent tag bound to a short acyl chain replaces the native acyl chain of gangliosides.

Key words 2-ATTO-ganglioside, NeuATTO-gangliosides, ATTO-lyso-gangliosides, 2-NBD-gangliosides, NBD-C6-gangliosides, Pyrene-gangliosides, DPH-gangliosides, Lyso-gangliosides

1 Introduction

Gangliosides are ubiquitous in vertebrate cells and are notably plentiful in neuronal plasma membranes [1]. Figure 1 shows the structures of the main gangliosides of mammalian brains.

The unraveling of ganglioside metabolism and function since their first discovery [2] depended on the extensive study of inherited human storage disease and specifically designed mouse models. For these studies, mainly isotope labeled gangliosides have been used. However, only few of the accomplishments made so far with respect to transport within cells, their distribution in membranes as well as their interaction with lipids and proteins would have been possible without other labeled gangliosides, such as fluorescent, photoactivatable, paramagnetic as well as affinity-tagged ganglioside derivatives. These derivatives have been studied after their uptake by and insertion into cultured cells or in model membranes. For a review *see* [3].

There are two general sites to add a label to a ganglioside molecule: (a) the lipophilic ceramide moiety or (b) the sugar part of

Sandro Sonnino and Alessandro Prinetti (eds.), *Gangliosides: Methods and Protocols*, Methods in Molecular Biology, vol. 1804, https://doi.org/10.1007/978-1-4939-8552-4_16, © Springer Science+Business Media, LLC, part of Springer Nature 2018

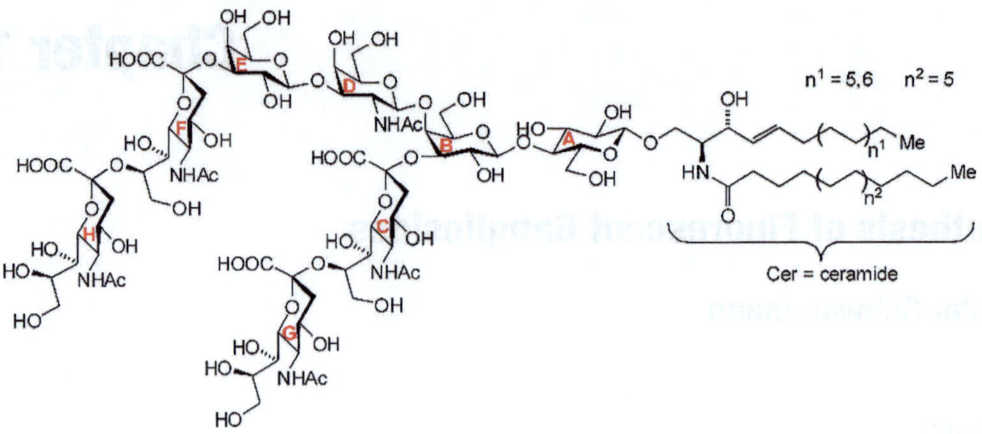

Fig. 1 The structures of mammalian brains major gangliosides are immanent in the tetrasialoganglioside GQ1b as shown above; In addition to the ceramide moiety (Cer) gangliosides contain the sugars indicated by bold letters in red. GM3—A to C, GD1a—A to F, GD1b—A to E + G; GM2—A to D, GT1b—A to G, GT1a—A to F + H; GM1—A to E, GQ1b—A to H; A = D-glucose B = E = D-galactose D = *N*-acetyl-D-galactosamine; C = F = G = H = *N*-acetylneuraminic acid

the molecule. The introduction of a label into the sugar part without destruction of the latter is mostly done for affinity-tagged gangliosides [4, 5] and in some cases also for fluorescent gangliosides [6]. In most cases fluorescent, paramagnetic, and photoactivatable tags are placed in the ceramide portion of the ganglioside molecule [7–20].

For the preparation of labeled gangliosides having the appropriate tag in their ceramide the usual procedure is to reacylate lyso-gangliosides, i.e., gangliosides lacking the genuine fatty acyl residue. This reacylation can be done with an activated labeled fatty acid or with an activated fatty acid bearing a functional group that subsequently can be modified with a desired tag. For the preparation of the necessary lyso-gangliosides four procedures have been developed over the last decades. These are alkaline deacylation, selective alkaline deacylation, enzymatic deacylation, partial and total synthesis from sugars and sphingoids. All these procedures have their benefits and drawbacks. I shall outline examples of these procedures except for the total synthesis of lyso-gangliosides as intermediates in the chemical synthesis of gangliosides, which is dealt with in another chapter of this volume.

2 Materials

2.1 Chemicals

1. Acetic anhydride (Ac_2O).
2. Aluminum oxide.
3. Ammonia (NH_3).
4. Ammonium chloride (NH_4Cl).

5. Anhydrous sodium sulfate (Na_2SO_4).
6. ATTO647N NHS ester (ATTO-TEC Siegen, Germany).
7. α-Azido stearic acid was prepared as described [18, 21].
8. *tert*-Butanol.
9. *tert*-Butylchlorodiphenylsilane (TBDPSCl).
10. Benzoyl chloride (BzCl).
11. Calcium chloride ($CaCl_2$).
12. Copper sulfate ($CuSO_4$).
13. Dichloroacetyl chloride ($CHCl_2COCl$).
14. *N,N′*-Dicyclohexylcarbodiimide (DCC).
15. *N,N*-Diisopropylethylamine (DIPEA).
16. 4-Dimethylaminopyridine (DMAP).
17. *N*-(9-Fluorenylmethoxycarbonyl chloride) (FmocCl).
18. Gangliosides were isolated from bovine brain and purified as described [22, 23].
19. Hexadecanoic acid NHS ester (Fluka).
20. Hydrochloric acid (HCl).
21. Hydrogen bromide in acetic acid.
22. *N*-Hydroxysuccinimide (NHSI).
23. KOH.
24. Mercury cyanide ($Hg(CN)_2$).
25. Ninhydrin.
26. 7-Nitrobenz-1,3-diazol-2-oxa-4-yl fluoride (NBDF) (Thermofisher).
27. Paraformaldehyde.
28. Platinum(IV)oxide (PtO_2).
29. 2-Propanol.
30. 3-Phenylpropionic acid was obtained (Acros Organics).
31. Sodium hydrogen carbonate ($NaHCO_3$).
32. Sphingolipid ceramide-*N*-deacylase (SCDase) (Sigma-Aldrich).
33. Sphingosine [(2S,3R,4E)-2-amino-4-octadecen-1,3-diol was prepared from human brain lipids as described [24].
34. Tetrabutylammonium fluoride 1 M in THF (TBAF).
35. Triethylamine (Et_3N).
36. Triethylsilane (Et_3SiH).
37. Trifluoroacetic acid (TFA).
38. Triphenylphosphane (Ph_3PH).
39. Triton™ X-100 (TX100) (Sigma-Aldrich).

2.2 Solvents

1. Acetic acid (HOAc).
2. Acetone.
3. Benzene.
4. 1-Butanol (n-BuOH) .
5. Chloroform ($CHCl_3$).
6. n-Decane.
7. Deionized water (DI H_2O).
8. Dichloromethane (CH_2Cl_2).
9. Dimethylformamide (DMF).
10. Ethyl acetate (EtOAc).
11. n-Hexane.
12. Methanol (MeOH).
13. 1-Propanol.
14. 2-Propanol.
15. Pyridine.
16. Tetrahydrofuran (THF).
17. Toluene.
18. Dry solvents are obtained immediately prior to use using filtration tubes equipped with a polyethylene frit and filled with either acidic, neutral or basic aluminum oxide to remove traces of water and peroxides from ethers, and amines from DMF similarly as reported [25].

2.3 Chromatographic Materials

1. LiChroprep RP-18 (40–63 μm) (Merck).
2. LiChroprep Si 60 (40–63 μm) (Merck).
3. Precoated TLC plates HPTLC Silica gel 60 F_{254} (Merck).

2.4 Solutions and Mixtures

1. 0.1 M $NaHCO_3$.
2. 0.02 M $NaHCO_3$.
3. 0.05 M $NaHCO_3$.
4. 0.1 M HCl.
5. 1 M NH_4Cl.
6. 1 M HCl in MeOH.
7. 0.05 M NH_3 in MeOH.
8. 0.2 M propanolic KOH.
9. 0.8 M KOH in MeOH.
10. 0.2 M methanolic NaOH.
11. 0.05 M NH_3 in $CHCl_3$–MeOH (1:1, v/v).
12. 0.05 M NH_3 in MeOH–H_2O (9:4, v/v).

13. 0.8% of Triton™ X-100 in 50 mM sodium acetate buffer (pH 6.0).
14. 33% hydrogen bromide in acetic acid.
15. $CHCl_3$–MeOH (1:1, v/v).
16. $CHCl_3$–MeOH (2:1, v/v).
17. $CHCl_3$–MeOH (4:1, v/v).
18. $CHCl_3$–MeOH/–H_2O (65:25:4, v/v/v).
19. $CHCl_3$–MeOH–H_2O (60:35:8, v/v/v).
20. $CHCl_3$–MeOH–H_2O (60:40:9, v/v/v).
21. $CHCl_3$–MeOH–H_2O (55:45:10, v/v/v).
22. $CHCl_3$–MeOH–15 mM $CaCl_2$ (60:40:9, v/v/v).
23. $CHCl_3$–MeOH–15 mM $CaCl_2$ (55:45:10, v/v/v).
24. $CHCl_3$–MeOH–15 mM $CaCl_2$ (60:35:8, v/v/v).
25. $CHCl_3$–MeOH–2 N NH_3, (50:50:12, v/v/v).
26. $CHCl_3$–MeOH–2 N NH_3, (50:50:14, v/v/v).
27. $CHCl_3$–MeOH–2.5 M NH_3 (60:40:9, v/v/v).
28. $CHCl_3$–MeOH–2.5 M NH_3 (60:40:10, v/v/v).
29. $CHCl_3$–MeOH–2.5 M NH_3 (50:50:16, v/v/v).
30. $CHCl_3$–2 M NH_3 in MeOH (10:1, v/v).
31. $CHCl_3$–2 M NH_3 in MeOH (50:1, v/v).
32. $CHCl_3$–MeOH–HOAc (190:10:1, v/v/v).
33. MeOH–H_2O (4:1, v/v).
34. MeOH–H_2O (85:15, v/v).
35. MeOH–H_2O (9:1, v/v).
36. Acetone–toluene (1:1, v/v).
37. Acetone–toluene (2:8, v/v).
38. Acetone–toluene (8:2, v/v).
39. n-hexane–EtOAc (4:1, v/v).
40. n-hexane–EtOAc, (8:1, v/v).
41. n-hexane–EtOAc (9:1, v/v).
42. n-hexane–EtOAc, (7:3, v/v).

3 Methods

3.1 Preparation of Lyso-Gangliosides Via Strong Alkaline Hydrolysis

This procedure is used for the synthesis of lyso-mono- and -disialo-gangliosides and leads in the first step to deAc-lyso-gangliosides. These lack the fatty acyl chain as well as the acetyl groups of their sialic acid residue(s) with very little, if any, deacetylation of their *N*-acetyl-galactosamine residue [16, 26]. The next step involves

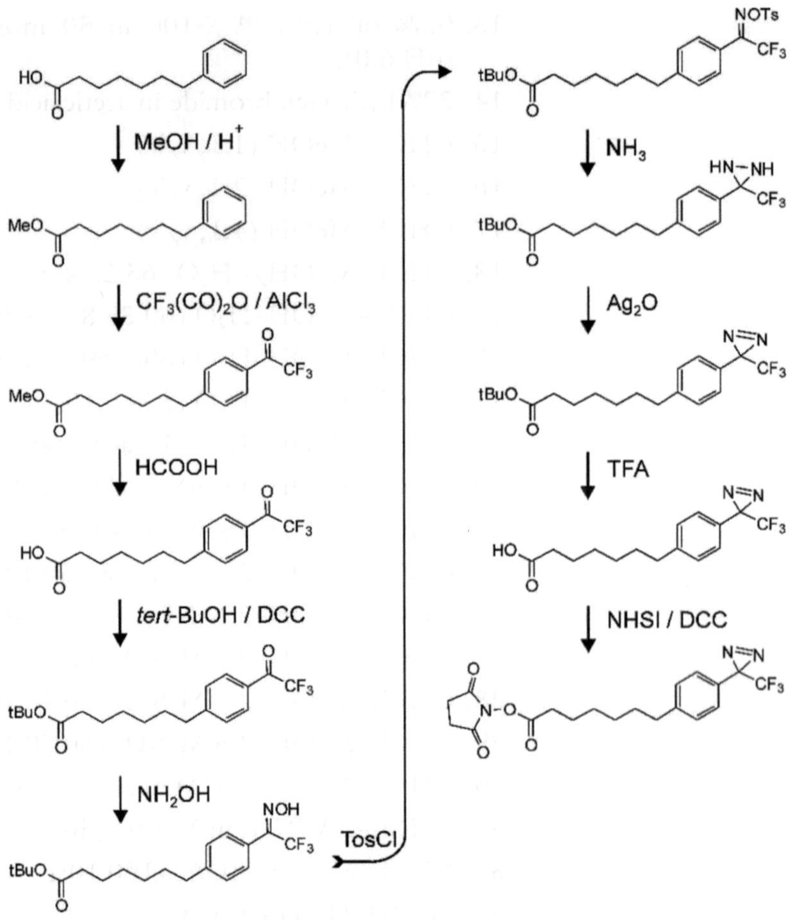

Fig. 2 Synthetic steps in the formation of lyso-GM1. Susceptible amide and amino groups are shown in red

protection of the sphingosine amino group with FmocCl followed by reacetylation of the neuraminyl group(s). Removal of the Fmoc residue furnishes the respective lyso-ganglioside in an overall yield of usually 35–40%. The methodology is illustrated in Fig. 2 for the synthesis of lyso-GM1.

In lieu of reacetylation a label can be introduced into the hydrophilic part of gangliosides prior to deprotection. Removal of Fmoc then allows the introduction of any kind of labeled acyl chain [4]. A similar procedure for alkaline hydrolysis using tetramethylammonium hydroxide instead of KOH was described by [27].

3.1.1 Hydrolysis

1. In a screw-capped vial equipped with a magnetic stir bar 50 mg ganglioside GM1 or 44.5 mg GM2 (32 μmol each) is dispersed in 16 mL 0.8 M KOH in MeOH, which has been purged with argon to remove oxygen. Under stirring the suspension is heated to 95–100 °C. After a short while GM1 or GM2 is completely dissolved. The solution is kept at this temperature

for 20–22 h (after this time it has turned yellow) and then cooled to room temperature.

2. Slowly add 1 M HCl in MeOH under stirring until pH 9 is reached as checked by wet indicator paper.
3. Remove the precipitated KCl by filtration and wash the precipitate with cold MeOH.
4. Dilute the filtrate with water to twice its volume and pass the solution over a small column of LiChroprep RP18. Wash with several column volumes of H_2O to remove salts.
5. Elute the lipophilic products containing released fatty acid and deAc-lyso-GM1 and -GM2 with 0.05 M NH_3 in MeOH (5 column volumes), followed by 0.05 M NH_3 in $CHCl_3$/MeOH (1:1, v/v) until no more ninhydrin-positive material is detected by TLC in the eluates.
6. Reduce the volume in a rotary evaporator care being taken to avoid foaming by the addition of small volumes of n-butanol and toluene.
7. Dissolve the wet residue in about 3 mL $CHCl_3$/MeOH/2 N NH_3, (50:50:12, v/v/v) and subject to chromatography on LiChroprep Si 60 (40–63 μm). Fractions containing the desired product (R_f 0.24 or 0.34 for deAc-lyso-GM1 or -GM2) as checked by TLC in $CHCl_3$/MeOH/2 N NH_3, (50:50:14, v/v/v) are pooled and dried in a nitrogen jet (*see* **Note 1**).
8. Redissolve the residue in a minimum of water and freeze-dry. The yield is 29.6 or 24.8 mg for deAc-lyso-GM1 or -GM2, respectively (23 μmol, 72%). ESI-MS (positive mode): m/z = 1266.6 $[M+H]^+$ and 1238.6 $[M+H]^+$ for deAc-lyso-GM1 with C20:1 and C18:1 sphingosine, respectively, and 1104.5 $[M+H]^+$and 1076.5 $[M+H]^+$ for deAc-lyso-GM2 with C20:1 and C18:1 sphingosine, respectively (*see* **Note 2**).

3.1.2 Protection of the Sphingoid Amino Group with FmocCl

1. To 22 μmol deAc-lyso-ganglioside add first 50 μL 0.1 M $NaHCO_3$ and 40 μL DIPEA. Then add 750 μL $CHCl_3$, followed by 750 μL MeOH. Stir until a clear solution is obtained.
2. Reduce the volume in a stream of nitrogen to one third to remove ammonia and other volatile amines, if present, and that otherwise will compete for FmocCl. Then replenish the volume with 1 mL $CHCl_3$/MeOH (1/1, v/v).
3. Cool the solution to 0 °C and slowly add with stirring a solution of 25 μmol FmocCl in 110 μL n-hexane. Continue stirring for 2 h. If TLC still shows a significant amount of educt further FmocCl is added. When TLC on silica gel 60 F_{254} developed with $CHCl_3$/MeOH/15 mM $CaCl_2$ (60:35:8, v/v/v) shows a main fluorescence quenching product of deAc-Fmoc-lyso-ganglioside (R_f 0.18 and 0.22 for the GM1 and GM2 derivative, respectively) besides little of educt and of traces of bis-Fmoc derivative (R_f 0.40 and 0.49 for the GM1

and GM2 derivative, respectively) the reaction is stopped by adding 1 M methanolic HCl to pH 7–8 as judged on a wet indicator paper.

4. Dry the mixture in a nitrogen jet. Dissolve the residue in 2 mL $CHCl_3$/MeOH (2:1, v/v) and subject to chromatography on LiChroprep Si 60 using $CHCl_3$/MeOH/H_2O (60:35:8, v/v/v).
5. Collect fractions containing pure products as revealed by TLC. Dry in a nitrogen stream, redissolve residue in little H_2O, and freeze-dry. Yield is 60 to 70%. ESI-MS (positive mode): m/z = 753.4 $[M+H+NH_4]^{2+}$ and 739.3 $[M+H+NH_4]^{2+}$ for deAc-Fmoc-lyso-GM1 with C20:1 and C18:1 sphingosine, respectively (*see* **Note 3**).

3.1.3 Reacetylation of Neuraminyl Residues

1. Disperse 15 µmol Fmoc-deAc-lyso-gangliosides first in 0.6 mL water, 0.3 mL $CHCl_3$, and 0.1 mL DIPEA. Then add 1.3 mL MeOH to solubilize the heterogeneous mixture.
2. After cooling in ice slowly add a total of 60 µL Ac_2O in instalments of 10 µL with vigorous stirring over 20 min. TLC should show complete conversion of the educt to Fmoc-lyso-gangliosides (R_f 0.30 and 0.44 for the GM1 and GM2 derivative, respectively).
3. Reduce the reaction volume in a nitrogen jet to near dryness. Dissolve the residue in 2 mL MeOH and 2 mL H_2O, and purify the products by chromatography on a small column of LiChroprep RP18 using a step gradient of 40 mL MeOH/H_2O (4:1, v/v), 10 mL MeOH/H_2O (85:15, v/v), and 40 mL MeOH/H_2O (9:1, v/v). TLC with shows a clean separation of Fmoc-lyso-gangliosides with C18 from C20 sphingosine moieties. Total yield is in the range from 70% to 80%. ESI-MS (negative mode): m/z = 1528.9 $[M-H]^-$ and 1500.9 $[M-H]^-$ for Fmoc-lyso-GM1 with C20:1 and C18:1 sphingosine, respectively.

3.1.4 Deprotection

1. Dissolve 10 µmol Fmoc-lyso-ganglioside in 300 µL MeOH followed by 300 µL piperidine. Keep the solution for 2 h at 20 °C. Check for complete conversion to lyso-gangliosides by TLC with $CHCl_3$/MeOH/15 mM $CaCl_2$ (60:40:9, v/v/v) (R_f 0.25 and 0.22 for lyso-GM1 (C20:1 and C18:1 sphingosine, respectively) and 0.30 for lyso-GM2 with C18:1 sphingosine).
2. Remove the solvents in a nitrogen jet with several additions of 200 µL benzene. Final purification of lyso-gangliosides is by column chromatography yielding 80–90% pure lyso-ganglioside (*see* **Note 4**). ESI-MS (positive mode): m/z = 1308.7 $[M+H]^+$ and 1280.6 $[M+H]^+$ for lyso-GM1 with C20:1 and C18:1 sphingosine, respectively, and m/z = 1118.6 $[M+H]^+$

and 1146.6 $[M+H]^+$ for lyso-GM2 with C18:1 and C20:1 sphingosine, respectively.

3.2 Preparation of Lyso-GM1 by Selective Alkaline Hydrolysis [22, 28]

1. Prepare oxygen-free 1-propanol by bubbling argon through it for 1 h. Then use this alcohol to prepare a 0.2 M KOH.
2. Place 24 mg (15 μmol) GM1 (containing both C18:1 and C20:1 sphingosine) in 15 mL 0.2 M propanolic KOH in a screw-cap flask and stir the mixture under argon atmosphere for 20 h at 90 °C (*see* **Note 5**).
3. Cool the reaction mixture to room temperature and add 1 M NH_4Cl to adjust the pH to about 9 as judged by wet indicator paper.
4. Remove the solvents in a nitrogen stream and dissolve the residue in 1 mL H_2O and 1 mL MeOH.
5. Purify the crude products by reverse phase (RP) chromatography using LiChroprep RP 18 (40–63 μm) in a small column (1 × 24 cm). Perform isocratic elution with 0.05 M NH_3 in MeOH/H_2O (9:4, v/v). Check fractions by TLC with $CHCl_3$/MeOH/15 mM $CaCl_2$ (60:40:9, v/v/v). Collect the two slightly overlapping ninhydrin-positive fractions containing lyso-GM1 with18:1 (R_f 0.22) and lyso-GM1 with 20:1 sphingosine (R_f 0.26) and dry in a nitrogen jet.
6. Redissolve the residues in 1 mL water, and freeze-dry. After lyophilization, lyso-GM1 (C18:1) and lyso-GM1·(C20:1) as well as the overlapping fraction are obtained as colorless powder (total 10.1 mg, 52%). Total yield is in close agreement with published data [28].
7. Further purification is achieved by flash chromatography on LiChroprep Si 60 (40–63 μm, 1.0 × 24 cm) with isocratic elution using $CHCl_3$/MeOH/2.5 M NH_3 (60:40:9, v/v/v). Yield of purified lyso-GM1 (C18:1) is about 2.9 μmol and that of lyso-GM1·(C20:1) is roughly 4.1 μmol. ESI-MS (positive mode): m/z = 1280.6 $[M+H]^+$, 640.8 $[M+2H]^{2+}$ for lyso-GM1 with C18:1 sphingosine. ESI-MS: m/z = 1308.6 $[M+H]^+$, 654.8 $[M+2H]^{2+}$ for lyso-GM1 with C20:1 sphingosine (*see* **Note 6**).

3.3 Preparation of Lyso-Gangliosides by Enzymatic Hydrolysis

The third method for preparing lyso-gangliosides is an enzymatic deacylation of gangliosides by using sphingolipid ceramide N-deacylase (SCDase). So far this method is the best to obtain the higher sialylated lyso-gangliosides that by alkaline hydrolysis of their parent gangliosides are difficult to obtain. The enzyme SCDase catalyzes the hydrolysis and also the reverse reaction of various glycosphingolipids, but not ceramides, in a reversible manner [29]. SCDase is reported to be regioselective for GM1 [30], GM2, GM3, GD3, GD1a, and GQ1b [31]. For gangliosides

GD1a, GD1b, GT1b, and GQ1b we obtained their lyso-gangliosides in reasonable yields after chromatographic purification [22]. The enzymatic preparation of lyso-gangliosides follows reported methods [31, 32] with, however, substituting Triton™ X-100 for sodium cholate. In our hands this leads to higher rates of deacylation [22].

1. Dissolve gangliosides (500 nmol) in 500 μL of 50 mM sodium acetate buffer (pH 6.0), which contains 0.8% of Triton™ X-100. Sonify the mixture in an ultrasonic bath for 15 s.

   To the clear solution add 3.6 μL (18 mU) of a SCDase solution (in 50 mM sodium acetate buffer, pH 6.0).
2. Then cover the aqueous solution with 5 mL of n-decane to extract the released fatty acid from the reaction mixture.
3. Incubated at 37 °C and replace the n-decane phase every few h.
4. Check the conversion of the ganglioside by TLC with $CHCl_3$/MeOH/2.5 M NH_3 (60:40:10, v/v/v). If considerable amounts of gangliosides remain add additional enzyme (up to 36 mU) in instalments over a period of 2 days.
5. Stop the reaction by the addition of a few drops of 2.5 M ammonia when only an insignificant amount of the remaining ganglioside persists.
6. Remove the solvent in a stream of nitrogen, dry the crude product and purify the respective lyso-ganglioside by column chromatography on LiChroprep Si 60 using $CHCl_3$/MeOH/2.5 M NH_3 (60:40:10, v/v/v) as eluent.
7. Collect fractions of pure lyso-gangliosides, dry in a nitrogen stream and freeze-dry from a minimum of water. After lyophilization, lyso-gangliosides are obtained as fluffy white powders. Yield of lyso-GD1a (C18:1 sphingosine) was 0.37 mg (47%), ESI-MS: m/z = 1571.7 $[M+H]^+$, 794.8 $[M+H+NH_4]^{2+}$, 786.4 $[M+2H]^{2+}$. R_f 0.24 on TLC in $CHCl_3$/MeOH/2.5 M NH_3 (60/44/12, v/v/v). Yield of lyso-GD1b (C20:1 sphingosine) was 0.43 mg (53%), ESI-MS: m/z = 1599.7 $[M+H]^+$, 808.8 $[M+H+NH_4]^{2+}$, 800.4 $[M+2H]^{2+}$. R_f 0.18 on TLC in $CHCl_3$/MeOH/2.5 M NH_3 (60:44:12, v/v/v). Yield of lyso-GT1b (C20:1 sphingosine) was 0.57 mg (61%), ESI-MS: m/z = 1890.8 $[M+H]^+$, 945.9 $[M+2H]^{2+}$. R_f 0.26 on TLC in $CHCl_3$/MeOH/2.5 M NH_3 (50/50/16, v/v/v).

Yield of lyso-GQ1b (C20:1 sphingosine) was 0.48 mg, (44%), ESI-MS: m/z = 1100.0 $[M+H+NH_4]^{2+}$, 1091.4 $[M+2H]^{2+}$. R_f 0.25 on TLC in $CHCl_3$/MeOH/2.5 M NH_3 (50/50/16, v/v/v).

3.4 Semisynthesis of Lyso-Gangliosides with Defined Sphingoid Residues

For metabolic and physical studies of gangliosides as well as for internal standards for MS quantification it is sometimes advantageous to apply labeled gangliosides that contain a sphingosine base different from their natural counterpart. Since several synthetic procedures for sphingosine synthesis are found in the literature and some of more rare sphingosines are already commercially available, semisynthetic approaches to gangliosides have become attractive. These rely on the glycosydation (*see* Fig. 3) of an appropriately protected sphingosine with derivatized ganglioside sugars [16, 33], which can be obtained from gangliosides by ozonolysis and alkaline fragmentation [34], as free oligosaccharides from milk sources [35] or by total synthesis of ganglioside sugars (*see* Chap. 14 and references cited therein).

The pure ganglioside GM1 sugar, monosialogangliotetraose, was prepared according to a modified procedure first described by Wiegandt and Baschang [34] starting from bovine brain ganglioside GM1 [16, 36].

Monosialolactose (GM3 sugar) was isolated from cow colostrum (*see* **Note 7**).

3.4.1 Pentadeca-O-Acetyl-Monosialogangliotetraose Sialoyl-II^2-Lactone

1. Disperse under an argon atmosphere a solution of monosialogangliotetraose, II [3]NeuAcGgOse4 (100 mg, 100 μmol) and 2.5 mg DMAP (20 μmol) in 2 mL pyridine and 1 mL Ac_2O. Stir the reaction mixture at room temperature for 48 h. The sugar dissolves only slowly.
2. Remove the solvents azeotropically in a nitrogen jet by several additions of toluene and dissolve the yellow-brownish residue in acetone/toluene (1:1, by vol.) and subject to column chromatography on silica gel LiChroprep Si 60 (40–63 μm) using acetone/toluene (1:1, v/v.) as eluent.
3. Monitor the elution profile by TLC with acetone-toluene (1:1, v/v.) as the mobile phase. Collect and dry all fractions containing pure product (R_f = 0.19).
4. Freeze-dry the residue from benzene. A colorless powder of a mixture of pure α- and β-pentadeca-*O*-acetyl-monosialogangliotetraose sialoyl-II^2-lactone (120 mg, 72%) is obtained. For ^{1}H NMR *see* [33]. ESI-MS (positive mode): m/z = 1628.5 $[M+NH_4]^+$ and 1611.5 $[M+H]^+$.

3.4.2 Tetradeca-O-Acetyl-α-Monosialogangliotetraosyl Bromide Sialoyl-II^2-Lactone

1. Dissolve pentadeca-O-acetyl-monosialogangliotetraose sialoyl-II^2-lactone (100 mg, 62 μmol) in 33% hydrogen bromide in acetic acid (2 mL) in a screw capped vial and keep for 30 min at 0 °C and then for further 30 min at room temperature.
2. Transfer the reaction mixture quickly to a separatory funnel containing ice-cold $CHCl_3$ (15 mL) and extract hydrogen bromide and acetic acid into ice water (10 mL). Wash the organic layer several times with ice water, 0.02 M $NaHCO_3$, and again with ice water.

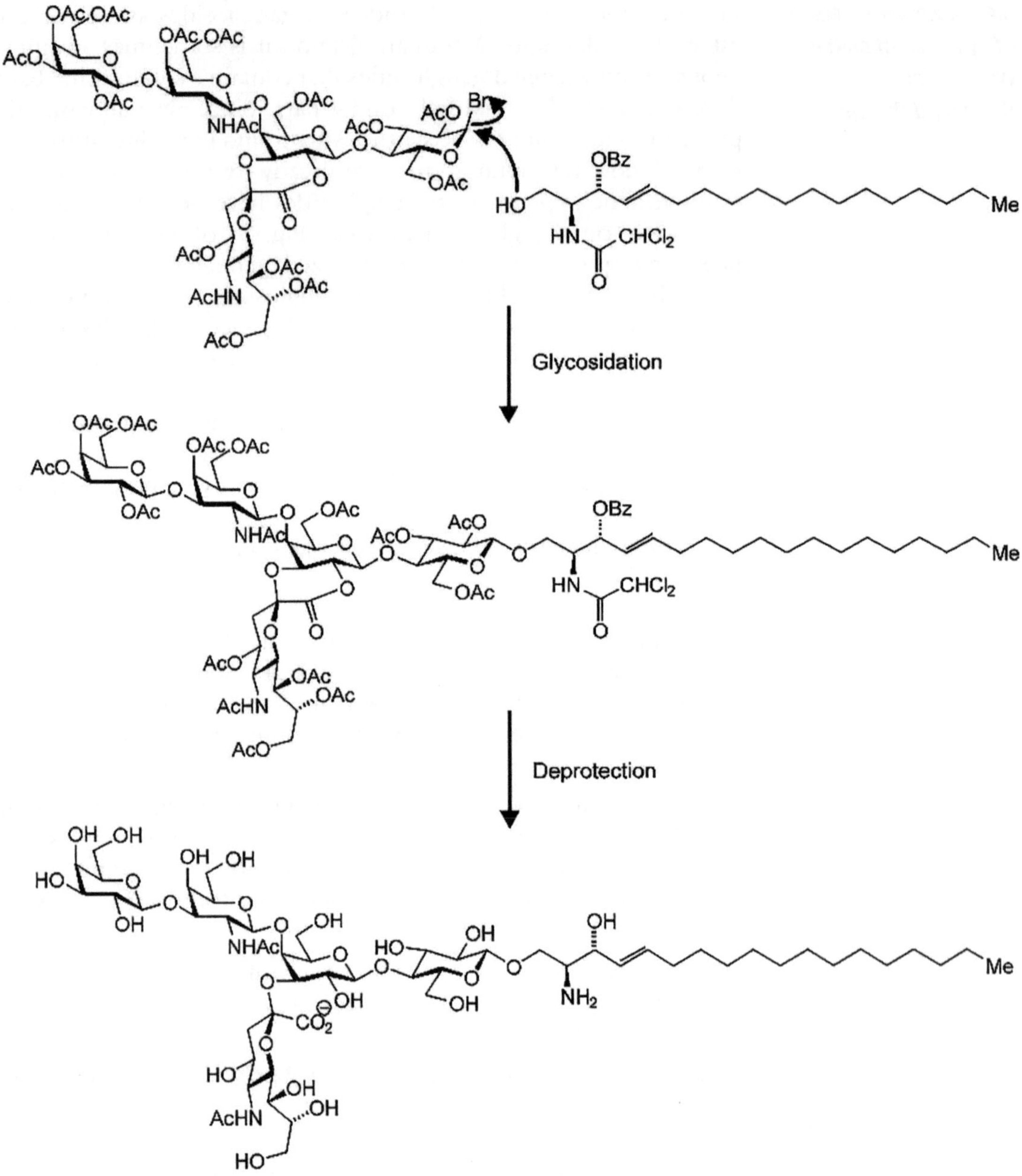

Fig. 3 Semisynthetic procedure for lyso-GM1

3. Pass the $CHCl_3$ solution over anhydrous Na_2SO_4 (1 g, in filtration tube equipped with a polyethylene frit). Dry the filtrate and $CHCl_3$ washes in a nitrogen stream with no heating.
4. Dissolve the residue in 1 mL benzene and freeze-dry. TLC in acetone/toluene (1:1, v/v) should reveal one single spot (R_f = 0.21) with no residual starting material (*see* **Note 8**).

3.4.3 Synthesis of Protected Sphingosine as Glycosyl Acceptor, i.e., (2S,3R,4E)-2-Dichloroacetamido-4-Octadecen-1,3-Diol

For synthetic scheme *see* Fig. 4.

1. Dissolve C18:1 sphingosine (300 mg, 1.0 mmol) in dry methanol (20 mL) and DIPEA (0.5 mL), then cool the solution to −20 °C.
2. To the cold solution add in increments of 125 μL dichloroacetyl chloride (0.75 mL, 8.2 mmol) in dry CH_2Cl_2 (4.25 mL) under vigorous stirring over a period of 1 h. Addition of dichloroacetyl chloride is continued until no more ninhydrin-positive material can be detected.
3. Pour this solution into a separatory funnel, add 50 mL H_2O and extract the title compound six times with ethyl acetate (50 mL). Dry the organic phase with anhydrous Na_2SO_4 and remove the solvent in a rotary evaporator. The crude product displays a single spot on TLC (R_f 0.4, ethyl acetate).
4. Freeze-dry from benzene to give *N*-dichloroacetyl-sphingosine (341 mg, 0.83 mmol, 83%) that is sufficiently pure for the next reaction step. For ^{1}HNMR and FABMS data *see* [37].

3.4.4 (2S,3R,4E)-1-O-tert-Butyldiphenylsilyl-2-Dichloroacetamido-4-Octadecen-1,3-Diol

1. Prepare a solution of 328 mg (0.8 mmol) *(2S,3R,4E)*-2-dichloroacetamido-4-octadecen-1,3-diol in dry CH_2Cl_2 (7 mL) and THF (1 mL), add DIPEA (200 μL), and DMPA (5 mg, 40 μmol).
2. Then add *tert*-butylchlorodiphenylsilane (123 μL, 0.9 mmol) and stir the solution for 7 h at room temperature. Stop the reaction by adding methanol (0.1 mL), and evaporate the solvent under nitrogen jet.
3. Purify the residue by flash chromatography on LiChroprep Si 60 (50 g) using n-hexane–EtOAc, (8:1, v/v) as eluent. Pool all fractions that on TLC with n-hexane–EtOAc, (7:3, v/v) appear pure (R_f 0.55). Remove solvents in a nitrogen jet and freeze-dry the residue from benzene. An yield of 0.47 g (0.72 mmol, 90%) of product is obtained.

3.4.5 (2S,3R,4E)-3-O-Benzoyl-1-O-tert-Butyldiphenylsilyl-2-Dichloroacetamido-4-Octadecen-1,3-Diol

1. To a cold (0 °C) solution of 0.45 g (0.69 mmol) of *(2S,3R,4E)*-1-*O*-*tert*-butyldiphenylsilyl-2-dichloroacetamido-4-octadecen-1,3-diol in pyridine (2 mL) add benzoyl chloride (0.16 mL, 1.38 mmol), and stir the suspension for 1 h at 0 °C, and then overnight at room temperature.
2. Stop the reaction by adding 0.1 M HCl (36 mL) and extract the product into ethyl acetate (20 mL). Wash the organic layer with H_2O, 0.1 M $NaHCO_3$, and again with H_2O.
3. Dry the EtOAc phase with anhydrous Na_2SO_4 and remove the solvent in a rotary evaporator. Purify the crude product on LiChroprep Si 60 (50 g) with n-hexane–EtOAc (4:1, v/v).

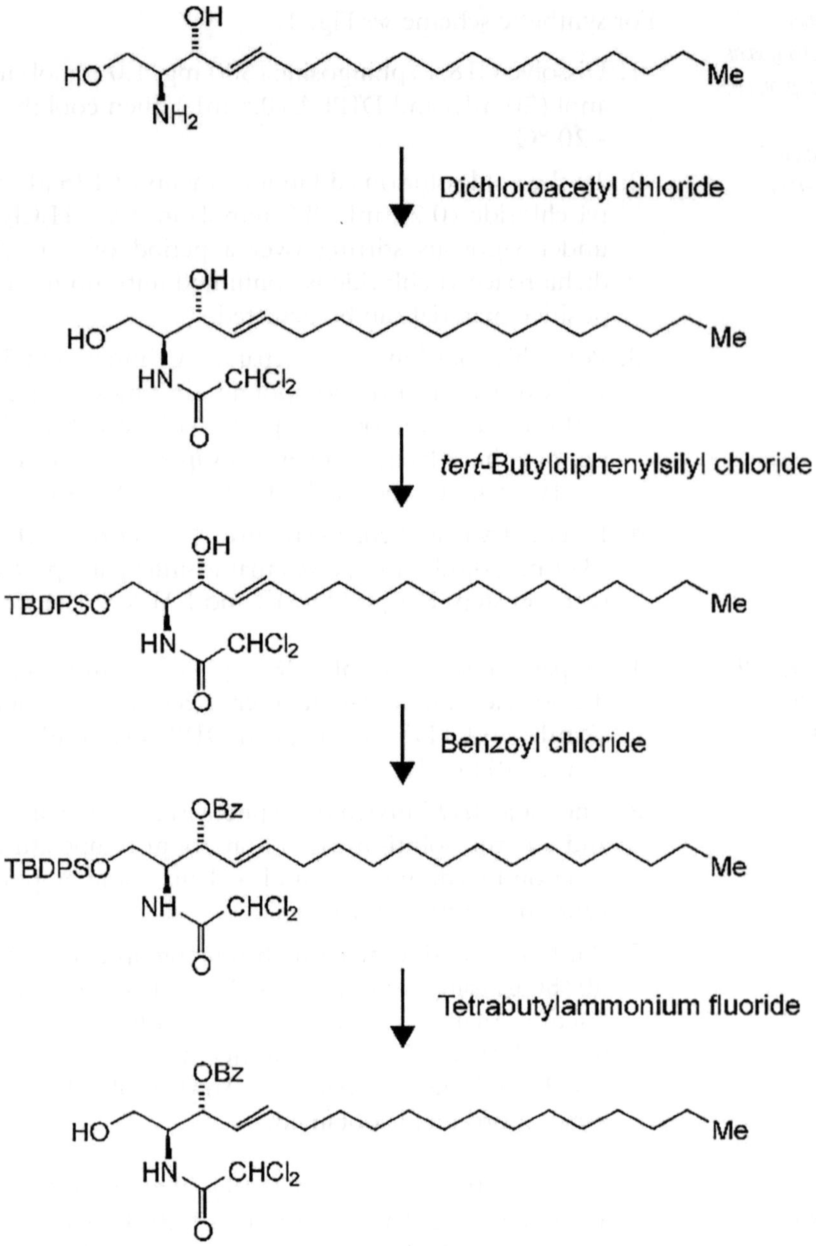

Fig. 4 Synthesis of a protected sphingosine as a glycosyl acceptor

Collect all fractions that on TLC with n-hexane–EtOAc, (4:1, v/v) appear pure (R_f 0.54). Yield is 0.5 g (0.66 mmol, 96%).

3.4.6 (2S,3R,4E)-3-O-Benzoyl-2-Dichloroacetamido-4-Octadecen-1,3-Diol

1. Dissolve 0.5 g (0.66 mmol) *(2S,3R,4E)*-3-*O*-*tert*-butyldiphenylsilyl-2-dichloroacetamido-4-octadecen-1,3-diol in 0.17 M TBAF in dry THF (5 mL) and stir overnight at room temperature.

2. Check progress of reaction by TLC with n-hexane–EtOAc, (7:3, v/v). The title compound (R_f 0.30) should have been formed on the expense of the educt (R_f = 0.65).
3. When the reaction is complete add 1 M HCl (2.5 mL) and extract the product into EtOAc (20 mL) and dry the organic phase with anhydrous Na_2SO_4. Remove the solvent in a nitrogen jet.
4. Purify the crude product by flash chromatography on LiChroprep Si 60 (50 g) with n-hexane–EtOAc (7:3, v/v). Collect pure fractions with R_f 0.30 (n-hexane–EtOAc, 7:3, v/v). Yield (238 mg, 0.46 mmol, 70%). For ^1H NMR (500 MHz, $CDCl_3$) *see* [37]. ESI-MS (positive mode): m/z = 536.2 $[M+Na]^+$ and 1049.4 $[2\ M+Na]^+$.

3.4.7 Glycosidation of (2S,3R,4E)-3-O-Benzoyl-2-Dichloroacetamido-4-Octadecen-1,3-Diol

1. Prepare a solution in a plastic vial of tetradeca-*O*-acetyl-α-monosialogangliotetraosyl bromide sialoyl-II^2-lactone (38 mg, 23 μmol) and 2-*N*-dichloroacetyl-3-*O*-benzoyl-D-erythro-sphingosine (13 mg, 25 μmol) in 0.4 mL of a mixture of anhydrous toluene/nitromethane (1/4, v/v).
2. Add finely powdered anhydrous $Hg(CN)_2$ (25 mg) under argon and stir the dispersion at 37 °C for 72 h.
3. Transfer the mixture to a screw-cap vial and dilute with benzene (3 mL) and centrifuge to remove most of the salts. Wash the supernatant successively with water containing H_2S (2 × 2 mL), 0.05 M $NaHCO_3$ (2 × 2 mL), and H_2O (2 × 2 mL). Dry the organic phase with anhydrous Na_2SO_4. Remove the solvent in a nitrogen jet.
4. Purify the residue by flash chromatography on LiChroprep Si 60 (0.9 × 100 cm) using a linear gradient of acetone/toluene (2:8, v/v) and acetone/toluene (8:2, v/v) (150 mL each). Monitor the elution profile by TLC in acetone/toluene (1:1, v/v). Collect all fractions containing the desired pure condensation product (R_f 0.31), evaporate the solvents and lyophilize from benzene. Yield of the protected lyso-GM1 is 20.5 mg (9.9 μmol, 43%).

3.4.8 Deprotection to Lyso-GM1

1. Dissolve the glycosidation product (20 mg, 9.6 μmol) in 0.2 M methanolic NaOH (1.0 mL) and stir for 2–3 h at 65 °C. Monitor the removal of the protecting groups by TLC with $CHCl_3$/MeOH/15 mM $CaCl_2$ (60:40:9, v/v/v). When a ninhydrin-positive spot (R_f 0.19) is prominent on the expense of the educt neutralize the reaction mixture with HOAc.
2. Remove the solvents in a nitrogen jet. Dissolve the residue in MeOH/H_2O (1:1, v/v) and purify by reverse phase chromatography as described for lyso-GM1 before.
3. Collect fractions of pure lyso-GM1, evaporate the solvents and freeze-dry from a minimum amount of H_2O. Yield of lyso-

GM1 is 9.9 mg (7.7 μmol, 80%). ESI-MS (positive mode): m/z = 1280.6 $[M+H]^+$ (*see* **Note 9**).

3.5 ATTO647N-Labeled Gangliosides

For the studies of ganglioside behavior in model or live cell membranes by fluorescence correlation spectroscopy (FCS) in either conventional microscopy or stimulated emission depletion (STED) nanoscopy analogues of gangliosides are required that bear a photostable fluorescent tag. Among others the commercially available fluorescent and lipophilic dye ATTO647N has proven suitable for STED and FCS experiments [6] and is also applicable in conventional high-resolution microscopy. This dye comes with a short C4 linker and may replace the acyl chain of gangliosides. On the other hand and for comparative studies this ATTO-tag can replace the N-acetyl group of the sialic acid residue in monosialogangliosides, thus labeling the hydrophilic portion of the molecule. And thirdly, the label may be hooked onto the α- or ω-position of the acyl chain leaving the native ganglioside skeleton formally intact (*see* Fig. 5 for the various position of the ATTO label in ganglioside GM1). The α-position of the long acyl chain residue is favorable for the attachment both of lipophilic and hydrophilic dyes as it is adjacent to the interface where the lipophilic core meets the aqueous environment, and thus likely causing little, if any, membrane perturbation [18]. On the other hand, derivatization of the α-position creates a stereogenic center causing two diastereomeric ganglioside derivatives to occur. This may complicate interpretation of analytical data, but so far has not been observed [18]. Attaching the label at the end of the acyl chain removes possible complications associated with the presence of two diastereomers [38], is, however limited to lipophilic labels as polar labels may cause looping of the acyl chain and placing the label to the interface, thereby causing concomitant increase of the off-rate of such labeled ganglioside [39].

3.5.1 Synthesis of Gangliosides GM1, GM2, and GM3 Where ATTO647N Substitutes for the Native Acyl Chain, i.e., ATTO-Lyso-GM1, -GM2, and -GM3

1. In a small screw-cap vial dissolve the respective lyso-ganglioside (about 1 μmol) in 150 μL DMF and 100 μL MeOH. Add 10 μL freshly distilled DIPEA and then solid ATTO647N-NHS ester (0.9 mg, 1.2 μmol). Protect from light and stir at room temperature for 20 h.
2. Dry a small aliquot (1–2 μL) of the reaction mixture in a nitrogen jet. Dissolve the residue in 20 μL $CHCl_3/MeOH/H_2O$ (60/35/8, v/v/v) and perform a TLC check with the same solvent mixture. Usually TLC reveals a 50–70% conversion of the lyso-ganglioside. Add another μmol of ATTO647N-NHS ester and continue stirring for 20 h. Thereafter, the formation of ATTO-lyso-ganglioside should be about or better than 90%.
3. Dry the reaction mixture in a nitrogen jet, dissolve the residue in solvent as above and subject to flash chromatography with step gradients of $CHCl_3/MeOH$ (4:1, v/v), $CHCl_3/MeOH/$

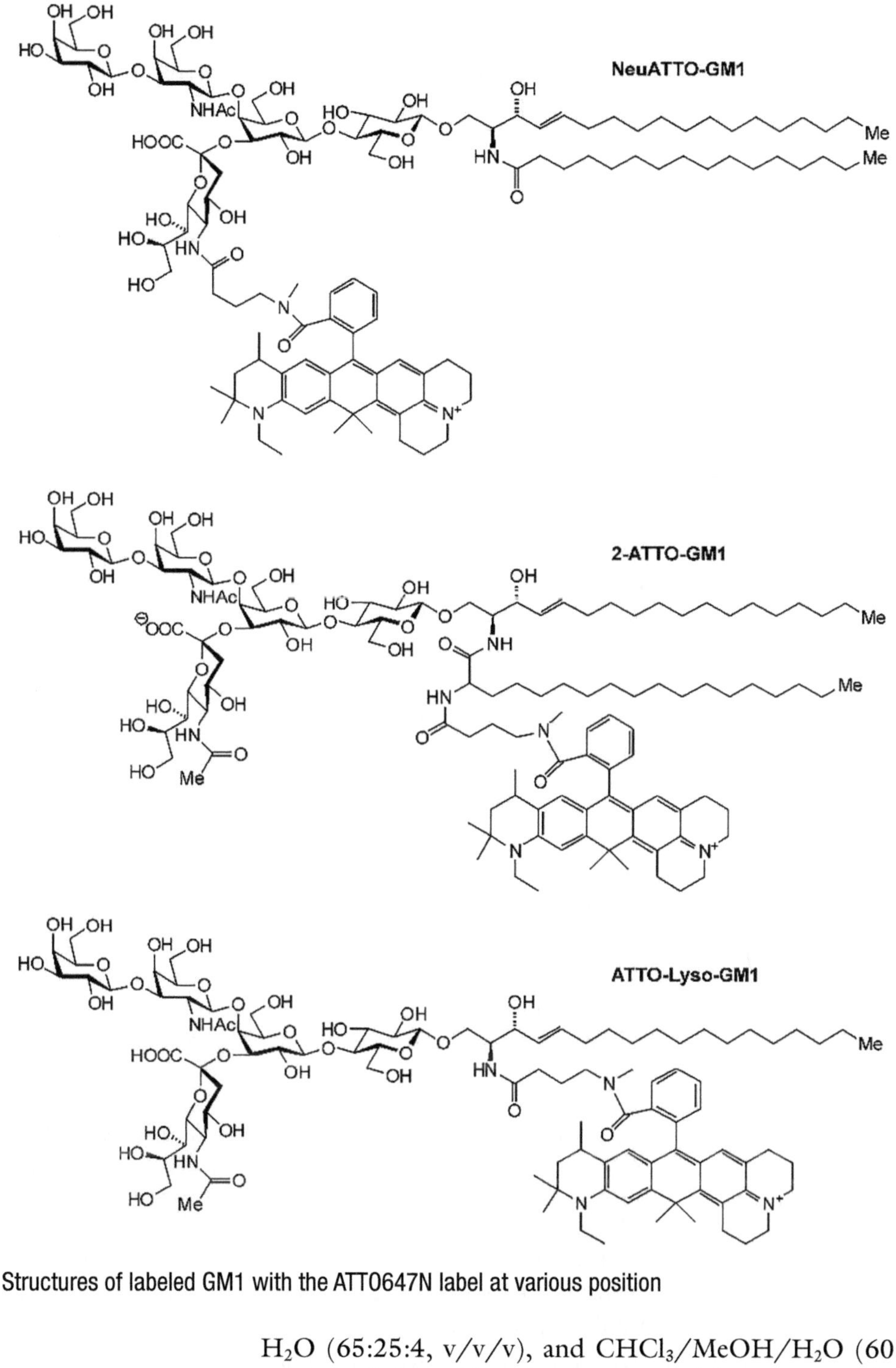

Fig. 5 Structures of labeled GM1 with the ATTO647N label at various position

H_2O (65:25:4, v/v/v), and $CHCl_3$/MeOH/H_2O (60:35:8, v/v/v). Collect blue fractions containing pure products as checked by TLC with $CHCl_3$/MeOH/H_2O (60:35:8, v/v/v); R_f 0.37, 0.49, and 0.52 for ATTO-lyso-GM1, –GM2, and -GM3, respectively (*see* **Note 1**).

4. Dry the pure fractions in a nitrogen jet, dissolve the residue in an amount as little as possible of water and lyophilize. Yield of purified products is in the range of 70%. ESI-MS confirms structure. ESI-MS (positive mode): *m/z* = 977.0 and 977.5 $[M+H+NH_4]^{2+}$ for the C20:1 and 963.5 and 963.0 $[M+H+NH_4]^{2+}$ for the C18:1 sphingosine-containing ATTO-lyso-GM1 (*see* **Note 10**). ESI-MS for ATTO-lyso-GM2 and -GM3 (containing C18:1 sphingosine) *m/z* = 882.0 and 882.5 $[M+H+NH_4]^{2+}$, and 781.0 and 781.5 $[M+H+NH_4]^{2+}$, respectively.

3.5.2 Synthesis of Gangliosides GM1 and GM3 with the Label Substituting for the Acetyl Residue in the Sialic Acid, i.e., NeuATTO-GM1 and -GM3

Reaction of Fmoc-deAc-Lyso-GM1 and -GM3 with ATTO647N-NHS Ester

1. In a screw-cap vial dissolve Fmoc-deAc-lyso-GM1 or -GM3 (3 μmol, 4.5 and 3.4 mg, respectively) in 250 μL DMF and 20 μL freshly distilled DIPEA. To the solution add solid ATTO647N-NHS ester (3.7 mg, 5 μmol) and stir at room temperature for 48 h. Check the progress of reaction by TLC with $CHCl_3/MeOH/H_2O$ (60:35:8, v/v/v); R_f 0.49 and 0.62 for NeuATTO-Fmoc-lyso-GM1, and -GM3, respectively. If the product formation is less than 80% add more solid ATTO647N-NHS ester.
2. Dry the mixture in a nitrogen jet and subject the residue to chromatographic purification on silica gel Si 60 with step gradients of $CHCl_3/MeOH$ (4:1, v/v), $CHCl_3/MeOH/H_2O$ (65:25:4, v/v/v), and $CHCl_3/MeOH/H_2O$ (60:35:8, v/v/v). Collect blue fractions containing pure products as checked by TLC with $CHCl_3/MeOH/H_2O$ (60:35:8, v/v/v).
3. Take pure fractions to dryness as before, and freeze-dry from a minimum of H_2O. Yield of pure product is about 60–68%. ESI-MS (positive mode): *m/z* = 2117.1 as well as 2116.1 $[M+H]^+$ and 1067.5 as well as 1067.0 $[M+H+NH_4]^{2+}$ for the C20:1, and 2089.1 as well as 2088.1 $[M+H]^+$ and 1053.5 as well as 1053.0 $[M+H+NH_4]^{2+}$ for the C18:1 sphingosine-containing NeuATTO-Fmoc-lyso-GM1, respectively (*see* **Note 10**).

Deprotection of NeuATTO-Fmoc-Lyso-Gangliosides

1. Dissolve NeuATTO-Fmoc-lyso-GM1 and -GM3 (1.5 μmol, 3.2 mg, and 2.6 mg, respectively) in 200 μL MeOH and 150 μL piperidine. Keep solution at room temperature for 2 h. Check for complete Fmoc removal by TLC with $CHCl_3/MeOH/H_2O$ (60:35:8, v/v/v). For the blue spot of NeuATTO-lyso-GM1 and -GM3 the R_f value is 0.46, and 0.50, respectively.
2. To completely remove piperidine evaporate solvents azeotropically by several additions of ethanol and benzene in a nitrogen jet. The respective lyso-compounds are purified by chromatography as above. Pure fractions are collected and dried in a nitrogen jet. Yield is in the range of 90%. ESI-MS (positive

mode): m/z = 948.0 as well as 947.5 $[M+2H]^{2+}$, and 934.0 as well as 933.5 $[M+2H]^{2+}$for NeuATTO-lyso-GM1 containing C20:1 and C18:1 sphingosine, respectively.

Reacylation of NeuATTO-Lyso-GM1 and -GM3

1. Dissolve dry residue in 110 μL DMF and 20 μL DIPEA. Add solid hexadecanoic acid succinimidyl ester (2 mg, 5.6 μmol) and stir the mixture for 40 h at room temperature.
2. Check for almost complete acylation by TLC with $CHCl_3$/ $MeOH/H_2O$ (60:35:8, v/v/v). The products display blue spots with R_f 0.58 and 0.66 for NeuATTO-GM1 and -GM3, respectively (*see* **Note 1**). Usually product formation is better than 90%. Take the reaction mixture to dryness in a nitrogen jet and subject the residue to chromatographic purification. Yield of purified NeuATTO-GM1 and -GM3 is about 1.1 and 1.3 μmol (73% and 86%, respectively). ESI-MS (positive mode): m/z = 1075.6 as well as 1075.1 $[M+H+NH_4]^{2+}$, and 1061.6 as well as 1061. $[M+H+NH_4]^{2+}$ for NeuATTO-GM1 containing C20:1 and C18:1 sphingosine, respectively (*see* **Note 10**).

3.6 Synthesis of GM1 with the ATTO Label Attached to the Alpha Position of Its Acyl Chain, i.e., 2-ATTO-GM1

3.6.1 Synthesis of 2-Fmoc-GM1

1. Dissolve lyso-GM1 (4.6 mg, 3.5 μmol) in 300 μL MeOH and 10 μL DIPEA. Add N-Hydroxysuccinimidyl (2R/S)-[(9-Fluorenylmethoxycarbonyl)amino]stearate, i.e., 2-Fmoc-aminostearic acid NHS ester, (3.1 mg, 5 μmol) dissolved in 100 μL DCM. Stir for 2 h at room temperature. Check reaction by TLC with $CHCl_3/MeOH/H_2O$ (60:35:8, v/v/v) against GM1 (R_f 0.30) (*see* **Note 1**). The product yields a fluorescent quenching spot (R_f 0.40). If less than 90% reaction add more NHS ester and keep the mixture overnight.
2. Dry the reaction mixture in a nitrogen jet and purify by flash chromatography with $CHCl_3/MeOH/H_2O$ (60:35:8, v/v/v). Collect pure fractions, dry, and freeze-dry the residue from a minimum of water. Yield of pure product 3.8 mg (2.1 μmol, 60%). ESI-MS (negative mode): m/z = 1782.0 and 1810.0 $[M-H]^-$ for the C18:1 and C20:1 sphingosine-containing products, respectively.

3.6.2 Deprotection of 2-Fmoc-GM1

1. Dissolve 2-Fmoc-GM1 (3.6 mg, 2 μmol) in 200 μL MeOH, add 100 μL piperidine and keep the solution for 2 h at room temperature. TLC with $CHCl_3/MeOH/H_2O$ (60:35:8, v/v/v) in comparison to GM1 (R_f 0.30) should show complete loss of the Fmoc educt and formation of two ninhydrin-positive new spots, representing the diastereomeric forms of 2-NH_2-GM1 (R_f 0.32 and 0.28).
2. Dry the reaction mixture azeotropically with repeated addition of benzene or toluene to assure complete removal of piperi-

dine. Purify by chromatography on silica gel as above. Collect and dry the pure fractions of diastereomeric 2-NH_2-GM1 and freeze-dry from a minimum of water. Yield 2.9 mg (1.84 μmol, 92%). ESI-MS (positive mode): m/z = 1589.9 and 1561.9 $[M+H]^+$ and m/z = 803.9 and 789.9 $[M+H+NH_4]^{2+}$ for the C20:1 and C18:1 sphingosine-containing products, respectively.

3.6.3 Synthesis of 2-ATTO-GM1

1. Dissolve 2.4 mg (1.5 μmol) 2-NH_2-GM1 in 0.3 mL DMF and 10 μL DIPEA with slight warming and sonication. Cool the solution in an ice-bath and add solid ATTO647N-NHS (2.2 mg, 2.9 μmol). Stir the solution at room temperature for 24 h and check reaction progress by TLC with $CHCl_3$/MeOH/15 mM $CaCl_2$ (60:35:8, v/v/v) in comparison to GM1 (R_f 0.30). The title compound displays a blue spot (R_f 0.45) (*see* **Note 1**). If copper spray and heating [40] reveals a formation of about 90% (as visualized by spot density) of the title compound the solvents are dried in a nitrogen jet.
2. Dissolve the blue residue in a minimum of $CHCl_3$/MeOH/H_2O (60:35:8, v/v/v) (*see* **Note 11**) and apply the solution as a streak onto a TLC plate (10 × 10 cm) and develop the plate in a vapor saturated chromatography chamber. A blue band is clearly separated from side products of ATTO647N-NHS ester. Using a razor blade scrape the band (*see* **Note 12**) of 2-ATTO-GM1 and elute the silica gel with $CHCl_3$/MeOH/H_2O (60:35:5, v/v/v). Centrifuge the blue solution and reextract the silica gel until it remains almost colorless. Dry the solvents and freeze-dry the residue from a minimum of water. Yield is 2.2 mg (1.0 μmol, 67%). ESI-MS (positive mode): m/z = 1120.1 and 1120.6 $[M+H+Na]^{2+}$ for the C20:1 and 1106.1 and 1106.6 $[M+H+Na]^{2+}$ for the C18:1 sphingosine-containing 2-ATTO-GM1 (*see* **Note 10**).

3.7 NBD-Labeled Gangliosides

To investigate the ganglioside binding and transfer activity of saposins apart from its interaction with the enzyme, a simple transfer assay based on Förster-resonance energy transfer (FRET) [41] was developed that allows real time measurements. So far a couple of sphingolipids including gangliosides GM2 and GM1 have been synthesized that in addition to their natural skeleton contain a 7-nitrobenz-2-oxa-1,3-diazol (NBD) group at the α-position of their acyl chain [18]. Although this introduces an additional stereogenic center in the molecule this is no shortcoming as the R and S forms are easily separated by chromatographic means. On the other hand placing this label at the ω-position would omit a new stereo center, however, would cause looping of the acyl chain and placing the NBD group to the membrane surface [39, 42]. This would render such labeled lipids inadequate for transfer assays because of their spontaneous transfer between membranes even in the absence of transfer proteins. Hence an uncompromised transfer assay requires transfer lipids with a low off-rate from membranes [43]. This criterion is met by lipids bearing the label at α-position of their long acyl chain but is not fulfilled with labels such as NBD or BODIPY placed at the terminal end of the acyl chain [44–46]. This may also not be fulfilled for a polar rhodamine label such as KK114 [47]. Furthermore, the quenching lipid must not be transferable at all. This requisite is met by bipolar membrane-spanning lipids [12]. The 7-nitrobenz-2-oxa-1,3-diazol (NBD) group was chosen as the donor fluorophore and a rhodamine group as a quenching fluorophore. The structure of the fluorescent ganglioside analogue β-D-galactopyranosyl-(1,3)-2-acetamido-2-deoxy-β-D-galactopyranosyl-(1,4)-[α-D-neuraminyl-(2,3)]-β-D-galactopyranosyl-(1,4)-β-D-glucopyranosyl-(1,1)-(2S,3R,4E)-2-[(2-N-(7-nitrobenz-1,3-diazol-2-oxa-4-yl)-amino)-octadecanamido]-4-octadecen-1,3-diol (2-NBD-GM1) is shown in Fig. 6.

3.7.1 Synthesis of Gangliosides GM1 and GM2 with the NBD-Label at the α-Position of Their Long Acyl Chain (2-NBD-GM1 and 2-NBD-GM2)

1. In a small screw-cap vial dissolve the respective lyso-ganglioside (about 2 μmol) in 150 μL DMF and 100 μL THF. Add 10 μL freshly distilled DIPEA and then a solution of 4 μmol 2-NBD-octadecanoic acid-NHS ester in 200 μL EtOAc. Remove about 200 μL solvent in a nitrogen jet and stir the reaction mixture in the dark at room temperature for 20 h.
2. Check progress of reaction by TLC with $CHCl_3$/MeOH/15 mM $CaCl_2$ (55:45:10, v/v/v). With vapor saturation in a horizontal chamber R_f values are found for 2-R-NBD-GM2 (0.45), 2-S-NBD-GM2 (0.42), 2-R-NBD-GM1 (0.37), and 2-S-NBD-GM1 (0.36) (*see* **Note 1**). If reaction is complete dry solvents and purify 2-NBD GM1 and -GM2 by chromatography using a step gradient of $CHCl_3$/MeOH/H_2O (65:25:4, v/v/v) and $CHCl_3$/MeOH/H_2O (60:35:8, v/v/v). Pool the respective pure fractions and dry (*see* **Note 13**).Yield of pure products is in the range of 80–90%. ESI-MS (negative mode): m/z = 874.9 $[M-2H]^{2-}$ for the C20:1 and 860.9 $[M-2H]^{2-}$ for the C18:1 sphingosine-containing 2-NBD-GM1, m/z = 779.8 $[M-2H]^{2-}$ for the C18:1 sphingosine-containing 2-NBD-GM2. ESI-MS (positive mode): m/z = 814.9 $[M+3Na]2^{+}$ for the sodium salt of 2-NBD-GM2.

3.7.2 Synthesis of Gangliosides with the NBD-Label at the ω-Position of a Short Acyl Chain

These syntheses use lyso-gangliosides and commercially available N-[6-[(7-Nitro-2-1,3-benzoxadiazol-4-yl)amino]hexanoic acid succinimidyl ester. A synthetic approach is given for NBD-C6-lyso-gangliosides that proved useful for the investigation of endocytosis, intracellular distribution and metabolism [11, 19].

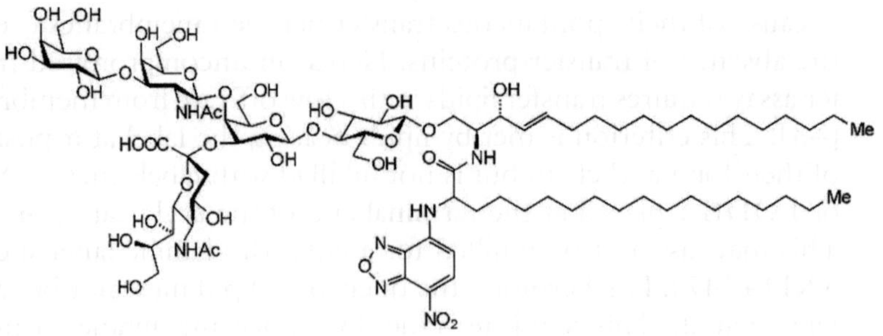

2-NBD-GM1

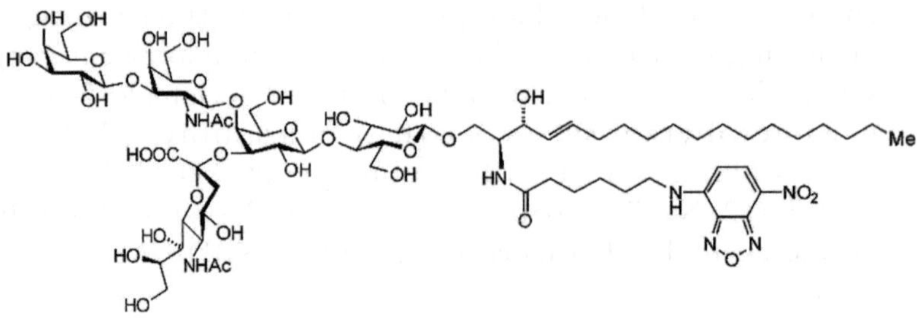

NBD-C6-GM1

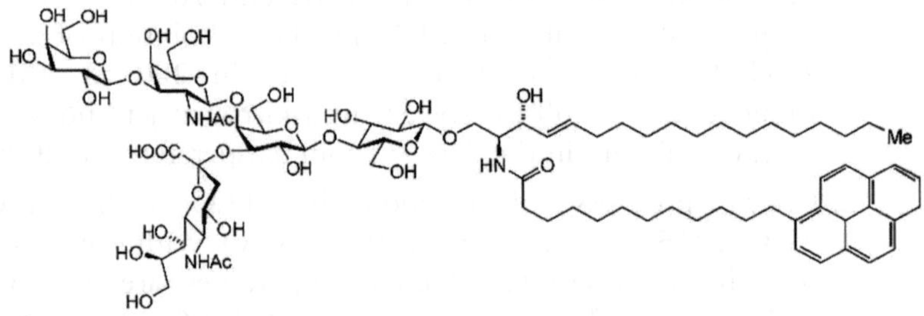

Pyr-GM1

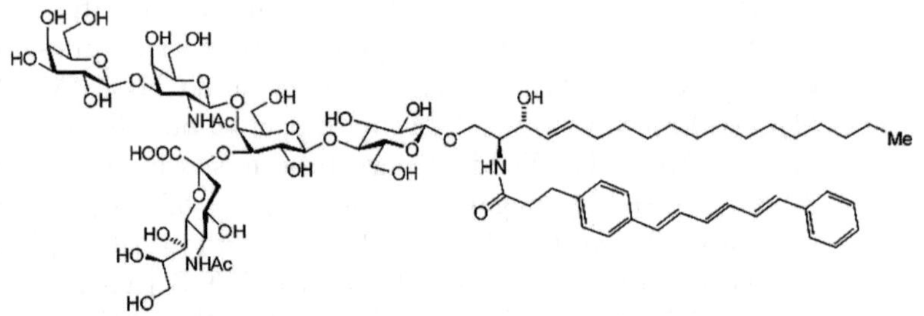

DPH-GM1

Fig. 6 Structures of fluorescently labeled GM1

1. Under stirring dissolve lyso-ganglioside (2 μmol) in 100 μL DMF and 50 μL DCM and add 10 μL DIPEA. A clear solution should result. Then add solid N-[6-[(7-Nitro-2-1,3-benzoxadiazol-4-yl)amino]hexanoic acid succinimidyl ester (1.6 mg, 4 μmol) and stir at room temperature for 20 h.

2. Pursue the reaction progress by TLC with $CHCl_3$/MeOH/15 mM $CaCl_2$ (55:45:10, v/v/v). The NBD-C6-lyso-gangliosides yield yellow spots (green fluorescent under 366 nm UV lamp) that give rise to dark brown spots when sprayed with $CuSO_4$ followed by charring [40]. For the title compounds with C18:1 sphingosine the following R_f-values are observed: 0.49, 0.37, 0.29, and 0.18 for NBD-C6-lyso-GM3, -GM2, -GM1, and GD1a, respectively.
3. To purify the NBD-C6-lyso-gangliosides from excess ester or side products thereof and from remaining lyso-ganglioside, if any, use either preparative TLC or flash chromatography. The purified compounds may be stored in MeOH in a freezer for years.

3.8 Pyrene-Labeled Gangliosides (Pyr-Gangliosides)

Pyrene-labeled gangliosides are well suited to study, by use of the excimer technique, the formation of ganglioside micelles as well as their distribution and lateral diffusion in model membranes. For these studies 12-(1-pyrenyl)dodecanoic acid was substituting for the native acyl chain in gangliosides GM3, GM2, GM1, GD1a, and GD1b [10, 48].

3.8.1 Synthesis of Pyr-Gangliosides

1. Prepare N-succinimidyl pyrenedodecanoate as described [16].
2. Prepare a solution of Iyso-GM3, -GM2, -GM1, -GD1a, or -GD1b (5 μmol) in DMF (0.4 mL) and DIPEA (20 μL).
3. Dissolve N-succinimidyl pyrenedodecanoate (25 mg, 50 μmol) in DMF (0.5 mL).
4. Add 0.1 mL of the above solution to each of the lyso-ganglioside solution.
5. Keep each of the resulting clear mixture under argon at 30 °C in the dark for 48 h.
6. Monitor the reaction progress by TLC with $CHCl_3$/MeOH/15 mM $CaCl_2$ (60:35:8, v/v/v) or $CHCl_3$/MeOH/15 mM $CaCl_2$ (60:40:9, v/v/v) for Pyr-GM3, -GM2, -GM1 or -GD1a and -GD1b, respectively. The Pyr-gangliosides (for structure of Pyr-GM1 *see* Fig. 6) have higher R_f values than their parent gangliosides and are visible both in UV light (366 nm) as well as after spraying with $CuSO_4$ [40]. If significant amounts of lyso-gangliosides should still be present in the reaction mixture, it may be helpful to add more of the active ester.
7. Remove the solvents in a nitrogen jet and purify Pyr-gangliosides by flash chromatography on LiChroprep Si 100 (0.6 × 90 cm). Pyr-GM3, -GM2, and -GM1 are eluted with a linear gradient of $CHCl_3$/MeOH/H_2O (60:35:8, v/v/v) and

$CHCl_3$/MeOH/H_2O (60:40:9, v/v/v) (200 mL each) and Pyr-GD1a and -GD1b are eluted with a linear gradient of $CHCl_3$/MeOH/H_2O (60:40:9, v/v/v) and $CHCl_3$/MeOH/H_2O (55:45:10, v/v/v) (200 mL each).

8. Collect fractions with pure Pyr-gangliosides as judged by TLC and remove solvents in a nitrogen jet.
9. Dissolve each residue in a minimum of H_2O and freeze-dry. The yield is generally about 4 μmol (80%) as determined by weight and sialic acid analysis.

3.9 Synthesis of 2-Aminostearic Acid Derivatives (See Fig. 7 for a Synthetic Scheme)

3.9.1 Synthesis of Racemic 2-NBD-Aminostearic Acid-N-Succinimidyl Ester

tert-Butyl 2-Azidostearate

1. Dissolve 2-azido-stearic acid (330 mg, 1 mmol) in 10 mL dry DCM and 0.2 mL *tert*-butanol into a 25 mL screw-cap flask containing a magnetic stirring bar. Start reaction by adding DCC (210 mg, 1 mmol) and as catalyst DMAP (20 mg).
2. Stir solution for one week at room temperature.
3. Dilute solution with 10 mL n-hexane–ethylacetate (4:1, v/v) to crystallize out most of dicyclohexylurea overnight.
4. Filter solution over a polyethylene frit in a filtration tube (Supelco). Wash crystals twice with 4 mL n-hexane–ethylacetate (4:1, v/v).
5. Reduce volume of filtrate in a nitrogen jet. Dissolve residue in 4 mL n-hexane–ethylacetate (4:1, v/v) and purify *tert*-butyl 2-azidostearate by flash chromatography on 84 g silica gel Si 60 using 160 mL n-hexane–ethylacetate (9:1, v/v) followed by 160 mL n-hexane/ethylacetate (4:1, v/v).
6. Collect fractions that on TLC show one pure spot (R_f 0.68, n-hexane–ethylacetate (9:1, v/v). Evaporate solvent in a nitrogen jet. The product crystallizes in needles. Yield of pure product (267–280 mg) 70–73%. ESI-MS (positive mode): $m/z = 399.3$ $[M+NH_4]^+$ and 780.7 $[2\ M+NH_4]^+$.

Synthesis of *tert*-Butyl 2-Aminostearate

1. In a 10 mL screw-cap tube equipped with a stir bar dissolve *tert*-butyl 2-azidostearate (200 mg, 520 μmol) in 2 mL 2-propanol. Replace air with argon, add 25 mg PtO_2 and connect to a hydrogen cylinder. Stir at 1 bar hydrogen overnight.
2. Centrifuge to sediment platinum. Transfer clear supernatant to new tube. Rinse the reaction tube with 2-propanol, centrifuge and add the clear solution to the first supernatant. Dry the solvents in a nitrogen jet.
3. Dissolve residue in $CHCl_3$/2 M NH_3 in MeOH (10/1, v/v) and purify product by flash chromatography on 84 g silica gel Si 60 using 140 mL $CHCl_3$/2 M NH_3 in MeOH (50:1, v/v) followed by 360 mL $CHCl_3$/2 M NH_3 in MeOH (10:1, v/v). Collect all fractions that display a single and ninhydrin-positive

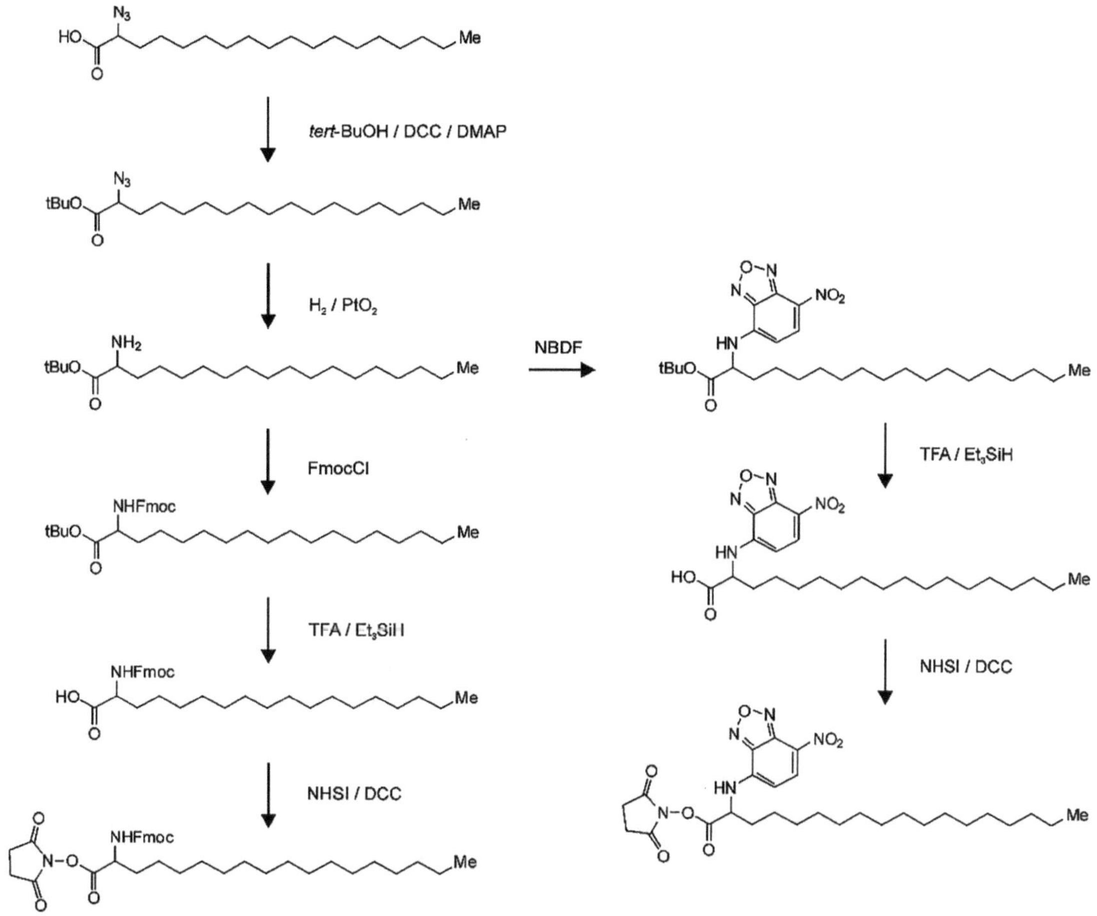

Fig. 7 Synthetic steps for the formation of 2-*N*-Fmoc-aminostearic acid hydroxysuccinimidyl ester and of 2-NBD-aminostearic acid hydroxysuccinimidyl ester

spot. Dry fractions azeotropically with the addition of two times 1 mL benzene. Residue crystallizes in colorless needles. Yield of pure product 148 mg (0.42 mmol, 80%). ESI-MS (positive mode): $m/z = 356.3$ $[M+H]^+$ and 711.7 $[2\ M+H]^+$.

Synthesis of *tert*-Butyl 2-NBD-Aminostearate

1. Dissolve *tert*-butyl 2-aminostearate (140 mg, 0.39 mmol) in 1 mL THF and 1 mL CH_2Cl_2, add 80 μL DIPEA (465 μmol). Then add 75 mg (0.41 mmol) NBDF and stir the solution at room temperature for 6 h. TLC in n-hexane–ethyl acetate (4:1, v/v) should show better than 95% production of the 2-NBD product. The solvent is reduced in a nitrogen jet.
2. Dissolve residue in n-hexane–ethyl acetate (4:1, v/v) and purify *tert*-butyl 2-NBD-aminostearate by flash chromatography on 84 g silica gel Si 60 using 40 mL n-hexane–ethylacetate (9:1, v/v) followed by 120 mL n-hexane–ethylacetate (4:1, v/v) and finally by 100 mL n-hexane/ethylacetate (7:3, v/v)

3. Collect fractions that on TLC show one single spot (R_f 0.52, n-hexane–ethylacetate (4:1, v/v) or R_f 0.80, $CHCl_3$–MeOH–HOAc (190:10:1, v/v/v). Evaporate solvent in a nitrogen jet. The product crystallizes in yellow-orange needles. Yield of pure product (180 mg, 0.35 mmol) 89%. ESI-MS (positive mode): m/z = 536.4 $[M+NH_4]^+$ and 1054.7 $[2M+NH_4]^+$.

Synthesis of 2-NBD-Aminostearic Acid *N*-Succinimidyl Ester

1. Dissolve *tert*-butyl 2-NBD-aminostearate (156 mg, 0.3 mmol) in 2 mL CH_2Cl_2 and 1 mL CF_3COOH. Add 100 μL Et_3SiH and keep overnight. TLC in n-hexane–ethyl acetate (4:1, v/v) shows complete conversion of the *tert*-butyl ester into free acid (R_f 0.12). TLC in $CHCl_3$/MeOH/HOAc (190:10:1, v/v/v) gives R_f 0.20. Dry product with a nitrogen jet. ESI-MS (positive mode): m/z = 463.3 $[M+H]^+$ as well as 480.3 $[M+Na]^+$.
2. Dissolve residue in anhydrous ethyl acetate (5 mL) in a screw-cap vial equipped with a stir bar. Add NHSI (50 mg, 0.43 mmol) and stir at room temperature for 1 h. Then add DCC (62 mg, 0.3 mmol) and stir overnight. Keep the orange-yellow solution under argon in a refrigerator until use. The success of this reaction can be checked by TLC in $CHCl_3/CH_3OH/CH_3COOH$ (190:10:1, v/v/v). The succinimidyl ester (R_f 0.72) is well separated from remaining free acid, if any. For acylation of lyso-gangliosides the ester can be used even in the presence of some free acid since the latter does not interfere with acylation and is easily separated from fluorescent gangliosides in their chromatographic purification.
3. For characterization the succinimidyl ester may be purified by chromatography. ESI-MS (positive mode) furnishes main signals at m/z = 479.3 $[M+NH_4]^+$ and 940.6 $[2\ M+NH_4]^+$ ascribable to 2-NBD-octadecanoylamide that spontaneously forms from the active ester in the presence of ammonium acetate.

3.9.2 Synthesis of *N*-Fmoc-2-Aminostearic acid *N*-Succinimidyl Ester

Synthesis of *tert*-Butyl 2-*N*-Fmoc-Aminostearate

1. Under an argon atmosphere dissolve *tert*-butyl 2-aminostearate (140 mg, 0.39 mmol) prepared as above (Subheading "Synthesis of *tert*-Butyl 2-Aminostearate") in 5 mL THF. Add FmocCl (110 mg, 0.42 mmol) and $NaHCO_3$ (84 mg, 1 mmol). Stir the mixture at room temperature in the dark overnight.
2. Slightly acidify the mixture by adding a few drops of 1 M HCl and remove most of THF in a nitrogen stream. To the residue add 2 mL water and 5 mL ethyl acetate. Vortex and centrifuge, collect the upper phase and reextract the aqueous phase twice with 2 mL ethyl acetate. Dry the combined upper phases azeotropically with the addition of two times 1 mL benzene in a nitrogen jet.
3. The oily clear residue is dissolved in 4 mL n-hexane–ethyl acetate (4:1, v/v) and chromatographed on 84 g silica gel Si 60.

Fractions containing a single and fluorescence quenching material (R_f 0.59) on TLC in n-hexane–ethyl acetate (4:1, v/v) are collected and dried in a nitrogen jet. The oily residue crystallizes overnight. Yield 180 mg (0.31 mmol, 80%). ESI-MS (positive mode): m/z = 600.4 $[M+Na]^+$ and 1177.8 $[2\ M+Na]^+$.

Synthesis of ***N***-Fmoc-2-Aminostearic Acid

1. Dissolve 180 mg *tert*-butyl 2-*N*-Fmoc-aminostearate in 2 mL DCM and 1 mL TFA and add 0.1 mL Et_3SiH. After 2 h at room temperature dilute reaction mixture with 5 mL benzene and remove solvents to near dryness in a nitrogen jet. Add further 5 mL benzene and dry.
2. Dissolve residue in 3 mL DCM and apply to chromatography on 84 g silica gel Si 60 with chloroform/methanol/acetic acid (190:10:1, v/v/v) as eluent. Collect all fractions that show a single fluorescence quenching spot (R_f 0.40) on TLC in the same solvent. Yield of colorless crystals 129 mg (0.25 mmol, 80%).

Synthesis of ***N***-Fmoc-2-Aminostearic Acid Succinimidyl Ester

1. Dissolve 126 mg (0.24 mmol) N-Fmoc-2-aminooctadecanoic acid in 6 mL DCM, add 1 mL benzene, 9 mg DMAP, and 35 mg (0.3 mmol) N-hydroxysuccinimide.
2. Stir mixture under mild warming to dissolve most of N-hydroxysuccinimide. Then reduce volume in a nitrogen jet to about 5 mL to remove traces of water, if any, azeotropically.
3. Add 50 mg (0.24 mmol) DCC and stir at room temperature overnight in an argon atmosphere.

Control reaction by TLC in chloroform/methanol/acetic acid (190:10:1, v/v/v). Product shows fluorescence quenching spot (R_f 0.84). If more than a trace of free acid is displayed, further 2–4 mg DCC may be added and stirring continued for a couple of h (*see* **Note 14**).

The clear supernatant containing the title compound can be used without further purification for acylation of lyso-gangliosides. If necessary for NMR characterization, purification can be achieved by flash chromatography on silica gel with EtOAc as eluent [38]. The NMR data are found in [38].

3.10 Synthesis of DPH-GM1

The diphenylhexatriene (DPH) is a valuable tool to study fluidity in lipid bilayer membranes. Since this is a very lipophilic probe lacking a polar end it moves freely within membrane in a lateral as well as in transversal direction. Thus, it is desirable to obtain a probe that is restricted in its movement. In the past a diphenylhexatrienyl propionic acid (DPH-propionic acid) has been employed with success in the study of bilayer thickness and fluidity [49]. On the other hand, a DPH-ganglioside in which part of the acyl chain is replaced by the DPH residue has shown to be suitable for FRET studies. These studies made use of the energy transfer of an exited DPH to NBD. We have thus synthesized a DPH-GM1 in which a DPH-propionyl residue substitutes for the native acyl chain thereby approaching its native chain length [8, 50]. As the DPH-propionic acid is no longer commercially available I shall outline a brief protocol for its synthesis that makes use of commercially available 3-phenylpropionic acid.

3.10.1 4½-Bromomethyl-3-Phenylpropionic Acid

Conversion of 3-phenylpropionic acid to 4′-bromomethyl-3-phenylpropionic acid follows, with slight modification, the procedure described for the synthesis of benzyl bromide from benzene [51].

1. Add to 60 g (400 mmol) 3-phenylpropionic acid in 30 mL glacial acetic acid 12.1 g (400 mmol) paraformaldehyde and 41 g (400 mmol) finely powdered dry NaBr.
2. To this mixture add slowly under vigorous stirring a mixture of 40 g conc. H_2SO_4 and 20 mL HOAc over a period of 3 h taking care not to let the temperature exceed 80 °C. Thereafter, stir the mixture for another 9 h at 80 °C, and then keep for 8 h at 80 °C without stirring (*see* **Note 15**).
3. As soon as the mixture has reached room temperature carefully mix with 200 mL cold water
4. Extract the product into 200 mL toluene. Repeat this extraction and combine the organic phases in a separatory funnel. Wash several times with water and aqueous $NaHCO_3$ and dry the organic phase over anhydrous Na_2SO_4.
5. Remove toluene with a rotary evaporator and crystallize the residue from n-hexane/acetone. Yield 30%, mp 173–177 °C (uncorrected) (*see* **Note 16**).

3.10.2 (4-(3-Carboxy-propyl)Benzyl) Triphenylphosphonium Bromide

1. Dissolve 12.1 g (50 mmol) of 3-(4-bromomethylphenyl)propionic acid and 14 g (53 mmol) Ph_3PH in 250 mL dry toluene and heat under reflux for 2–3 h until a white precipitate has formed.
2. After cooling in an ice bath for 1 h collect the solid on a filter and wash with ether. Reduce the filtrate to one third in a rotary evaporator and cool again to 0 °C. Isolate the crystals and combine with the first crop of solid phosphonium bromide.
3. Recrystallize the phosphonium bromide from Et_2O/EtOH. Total yield is about 20.2 g (80%).

3.10.3 Synthesis of DPH-Propionic Acid (3-(4-((1E,3E,5E)-6-Phenylhexa-1,3,5-trienyl) Phenyl)Propionic Acid)

1. Under argon atmosphere dissolve 5.1 g (10 mmol) phosphonium bromide and 1.59 g (10 mmol) (2E,4E)-5-phenylpenta-2,4-dienal (prepared in a one pot synthesis as described [52]) in 50 mL MeOH. Add 20 mmol $NaOCH_3$ in 20 mL MeOH and stir at room temperature for 2 h.

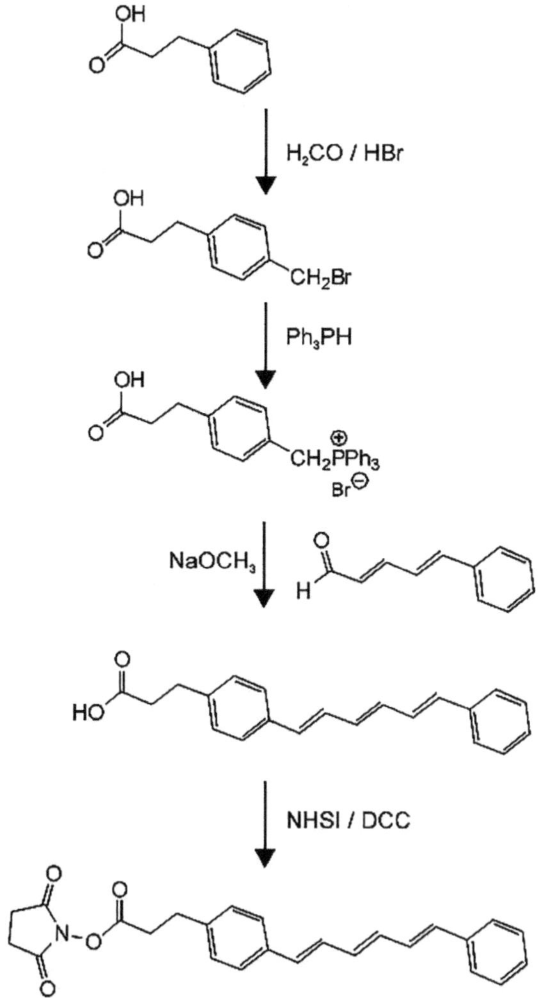

Fig. 8 Synthetic steps in the formation of DPH-propionic acid hydroxysuccinimidyl ester

2. Isolate the solid precipitate formed in this Wittig reaction [53], which is the sodium salt of the product, by filtration and wash once with cold 50% aqueous EtOH.
3. Suspend this salt in H_2O, acidify with 3 N HCl and extract the free DPH-propionic acid into EtOAc using a separatory funnel.
4. Remove the solvent in a rotary evaporator and crystallize the residue from aqueous acetone. The yield is in the range of 50%, mp 250–255 °C (*see* **Note 17**) (Fig. 8).

3.10.4 Synthesis of 3-(4-((1E,3E,5E)-6-Phenylhexa-1,3,5-Trienyl)Phenyl)Propionyl)-lyso-GM1 (DPH-GM1)

DPH-GM1 is synthesized as described for Pyr-GM1 starting from lyso-GM1 and the NHS ester of DPH-propionic acid [50]. ESI-MS (negative mode): m/z = 1592.8 and 1564.8 $[M-H]^-$ for DPH-GM1 with C20:1 and C18:1 sphingosine, respectively. TLC with $CHCl_3$/MeOH/15 mM $CaCl_2$ demonstrates R_f 0.34.

4 Notes

1. R_f values may vary since some parameters such as room temperature, activation of precoated TLC plates, vapor saturation in TLC chamber, and relative humidity during sample application are no constants. Thus, it is advisable to run a reference sample, if available, or a sample of similar structure on the same TLC plate and judge R_f values accordingly.
2. GM3, GD1a and GD1b were treated as for GM1. Similar yields were obtained for lysoGM3, -GD1a, and -GD1b after chromatographic purification. Data for GM3 derivatives are: ESI-MS (positive mode): m/z = 915.5 $[M+H]^+$ and 937.5 $[M+Na]^+$ for lyso-GM3 with C18:1 sphingosine, and m/z = 873.4 $[M+H]^+$ and 895.4 $[M+Na]^+$ for de-Ac-lyso-GM3. TLC with $CHCl_3$-MeOH-15 mM $CaCl_2$ (60:40:9, v/v/v) yields R_f 0.41 for lyso-GM3.
3. The deAc-Fmoc-lyso-gangliosides are isolated when a label is to be introduced into the sugar portion of gangliosides such as a fluorescent or an affinity tag [3–6, 54–57]. If this is not required then the deAc-lyso-gangliosides can be readily converted to Fmoc-lyso-ganglioside in a one-pot reaction by forming first the deAc-Fmoc-lyso-gangliosides prior to adding Ac_2O.
4. This purification step can be omitted if TLC shows a purity of better than 95% and reacylation follows with an active ester of a labeled fatty acid yielding a labeled ganglioside that is finally purified by chromatography.
5. It is imperative to exclude oxygen in alkaline hydrolysis. If not, the solution will darken and gangliosides will be destroyed.
6. This method, for unknown reason, is much to our regret not efficiently applicable to gangliosides GM3 [58], GD1a [22], and to gangliosides GM2, GD1b, GT1b, and GQ1b (unpublished observation by the author).
7. A protocol for the synthesis of labeled GM1 and GM3 as well as a GM1 with a thioglycosidic bond of its sugar to its ceramide residue [33] is described in [16].
8. The crude α-glycosyl bromide (96 mg, 94%) can be stored in a freezer for several weeks and be used without further purification for glycosidation.
9. The synthesis of lyso-GM3 starting from monosialolactose and following glycosidation as described for protected lyso-GM1 furnishes similar yields [16].
10. Two main masses owing to the isotope distribution are prominent in the mass spectra.

11. From the author's experience dissolution of gangliosides and their derivatives is facilitated by adding first water followed by chloroform and then breaking the two phases by methanol.
12. Scraping of streaks from TLC plates is aided when the plates are moistened with a fine spray of MeOH–H_2O (1:1, v/v) prior to scraping with either a razor blade or a small spatula.
13. Due to the R and S configuration of the NBD group the two isomers are mostly separated with little overlapping by chromatography if pure C18:1 or C20:1 containing lyso-ganglioside is used for acylation. Separation into the two isomeric forms is difficult when a mixture of both sphingoid chain length are used for acylation. More clearly separated are the respective isomers of 2-NH_2-gangliosides.
14. The free acid can be transformed into the N-hydroxysuccinimidyl ester using bis(*N*-succinimidyl) carbonate as described [38].
15. Synthesis has to be performed in an efficient hood and wearing protective gloves because of the corrosiveness of the reagents and the lachrymatory properties of the bromomethylated products.
16. A simpler method for bromomethylation was devised by [59, 60] using hydrogen bromide in acetic acid rather than generating it in situ from sodium bromide and sulfuric acid.
17. Following this protocol the corresponding DPH-butyric acid is synthesized from commercially available 4-phenylbutyric acid

Acknowledgments

The author's work was supported by the Deutsche Forschungsgemeinschaft. Grants Schw 143/5-3 and SFB 284/B5, which hereby is gratefully acknowledged.

References

1. Ledeen R (1985) Gangliosides of the neuron. Trends Neurosci 8(4):169–174. https://doi.org/10.1016/0166-2236(85)90064-5
2. Klenk E (1942) Über die Ganglioside, eine neue Gruppe von zuckerhaltigen Gehirnlipoiden. Hoppe-Seyler's Zeitschr Physiol Chem 273(1-2):76–86
3. Schwarzmann G, Arenz C, Sandhoff K (2014) Labeled chemical biology tools for investigating sphingolipid metabolism, trafficking and interaction with lipids and proteins. Biochim Biophys Acta 1841(8):1161–1173. https://doi.org/10.1016/j.bbalip.2013.12.011
4. Albrecht B, Pohlentz G, Sandhoff K, Schwarzmann G (1997) Synthesis and mass spectrometric characterization of digoxigenin and biotin labeled ganglioside GM1 and their uptake by and metabolism in cultured cells. Chem Phys Lipids 86(1):37–50
5. Möbius W, Herzog V, Sandhoff K, Schwarzmann G (1999) Intracellular distribution of a biotin-labeled ganglioside, GM1, by immunoelectron microscopy after endocytosis in fibroblasts. J Histochem Cytochem 47(8):1005–1014

6. Eggeling C, Ringemann C, Medda R, Schwarzmann G, Sandhoff K, Polyakova S, Belov VN, Hein B, von Middendorff C, Schönle A, Hell SW (2009) Direct observation of the nanoscale dynamics of membrane lipids in a living cell. Nature 457(7233):1159–1162. https://doi.org/10.1038/nature07596
7. Acquotti D, Sonnino S, Masserini M, Casella L, Fronza G, Tettamanti G (1986) A new chemical procedure for the preparation of gangliosides carrying fluorescent or paramagnetic probes on the lipid moiety. Chem Phys Lipids 40(1):71–86
8. Antes P, Schwarzmann G, Sandhoff K (1992) Distribution and metabolism of fluorescent sphingosines and corresponding ceramides bearing the diphenylhexatrienyl (DPH) fluorophore in cultured human fibroblasts. Eur J Cell Biol 59(1):27–36
9. Müller E, Giehl A, Schwarzmann G, Sandhoff K, Blume A (1996) Oriented 1,2-dimyristoyl-sn-glycero-3-phosphorylcholine/ganglioside membranes: a Fourier transform infrared attenuated total reflection spectroscopic study. Band assignments; orientational, hydrational, and phase behavior; and effects of Ca2+ binding. Biophys J 71(3):1400–1421. https://doi.org/10.1016/S0006-3495(96)79342-6
10. Ollmann M, Robitzki A, Schwarzmann G, Galla HJ (1988) Minor effects of bulk viscosity on lipid translational diffusion measured by the excimer formation technique. Eur Biophys J 16(2):109–112
11. Schwarzmann G (2001) Uptake and metabolism of exogenous glycosphingolipids by cultured cells. Semin Cell Dev Biol 12(2):163–171. https://doi.org/10.1006/scdb.2000.0233
12. Schwarzmann G, Breiden B, Sandhoff K (2015) Membrane spanning lipids for an uncompromised monitoring of membrane fusion and intermembrane lipid transfer. J Lipid Res 56(10):1861–1879. https://doi.org/10.1194/jlr.M056929
13. Schwarzmann G, Hinrichs U, Sonderfeld S, Marsh D, Sandhoff K (1986) Metabolism of exogenous gangliosides in cultured fibroblasts and cerebellar cells. In: Freysz L, Dreyfus H, Massarelli R, Gatt S (eds) Enzymes of lipid metabolism II. Springer, Boston, pp 553–562. https://doi.org/10.1007/978-1-4684-5212-9_70
14. Schwarzmann G, Hoffmann-Bleihauer P, Schubert J, Sandhoff K, Marsh D (1983) Incorporation of ganglioside analogues into fibroblast cell membranes. A spin-label study. Biochemistry 22(21):5041–5048
15. Schwarzmann G, Marsh D, Herzog V, Sandhoff K (1987) In vitro incorporation and metabolism of gangliosides. In: Rahmann H (ed) Gangliosides and modulation of neuronal functions, vol 7. NATO ASI Series. Springer, Berlin, pp 217–229. https://doi.org/10.1007/978-3-642-71932-5_18
16. Schwarzmann G, Sandhoff K (1987) Lysogangliosides: synthesis and use in preparing labeled gangliosides. Methods Enzymol 138:319–341
17. Schwarzmann G, Sandhoff K (1990) Metabolism and intracellular transport of glycosphingolipids. Biochemistry 29(49):10865–10871
18. Schwarzmann G, Wendeler M, Sandhoff K (2005) Synthesis of novel NBD-GM1 and NBD-GM2 for the transfer activity of GM2-activator protein by a FRET-based assay system. Glycobiology 15(12):1302–1311. https://doi.org/10.1093/glycob/cwj018
19. Sofer A, Schwarzmann G, Futerman AH (1996) The internalization of a short acyl chain analogue of ganglioside GM1 in polarized neurons. J Cell Sci 109(Pt 8):2111–2119
20. Wendeler M, Hoernschemeyer J, Hoffmann D, Kolter T, Schwarzmann G, Sandhoff K (2004) Photoaffinity labelling of the human GM2-activator protein. Mechanistic insight into ganglioside GM2 degradation. Eur J Biochem 271(3):614–627
21. Mraz W, Schwarzmann G, Sattler J, Momoi T, Seemann B, Wiegandt H (1980) Aggregate formation of gangliosides at low concentrations in aqueous media. Hoppe Seylers Z Physiol Chem 361(2):177–185
22. Gantner M, Schwarzmann G, Sandhoff K, Kolter T (2014) Partial synthesis of ganglioside and lysoganglioside lipoforms as internal standards for MS quantification. J Lipid Res 55(12):2692–2704. https://doi.org/10.1194/jlr.D054734
23. Schwarzmann G (1978) A simple and novel method for tritium labeling of gangliosides and other sphingolipids. Biochim Biophys Acta 529(1):106–114
24. Sarmientos F, Schwarzmann G, Sandhoff K (1985) Direct evidence by carbon-13 NMR spectroscopy for the erythro configuration of the sphingoid moiety in Gaucher cerebroside and other natural sphingolipids. Eur J Biochem 146(1):59–64
25. Pangborn AB, Giardello MA, Grubbs RH, Rosen RK, Timmers FJ (1996) Safe and convenient procedure for solvent purification. Organometallics 15(5):1518–1520. https://doi.org/10.1021/Om9503712

26. Neuenhofer S, Schwarzmann G, Egge H, Sandhoff K (1985) Synthesis of lysogangliosides. Biochemistry 24(2):525–532
27. Sonnino S, Kirschner G, Ghidoni R, Acquotti D, Tettamanti G (1985) Preparation of GM1 ganglioside molecular species having homogeneous fatty acid and long chain base moieties. J Lipid Res 26(2):248–257
28. Sonnino S, Acquotti D, Kirschner G, Uguaglianza A, Zecca L, Rubino F, Tettamanti G (1992) Preparation of lyso-GM1 (II3Neu5AcGgOse4-long chain bases) by a one-pot reaction. J Lipid Res 33(8):1221–1226
29. Kita K, Kurita T, Ito M (2001) Characterization of the reversible nature of the reaction catalyzed by sphingolipid ceramide N-deacylase. A novel form of reverse hydrolysis reaction. Eur J Biochem 268(3):592–602
30. Kurita T, Izu H, Sano M, Ito M, Kato I (2000) Enhancement of hydrolytic activity of sphingolipid ceramide N-deacylase in the aqueous-organic biphasic system. J Lipid Res 41(5):846–851
31. Ito M, Kita K, Kurita T, Sueyoshi N, Izu H (2000) Enzymatic N-deacylation of sphingolipids. Methods Enzymol 311:297–303
32. Tagawa Y, Laroy W, Nimrichter L, Fromholt SE, Moser AB, Moser HW, Schnaar RL (2002) Anti-ganglioside antibodies bind with enhanced affinity to gangliosides containing very long chain fatty acids. Neurochem Res 27(7):847–855. https://doi.org/10.1023/a:1020221410895
33. Schwarzmann G, Hofmann P, Pütz U (1997) Synthesis of ganglioside GM1 containing a thioglycosidic bond to its labeled ceramide(s). A facile synthesis starting from natural gangliosides. Carbohydr Res 304(1):43–52
34. Wiegandt H, Baschang G (1965) The Isolation of the sugar portion of glycosphingolipids by ozonolysis and fragmentation. Zeitschr Naturforsch Teil B Chem Biochem Biophys Biol Gebiete 20:164–166
35. von Nicolai H, Müller HE, Zilliken F (1978) Substrate specificity of neuraminidase from Erysipelothrix rhusiopathiae. Hoppe Seylers Z Physiol Chem 359(3):393–398
36. Sattler J, Schwarzmann G, Staerk J, Ziegler W, Wiegandt H (1977) Studies of the ligand binding to cholera toxin, II. The hydrophilic moiety of sialoglycolipids. Hoppe Seylers Z Physiol Chem 358(2):159–163
37. Albrecht B, Pütz U, Schwarzmann G (1995) Synthesis of fluorescent and radioactive analogues of two lactosylceramides and glucosylceramide containing beta-thioglycosidic bonds that are resistant to enzymatic degradation. Carbohydr Res 276(2):289–308
38. Polyakova SM, Belov VN, Yan SF, Eggeling C, Ringemann C, Schwarzmann G, de Meijere A, Hell SW (2009) New GM1 ganglioside derivatives for selective single and double labelling of the natural glycosphingolipid skeleton. Eur J Org Chem 2009(30):5162–5177. https://doi.org/10.1002/ejoc.200900645
39. Chattopadhyay A, London E (1987) Parallax method for direct measurement of membrane penetration depth utilizing fluorescence quenching by spin-labeled phospholipids. Biochemistry 26(1):39–45
40. Touchstone JC, Levin SS, Dobbins MF, Carter PJ (1981) Differentiation of saturated and unsaturated phospholipids on thin layer chromatograms. J High Resolut Chromatogr 4(8):423–424. https://doi.org/10.1002/jhrc.1240040816
41. Förster T (1949) Experimentelle und theoretische Untersuchung des zwischenmolekularen Übergangs von Elektronenanregungsenergie. Z Naturforsch 4a(5):321–327
42. Chattopadhyay A (1990) Chemistry and biology of N-(7-nitrobenz-2-oxa-1,3-diazol-4-yl)-labeled lipids: fluorescent probes of biological and model membranes. Chem Phys Lipids 53(1):1–15
43. Brown RE, Mattjus P (2007) Glycolipid transfer proteins. Biochim Biophys Acta 1771(6):746–760. https://doi.org/10.1016/j.bbalip.2007.01.011
44. Bai JN, Pagano RE (1997) Measurement of spontaneous transfer and transbilayer movement of BODIPY-labeled lipids in lipid vesicles. Biochemistry 36(29):8840–8848
45. Nichols JW, Pagano RE (1981) Kinetics of soluble lipid monomer diffusion between vesicles. Biochemistry 20(10):2783–2789
46. Nichols JW, Pagano RE (1982) Use of resonance energy transfer to study the kinetics of amphiphile transfer between vesicles. Biochemistry 21(8):1720–1726
47. Sezgin E, Levental I, Grzybek M, Schwarzmann G, Mueller V, Honigmann A, Belov VN, Eggeling C, Coskun U, Simons K, Schwille P (2012) Partitioning, diffusion, and ligand binding of raft lipid analogs in model and cellular plasma membranes. Biochim Biophys Acta 1818(7):1777–1784. https://doi.org/10.1016/j.bbamem.2012.03.007
48. Ollmann M, Schwarzmann G, Sandhoff K, Galla HJ (1987) Pyrene-labeled gangliosides: micelle formation in aqueous solution, lateral diffusion, and thermotropic behavior in phosphatidylcholine bilayers. Biochemistry 26(18):5943–5952

49. Trotter PJ, Storch J (1989) 3-[p-(6-Phenyl)-1,3,5-hexatrienyl]phenylpropionic acid (PA-DPH): characterization as a fluorescent membrane probe and binding to fatty acid binding proteins. Biochim Biophys Acta 982(1):131–139
50. Antes P, Schwarzmann G, Sandhoff K (1992) Detection of protein mediated glycosphingolipid clustering by the use of resonance energy transfer between fluorescent labelled lipids. A method established by applying the system ganglioside GM1 and cholera toxin B subunit. Chem Phys Lipids 62(3):269–280
51. Kubiczek G, Neugebauer L (1950) Experimenteller Beitrag Zur Brommethylierung Aromatischer Verbindungen. Monatsh Chem 81(6):917–920. https://doi.org/10.1007/Bf00899333
52. Bellassoued M, Salemkour M (1996) Exclusive γ-regio functionalization of crotonaldehyde using γ-trimethylsilyl crotonaldimine. Application to the one pot synthesis of conjugated dienals. Tetrahedron 52(13):4607–4624. https://doi.org/10.1016/0040-4020(96)00165-2
53. Schöllkopf U (1959) Neuere präparative methoden der organischen chemie II 13. Carbonyl-Olefinierung mit Triphenyl-phosphin-methylenen Wittig-Reaktion. Angew Chem 71(8):260–273. https://doi.org/10.1002/ange.19590710803
54. Möbius W, Herzog V, Sandhoff K, Schwarzmann G (1999) Gangliosides are transported from the plasma membrane to intralysosomal membranes as revealed by immuno-electron microscopy. Biosci Rep 19(4):307–316
55. Mueller V, Honigmann A, Ringemann C, Medda R, Schwarzmann G, Eggeling C (2013) FCS in STED microscopy: studying the nanoscale of lipid membrane dynamics. Methods Enzymol 519:1–38. https://doi.org/10.1016/B978-0-12-405539-1.00001-4
56. Mueller V, Ringemann C, Honigmann A, Schwarzmann G, Medda R, Leutenegger M, Polyakova S, Belov VN, Hell SW, Eggeling C (2011) STED nanoscopy reveals molecular details of cholesterol- and cytoskeleton-modulated lipid interactions in living cells. Biophys J 101(7):1651–1660. https://doi.org/10.1016/j.bpj.2011.09.006
57. Schwarzmann G, von Coburg A, Möbius W (2000) Using biotinylated gangliosides to study their distribution and traffic in cells by immunoelectron microscopy. Methods Enzymol 312:534–562
58. Valiente O, Mauri L, Casellato R, Fernandez LE, Sonnino S (2001) Preparation of deacetyl-, lyso-, and deacetyl-lyso-GM(3) by selective alkaline hydrolysis of GM3 ganglioside. J Lipid Res 42(8):1318–1324
59. Gates M (1982) Condensation of naphthoquinones with polar ethylenes. A reexamination. J Org Chem 47(3):578–582. https://doi.org/10.1021/jo00342a045
60. Van der Made AW, Van der Made RH (1993) A convenient procedure for bromomethylation of aromatic compounds. Selective mono-, bis-, or trisbromomethylation. J Org Chem 58(5):1262–1263. https://doi.org/10.1021/jo00057a046

Chapter 17

Synthesis of Photoactivatable and Paramagnetic Gangliosides

Günter Schwarzmann

Abstract

For the physicochemical studies aimed at elucidating the dynamic of gangliosides in membranes and their interaction with proteins and membrane lipids, photoactivatable and paramagnetic ganglioside derivatives have proved to be invaluable tools. Here, protocols for the synthesis of such ganglioside derivatives are described. These derivatives bear in their ceramide portion either a highly photoreactive (3-trifluoromethyl) phenyldiazirinyl- or a spin active doxyl-labeled acyl chain in place of their natural acyl chain.

Key words Photoactivatable gangliosides, C7-TPD-GM2, C10-TPD-GM1, (3-Trifluoromethyl) phenyldiazirine, Paramagnetic gangliosides, 9-Doxyl-GM1, 9-Doxyl-GM3, Spin-labeled stearic acid

1 Introduction

1.1 Photoactivatable Gangliosides

Apart from the phospholipids that constitute the majority of lipids gangliosides are characteristic components of the plasma membrane of vertebrate cells and are especially abundant in neuronal membranes [1]. Gangliosides of the plasma membrane are generally known to be anchored in the outer leaflet of the membrane bilayer by their hydrophobic ceramide moiety with their hydrophilic oligosaccharide residues protruding into the extracellular space (for a review *see* [2].

Physicochemical studies on model membranes and membranes of cultured cells suggested that gangliosides of natural membranes are not randomly distributed but are, rather, hierarchically concentrated in membrane domains, i.e., the so-called lipid rafts. The notion that lipid segregation occurs at the cell surface, leading to the dynamic separation of lipid rafts is still a matter of intense debate (*see* [3] and references cited therein). It has been shown that besides paramagnetic gangliosides (see below) photoactivatable gangliosides are useful tools to study ganglioside–protein [4–7] and ganglioside–lipid interaction [8]. A photoreactive GM1 has also been

Sandro Sonnino and Alessandro Prinetti (eds.), *Gangliosides: Methods and Protocols*, Methods in Molecular Biology, vol. 1804, https://doi.org/10.1007/978-1-4939-8552-4_17,

© Springer Science+Business Media, LLC, part of Springer Nature 2018

applied to titrate the water content of the hydrophobic core of model membranes [9].

Mainly three chemical residues such as phenyldiazirine, arylazide, and benzophenone are used as photoactivatable tags. Of these (3-trifluoromethyl)phenyldiazirine is the most promising highly active carbene precursor. Upon irradiation the diazirinyl group does not rearrange or isomerize to a less reactive electrophilic species. This is clearly an advantage over azides as nitrene precursors for the latter easily isomerize to inactive species [10, 11]. Furthermore, the (3-trifluoromethyl)phenyldiazirinyl group is chemically quite stable and the carbene generated upon irradiation is highly reactive and inserts even into chemically inert C-H bonds within nanoseconds. Thus, photocoupling furnishes stable covalent bonds that persist isolation and purification processes. Photolysis is nicely achieved within few seconds at around 360 nm, which for example is provided by a 2000 W Osram Ultramed lamp with an output of about 400 W in this UV range.

1.2 Paramagnetic Gangliosides

With spin-labeled gangliosides having the nitroxyl radical at various positions of their acyl chain it is possible to examine their motion at various depth of the membrane and the interaction with membrane proteins by electron spin resonance (ESR) spectroscopy [12–14]. Therefore, replacement of the native acyl chain by spin-labeled acyl chain in gangliosides is attractive and easily achieved by acylation of a lyso-ganglioside with the NHS ester of a spin-labeled fatty acid [15, 16]. For example, stearic acid labeled with a nitroxyl spin label moiety (doxyl group) at various carbons (5, 7, 9, 12, 13, 14 and 16) with respect to the carbonyl group have been used to study the flexibility gradient of membrane lipids to understand membrane fluidity conditions at different depths of their lipid bilayer organization. ESR experiments indicate that the labeled gangliosides behave similarly to natural GMI, in both the aggregation properties and the capability to be transferred from micelles to membranes of living cells [17, 18]. With doxyl-ganglioside derivatives we have been able to demonstrate unambiguously for the first time the insertion into the plasma membrane of cultured cells of ganglioside molecules that have been applied exogenously from their micellar state [18, 19]. Few spin-labeled fatty acids are commercially available such as 5- and 16-doxyl-stearic acid, others have to be synthesized. Here, I provide a protocol for the synthesis of a 9-doxyl-stearic acid NHS ester. This synthesis follows published procedures with minor modifications [20–24].

2 Materials

2.1 Chemicals

1. Acetic anhydride [1-^{14}C] ([1-^{14}C]-Ac_2O) (Biotrend, ARC).
2. Aluminum chloride ($AlCl_3$).

3. 2-amino-2-methyl-1-propanol.
4. Ammonia (NH_3).
5. Anhydrous sodium sulfate (Na_2SO_4).
6. Barium oxide (BaO) 97%.
7. Bromononane.
8. *tert*-butanol (*tert*-BuOH).
9. Calcium chloride ($CaCl_2$).
10. m-Chloroperbenzoic acid (MCPBA).
11. Copper sulfate ($CuSO_4$).
12. *N,N'*-dicyclohexylcarbodiimide (DCC).
13. *N,N*-diisopropylethylamine (DIPEA).
14. 4-Dimethylaminopyridine (DMAP).
15. Formic acid (HCOOH).
16. Hydrochloric acid (HCl).
17. Hydroxylamine hydrochloride ($NH_2OH{\cdot}HCl$).
18. *N*-Hydroxysuccinimide (NHSI).
19. Magnesium turnings (Mg).
20. Nonanedioic acid.
21. 10-phenyldecanoic acid.
22. 7-phenylheptanoic acid.
23. Potassium carbonate (K_2CO_3).
24. Potassium hydroxide (KOH).
25. Silver nitrate ($AgNO_3$).
26. Sodium hydrogen carbonate ($NaHCO_3$).
27. Sodium hydrogen carbonate ($NaHCO_3$).
28. Sodium hydroxide (NaOH).
29. Sodium sulfate anhydrous (Na_2SO_4).
30. Sulfuric acid (H_2SO_4).
31. Thionyl chloride ($OSCl_2$).
32. Toluene sulfonic acid ($TsOH{\cdot}H_2O$).
33. Toluene sulfonyl chloride (TsCl).
34. Trifluoroacetic acid (TFA).
35. Trifluoroacetic anhydride $(CF_3CO)_2O$.
36. Zinc chloride ($ZnCl_2$).

2.2 Solvents

1. Acetic acid (HOAc).
2. Benzene.
3. Chloroform ($CHCL_3$).

4. Deionized water (DI H_2O).
5. Dichloromethane (CH_2Cl_2).
6. Diethylether (Et_2O).
7. Dimethylformamide (DMF).
8. Dioxane.
9. Ethanol.
10. Ethyl acetate (EtOAc).
11. n-Hexane.
12. Methanol (MeOH).
13. Tetrahydrofuran (THF)
14. Toluene.
15. Dry solvents are obtained immediately prior to use using filtration tubes equipped with a polyethylene frit and filled with either acidic, neutral or basic aluminum oxide to remove traces of water and peroxides from ethers, and amines from DMF, respectively, similarly as reported [25].

2.3 Chromatographic Materials

1. LiChroprep Si 60 (40–63 μm) (Merck).
2. Precoated TLC plates HPTLC Silica gel 60 F_{254} (Merck).

2.4 Solutions and Mixtures

1. 1 M $NaHCO_3$.
2. 1 M NaOH.
3. 4 M NaOH.
4. HCl (37%).
5. 1 M HCl.
6. 0.1 N H_2SO_4.
7. n-hexane–EtOAc (7:3, v/v).
8. n-hexane–EtOAc (9:1, v/v).
9. n-hexane–$CHCl_3$ (3:7, v/v).
10. n-hexane–CH_2Cl_2 (2:3, v/v).
11. n-hexane–CH_3COOH (9:1, v/v).
12. $CHCl_3$–EtOAc (9:1, v/v).
13. CH_2Cl_2–Et_2O (10:1, v/v).
14. $CHCl_3$–MeOH–2 M NH_3 (60:40:9, v/v/v).
15. $CHCl_3$–MeOH–2 M NH_3 (60:35:8, v/v/v).
16. $CHCl_3$–MeOH–HOAc (190:10:1, v/v/v).
17. $CHCl_3$–MeOH–15 mM $CaCl_2$ (60:40:9, v/v/v)

3 Methods

3.1 Synthesis of Photoactivatble Gangliosides (TPD Gangliosides) (for Structure See Fig. 1)

The multistep synthesis involves esterification, Friedel-Crafts acylation, transesterification, and cyclization as well as oxidation. For synthetic steps *see* Fig. 2.

3.1.1 Synthesis of 7-(p-(3-(Trifluoromethyl) Diazirinyl)Phenyl)Heptanoic Acid N-Succinimidyl Ester

Synthesis of 7-Phenylheptanoic Acid Methyl Ester

1. Suspend 20.64 g (0.1 mol) 7-phenylheptanoic acid in 25 mL MeOH. Then add 1 mL conc. H_2SO_4 and heat the mixture under reflux for 20 h.
2. Remove the excess MeOH in a rotary evaporator and pour the oily residue into a sixfold amount of ice-cold H_2O in an appropriate separatory funnel.
3. Extract the methyl ester into Et_2O (20 mL) and repeat this extraction twice.
4. Wash the combined organic phases with H_2O (15 mL), then with 1 M $NaHCO_3$ (15 mL), and again twice with H_2O (15 mL). Dry the washed ether phase with anhydrous Na_2SO_4.
5. Remove Et_2O in a rotary evaporator under normal pressure and distill the oily residue in vacuo (at 0.07 hPa). Collect fractions boiling from 82 to 89 ° C. The yield is about 19.83 g (90 mmol, 90%). TLC in n-hexane–EtOAc (7:3, v/v) shows R_f 0.63 and 0.45 for the ester and acid, respectively (*see* **Note 1**).
   Analytical data [26]:

   ^{1}H-NMR (400 MHz, $CDCl_3$): δ = 1.27 (m, 4 H, H-4, H-5), 1.55 (m, 4 H, H-3, H-6), 2.22 (t, 2 H, $J_{2,3}$ 7.5 Hz, H-2), 2.52 (t, 2 H, $J_{6,7}$ 7.7 Hz, H-7), 3.60 (s, 3 H, OCH_3), 7.10 (m, 3 H, Bz), 7.19 (d, 2 H, J_{HH} 7.6 Hz, Bz).

   ^{13}C-NMR (100 MHz, $CDCl_3$): δ = 24.83 (1 C, CH_2, C-3), 28.85 (1 C, CH_2, C-4), 28.95 (1 C, CH_2, C-5), 31.23 (1 C, CH_2, C-6), 34.03 (1 C, CH_2, C-2), 35.83 (1 C, CH_2, C-7), 51.41 (1 C, OCH_3), 125.58 (1 C, CH-Bz_4), 128.20 (2 C, CH-Bz), 128.35 (2 C, CH-Bz), 142.65 (1 C, C-Bz_1), 174.22 (1 C, COOR).

Synthesis of 7-(p-(2,2,2-Trifluoroacetyl)Phenyl) Heptanoic Acid Methyl Ester

1. Suspend 50 g (375 mmol) anhydrous $AlCl_3$ in about 100 mL $CHCl_3$. Cool this suspension in an ice-bath and slowly add 14 mL $(CF_3CO)_2O$ (21 g, 100 mmol) with stirring. Stir the suspension for further 0.5 h.
2. To the cold suspension add drop-wise 19 g (86 mmol) of the methyl ester diluted with 10 mL $CHCl_3$ with stirring. Continue stirring at room temperature for 2 h.
3. Pour the mixture into a separatory funnel containing 600 g ice. Add 5 mL HCl (37%) to avoid precipitation of aluminum

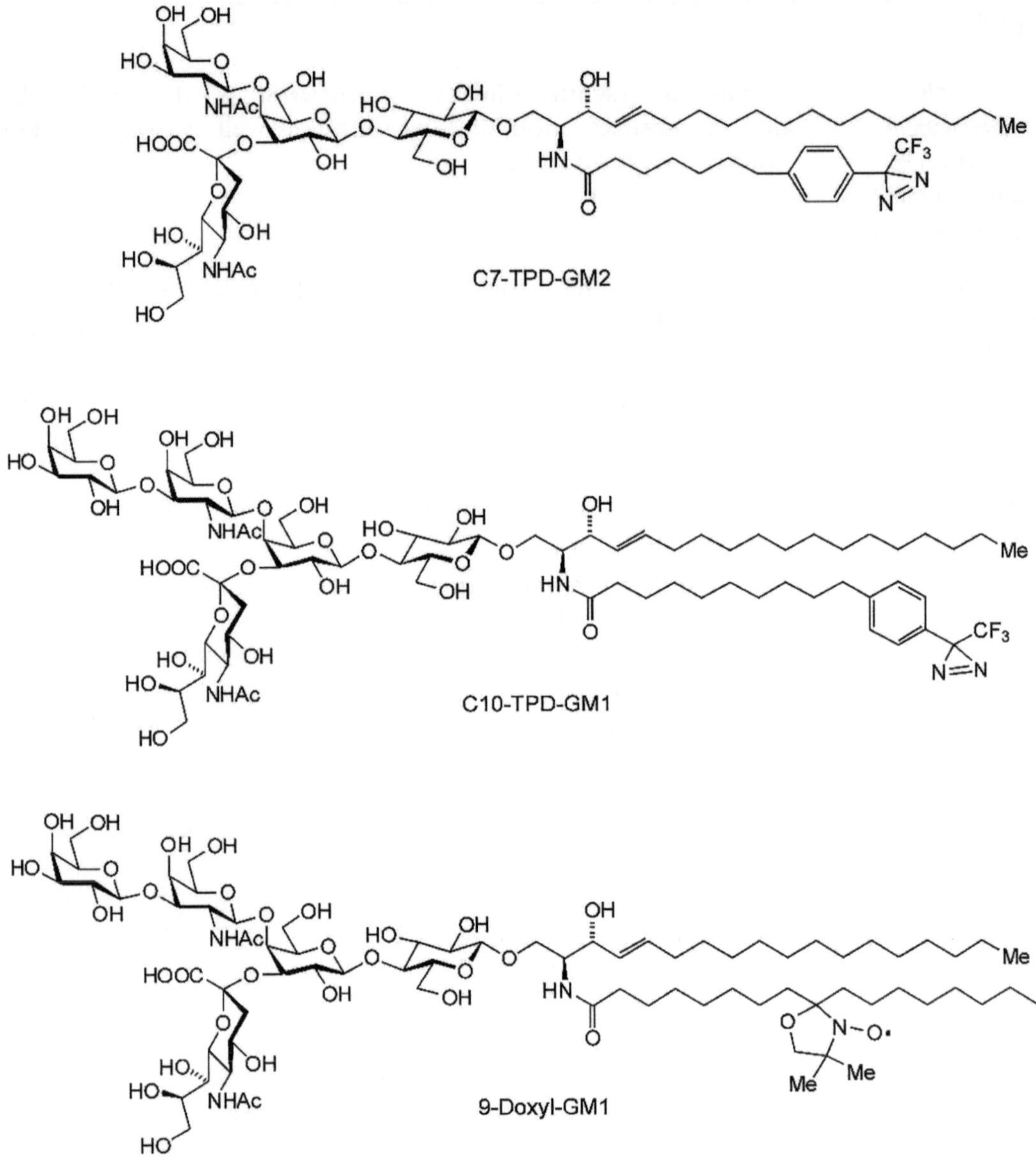

Fig. 1 Structures of photoactivatable Gangliosides C7-TPD-GM2, C10-TPD-GM1, and of the paramagnetic ganglioside 9-Doxyl-GM1

hydroxide. Remove the organic phase and extract the aqueous phase three times with 50 mL $CHCl_3$.

4. Wash the combined organic phases with H_2O (40 mL), then with 1 M $NaHCO_3$ (40 mL), and again twice with H_2O (40 mL). Dry the washed organic phase with anhydrous Na_2SO_4.
5. Remove $CHCl_3$ in a rotary evaporator. Distill the oily residue in oil pump vacuum. Combine fractions that on TLC developed with n-hexane–EtOAc (9:1, v/v) appear pure ($R_f = 0.23$)

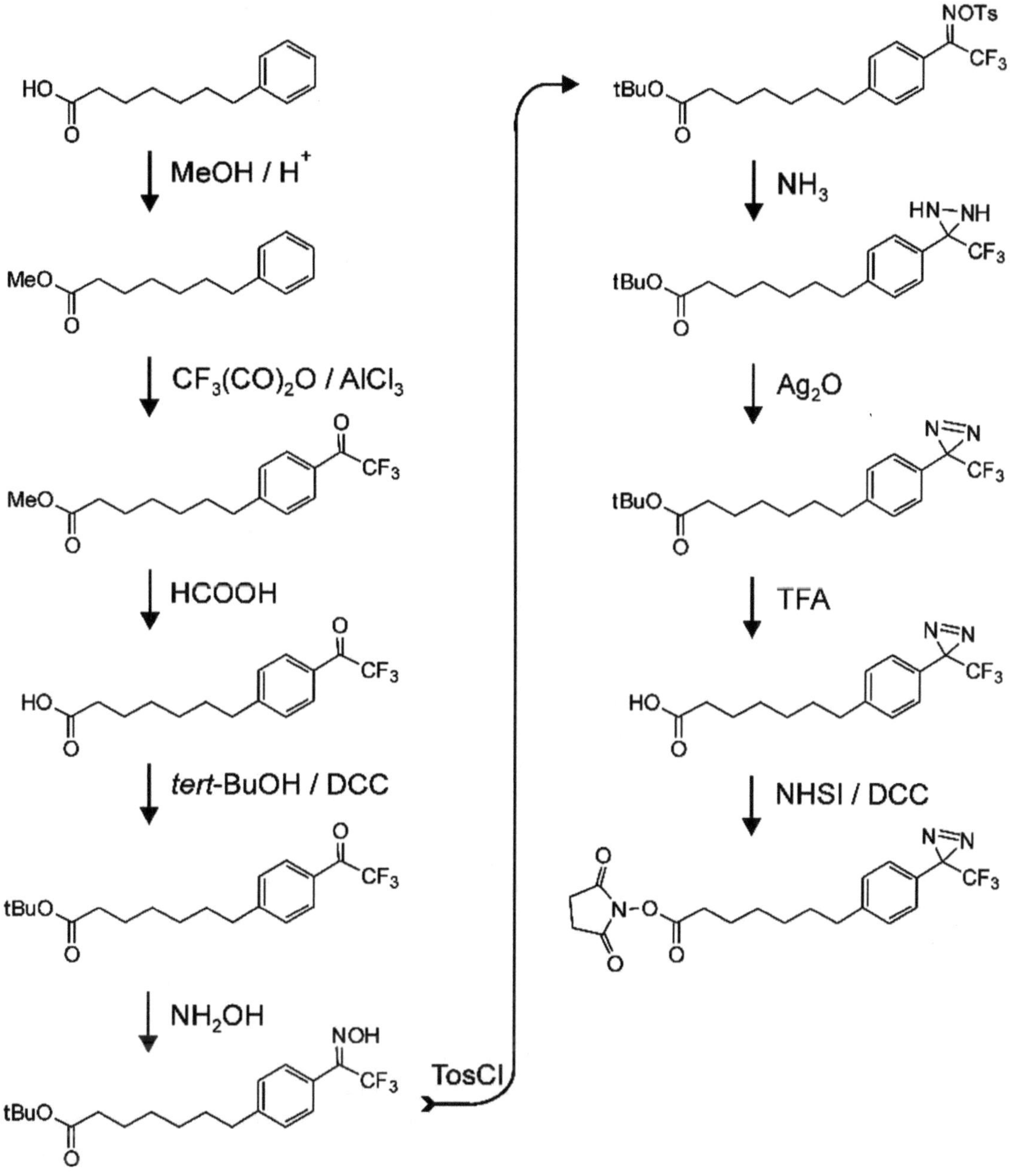

Fig. 2 Scheme for the synthesis of C7-TPD-NHS ester

as easily detected by fluorescence quenching under UV light at 254 nm. Yield is about 20.2 g (64 mmol, 74%).
Analytical data [26]:

^{1}H-NMR (400 MHz, $CDCl_3$): δ = 1.28 (m, 4 H, H-4, H-5), 1.56 (m, 4 H, H-3, H-6), 2.22 (t, 2 H, $J_{2,3}$ 7.5 Hz, H-2,), 2.61 (t, 2 H, $J_{6,7}$ 7.7 Hz, H-7), 3.57 (s, 3 H, OCH_3), 7.26 (d, 2 H, J_{HH} 7.6 Hz, $Bz_{2,6}$), 7.90 (d, 2 H, J_{HH} 7.6 Hz, $Bz_{3,5}$).

^{13}C-NMR (100 MHz, $CDCl_3$): $\delta = 24.67$ (1 C, C-3), 28.73 (1 C, C-4), 28.77 (1 C, C-5), 30.56 (1 C, C-6), 33.85 (1 C, C-2), 35.98 (1 C, C-7), 51.35 (1 C, OCH_3), 116.69 (q, 1 C, $^1J_{CF}$ 291.4 Hz, CF_3), 127.53 (1 C, C-Bz_4), 129.09 (2 C, CH-$Bz_{2,6}$), 130.21 (1 C, CH-Bz), 130.23 (1 C, CH-Bz), 151.58 (1 C, C-Bz_1), 174.06 (1 C, COOR), 179.96 (q, 1 C, $^2J_{CF}$ 34.8 Hz, $CF_3\underline{C}O$).

Synthesis of 7-(p-(2,2,2-Trifluoroacetyl)Phenyl) Heptanoic Acid

1. Dissolve 20 g of methyl ester (63 mmol) in 100 mL HCOOH. Add 1.2 mL conc. H_2SO_4 and heat this mixture 5 h under reflux with as little as possible cooling that will allow methyl formiate (bp 32 °C) to escape.
2. Thereafter remove most of the HCOOH in a rotatory evaporator. To the residue add ice-cold H_2O under stirring.
3. Collect the white precipitate by filtration and dry.
4. Recrystallize from a mixture of acetone and H_2O to obtain 16.9 g (56 mmol, 89%) colorless needles. TLC developed with $CHCl_3$ shows $R_f = 0.33$.
   Analytical data [26]:

   ^{1}H-NMR (400 MHz, $CDCl_3$): $\delta = 1.29$ (m, 4 H, H-4, H-5), 1.56 (m, 4 H, H-3, H-6), 2.27 (t, 2 H, $J_{2,3}$ 7.4 Hz, H-2), 2.61 (t, 2 H, $J_{6,7}$ 7.7 Hz, H-7), 7.26 (d, 2 H, J_{HH} 8.4 Hz, $Bz_{2,6}$), 7.91 (d, 2 H, J_{HH} 7.6 Hz, $Bz_{3,5}$).

   ^{13}C-NMR (100 MHz, $CDCl_3$): $\delta = 24.42$ (1 C, C-3), 28.72 (1 C, C-4), 28.74 (1 C, C-5), 30.57 (1 C, C-6), 33.93 (1 C, C-2), 36.00 (1 C, C-7), 116.71 (q, 1 C, $^1J_{CF}$ 291.5 Hz, CF_3), 127.56 (1 C, C-Bz_4), 129.12 (2 C, CH-$Bz_{2,6}$), 130.27 (1 C, CH-Bz), 130.29 (1 C, CH-Bz), 151.58 (1 C, C-Bz_1), 180.23 (1 C, COOH), 180.25 (q, 1 C, $^2J_{CF}$ 34.3 Hz, $CF_3\underline{C}O$).

Synthesis of 7-(p-(2,2,2-Trifluoroacetyl)Phenyl) Heptanoic Acid ***tert*-Butyl** Ester

1. Dissolve 16.6 g (55 mmol) acid in a mixture of 100 mL dry EtOAc and 100 mL dry Et_2O.
2. To the solution add 10 mL (104 mmol) *tert*-BuOH and 0.61 g (5 mmol) DMAP. Then add 12.4 g (60 mmol) DCC. Stir the mixture for 3 days at room temperature.
3. Remove precipitated dicyclohexylurea by filtration. Reduce the volume of the clear solution by rotary evaporation and remove any precipitated dicyclohexylurea by either filtration or centrifugation.
4. Purify the *tert*-butyl ester by flash chromatography on silica gel with n-hexane–$CHCl_3$ (3:7, v/v). Collect all fractions that on TLC with $CHCl_3$ will show fluorescence quenching spots ($R_f = 0.6$) with no contamination of the acid ($R_f = 0.33$). Dry combined pure fractions in a stream of N_2 to obtain 11.82 g (33 mmol, 60%) of *tert*-butyl ester as a faint yellowish oil.
   Analytical data [26]:

^{1}H-NMR (400 MHz, $CDCl_3$): δ = 1.30 (m, 4 H, H-4, H-5), 1.39 (s, 9 H, tBu), 1.53 (m, 2 H, H-3), 1.60 (m, 2 H, H-6), 2.15 (t, 2 H, $J_{2,3}$ 7.3 Hz, H-2), 2.64 (t, 2 H, $J_{6,7}$ 7.7 Hz, H-7), 7.29 (d, 2 H, J_{HH} 8.6 Hz, $Bz_{2,6}$), 7.93 (d, 2 H, J_{HH} 7.6 Hz, $Bz_{3,5}$).

^{13}C-NMR (100 MHz, $CDCl_3$): δ = 24.81 (1 C, C-3), 27.94 (3 C, tBu-CH_3), 28.71 (1 C, C-4), 28.75 (1 C, C-5), 30.56 (1 C, C-6), 35.32 (1 C, C-2), 35.97 (1 C, C-7), 79.81 (1 C, tBu-C_q), 116.69 (q, 1 C, $^1J_{CF}$ 291.4 Hz, CF_3), 127.51 (1 C, C-Bz_4), 129.07 (2 C, CH-$Bz_{2,6}$), 130.16 (1 C, CH-Bz), 130.19 (1 C, CH-Bz), 151.62 (1 C, C-Bz_1), 172.98 (1 C, COOR), 179.92 (q, 1 C, $^2J_{CF}$ 34.3 Hz, $CF_3\underline{C}O$).

Synthesis of 7-(p-(1-Hydroxyimino-2,2,2-Trifluoroethyl)Phenyl) Heptanoic Acid ***tert***-Butyl Ester (Oxime)

1. Dissolve 11.2 g (31.25 mmol) *tert*-butyl ester and 2.4 g (34 mmol) $NH_2OH{\cdot}HCl$ in 150 mL dry EtOH. Reflux this solution for 2 h while carefully and continously adjusting the pH between 6 and 7 with 4 M NaOH.
2. Cool to room temperature and remove precipitated NaCl by filtration. Wash precipitated NaCl with EtOH and remove combined solvents by rotary evaporation.
3. Dissolve the residue in 150 mL Et_2O, wash three times with 20 mL H_2O and dry the organic phase with anhydrous Na_2SO_4.
4. Remove Et_2O in a rotary evaporator and purify the crude product by flash chromatography on silica gel using $CHCl_3$ as eluent. Collect pure fractions that on TLC with $CHCl_3$ show R_f = 0.2. Yield is 8.75 g (23.4 mmol, 75%).
   Analytical data [26]:

   ^{1}H-NMR (400 MHz, $CDCl_3$): δ = 1.37 (m, 4 H, H-4, H-5), 1.46 (s, 9 H, tBu), 1.63 (m, 4 H, H-3, H-6), 2.24 (t, 2 H, $J_{2,3}$ 7.5 Hz, H-2), 2.64 (t, 2 H, $J_{6,7}$ 7.7 Hz, H-7), 7.27 (d, 2 H, J_{HH} 8.4 Hz, Bz), 7.48 (d, 2 H, J_{HH} 8.1 Hz, Bz), 10.35 (s, 1 H, N-OH).

   ^{13}C-NMR (100 MHz, $CDCl_3$): δ = 24.91 (1 C, C-3), 27.91 (3 C, tBu-CH_3), 28.74 (1 C, C-4), 28.79 (1 C, C-5), 30.79 (1 C, C-6), 35.56 (1 C, C-2), 35.67 (1 C, C-7), 80.59 (1 C, tBu-C_q), 120.81 (q, 1 C, $^1J_{CF}$ 274.7 Hz, CF_3), 123.52 (1 C, C-Bz_4), 128.38 (2 C, CH-Bz), 128.57 (2 C, CH-Bz), 145.32 (1 C, Bz_1), 146.81 (q, 1 C, $^2J_{CF}$ 32.0 Hz, C=N), 174.33 (1 C, COOR).

Synthesis of 7-(p-(1-(***O***-(p-Toluene Sulfonyl) Hydroxyimino)-2,2,2-Trifluoroethyl)Phenyl) Heptanoic Acid ***tert***-Butyl Ester (Tosylate)

1. Dissolve 8.6 g (23 mmol) of the above oxime and 6 g (31.4 mmol) TsCl in 200 mL THF.
2. To this solution add 12 g (86.8 mmol) finely powdered K_2CO_3. Heat the suspension with vigorous stirring for 2 h under reflux.
3. Cool the mixture to room temperature and remove KCl and remaining K_2CO_3 by filtration. Remove the solvent by evaporation.

4. Purify the tosylate by flash chromatography on silica gel using n-hexane–CH_2Cl_2 (2:3, v/v) as eluent. Collect pure fractions of tosylat with $R_f = 0.2$ in TLC developed with n-hexane–CH_2Cl_2 (2:3, v/v) to obtain 10.8 g (20.4 mmol, 89%) tosylate as a colorless oil.
   Analytical data [26]:

   ^{1}H-NMR (400 MHz, $CDCl_3$): (*see* **Note 2**): δ = 1.33 (m, 4 H, H-4, H-5), [1.41, 1.42 (s, 9 H, tBu)], 1.58 (m, 4 H, H-3, H-6), [2.17, 2.18 (t, 2 H, H-2, J_{HH} 7.4 Hz)], [2.43, 2.44 (s, 3 H, $TosCH_3$)], 2.61 (m, 2 H, H-7), {[7.19 (d, 1 H, J_{HH} 8.4 Hz, Bz), 7.24 (d, 1 H, J_{HH} 8.6 Hz, Bz)], 7.34 (m, 2 H, TosCH), [7.86 (d, 2 H, J_{HH} 8.4 Hz, TosCH), 7.88 (d, 2 H, J_{HH} 8.4 Hz, TosCH)].

   ^{13}C-NMR (100 MHz, $CDCl_3$): δ = [21.70, 21.72 (1 C, $TosCH_3$)], [24.90, 24.92 (1 C, C-3)], 28.06 (3 C, tBu-CH_3), 28.80, (1 C, C-4), 28.90 (1 C, C-5), [30.78, 30.82 (1 C, C-6)], [35.44, 35.47 (1 C, C-2)], [35.67, 35.78 (1 C, C-7)], 79.92 (1 C, tBu-C_q), [119.67 (q, 1 C, $^1J_{CF}$ 277.7 Hz, CF_3), 120.19 (q, 1 C, $^1J_{CF}$ 283.8 Hz, CF_3)], [121.77, 125.02 (1 C, C-Bz_4)], [128.48 (2 C, CH-Bz)], [128.73, 128.75 (1 C, CH-Bz)], [128.79, 128.81 (1 C, CH-Bz)], [129.07, 129.22 (2 C, CH-Bz)], [129.81 (2 C, CH-Bz)], [131.21, 131.49 (1 C, Tos-C_4)], [145.88, 146.03 (1 C, Tos-C_1)], [147.01, 147.19 (1 C, Bz_1)], [153.88 (q, $^2J_{CF}$ 32.8 Hz, C=N), 153.98 (q, 1 C, $^2J_{CF}$ 32.0 Hz, C=N)], 173.15 (1 C, COOR).

Synthesis of 7-(p-(3-(Trifluoromethyl) Diaziridinyl)Phenyl) Heptanoic Acid *tert*-Butyl Ester (Diaziridine)

1. For the formation of the diaziridinyl compound use pressure resistant screw-capped Pyrex vials (25 mL). Dissolve 10.6 g (20 mmol) of the above tosylate in dry Et_2O. Distribute this solution in 10 mL portions to 5 vials.
2. Solidify each solution by immersing the closed vials in liquid nitrogen. To each vial add about 2 mL liquid NH_3 (1 g, 58 mmol) (*see* **Note 3**).
3. Tighten the vials and stir the solutions overnight at room temperature.
4. Cool the vials again and open the caps to let the excess of NH_3 evaporate slowly without splashing.
5. Remove precipitated $TsNH_2$ by filtration, wash the amide with Et_2O. Wash the combined filtrates twice with H_2O, dry over anhydrous Na_2SO_4 and evaporate the solvent.
6. Purify the diaziridine by flash chromatography on silica gel using $CHCl_3$–EtOAc (9:1, v/v) as eluent. Collect pure fractions which on TLC developed with $CHCl_3$–EtOAc (9:1, v/v) show $R_f = 0.56$ to obtain 6.1 g (16.2 mmol, 81%) product as a colorless oil.

Analytical data [26]:

^{1}H-NMR (400 MHz, $CDCl_3$): δ = 1.30 (m, 4 H, H-4, H-5), 1.39 (s, 9 H, tBu), 1.55 (m, 4 H, H-3, H-6), 2.15 (t, 2 H, $J_{2,3}$ 7.5 Hz, H-2), 2.31 (d, 1 H, J_{HH} 8.6 Hz, NH), 2.58 (t, 2 H, $J_{6,7}$ 7.7 Hz, H-7), 2.75 (d, 1 H, J_{HH} 8.6 Hz, NH), 7.17 (d, 2 H, J_{HH} 8.4 Hz, Bz), 7.46 (d, 2 H, J_{HH} 8.1 Hz, Bz).

^{13}C-NMR (100 MHz, $CDCl_3$): δ = 24.85 (1 C, C-3), 27.96 (3 C, tBu-CH_3), 28.75 (2 C, C-4,5), 30.94 (1 C, C-6), 35.37 (1 C, C-2), 35.49 (1 C, C-7), 57.74 (q, $^{2}J_{CF}$ 35.9 Hz, diaziridine-C), 79.81 (1 C, tBu-C_q), 123.52 (q, 1 C, $^{1}J_{CF}$ 278.5 Hz, CF_3), 127.91 (2 C, CH-Bz), 128.61 (2 C, CH-Bz), 128.91 (1 C, C-Bz_4), 144.85 (1 C, C-Bz_1), 173.06 (1 C, COOR).

Synthesis of 7-(p-(3-(Trifluoromethyl)Diazirinyl) Phenyl)Heptanoic Acid *tert*-Butyl Ester

For the oxidation of the diaziridine to diazirine freshly prepared Ag_2O is necessary.

1. Dissolve 8.5 g (50 mmol) $AgNO_3$ in 50 mL H_2O. Heat until the solution boils.
2. To the boiling solution add 2.0 g (50 mmol) NaOH dissolved in 80 mL H_2O under stirring.
3. Under dim light isolate precipitated Ag_2O by filtration, wash the solid successively with H_2O, acetone and Et_2O. Dry Ag_2O in vacuo. Keep the solid light-protected until use.
4. Under dim light dissolve 3 g (8 mmol) diaziridine in 60 mL Et_2O. Add to this solution 9.3 g (40 mmol) Ag_2O. Stir the suspension for 5 h in the dark.
5. Remove precipitated silver and Ag_2O by filtration. Wash the solids once with Et_2O and remove the solvents by evaporation. TLC developed with $CHCL_3$ shows one prominent spot (R_f = 0.65). Yield 2.9 g (7.8 mmol, 97%). The faint yellowish oil can be used in the next step without further purification.

Analytical data [26]:

^{1}H-NMR (400 MHz, $CDCl_3$): δ = 1.31 (m, 4 H, H-4, H-5), 1.42 (s, 9 H, tBu), 1.57 (m, 4 H, H-3, H-6), 2.17 (t, 2 H, $J_{2,3}$ 7.4 Hz, H-2), 2.58 (t, 2 H, $J_{6,7}$ 7.7 Hz, H-7), 7.07 (d, 2 H, J_{HH} 8.4 Hz, Bz), 7.17 (d, 2 H, J_{HH} 8.1 Hz, Bz).

^{13}C-NMR (100 MHz, $CDCl_3$): δ = 24.92 (1 C, C-3), 28.05 (3 C, tBu-CH_3), 28.33 (q, 1 C, $^{2}J_{CF}$ 40.4 Hz, diazirine-C), 28.80 (1 C, C-4), 28.82 (1 C, C-5), 30.96 (1 C, C-6), 35.46 (2 C, C-2, C-7), 79.90 (1 C, tBu-C_q), 122.19 (q, 1 C, $^{1}J_{CF}$ 274.7 Hz, CF_3), 126.36 (1 C, C-Bz_4), 126.36 (2 C, CH-Bz), 128.83 (2 C, CH-Bz), 144.57 (1 C, C-Bz_1), 173.14 (1 C, COOR).

Synthesis of 7-(p-(3-(Trifluoromethyl)Diazirinyl) Phenyl)Heptanoic Acid (C7-TPD-Acid)

1. Dissolve 2.22 g (6 mmol) of the diazirine in 10 mL CH_2Cl_2.
2. Add 4 mL TFA and 0.1 mL Et_3SiH. Keep the solution at 20 °C overnight.
3. Remove solvents including TFA in a nitrogen jet.
4. After drying in high vacuum 1.87 g (5.9 mmol, 98%) colorless crystals are obtained.
   Analytical data [26]:

   ^{1}H-NMR (400 MHz, $CDCl_3$): δ = 1.33 (m, 4 H, H-4, H-5), 1.60 (m, 4 H, H-3, H-6), 2.33 (t, 2 H, $J_{2,3}$ 7.4 Hz, H-2), 2.59 (t, 2 H, $J_{6,7}$ 7.7 Hz, H-7), 7.08 (d, 2 H, J_{HH} 8.4 Hz, Bz), 7.18 (d, 2 H, J_{HH} 8.4 Hz, Bz), 10.8-11.2 (b, 1 H, OH).

   ^{13}C-NMR (100 MHz, $CDCl_3$): δ = 24.49 (1 C, C-3), 28.36 (q, $^2J_{CF}$ 39.7 Hz, diazirine-C), 28.75 (1 C, C-4), 28.78 (1 C, C-5), 30.92 (1 C, C-6), 33.98 (1 C, C-2), 35.46 (1 C, C-7), 122.20 (q, 1 C, $^1J_{CF}$ 274.7 Hz, CF_3), 126.40 (1 C, C-Bz_4), 126.40 (2 C, CH-Bz), 128.84 (2 C, CH-Bz), 144.49 (1 C, C-Bz_1), 180.37 (1 C, COOH).

Preparation of 7-(p-(3-(Trifluoromethyl)Diazirinyl) Phenyl)Heptanoic Acid *N*-Succinimidyl Ester (C7-TPD-NHS Ester)

1. Dissolve 1.6 g (5.1 mmol) C7-TPD acid in 30 mL THF and 20 mL EtOAc. Add 0.81 g (7 mmol) solid NHSI and stir at room temperature until most NHSI is dissolved.
2. To this solution add 1.07 g (5.2 mmol) DCC. Stir the mixture overnight.
3. Remove precipitated dicyclohexylurea by filtration. Remove solvents by evaporation and purify the residue by flash chromatography on silica gel using CH_2Cl_2–Et_2O (10:1, v/v) as eluent. Collect fractions that on TLC developed with the same solvent mixture display single spots (R_f = 0.59) to yield 1.74 g C7-TPD-NHS ester (4.23 mmol, 83%) (*see* **Note 4**).
   Analytical data [26]:

   ^{1}H-NMR (400 MHz, $CDCl_3$): δ = 1.32 (m, 2 H, H-5), 1.40 (m, 2 H, H-4), 1.58 (dt, 2 H, J_{HH} 7.4 Hz, J_{HH} 15.1 Hz, H-6), 1.71 (dt, 2 H, J_{HH} 7.7 Hz, J_{HH} 15.0 Hz, H-3), 2.57 (t, 2 H, J_{HH} 7.4 Hz, H-2), 2.59 (t, 2 H, J_{HH} 7.7 Hz, H-7), 2.75 (s, 4 H, Succ-CH_2), 7.07 (d, 2 H, J_{HH} 8.1 Hz, $Bz_{2,6}$), 7.17 (d, 2 H, J_{HH} 8.6 Hz, $Bz_{3,5}$).

   ^{13}C-NMR (100 MHz, $CDCl_3$): δ = 24.39 (1 C, C-3), 25.52 (2 C, Succ-CH_2), 28.36 (q, 1 C, $^2J_{CF}$ 40.4 Hz, diazirine-C), 28.43 (1 C, C-4), 28.51 (1 C, C-5), 30.78 (1 C, C-6), 30.81 (1 C, C-2), 35.32 (1 C, C-7), 122.16 (q, 1 C, $^1J_{CF}$ 274.7 Hz, CF_3), 126.33 (1 C, C-Bz_4), 126.35 (2 C, CH-$Bz_{2,6}$), 128.82 (2 C, CH-$Bz_{3,5}$), 144.40 (1 C, C-Bz_1), 168.56 (1 C, COOR), 169.15 (2 C, Succ-CO).

*3.1.2 Synthesis of 10-(p-(3-(Trifluoromethyl)Diazirinyl) Phenyl)Decanoic Acid **N**-Succinimidyl Ester (C10-TPD-NHS Ester)*

This synthesis uses 10-phenyldecanoic acid as starting material and is performed similarly to the C7-TPD-NHS ester formation. Due to the higher boiling point than those of the phenylheptanoic acid derivatives the respective phenyldecanoic acid derivatives of the first two synthetic steps are purified by flash chromatography rather than by distillation.

Formation of 10-Phenyldecanoic Acid Methyl Ester

1. Suspend 22 g (88.6 mmol) 10-phenyldecanoic acid in 30 mL MeOH. Add 1 mL conc. H_2SO_4 and boil the mixture under reflux for 20 h.
2. Remove excess MeOH and Me_2O in a rotary evaporator.
3. Disperse the oily residue into a sixfold amount of ice-cold H_2O in an appropriate separatory funnel.
4. Extract the methyl ester into Et_2O (20 mL) and repeat this extraction twice.
5. Wash the combined organic phases with H_2O (15 mL), then with saturated $NaHCO_3$ (15 mL), and again twice with H_2O (15 mL). Dry the washed ether phase with anhydrous Na_2SO_4.
6. Remove Et_2O in a rotary evaporator under normal pressure and purify the oily residue by flash chromatography on silica gel using n-hexane–CH_3COOH (9:1, v/v) as eluent. Collect fractions with R_f = 0.52 on TLC developed with n-hexane–CH_3COOH (9:1, v/v) to obtain 23.1 g (88 mmol, 90%) methyl ester.

Analytical data correspond to those of 7-phenylheptanoic acid derivatives and are found elsewhere [8, 26].

Synthesis of 10-(p-(2,2,2-Trifluoroacetyl)Phenyl) Decanoic Acid Methyl Ester

1. Suspend 50 g (375 mmol) anhydrous $AlCl_3$ in about 100 mL $CHCl_3$. Cool this suspension in an ice-bath and slowly add 14 mL $(CF_3CO)_2O$ (21 g, 100 mmol) with stirring. Stir the suspension for further 0.5 h.
2. To the cold suspension add dropwise with stirring 22.5 g (86 mmol) of the methyl ester diluted with 10 mL $CHCl_3$. Continue stirring at room temperature for 2 h.
3. Pour the mixture into a separatory funnel containing 600 g ice. Add 5 mL HCl (37%) to avoid precipitation of aluminum hydroxide. Remove the organic phase and extract the aqueous phase three times with 50 mL $CHCl_3$.
4. Wash the combined organic phases with H_2O (40 mL), then with saturated $NaHCO_3$ (40 mL), and again twice with H_2O (40 mL). Dry the washed organic phase with anhydrous Na_2SO_4.
5. Remove $CHCl_3$ in a rotary evaporator and purify the oily residue by flash chromatography on silica gel using n-hexane–EtOAc (9:1, v/v) as eluent. Combine fractions that on TLC developed with n-hexane–EtOAc (9:1, v/v) appear pure

($R_f = 0.32$) as easily detected by fluorescence quenching under UV light at 254 nm. Yield is about 22.8 g (63.6 mmol, 74%).

Analytical data correspond to those of 7-(p-(2,2,2-Trifluoroacetyl)phenyl)heptanoic acid methyl ester [8, 26] (*see* **Note 5**).

3.1.3 Synthesis of Photoactivatable Gangliosides C7-TPD-GM2 and C10-TPD-GM1

For the study of ganglioside–lipid and ganglioside–protein interaction by photoactivatable gangliosides it is indispensable to introduce a further label into the molecule to facilitate the characterization of the photocoupled products as well as their quantity. This is best achieved by introducing a radioactive isotope rather than a fluorescent tag or an affinity tag [5] which may lead to a significant deviation from the overall structure of the ganglioside under study. For isotope labeling of photoactivatable gangliosides deAc-lyso-gangliosides (*see* Chapter 16, Subheading 3.1.1) are perfectly suited as starting material. Owing to the strong basicity of the sphingoid amino group compared to the neuraminyl amino group a selective *N*-acylation of the former with equimolar amounts of C7-TPD-NHS ester or C10-TPD-NHS ester, respectively, can be accomplished prior to re-*N*-acetylation of the neuraminyl residue. Here, radiocarbon- or tritium-labeled acetic anhydride is the reagent of choice. This procedure will circumvent the protection of the sphingoid amino group with FmocCl [27]. Unlike re-*N*-acetylation of deAc-Fmoc-lyso-gangliosides with excess of unlabeled acetic anhydride in protic solvents to avoid *O*-acetylation of Fmoc-lyso-gangliosides, here re-*N*-acetylation has to be performed with equimolar amounts and with extreme care in an aprotic solvent because radioactive acetic anhydride is too expensive and too hazardous for use in huge excess.

Synthesis of Radiocarbon-Labeled C7-TPD-GM2

The synthetic steps are displayed in Fig. 3.

1. in a screw cap vial dissolve 86 mg (80 μmol) deAc-lyso-GM2 (*see* Chapter 16, Subheading 3.1.1) in 1 mL DMF. Add 20 μL DIPEA and reduce the volume to about 0.8 mL in a nitrogen jet to remove any volatile amine that otherwise may interfere with *N*-acylation.
2. Under argon add further 10 μL DIPEA. Cool the mixture in an ice bath and add 33 mg (80 μmol) C7-TPD-NHS ester dissolved in 200 μL THF with stirring. Keep at room temperature overnight.
3. Check the progress of reaction by TLC with $CHCl_3$–MeOH–2 M NH_3 (60:40:9, v/v/v). Visualize products by fluorescence quench and with $CuSO_4$ spray and heating according to [28] A new product (about 80% or more) at the expense of deAc-lyso-GM2 should have formed. If yield seems to be too low add a litle more C7-TPD-NHS ester with care to avoid acylation of the neuraminyl residue.

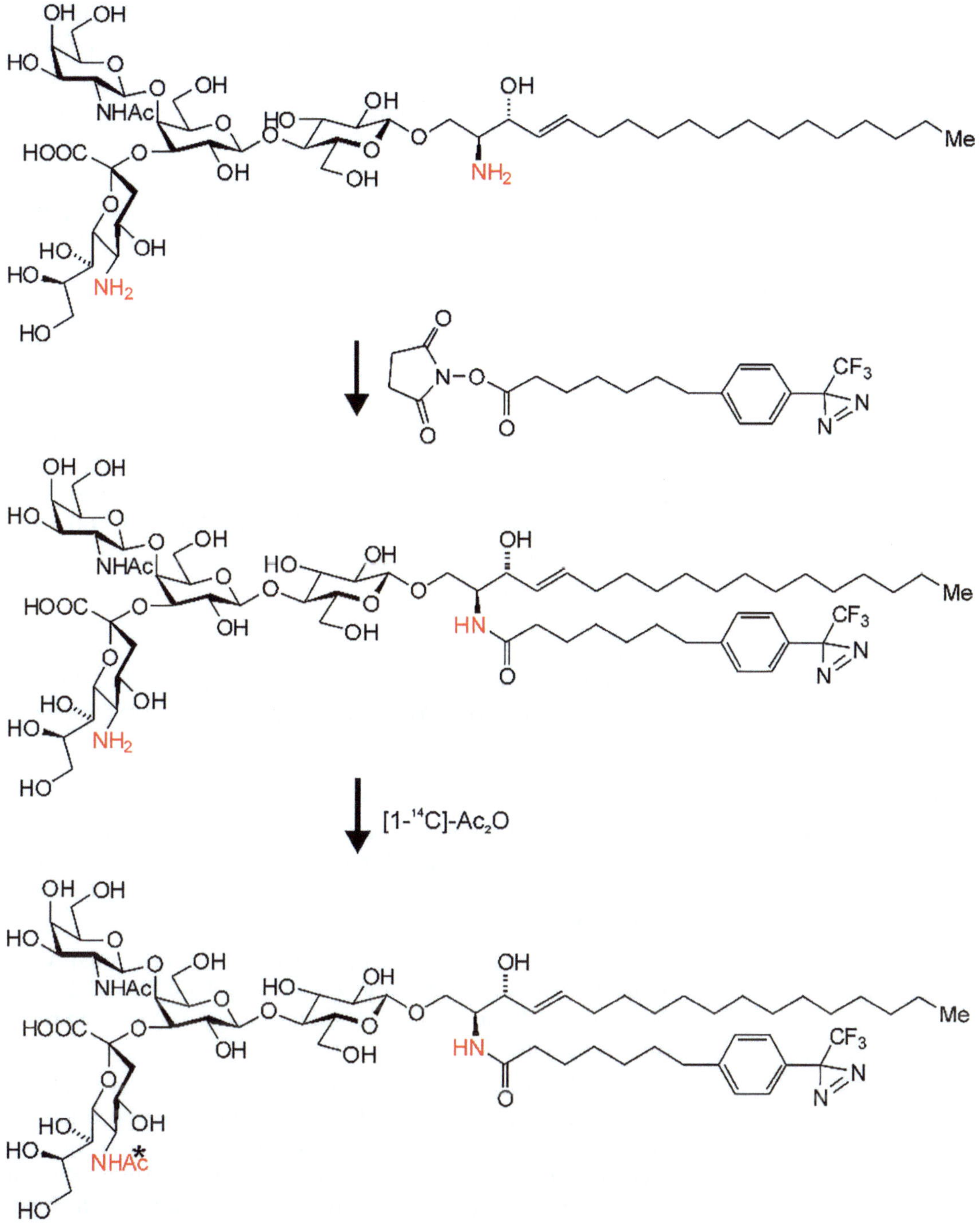

Fig. 3 Scheme for the synthesis of radioactive C7-TPD-GM2 with susceptible amino groups shown in red. Asterisk denotes residue bearing the ^{14}C-label

4. Purify the deAc-C7-TPD-GM2 by flash chromatography on silica gel using $CHCl_3$–MeOH–2 M NH_3 (60:40:9, v/v/v) as eluent.
5. Collect and combine all pure fractions and remove solvent in a nitrogen jet to obtain 66 mg (48 µmol, 60%) of deAc-C7-TPD-GM2.
6. In a plastic vial (Eppendorf) dissolve 11.7 mg (8.5 µmol) deAc-C7-TPD-GM2 in 0.2 mL DMF. Add 10 µL DIPEA. Reduce volume by nearly 15% in an argon jet to eliminate volatile amines and imines, if any.
7. To the above solution add 8.6 µmol (37 MBq) [1-^{14}C]-Ac_2O (specific activity 4.29 TBq per mol) in toluene (as delivered). Shake the vial overnight at room temperature.
8. Add more DIPEA and 5 µL H_2O to destroy excess of [1-^{14}C]-Ac_2O, if any, and to bind released [1-^{14}C]-AcOH as amine salt. Reduce the volume in a well vented hood by 80%. Dilute the remainder with $CHCl_3$–MeOH–2 M NH_3 (60:35:8, v/v/v) to about 1 mL.
9. Purify the raw product by flash chromatography on silica gel using $CHCl_3$–MeOH–2 M NH_3 (60:35:8, v/v/v) as eluent. Analyze and quantify aliquots of fractions by liquid scintillation counting and by phosphoimaging of films exposed to TLC of the product fractions. Yield of pure product is in the range of 3.7 mg (2.6 µmol, 30%) (*see* **Notes 6** and **7**).

Synthesis of Radiocarbon-Labeled C10-TPD-GM1

1. In a screw cap vial dissolve 99 mg (80 µmol) deAc-lyso-GM1 (*see* Chapter 16, Subheading 3.1.1) in 1 mL DMF. Add 20 µL DIPEA and reduce the volume to about 0.8 mL in a nitrogen jet to remove any volatile amine that otherwise may interfere with *N*-acylation.
2. Under argon add further 10 µL DIPEA. Cool the mixture in an ice bath and add 36.3 mg (80 µmol) C10-TPD-NHS ester dissolved in 200 µL THF with stirring. Keep at room temperature overnight.
3. Check the progress of reaction by TLC with $CHCl_3$–MeOH–2 M NH_3 (60:40:9, v/v/v). Visualize products by fluorescence quench and with $CuSO_4$ spray and heating according to [28] A new product (R_f = 0.28, about 80% or more) at the expense of deAc-lyso-GM1 should have formed. If yield seems to be too low add a litle more C10-TPD-NHS ester with care to avoid acylation of the neuraminyl residue.
4. Purify the deAc-C10-TPD-GM1 by flash chromatography on silica gel using $CHCl_3$–MeOH–2 M NH_3 (60:35:8, v/v/v) as eluent.
5. Collect and combine all pure fractions and remove solvent in a nitrogen jet to obtain 94 mg (56 µmol, 70%) of deAc-C10-TPD-GM1.

6. In a plastic vial (Eppendorf) dissolve 14.2 mg (8.5 μmol) deAc-C10-TPD-GM1 in 0.2 mL DMF. Add 10 μL DIPEA. Reduce volume by nearly 15% in an argon jet to eliminate volatile amines and imines, if any.
7. To the above solution add 8.6 μmol (37 MBq) [1-^{14}C]-Ac_2O (specific activity 4.29 TBq per mol) in toluene (as delivered). Shake the vial overnight at room temperature.
8. Add more DIPEA and 5 μL H_2O to destroy excess of [1-^{14}C]-Ac_2O, if any, and to bind released [1-^{14}C]-AcOH as amine salt. Reduce the volume in a well vented hood by 80%. Dilute the remainder with $CHCl_3$–MeOH–2 M NH_3 (60:35:8, v/v/v) to about 1 mL.
9. Purify the raw product by flash chromatography on silica gel using $CHCl_3$–MeOH–2 M NH_3 (60:40:9, v/v/v) as eluent. TLC in the above solvent displays R_f = 0.40. Analyze and quantify aliquots of fractions by liquid scintillation counting and by phosphoimaging of films exposed to TLC of the product fractions. Yield of pure product is in the range of 3.8 mg (2.2 μmol, 26%). Analytical data are found elsewhere [8] (*see* **Notes 6** and 7).

3.2 Synthesis of Paramagnetic (Doxyl) Gangliosides (for Structure See Fig. 1)

Figure 4 shows the multistep convergent synthesis starting from nonanedioic acid (azelaic acid) and 1-bromononane (n-nonyl bromide). The synthetic steps as outlined in Fig. 4 apply also to the other doxyl-stearic acids with, however, other corresponding starting chemicals such as dodecanedioic acid and 1-bromohexane for 12-doxyl-stearic acid.

3.2.1 Synthesis of 9-Doxyl-Stearic Acid NHS Ester

Synthesis of Dimethyl Nonanedioate

1. Suspend 47 g (250 mmol) nonanedioic acid in 250 mL dry MeOH. To the suspension add 6 mL conc. H_2SO_4 and heat the mixture for 5 h under reflux.
2. Evaporate excess MeOH and some Me_2O formed during this process under reduced pressure in a rotary evaporator.
3. Pour the residue onto 300 mL ice water in a separatory funnel and extract the crude dimethyl ester three times with 50 mL Et_2O.
4. Wash the combined ether phases with saturated $NaHCO_3$ to remove residual acids. Then wash with water and dry the organic phase with anhydrous Na_2SO_4.
5. Remove Et_2O by distillation and purify the dimethyl ester by distillation under high vacuum. The yield is 32.5–43.3 g (150–200 mmol, 60–80%) depending on the fractionation (*see* **Note 8**).

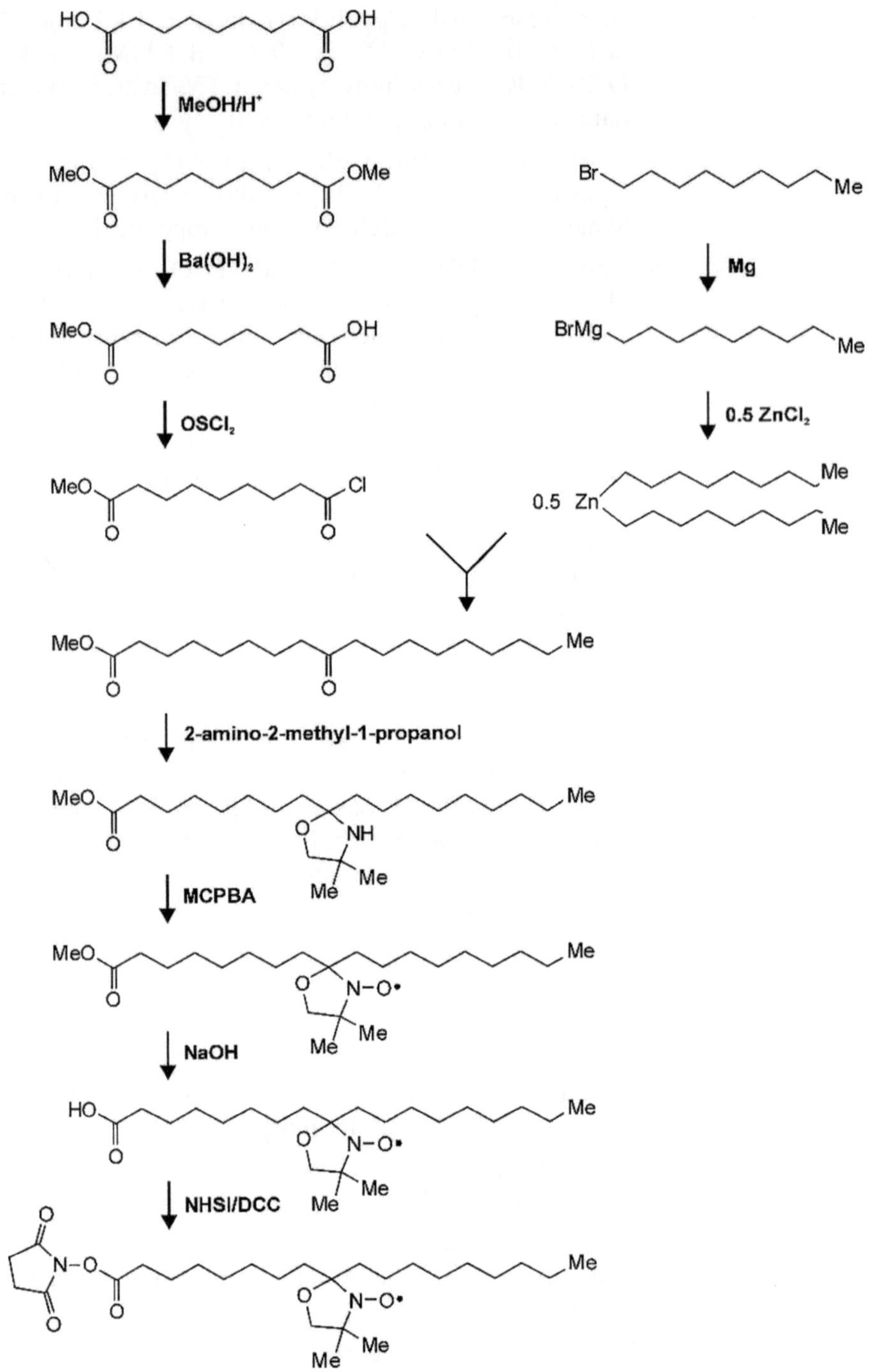

Fig. 4 Scheme of the convergent synthesis of 9-doxyl-stearic acid NHS ester

Synthesis of Nonanedioic Acid Monomethyl Ester

This synthesis relies on partial saponification of the dimethyl ester with $Ba(OH)_2$ instead of transesterification [29].

1. Suspend 22.5 g ((146 mmol) anhydrous BaO in 250 mL MeOH. To the suspension add 2.52 g (140 mmol) H_2O and stir the mixture vigorously until most of BaO has been converted into soluble $Ba(OH)_2$.
2. Remove any solid by filtration and titrate the content with 0.1 M HCl using an aliquot of $Ba(OH)_2$.
3. Dissolve 41.5 g (190 mmol) of the dimethyl ester in 190 mL toluene and add 100 mmol of the freshly prepared methanolic $Ba(OH)_2$.
4. Shake this mixture vigorously for 6 h at 20 °C.
5. Isolate the precipitate formed by filtration. Wash the precipitate with 80 mL Et_2O, 80 mL toluene and again with 80 mL Et_2O to remove remaining diester.
6. In a separatory funnel suspend the wet filter cake, which is mainly the Ba-salt of the monomethyl ester, into 100 mL Et_2O and carefully acidify the suspension with 1 M HCl to free the monomethyl ester from its salt.
7. Wash the Et_2O phase with water to remove surplus of HCl and dry the Et_2O phase with anhydrous Na_2SO_4.
8. Remove Et_2O by evaporation and purify the monomethyl ester by vacuum distillation. Yield is in the range of 23.1–26.9 g (114–133 mmol, 60–70%) (*see* **Note 9**).

Synthesis of Methyl 9-Chloro-9-Oxononanoate

1. To 24.3 g (120 mmol) monomethyl ester add 14.5 mL (23.8 g, 200 mmol) $OSCl_2$ and boil this mixture with stirring until no more HCl evolves.
2. Distill off excess $OSCl_2$ in a rotary evaporator.
3. Purify methyl 9-chloro-9-oxononanoate by crystallization from n-hexane. Yield is in the range from 19.9 to 21.2 g (90–96 mmol, 75–80%).

Synthesis of 9-Oxo-Stearic Acid Methyl Ester

1. Dissolve 25 g (120 mmol) nonyl bromide in 180 mL Et_2O. To the solution add 3 g (0.123 g-atoms) Mg turnings.
2. Reflux the mixture until almost all Mg has disappeared and cool the mixture in an ice bath.
3. To the ice-cold mixture (Grignard solution) add 8.8 g (64.5 mmol) finely powdered anhydrous $ZnCl_2$ in one portion under vigorous stirring (*see* **Note 10**).
4. Remove the ice bath and stir the mixture at room temperature for 30 min.

5. Distill off Et_2O almost completely in a rotary evaporator and dissolve the zink dinonyl in 160 mL dry benzene and cool the solution in an ice bath.
6. Dissolve 21 g (95 mmol) methyl 9-chloro-9-oxononanoate in 50 mL dry benzene. Add this solution slowly and with vigorous stirring to the cold solution of zink dinonyl over a period of 5 min.
7. Boil this mixture under reflux for 1 h and cool again in an ice bath.
8. Add to the mixture approximately 50 mL H_2O and then slowly a large excess of 0.1 N H_2SO_4 until two distinct phases form.
9. Separate the organic phase and reextract the aqueous phase with 50 mL benzene. Wash the combined organic phases with H_2O and dry over anhydrous Na_2SO_4.
10. Remove benzene by rotary evaporation. Purify the residue by twice recrystallization from n-hexane in a refrigerator to furnish 18.5 g (59 mmol, 62%) of methyl 9-ketostearate.

Synthesis of 2-(7-Methoxycarbonylheptyl)-2-Nonyl-4,4-Dimethyl-3-Oxazolidine (4′,4′-Dimethyloxazolidine Derivative of the Methyl 9-Ketostearateate)

1. Dissolve 18 g (57.6 mmol) methyl 9-ketostearate in 300 mL toluene. Then add 56 mL (51.7 g, 580 mmol) 2-amino-2-methyl-1-propanol and 63 mg (0.33 mmol) $TsOH{\cdot}H_2O$ as a catalyst (*see* **Note 11**).
2. Boil this mixture under reflux for 8 days using a Dean-Stark trap for water removal.
3. Wash the toluene solution six times with 100 mL saturated $NaHCO_3$ and four times with 100 mL H_2O. Dry the toluene solution with anhydrous Na_2SO_4 and remove toluene in a rotary evaporator to obtain the oxazolidine derivative as a colorless oil, which at this stage, is not further purified.

Synthesis of 2-(7-Methoxycarbonylheptyl)-2-Nonyl-4,4-Dimethyl-Oxazolidine-3-Oxyl

1. Dissolve the crude oil from the last step above in 300 mL Et_2O and cool the solution in an ice bath.
2. Over a period of 2 h add a solution of 14.8 g (66 mmol) of (77%) MCPBA in 60 mL Et_2O.
3. Allow the mixture to stand for 12 h.
4. Wash the ether phase four times with 110 mL saturated $NaHCO_3$ and four times with 110 mL H_2O. Dry the Et_2O solution with anhydrous Na_2SO_4 and remove Et_2O in a rotary evaporator to obtain a yellow viscous oil.
5. Purify the oxazolidine-3-oxyl methyl ester by flash chromatography on silica gel using benzene–Et_2O (7:3, v/v) as eluent. Collect most of the fast moving yellow band to obtain after evaporation of the solvent 6.9 g (17.3 mmol, 30%) of 2-(7-methoxycarbonylheptyl)-2-nonyl-4,4-dimethyl-oxazolidine-3-oxyl (the methyl ester of 9-doxyl-stearic acid).

Hydrolysis of 2-(7-Methoxy-carbonylheptyl)-2-Nonyl-4,4-Dimethyl-Oxazolidine-3-Oxyl

1. Dissolve 6.8 g (17 mmol) of the nitroxide ester in 60 mL dioxane. Then add 25 mL 1 M NaOH and stir the mixture for 6 h in the dark.
2. Acidify the mixture with 30 mL 1 M HCl and immediately extract the free acid into 100 mL EtOAc. Reextract the aqueous phase with 100 mL EtOAc. Dry the organic phase with anhydrous Na_2SO_4 and remove dioxane and EtOAc in a rotary evaporator.
3. Purify the free acid by flash chromatography on silica gel using $CHCl_3$–MeOH–HOAc (190:10:1, v/v/v) as eluent. Collect the center fraction of the yellow material and evaporate the solvents to obtain 4.6 g (11.9 mmol, 70%) free 9-doxyl-stearic acid.

Synthesis of the N-Succinimidyl Ester of the Spin-Labeled Stearic Acid (2-(7-***N***-Succinimidyloxy-carbonylheptyl)-2-Nonyl-4,4-Dimethyl-Oxazolidine-3-Oxyl) (9-Doxyl-Stearic Acid NHS Ester)

1. Dissolve 193 mg (500 μmol) of the spin-labeled stearic acid in 10 mL dry EtOAc. To this solution add solid 64 mg (550 μmol) NHSI and stir to bring most of the NHSI into solution.
2. Add solid 104 mg (500 μmol) DCC and stir the mixture at room temperature overnight.
3. Remove the precipitated dicyclohexylurea by centrifugation. Wash the precipitate once with 5 mL EtOAc and evaporate the solvent in a nitrogen jet. Keep the solid refrigerated until use. Usually the yield is in the range of 75–80%.

   Alternatively, the solid can be crystallized from aqueous EtOH (*see* **Note 12**).

3.2.2 Synthesis of Spin-Labeled Gangliosides GM1 and GM3 (9-doxyl-GM1, 9-Doxyl-GM3)

1. In a screw-cap vial dissolve 7.8 mg (6 μmol) lyso-GM1 (*see* Chapter 16, Subheading 3.1.1) in 0.4 mL DMF. Add 10 μL DIPEA and reduce the volume to about 0.3 mL in a nitrogen jet to remove any volatile amine that otherwise may interfere with *N*-acylation.
2. Under argon add further 5 μL DIPEA and 4.9 mg (10 μmol) 9-doxyl-stearic acid NHS ester dissolved in 150 μL THF with stirring. Keep at room temperature overnight.
3. Check the progress of reaction by TLC with $CHCl_3$–MeOH–15 mM $CaCl_2$ (60:40:9, v/v/v) besides GM1 as a reference compound. Visualize products with $CuSO_4$ spray and heating according to [28] A new product (about 80% or more) at the expense of lyso-GM1 should have formed (R_f = 0.45) (*see* **Note 13**). If yield seems to be too low add a litle more 9-doxyl-stearic acid NHS ester.
4. Purify the 9-doxyl-GM1 by flash chromatography on silica gel using $CHCl_3$–MeOH–2 M NH_3 (60:40:9, v/v/v) as eluent.
5. Collect and combine all pure fractions and remove solvent in a nitrogen jet to obtain 7.1 mg (4.2 μmol, 70%) of 9-doxyl-

GM1. Characterization can also be performed by quantifying the spin signals in ESR spectroscopy.

6. The synthesis of 9-doxyl-GM3 is similar to the synthesis of 9-doxyl-GM1 except for use of less polar solvent mixtures in TLC and flash chromatography on silica gel (*see* **Note 14**).

4 Notes

1. 7-phenylheptanoic acid methyl ester is also currently available from Sigma-Aldrich.
2. NMR-spectroscopy showed two isomers of the product (Tos-group in cis/trans-conformation to the benzene-ring), which are chemically identical. The correlating signals are displayed in brackets.
3. Liquid NH_3 is obtained by condensing NH_3 from a pressure bottle in a vial immersed in dry ice/acetone. Alternatively, NH_3 can also be solidified in the reaction vial placed in in liquid nitrogen.
4. In most cases flash chromatography may be omitted if a little less than equimolar amount of DCC is used because any remaining free acid does not interfere with the acylation reaction with lyso-gangliosides.
5. For synthesis of C10-TPD-NHS ester pursue Subheadings "Synthesis of 7-Phenylheptanoic Acid Methyl Ester," "Synthesis of 7-(p-(2,2,2-Trifluoroacetyl)phenyl)heptanoic Acid Methyl Ester," "Synthesis of 7-(p-(2,2,2-Trifluoroacetyl) phenyl)heptanoic Acid," "Synthesis of 7-(p-(2,2,2-Trifluoroacetyl)phenyl)heptanoic Acid *tert*-Butyl Ester," "Synthesis of 7-(p-(1-Hydroxyimino-2,2,2-trifluoroethyl)phenyl)heptanoic Acid *tert*-Butyl Ester (Oxime)," "Synthesis of 7-(p-(1-(*O*-(p-Toluene Sulfonyl)hydroxyimino)-2,2,2-trifluoroethyl)phenyl)heptanoic Acid *tert*-Butyl Ester (Tosylate)," "Synthesis of 7-(p-(3-(Trifluoromethyl)diaziridinyl)phenyl) heptanoic Acid *tert*-Butyl Ester (Diaziridine)," "Synthesis of 7-(p-(3-(Trifluoromethyl)diazirinyl)phenyl)heptanoic Acid *tert*-Butyl Ester," Synthesis of 7-(p-(3-(Trifluoromethyl) diazirinyl)phenyl)heptanoic Acid (C7-TPD-Acid)," and "Preparation of 7-(p-(3-(Trifluoromethyl)diazirinyl)phenyl) heptanoic Acid *N*-Succinimidyl Ester (C7-TPD-NHS Ester)." The analytical data for these last eight steps are found in [8].

6. Except for the collected pure product carefuly dispose of all radioactive fractions to radioactive waste. Notify your radiation officer.
7. To protect the product from radioautolysis keep refrigerated dissolved in a mixture of EtOH–toluene–H_2O (10:5:2, v/v/v) in the dark until use.
8. Dimethyl azelate, 97% (nonanedioic acid dimethyl ester) and monomethyl azelate, 96% (nonanedioic acid monomethyl ester) are also available from Alfa Aesar.
9. Alternatively this monomethyl ester can be purified by crystallization from EtOAc in a refrigerator yielding crystals with mp 23 °C. The method using $Ba(OH)_2$ is not satisfactory for lower-molecular-weight half esters, for their Ba salts are too soluble in MeOH.
10. To obtain anhydrous $ZnCl_2$ melt the commercially available salt and let it recrystallize in a desiccator before grinding in an inert dry atmosphere.
11. Formerly, 2-amino-2-methyl-1-propanol from Fluka gave better results than amino-methyl-propanol from other brands.
12. The raw NHS ester may contain the excess of NHSI and some free acid. Both do not interfere with acylation of lyso-gangliosides and are removed during purification of the doxyl-gangliosides.
13. The spin-labeled (doxyl-)gangliosides and their parent gangliosides have similar R_f values.
14. Sometimes it is useful to incorporate an isotope label in addition to the spin label into doxyl-gangliosides. This will facilitate quantification of the rate of incorporation of doxyl-gangliosides into membranes of living cells [19]. For the synthesis of such double labeled gangliosides start from deAc-lyso-gangliosides as described under Subheading "Synthesis of Radiocarbon-Labeled C7-TPD-GM2" using n-doxyl-stearic acid NHS ester for re-*N*-acylation prior to re-*N*-acetylation of the neuraminyl residue with radiolabeled Ac_2O.

Acknowledgments

The author's work was supported by the Deutsche Forschungsgemeinschaft. Grants Schw 143/5-4 and SFB 284/B5, which hereby is gratefully acknowledged.

References

1. Ledeen R (1985) Gangliosides of the Neuron. Trends Neurosci 8(4):169–174. https://doi.org/10.1016/0166-2236(85)90064-5
2. Schwarzmann G, Sandhoff K (1990) Metabolism and intracellular transport of glycosphingolipids. Biochemistry 29(49): 10865–10871
3. Eggeling C, Ringemann C, Medda R, Schwarzmann G, Sandhoff K, Polyakova S, Belov VN, Hein B, von Middendorff C, Schönle A, Hell SW (2009) Direct observation of the nanoscale dynamics of membrane lipids in a living cell. Nature 457(7233):1159–1162. https://doi.org/10.1038/nature07596
4. Wendeler M, Hoernschemeyer J, Hoffmann D, Kolter T, Schwarzmann G, Sandhoff K (2004) Photoaffinity labelling of the human GM2-activator protein. Mechanistic insight into ganglioside GM2 degradation. Eur J Biochem 271(3):614–627
5. Hashimoto M, Hatanaka Y (2008) A novel biotinylated diazirinyl ceramide analogue for photoaffinity labeling. Bioorg Med Chem Lett 18(2):650–652. https://doi.org/10.1016/j.bmcl.2007.11.071
6. Sonnino S, Chigorno V, Acquotti D, Pitto M, Kirschner G, Tettamanti G (1989) A photoreactive derivative of radiolabeled GM1 ganglioside: preparation and use to establish the involvement of specific proteins in GM1 uptake by human fibroblasts in culture. Biochemistry 28(1):77–84
7. Sonnino S, Chigorno V, Valsecchi M, Pitto M, Tettamanti G (1992) Specific ganglioside-cell protein interactions: a study performed with GM1 ganglioside derivative containing photoactivable azide and rat cerebellar granule cells in culture. Neurochem Int 20(3):315–321
8. Von Coburg A (2004) Untersuchung der Lipidnachbarschaft von derivatisiertem GM1 in Modellmembranen und kultivierten Zellen. http://hss.ulb.uni-bonn.de/2004/0304/0304.htm. http://nbn-resolving.de/urn:nbn:de:hbz:5n-03048
9. Meier EM, Schummer D, Sandhoff K (1990) Evidence for the presence of water within the hydrophobic core of membranes. Chem Phys Lipids 55(2):103–113
10. Brunner J (1993) New photolabeling and crosslinking methods. Annu Rev Biochem 62:483–514. https://doi.org/10.1146/annurev.bi.62.070193.002411
11. Brunner J, Senn H, Richards FM (1980) 3-Trifluoromethyl-3-phenyldiazirine. A new carbene generating group for photolabeling reagents. J Biol Chem 255(8):3313–3318
12. Esmann M, Marsh D, Schwarzmann G, Sandhoff K (1988) Ganglioside-protein interactions: spin-label electron spin resonance studies with (Na+,K+)-ATPase membranes. Biochemistry 27(7):2398–2403
13. Mantipragada SB, Horvath LI, Arias HR, Schwarzmann G, Sandhoff K, Barrantes FJ, Marsh D (2003) Lipid-protein interactions and effect of local anesthetics in acetylcholine receptor-rich membranes from Torpedo marmorata electric organ. Biochemistry 42(30):9167–9175. https://doi.org/10.1021/bi034485q
14. Marsh D (2008) Electron spin resonance in membrane research: protein-lipid interactions. Methods 46(2):83–96. https://doi.org/10.1016/j.ymeth.2008.07.001
15. Schwarzmann G, Sandhoff K (1987) Lysogangliosides: synthesis and use in preparing labeled gangliosides. Methods Enzymol 138:319–341
16. Acquotti D, Sonnino S, Masserini M, Casella L, Fronza G, Tettamanti G (1986) A new chemical procedure for the preparation of gangliosides carrying fluorescent or paramagnetic probes on the lipid moiety. Chem Phys Lipids 40(1):71–86
17. Mraz W, Schwarzmann G, Sattler J, Momoi T, Seemann B, Wiegandt H (1980) Aggregate formation of gangliosides at low concentrations in aqueous media. Hoppe Seylers Z Physiol Chem 361(2):177–185
18. Schwarzmann G, Hinrichs U, Sonderfeld S, Marsh D, Sandhoff K (1986) Metabolism of exogenous gangliosides in cultured fibroblasts and cerebellar cells. In: Freysz L, Dreyfus H, Massarelli R, Gatt S (eds) Enzymes of lipid metabolism II. Springer, Boston, pp 553–562. https://doi.org/10.1007/978-1-4684-5212-9_70
19. Schwarzmann G, Hoffmann-Bleihauer P, Schubert J, Sandhoff K, Marsh D (1983) Incorporation of ganglioside analogues into fibroblast cell membranes. A spin-label study. Biochemistry 22(21):5041–5048
20. Hubbell WL, McConnell HM (1969) Motion of steroid spin labels in membranes. Proc Natl Acad Sci U S A 63(1):16–22
21. Hubbell WL, McConnell HM (1971) Molecular motion in spin-labeled phospholipids and membranes. J Am Chem Soc 93(2): 314–326

22. Keana JFW (1978) Newer aspects of synthesis and chemistry of nitroxide spin labels. Chem Rev 78(1):37–64. https://doi.org/10.1021/Cr60311a004
23. Keana JFW, Keana SB, Beetham D (1967) A new versatile ketone spin label. J Am Chem Soc 89(12):3055. https://doi.org/10.1021/Ja00988a049
24. Waggoner AS, Kingzett TJ, Rottschaefer S, Griffith OH, Keith AD (1969) A spin-labeled lipid for probing biological membranes. Chem Phys Lipids 3(3):245–253
25. Pangborn AB, Giardello MA, Grubbs RH, Rosen RK, Timmers FJ (1996) Safe and convenient procedure for solvent purification. Organometallics 15(5):1518–1520. https://doi.org/10.1021/Om9503712
26. Hörnschemeyer J (2001) Examinations of the lipid neighbourhoodof sphingolipids in model membranes and membranes ofcultured cells. http://worldcat.org. http://nbn-resolving.de/urn:nbn:de:hbz:5n-00506
27. Sonnino S, Kirschner G, Ghidoni R, Acquotti D, Tettamanti G (1985) Preparation of GM1 ganglioside molecular species having homogeneous fatty acid and long chain base moieties. J Lipid Res 26(2):248–257
28. Touchstone JC, Levin SS, Dobbins MF, Carter PJ (1981) Differentiation of saturated and unsaturated phospholipids on thin layer chromatograms. J High Resolut Chromatogr 4(8):423–424. https://doi.org/10.1002/jhrc.1240040816
29. Durham LJ, McLeod DJ, Cason J (1963) Methyl hydrogen hendecanedioic acid. Org Syn Coll 4:635–638. https://doi.org/10.1002/0471264180.os038.14

Chapter 18

Methods for Assay of Ganglioside Catabolic Enzymes

Nicoletta Loberto, Giulia Lunghi, Domitilla Schiumarini, Maura Samarani, Elena Chiricozzi, and Massimo Aureli

Abstract

Cell plasma membrane gangliosides content and pattern is finely regulated by a complex metabolic machinery. Among different pathways, past and new evidence suggests an important role of the sphingolipid catabolic enzymes associated with both lysosomes and plasma membrane. Over the years, several cell-free and cell-based assays are developed in order to evaluate the activities both intracellularly and at the cell surface. Here, we propose a selection of the most efficient, sensitive, and specific assays that allow determining the activity of the main gangliosides-glycohydrolases such as sialidases, ß-hexosaminidases, ß-galactosidases, and ß-glucosidases in both cell/tissue lysates and directly in living cells.

Key words Glycohydrolases, Plasma membrane, Lysosome, Sialidases, Fluorigenic substrates, Radioactive substrates, Assays on living cells

1 Introduction

Qualitative and quantitative changes in gangliosides expression occur in response to different cell stimuli [1, 2].

A complex metabolic machinery is involved in the control of cell plasma membrane (PM) ganglioside content, consisting in neobiosynthesis, catabolism, and trafficking directed into and out the cell [3]. The fine control of the ganglioside homeostasis is fundamental to preserve cell physiology. For example, loss-of-function mutations in gene coding for enzymes involved in ganglioside lysosomal catabolism result in an aberrant accumulation of not-degraded compounds leading to the onset of severe clinical phenotypes [4].

However, lysosomes are not the only site involved in the ganglioside catabolism. Recent lines of evidence suggest that changes in PM gangliosides could be triggered directly in situ by the action of the so-called PM-associated glycohydrolases. Interestingly, among these enzymes, sialidase NEU3 and the β-glucosylceramidase GBA2 are glycohydrolases mainly resident at PM, while

Sandro Sonnino and Alessandro Prinetti (eds.), *Gangliosides: Methods and Protocols*, Methods in Molecular Biology, vol. 1804, https://doi.org/10.1007/978-1-4939-8552-4_18, © Springer Science+Business Media, LLC, part of Springer Nature 2018

β-galactosidase, β-galactocerebrosidase, β-hexosaminidase A, and β-glucocerebrosidase GBA1 found at the cell surface are the same enzymes present in lysosomes [5, 6].

In literature, several assays are available to measure the activity of both lysosomal and PM-associated glycohydrolases. The most common, used also for the sphingolipidoses diagnosis, are based on the use of cell/tissue homogenates (cell-free assays) as enzymatic sources and fluorigenic substrates represented by glycosylated derivatives of methylumbelliferone, as artificial substrates [5]. Despite the wide use and the strength of these assays, they allow highlighting only an impairment in the activity of specific enzymes and do not give any indication on the entire endolysosomal compartment functionality. To this purpose, several years ago, it has been developed an assay based on the use of living cells fed with GM3 and GM1 gangliosides tritium labeled on the sphingosine moiety [7]. The cell administration of these substrates allows following the ganglioside catabolism in the lysosomes of living cells, by a simple lipid extraction and a subsequent thin layer chromatography. Using the same strategy but in the presence of chloroquine, a drug able to block the lysosomal activity, it is possible to follow the PM degradation of the radioactive gangliosides triggered directly at the cell surface of living cells by the action of the PM-associated glycohydrolases [8].

However, one of the open questions related to the PM-enzymes concerns their ability to degrade substrates belonging to the same membrane (*cis* activity) and/or those associated with neighboring membranes (*trans* activity). For this reason, an assay based on the immobilization of radioactive lactosylceramide has been developed in order to establish the *trans* activity of PM-associated β-galactosidase and β-glucocerebrosidases [9].

Nevertheless, since the use of radioactive substrates is very expensive and needs specific facilities for their handling, it has been established an assay for the detection of the PM-associated glycohydrolases activity directly on living cells or in fraction of PM based on the use of specific fluorigenic substrates [9, 10].

Here, we intend to summarize the entire scenario of enzymatic assays available for testing both lysosomal and PM-associated glycohydrolases involved in the ganglioside catabolism, such as sialidases, β-hexosaminidases, β-galactosidases, and β-glucosidases.

2 Materials

Prepare all the solutions using ultrapure water (doubly distilled or deionized water to attain a resistivity of 18 MΩ cm at 25°C). Commercial chemicals are of the highest purity available.

2.1 Materials for Cell-Free Glycohydrolases Assays Based on the Use of Artificial Substrates: Enzymes Associated with Cell Lysate

1. Phosphate Buffered Saline (PBS): 137 mM NaCl, 2.7 mM KCl, 10 mM Na_2HPO_4, 2 mM KH_2PO_4. To prepare 1 l of PBS pH 7.4 dissolve 8 g of NaCl, 0.2 g of KCl, 1.44 g of Na_2HPO_4, 0.24 g of KH_2PO_4 in 800 ml H_2O. Adjust pH to 7.4 with HCl and adjust volume to 1 l with additional H_2O using a graduated cylinder.
2. McIlvaine buffer 4×: Two different aqueous solutions 0.4 M citric acid (solution A) and 0.8 M Na_2HPO_4 (solution B) need to be prepared. For the solution A, dissolve 21.01 g of citric acid monohydrate in 800 ml of water and adjust volume to 1 l with additional H_2O using a graduated cylinder. For the solution B, dissolve 28.40 g of Na_2HPO_4 in 800 ml of H_2O. After the complete solubilization, adjust volume to 1 l with additional water using a graduated cylinder. To prepare 100 ml of McIlvaine buffer 4× (pH 6) mix 36.85 ml of A with 63.15 ml of B. To prepare 100 ml of McIlvaine buffer 4× (pH 5.2) mix 46.40 ml of A with 53.60 ml of B. For 100 ml of McIlvaine buffer 4× (pH 4.4) mix 55.90 ml of A with 44.10 of B (*see* **Note 1**).
3. Substrates stock solution in water: 24 mM 4-methylumbelliferyl β-D-glucopyranoside (MUB-Glc); 1 mM 4-methylumbelliferyl β-D-galactopyranoside (MUB-Gal); 1 mM 4-methylumbelliferyl-N-acetyl-β-D-glucosaminide (MUG); 1 mM 4-methylumbelliferyl-*N*-acetyl-β-D-glucosaminide-6-sulfate (MUGS), 1 mM 2′-(4-methylumbelliferyl)-*N*-acetyl-α-D-neuraminic acid (MUB-NANA). Store at −20°C up to 1 year.
4. Lysis buffer: water containing commercial available protease inhibitors and (phenylmethylsulfonyl fluoride) PMSF 1 mM, freshly made (*see* **Note 1**).
5. Inhibitors stock solutions in water: 200 nM adamantane-pentyl-dNM (*N*-(5-adamantane-1-yl-methoxy-pentyl)-deoxynojirimycin (AMP-DNM)) and 40 mM conduritol B epoxide (CBE). Store at −20°C up to 1 year.
6. 4-methylumbelliferone standard solution in water: 10 μM MUB. Store at −20°C up to 1 year.
7. DC protein assay (Bio-Rad).
8. Blocking solution: 0.25 M glycine pH 10.7 in water. To prepare 1 l of blocking solution dissolve 18.8 g of glycine in 700 ml H_2O. Adjust pH to 10.7 with NaOH 6 M and adjust volume to 1 l with additional H_2O using a graduated cylinder.
9. Black, 96-well plate, OptiPlate-96 F (PerkinElmer).
10. Transparent, polystyrene 96-well plate for cell culture.

2.2 Materials for Cell-Free Glycohydrolases Assays Based on the Use of Artificial Substrates: Enzymes Associated with the Cell Surface

1. Phosphate buffered saline (PBS): 137 mM NaCl, 2.7 mM KCl, 10 mM Na_2HPO_4, 2 mM KH_2PO_4. To prepare 1 l of PBS pH 7.4 dissolve 8 g of NaCl, 0.2 g of KCl, 1.44 g of Na_2HPO_4, 0.24 g of KH_2PO_4 in 800 ml H_2O. Adjust pH to 7.4 with HCl and adjust volume to 1 l with additional H_2O using a graduated cylinder.
2. Biotinylation solution: 0.5 mg/ml of Sulfo-NHS-LC-Biotin (Pierce) in PBS.
3. Quenching solution: 100 mM Tris–HCl (pH 7.4). 12.14 g of tris(hydroxymethyl)aminomethane (Trizma) is dissolved in 600 ml of water and completely solubilized under stirring. Adjust pH to 7.4 with HCl and adjust volume to 1 l with additional H_2O using a graduated cylinder.
4. TBS-T 0.05% buffer: 10 mM Tris–HCl (pH 8) and 0.05% Tween 20. 121 mg of tris(hydroxymethyl)aminomethane (Trizma) is dissolved in 70 ml of water and completely solubilized under stirring. Adjust pH to 8 with HCl and add 46 μl of Tween 20 with solvent pipette. Finally, achieve the volume of 100 ml by adding H_2O using a graduated cylinder.
5. Lysis buffer: water solution containing 0.5% Triton X-100 (w/v) with a cocktail of protease inhibitors (Roche) and 1 mM PMSF.
6. Monomeric Avidin UltraLink Resin (Pierce).
7. Elution buffer: 2 mM D-Biotin in PBS.
8. Regenerating buffer: 0.1 M glycine, pH 2.8. 75 mg of glycine are dissolved in 70 ml of water. Adjust pH to 2.8 with HCl. Finally, achieve the volume of 100 ml by adding H_2O using a graduated cylinder.
9. Binding buffer: 0.1% Triton X-100 in PBS.
10. Polyvinylidene fluoride membrane (PVDF).
11. PVDF blocking solution: 5% milk in TBS-T 0.05%. 5 g of lyophilized milk are dissolved in 100 ml of TBS-T 0.05% (*see* **Note 2**).
12. HRP–streptavidin incubation buffer: dilute HRP–streptavidin at the final concentration of 0.5 μl/ml in PVDF blocking solution.
13. Dot blot washing buffer: 10 mM Tris–HCl (pH 8) and 0.1% Tween 20. 121 mg of tris(hydroxymethyl)aminomethane (Trizma) is dissolved in 70 ml of water and completely solubilized under stirring. Adjust pH to 8 with HCl and add 92 μl of Tween 20 with solvent pipette. Finally, achieve the volume of 100 ml by adding H_2O using a graduated cylinder.
14. DC protein assay (Bio-Rad)

2.3 Materials for Cell-Free Glycohydrolases Assays Based on the Use of Radioactive Substrates

1. Substrates stock solution:
   (a) [3-^{3}H-Sph]GM3 is solubilized in propan-1-ol–water, 7:3, by volume and stored at −20°C. Specific radioactivity is comprising 1 Ci/mmole and 10 Ci/mmole, and radioactive concentration is 23 nCi/ml.
   (b) Methanol solution of 30 μM GM3 stored at −20°C.
   (c) Methanol solution stored at −20°C of [^{3}H-GalNAc]GM2 with a specific radioactivity comprising 1 mCi/mmole and 10 mCi/mmole.
2. Reaction buffers:
   (a) Sialidase Neu3 reaction buffer 10×: 1 M sodium-acetate buffer (pH 4.2) containing 0.4% Triton X-100 (w/v). Two different aqueous solutions need to be prepared: 1 M sodium acetate trihydrate (solution A) and 1 M acetic acid (solution B). For solution A dissolve 27.22 g of sodium acetate trihydrate in 800 ml of water and adjust volume to 1 l with additional H_2O using a graduated cylinder. To prepare 1 l of solution B add 57.15 ml of glacial acetic acid to 942.85 ml of water. To prepare 100 ml of 1 M sodium-acetate buffer (pH 4.2) mix 26.5 ml of A with 73.5 ml of B. Finally add to 100 ml of buffer 0.4 g of Triton X-100 (*see* **Note 3**).
   (b) Hexosaminidase reaction buffer 10×: 1 M sodium-citrate buffer (pH 4.2) containing 0.6% BSA. Two different aqueous solutions need to be prepared: 1 M citric acid monohydrate (solution A) and 1 M trisodium citrate dehydrate (solution B). For solution A, dissolve 21.01 g of citric acid monohydrate in 800 ml of water and adjust volume to 1 l with additional H_2O using a graduated cylinder. To prepare 1 l of solution B solubilize 29.41 g of trisodium citrate dehydrate in water. To prepare 100 ml of sodium-citrate buffer (pH 4.2) mix 54 ml of A with 46 ml of B. Finally add to 100 ml of buffer 0.6 g of BSA (*see* **Note 3**).
3. GM2 activator protein (GM2AP): 3 μg for each assay.
4. Neu3 blocking solution: chloroform–methanol, 2:1, by volume.
5. High performance thin layer chromatography (HPTLC).
6. HPTLC solvent system: chloroform–methanol–0.2% aqueous $CaCl_2$, 55:45:10 (v:v:v).
7. [3-^{3}H-Sph]Lactosylceramide with a specific radioactivity comprising 1 Ci/mmole and 10 Ci/mmole, and a radioactive concentration of 23 nCi/ml. Methanol solution stored at −20°C.

8. Isolation and quantitation of liberated [^{3}H]GalNAc:
   (a) Silanized glass wool.
   (b) Glass Pasteur pipettes.
   (c) Silica gel LiChroprep RP18.
   (d) Washing solvent: chloroform–methanol 1:1 (v:v).
   (e) Equilibrating solvent: chloroform–methanol–0.1 M KCl 3:48:47 (v:v:v).
9. DC protein assay (Bio-Rad).
10. Scintillation fluid.

2.4 Materials for Cell-Live Based Assays Using Artificial Substrates: Detection of Enzymes Associated with the Cell Surface

1. Phosphate Buffered Saline (PBS): 137 mM NaCl, 2.7 mM KCl, 10 mM Na_2HPO_4, 2 mM KH_2PO_4. To prepare 1 l of PBS (pH 7.4) dissolve 8 g of NaCl, 0.2 g of KCl, 1.44 g of Na_2HPO_4, 0.24 g of KH_2PO_4 in 800 ml H_2O. Adjust pH to 7.4 with HCl and adjust volume to 1 l with additional H_2O using a graduated cylinder.
2. Reaction buffer: DMEM F-12 without phenol red containing 100 mM HEPES and 2 mM L-glutamine. The pH is adjusted to 6 by 6 M HCl.
3. Substrates solution in Reaction buffer: 6 mM 4-methylumbelliferyl β-D-glucopyranoside (MUB-Glc); 250 μM 4-methylumbelliferyl β-D-galactopyranoside (MUB-Gal); 1 mM 4-methylumbelliferyl-N-acetyl-β-D-glucosaminide (MUG); 1 mM 4-methylumbelliferyl-*N*-acetyl-β-D-glucosaminide-6-sulfate (MUGS). Store at −20°C up to 1 year.
4. Inhibitors stock solutions in water: 200 nM adamantane-pentyl-dNM (N-(5-adamantane-1-yl-methoxy-pentyl)-deoxynojirimycin (AMP-DNM)) and 40 mM conduritol B epoxide (CBE). Store at −20°C up to 1 year.
5. 4-methylumbelliferone standard solution in water: 10 μM MUB. Store at −20°C up to 1 year.
6. Blocking solutions: 0.25 M glycine (pH 10.7) in water. To prepare 1 l of blocking solutions dissolve 18.8 g of glycine 700 ml H_2O. Adjust pH to 10.7 with NaOH 6 M and adjust volume to 1 l with additional H_2O using a graduated cylinder.
7. Black, 96-well plate, OptiPlate-96 F(PerkinElmer).
8. Transparent, polystyrene 96-well plate for cell culture.

2.5 Materials for Cell-Live Based Assays Using Radioactive Gangliosides: Detection of trans Activity of PM-Associated β-Galactosidases and β-Glucocerebrosidases

1. [3-^{3}H-Sph]Lactosylceramide with a specific radioactivity comprising 1 Ci/mmole and 10 Ci/mmole, and radioactive concentration of 23 nCi/ml. Methanol solution stored at −20°C.
2. Washing buffer: EMEM supplemented with 10% fetal calf serum (FCS).
3. Phosphate buffered saline (PBS): 137 mM NaCl, 2.7 mM KCl, 10 mM Na_2HPO_4, 2 mM KH_2PO_4. To prepare 1 l of PBS pH 7.4 dissolve 8 g of NaCl, 0.2 g of KCl, 1.44 g of Na_2HPO_4, 0.24 g of KH_2PO_4 in 800 ml H_2O. Adjust pH to 7.4 with HCl and adjust volume to 1 l with additional distilled H_2O using a graduated cylinder.
4. Detachment solution: PBS solution composed by 0.5 g/l porcine trypsin and 0.2 g/l EDTA.
5. High performance thin layer chromatography (HPTLC).
6. 1st solvent: chloroform–methanol–water, 110:40:6 (v:v:v).
7. 2nd solvent system: chloroform–methanol–0.2% aqueous solution of $CaCl_2$, 50:42:11 (v:v:v).
8. [3-^{3}H-Sph]Glucosylceramide with a specific radioactivity comprising 1 Ci/mmole and 10 Ci/mmole, and radioactive concentration of 23 μCi/ml. Methanol solution stored at −20°C.
9. [3-^{3}H-Sph]Ceramide with a specific radioactivity comprising 1 Ci/mmole and 10 Ci/mmole, and radioactive concentration of 23 nCi/ml. Methanol solution stored at −20°C.

2.6 Materials for Cell-Live Based Assay by Cell Feeding with Radioactive Gangliosides

1. [3-^{3}H-Sph]GM3 and [3-^{3}H-Sph]GM1 are solubilized in propan-1-ol–water, 7:3 (v:v) and stored at −20°C. Specific radioactivity comprising 1 Ci/mmole and 10 Ci/mmole, and radioactive concentration of 23 nCi/ml.
2. Chloroquine stock solution: 50 mM in sterile water.
3. Washing buffer: EMEM supplemented with 10% fetal calf serum (FCS).
4. Incubation buffer: cell culture medium without serum.
5. Phosphate buffered saline (PBS): 137 mM NaCl, 2.7 mM KCl, 10 mM Na_2HPO_4, 2 mM KH_2PO_4. To prepare 1 l of PBS (pH 7.4) dissolve 8 g of NaCl, 0.2 g of KCl, 1.44 g of Na_2HPO_4, 0.24 g of KH_2PO_4 in 800 ml H_2O. Adjust pH to 7.4 with HCl and adjust volume to 1 l with additional distilled H_2O using a graduated cylinder.
6. Lyophilizator.
7. High performance thin layer chromatography (HPTLC).
8. [3-^{3}H-Sph]Glucosylceramide with a specific radioactivity comprising 1 Ci/mmole and 10 Ci/mmole, and radioactive concentration of 23 nCi/ml. Methanol solution stored at −20°C.

9. [3-^{3}H-Sph]Ceramide with a specific radioactivity comprising 1 Ci/mmole and 10 Ci/mmole, and radioactive concentration of 23 nCi/ml. Methanol solution stored at −20°C.
10. [3-^{3}H-Sph]Lactosylceramide with a specific radioactivity comprising 1 Ci/mmole and 10 Ci/mmole, and radioactive concentration of 23 nCi/ml. Methanol solution stored at −20°C.
11. Solvent system for aqueous phase: chloroform–methanol–0.2% aqueous solution of $CaCl_2$, 50:42:11 (v:v:v).
12. Solvent system for organic phase: chloroform–methanol–water, 110:40:6 (v:v:v).
13. Solvent system for ceramide detection: hexane–chloroform–acetone–acetic acid, 20:70:20:2 (v:v:v:v).
14. DC protein assay (Bio-Rad).
15. Extraction solvent: chloroform–methanol–water, 20:10:1 (v:v:v).

3 Methods

3.1 Method for Cell-Free Glycohydrolases Assays Based on the Use of Artificial Substrates: Enzymes Associated with Cell Lysate

The following method allows to detect the total cell activity of sialidases, β-hexosaminidases, β-galactosidases, and β-glucocerebrosidases using derivatives of the methylumbelliferone as artificial substrates [5, 11].

1. Sample preparation
   (a) Cell lysate. Cells are washed twice with 1 ml of PBS/10cm^2 cells growth area. Cells are harvested using the same quantity of PBS, previously indicated, by mechanical scraping and collected in falcon tubes of appropriate volume. After centrifugation at 450 × *g* for 10 min at 4 °C the supernatant is discarded and the volume of the cell pellet is estimated. Cells are lysed in 5 volumes (with respect to their volume) of lysis buffer and by ultrasonic homogenizer, using three rounds at 20 db for 10 s. During all the procedures, keep the sample at 4°C.
   (b) Tissue lysate. Explanted organ is washed with PBS and weighted. The organ is then mechanically homogenized with blade and suspended in 10 volumes (respect to their weight) of McIlvaine buffer 1× at pH 6. After homogenization with a Potter homogenizer at 4°C, 50 μl of Complete Protease Inhibitor Cocktail Tablets (Roche), and 5 μl of PMSF 1 mM were added to each ml of homogenate. After three rounds at 20 db for 10 s of sonication and centrifugation at 300 × *g* for 10 min at 4°C, supernatant is collected for protein concentration measurement.
   (c) Determination of the protein concentration.

For both cell lysate and tissue lysate, the protein concentration is evaluated by DC protein assay (Bio-Rad), accordingly to manufacturer's instruction (*see* **Note 1**).

2. Sialidases assay.

   Total sialidase activity is determined with MUB-NANA [12]. Ten to thirty micrograms of cell/tissue lysate proteins is transferred to a well of the transparent 96-well plate. Twenty-five microliters of McIlvaine buffer 4× (pH 4.4) is added and gently shaken and volume adjusted to 95 μl with water. The reaction starts by the addition of 5 μl of MUB-NANA.

   The reaction is performed at 37°C in thermostatic stirrer under gently shaking for 60–120 min.

3. β-glucocerebrosidase GBA1 and nonlysosomal β-glucosylceramidase GBA2 assays.

   Ten to thirty micrograms of cell/tissue lysate proteins are transferred to a well of the transparent 96-well plate and preincubated for 30 min at room temperature with a reaction mixture composed by: 25 μl of McIlvaine buffer 4× (pH 6), the specific inhibitors and water to a final volume of 75 μl. In par ticular, 2.5 μl of AMP-DNM to inhibit GBA2 or 2.5 μl of CBE to inhibit GBA1 are used. At the end of preincubation, the enzymatic reaction starts after the addition of 25 μl of MUB-Glc and the incubation of the 96-well plate at 37°C in thermostatic stirrer under gently shaking for 60–120 min.

4. β-galactosidase and β-hexosaminidase assays.

   Total β-galactosidase and β-hexosaminidase activity are determined with MUB-Gal and MUG respectively. One to ten micrograms of cell/tissue lysate proteins is transferred to a well of the transparent 96-well plate. Twenty-five microliters of McIlvaine buffer 4× (pH 5.2) is added and gently shaken and volume adjusted to 50 μl with water. The reaction starts by the addition of 50 μl of MUB-Gal or MUG.

   The reaction is performed at 37°C in thermostatic stirrer under gently shaking for 30–60 min.

   β-hexosaminidase activity, due to the enzymatic isoforms A and S, is determined by MUGS. 5–20 μg of cell/tissue lysate proteins are transferred to a well of the transparent 96-well plate. Twenty-five microliters of McIlvaine buffer 4× (pH 5.2) is added and gently shaken and volume adjusted to 50 μl with water. The reaction starts by the addition of 50 μl of MUGS.

   The reaction is performed at 37°C in thermostatic stirrer under gently shaking for 30–120 min.

5. Evaluation of the enzymatic activities.

   Assays are performed in triplicate. Negative controls for all the assays are performed using heat-inactivated cell homogenates (100°C for 5 min).

At different time points, 10 μl of the reaction mixtures are transferred to a black 96-well plate and 190 μl of blocking solution is added. Three different doses of MUB are used to obtain a standard curve. The fluorescence, due to the 4-methylumbelliferone released by the action of the enzyme on the specific fluorigenic substrate, is detected by a microplate reader (λex: 355 nm; λem: 460 nm). The standard curve allows determining the nmoles of converted substrate. The specific enzymatic activity is expressed as nmoles of converted substrate/mg cellular proteins x hour.

3.2 Method for Cell-Free Glycohydrolases Assays Based on the Use of Artificial Substrates: Enzymes Associated with the Cell Surface

The following method allows detecting the activity of sialidases, β-hexosaminidases, β-galactosidases, and β-glucocerebrosidase GBA1 associated with the external side of the cell PM using a derivative of the methylumbelliferone as artificial substrates [9, 13].

1. Biotinylation of Cell Surface Proteins.

   Cells are incubated for 1 h in medium without serum and then washed twice with ice-cold PBS (1 ml of PBS/10cm^2 cells growth area). Cells are incubated with biotinylation solution for 30 min at 4°C under gentle stirring. After the incubation, cells are maintained at 4°C, washed twice with PBS and the excess of biotin is blocked using quenching solution. Cells are harvested in PBS by mechanical scraping and collected in a falcon tube of appropriate volume. After centrifugation at 450 × *g* for 10 min at 4°C the supernatant is discarded and the volume of the cell pellet is estimated. Cells are lysed in 5 volumes of lysis buffer (with respect to their volume) and by ultrasonic homogenizer using three rounds at 20 db for 10 s. During all the procedure, keep the sample at 4°C. Cell lysates are centrifuged at 1300 *g* for 5 min at 4°C. Post-nuclear supernatants (PNS) are collected and protein content is evaluated by DC protein assay (*see* **Note 2**).

2. Affinity Purification.

   A glass column is carefully packed with 250 μl of Monomeric Avidin UltraLink Resin and washed with five resin-bed volumes of PBS. The column is then washed with three volumes of 2 mM D-Biotin in PBS. The reversible bound biotin is removed with five volumes of 0.1 M glycine, pH 2.8. Five resin-bed volumes of PBS are used to reequilibrate column for sample binding. From 100 to 300 μg of total proteins from cell lysate are applied to the column and the column flow is stopped for 30 min. To elute fractions containing non-biotinylated proteins (BB fractions), column is washed with 10 volumes of Binding Buffer, then biotinylated proteins (EB fractions) are recovered with five resin-bed volumes of Elution Buffer.

3. Dot Blot Assay

   Two microliters of each fraction is spotted on PVDF membrane. After drying at room temperature, PVDF is incubated in PVDF blocking solution for 1 h at room temperature and incubated with HRP–streptavidin for 30 min. After three washes with dot blot washing buffer, the presence of biotinylated proteins is assessed by chemiluminescence. Signals are visualized using a chemiluminescent kit (WESTAR ηC, Cyanagen). Digital images are obtained by the chemiluminescence system Alliance Mini HD9 (UVItec) (*see* **Note 2**).

4. Enzymatic Assays

The enzymatic assays are carried out as described under Subheading 3.1, **step 5**. The presence of Triton X-100 used to isolate PM proteins does not allow the detection of GBA2 (*see* **Note 2**).

3.3 Method for Cell-Free Glycohydrolases Assays Based on the Use of Radioactive Substrates

The following method allows detecting sialidases Neu3 activity using [3-^{3}H-Sph]GM3 as radioactive natural substrate and that of β-hexosaminidase using [^{3}H-GalNac]GM2 [14, 15].

1. Sialidase Neu3 assay using ganglioside GM3.

   Ganglioside GM3, radiolabeled in position 3 of the sphingosine, is incubated with cell/tissue homogenate obtained as described in the Subheading 3.1, **step 1**. The released radioactive lactosylceramide (LacCer) can easily be separated by GM3 using HPTLC with an appropriate solvent system. Radioactive lipids are visualized by digital-autoradiography. The experimental conditions used, strongly reduce the activity of the other sialidases.

   In detail, the assay contains, in a final volume of 50 μl, the following components: 50–100 μg of cell/tissue lysate, 5 μl of Neu3 reaction buffer 10×, 300 pmoles of GM3, 25 nCi of [3-^{3}H-Sph]GM3. In a 2-ml Eppendorf tube, 10 μl of GM3 and the appropriate amount of [3-^{3}H-Sph]GM3 are evaporated to dryness under a nitrogen stream. The lipid is then suspended in 5 μl of Neu3 reaction buffer 10× and sonicated for 5 min in a bath sonicator. Cell/tissue lysates are added and the final volume adjusted with water to 50 μl. Reactions are performed on ice. After addition of the enzyme, the samples are incubated at 37°C for 2 h under gentle shaking. Reactions are stopped by adding 1.5 ml of Neu3 blocking solution.

   20 μl of Re action mixture and 0.1 nCi of standard LacCer are loaded on a HPTLC and radioactive lipids are separated using the solvent system chloroform–methanol–0.2% aqueous $CaCl_2$, 55:45:10 (v:v:v). The radioactive lipids, separated by chromatography, are detected by digital autoradiography and quantified by a provided software.

All assays are performed at least in duplicate. The blank are carried out using the same reaction mixture where the homogenate are boiled for 10 min at 100°C. The enzymatic activity is expressed as pmoles of LacCer/mg of proteins × hour.

2. β-hexosaminidase assay using ganglioside GM2.

Ganglioside G M2, radiolabeled in its terminal GalNAc moiety, is incubated with cell/tissue homogenate obtained as described in the Subheading 3.1, **step 1**. The released radioactive, water-soluble GalNAc can easily be separated by reversed phase chromatography on single-use RP18 columns and quantitated by scintillation counting. For each assay system measuring the degradation of glycolipid GM2, either GM2AP or a detergent like a crude sodium taurodeoxycholate preparation (2.5 mM) have to be added. In detail, the assay contains, in a final volume of 40 μl, the following components: from 50 to 100 μg of cell/tissue lysate, 4 μl of Hex reaction buffer 10×, [^{3}H-GalNAc]GM2 10 nmol, 3 μg GM2AP (or sodium taurodeoxycholate 2.5 mM). The appropriate amount of [^{3}H-GalNAc]GM2 (10 nmol per assay) is evaporated to dryness under a nitrogen stream. The lipid is then resuspended in 4 μl of Hex reaction buffer 10× and sonicated for 5 min. Three micrograms of GM2AP and cell/tissue lysate are added and the final volume adjusted with water to 40 μl. Reactions are assembled on ice. After addition of the enzyme, the samples are incubated at 37°C for 1 h under gentle shaking. Reactions are stopped by adding 40 μl methanol. The enzymatic activity is evaluated as amount of [^{3}H]GalNAc released, which is quantified as follow: a self-packed RP18 cartridge is prepared using a small amount of silanized glass wool introduced into glass Pasteur pipettes and 0.5 ml of silica gel LiChroprep RP18 (Merck). The column is equilibrated by washing sequentially with 2 × 1 ml chloroform–methanol, 1:1 (v:v), 2 × 1 ml methanol and 2 × 1 ml chloroform–methanol–0.1 M KCl, 3:48:47 (v:v:v). The assay solution is then applied and the flow-through collected. Soluble [^{3}H]GalNAc is subsequently eluted with 2 × 1 ml water, and the eluate combined with the flow-through and 10 ml of scintillation fluid (Ultima Gold, Packard). Radioactivity is then measured by a scintillation counter, with correction for quenching. All assays are performed at least in duplicate, and blank values (from incubations without enzyme and/or GM2AP) are subtracted. The enzymatic activity is expressed as pmoles of [^{3}H-GalNAc]GM2/mg of proteins x hour.

3.4 Method for Cell-Live Based Assays Using Artificial Substrates: Detection of Enzymes Associated with the Cell Surface

The following method allows detecting the activity of β-hexosaminidases, β-galactosidases, and β-glucocerebrosidases directly on living cells in culture using derivatives of the methylumbelliferone as artificial substrates [5, 8].

1. Plate the cells in 96-well microplate at density not less than 15×10^3 cells/cm^2 24 h before the assay. The day after, cell medium is removed and cells are washed twice with 100 μl of reaction buffer. For GBA1 and GBA2 assays, cells are preincubated for 30 min at room temperature with 50 μl of reaction buffer containing respectively 5 nM AMP-DNM and 1 mM CBE. The well dedicated for the evaluation of the other enzymatic activity is left in reaction buffer. After, the incubation buffer is removed from all wells and 100 μl of substrate solutions is added. In particular, MUB-Gal is added for the detection of the β-galactosidase, MUG for the detection of the β-hexosaminidases, and MUGS for the detection of β-hexosaminidase A and S. For GBA1 and GBA2 assays, after remotion of the preincubation buffers, MUB-Glc containing 5 nM AMP-DNM or 1mM CBE is added. As blank the same volume of substrate solutions is loaded on wells without cells.

   The incubation is performed at 37°C under very gentle shaking (not in the presence of 5% CO_2 in order to maintain the pH of the reaction buffer). At different time points, 10 μl of the enzymatic assay mixture is transferred to a black microplate (Black, 96-well plate, OptiPlate-96 F, PerkinElmer) and 20 volumes of 0.25 M glycine (pH 10.7) are added. The fluorescence is evaluated using a microplate reader (λex: 355 nm/λem: 460 nm). The titration curve obtained by the MUB-standard allows determining the nmoles of converted substrate. All assays are performed at least in triplicate, and blank values are subtracted. The specific enzymatic activity is expressed as nmoles of converted substrate/10^6 cells × hour (*see* **Note 4**).

3.5 Methods for Cell-Live Based Assays Using Radioactive Gangliosides: Detection of trans Activity of PM-Associated β-Galactosidases and β-Glucocerebrosidases

The following method allows for evaluating the *trans* activity of the PM-associated β-galactosidases and β-glucosidases on substrates belonging to neighboring cell membrane, using natural substrates [9].

1. 300 pmoles of [3-^{3}H-Sph]Lactosylceramide is dried in a glass tube under nitrogen stream, solubilized in 30 μl ethanol and loaded on 0.3-cm^2 plastic polystyrene 96-wells plate and dried under a gentle stream of nitrogen. The well is then washed five times with 100 μl of washing buffer and from 3×10^3 to 3×10^4 cells are added and cultured in normal cell culture medium (pH 7.4) up to 48 h. Cells are then removed using the detachment solution followed by two washes with PBS. Radioactive lipids attached to the plate were recovered by five washes with

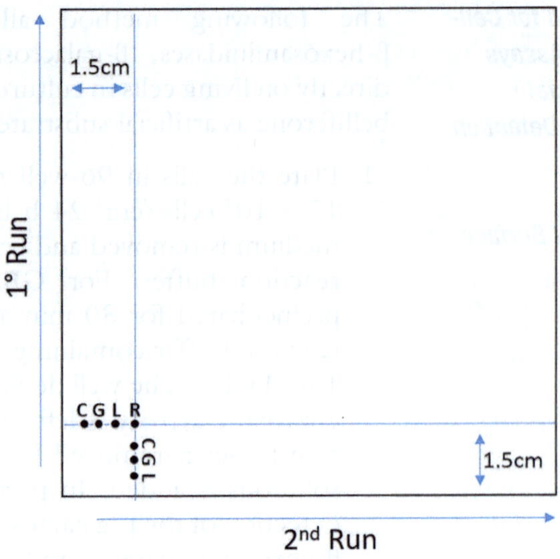

Fig. 1 Schematic representation of a two-dimensional HPTLC. A square HPTLC of 100 cm^2 is loaded with the 1 nCi of lipids extracted (R) and 0.2 nCi of radioactive standards such as lactosylceramide (L), glucosylceramide (G), and ceramide (C) as indicated. For the first run is used the solvent system chloroform-methanol-water, 110:40:6 (v:v:v) and for the second run chloroform-methanol–0.2% aqueous solution of $CaCl_2$, 50:42:11 (v:v:v)

100 μl of ethanol. Radioactivity is then measured by a scintillation counter. Radioactive lipids were analyzed by two-dimensional HPTLC. A square HPTLC of 10 cm of side is loaded with the 1 nCi of lipids extracted by the well (R) and 0.2 nCi of radioactive standards such as lactosylceramide (L), glucosylceramide (G), and ceramide (C) as indicated in Fig. 1 (*see* **Note 5**).

For the first run is used the solvent system chloroform-methanol-water, 110:40:6 (v:v:v) and for the second run chloroform-methanol–0.2% aqueous solution of $CaCl_2$, 50:42:11 (v:v:v). The radioactive lipids are identified using radioactive authentic standards and visualized by digital autoradiography. The radioactivity associated with individual lipids was determined with the specific software. The percentage of GlcCer formed represents the trans-activity of β-galactosidases whereas the percentage of ceramide formed reveals the *trans* activity of β-glucocerebrosidases (*see* **Note 5**).

3.6 Method for Cell-Live Based Assay by Cell Feeding with Radioactive Gangliosides

This methodology allows defining the hydrolytic activity of both lysosome and cell-PM enzymes directly on living cells, using radioactive gangliosides. The results obtained allow determining the efficacy of the catabolic pathways between different cell lines as well as the effect of specific cell treatments on these compartments [7, 16].

1. One day before the feeding, cells are plated in ϕ100 Petri culture dishes at the concentration of 10^4 cell/cm^2. The day after, isotopically labeled [3-^{3}H-Sph]GM3 or alternatively [3-^{3}H-Sph]GM1 is administered to the cells.

   [3-^{3}H-Sph]GM3 or [3-^{3}H-Sph]GM1 dissolved in propanol–water, 7:3 (v:v) is transferred to a sterile glass tube and dried under nitrogen flow. The residue is solubilized in cell culture medium without serum at the final concentration of 4.5×10^{-6} M. Complete solubilization of the radioactive lipids is evaluated by scintillation counter. To follow the catabolism of [3-^{3}H-Sph]gangliosides due only to the PM-associated glycohydrolases, dedicated cells are preincubated with 100 μM chloroquine for 1 h in cell medium without serum. After removal of the medium and rapid washing of cells with 5 ml of PBS, 6 ml of the medium containing the radioactive lipid is added to each dish and the cells are incubated at 37°C in the presence or absence of 100 μM chloroquine for 4 h. At the end of incubation, cells are washed three times with complete cell culture medium and incubated in the same medium for 30 min. Then, cells are washed three times with 5 ml of PBS, harvested in the same volume of PBS by mechanical scraping and collected in a falcon tube of appropriate volume. After centrifugation at $450 \times g$ for 10 min at 4°C the supernatant is discarded and cells pellet is suspended in 500 μl of water, transferred to a 2-ml Eppendorf tube and lyophilized. An aliquot of the cell lysate in water is preserved from the lyophilization in order to perform the DC protein assay.

   After lyophilization, lipid extraction and sphingolipid analyses are performed. Total lipids from lyophilized cells are extracted using the extraction solvent. The total lipid extracts are subjected to a two-phase partitioning by adding 20% water to the lipid extract. The radioactivity associated with the total lipid extract and with aqueous and organic phases is evaluated by scintillation counter. [^{3}H]sphingolipids of total extracts and organic phases are separated by HPTLC using the solvent system chloroform–methanol–water, 110:40:6 (v:v:v). For the specific determination of ceramide content the radioactive lipids of the organic phases are separated using the solvent system hexane–chloroform–acetone–acetic acid, 20:70:20:2 (v:v:v). Radioactive gangliosides contained in aqueous phases are separated by chloroform–methanol–0.2% aqueous solution of $CaCl_2$, 50:42:11 (v:v:v). [^{3}H]sphingolipids are identified by referring to the position of standards in the chromatogram, visualized by digital autoradiography and quantified by specific software (*see* **Note 6**). The amount of radioactivity to be loaded on the HPTLC depends on the sensitivity of the digital autoradiograph used.

4 Notes

1. Notes for cell-free glycohydrolases assays based on the use of artificial substrates: enzymes associated with cell lysate

   (a) Use only anhydrous salts.

   (b) Check every time the pH of the McIlvaine buffer; in case of differences adjust the pH adding small volume of solution A for acidification and B to basify.

   (c) To allow complete solubilization, heat the stock solution to 50°C.

   (d) Protein evaluation can be made also using other assays.

2. Notes for cell-free glycohydrolases assays based on the use of artificial substrates: enzymes associated with the cell surface

   (a) To better solubilize the milk, warm the solution at 37°C.

   (b) Protein evaluation can be made also using other assays.

   (c) Chemiluminescent signals can be revealed also by the classical methods employing the photographic film sheets.

   (d) To exclude the contamination with intracellular components of the fraction collected from the affinity column and containing the PM-biotinylated proteins, is suggested to perform western blot analysis using specific antibodies for endoplasmic reticulum, Golgi, and lysosomes protein markers.

3. Notes for cell-free glycohydrolases assays based on the use of radioactive substrates

   (a) Check every time the pH of the Neu3 and Hex reaction buffers; in case of differences adjust the pH adding small volume per time of solution A for acidification and B to basify.

4. Notes for cell-live based assays using artificial substrates: detection of enzymes associated with the cell surface

   (a) Control experiments are performed in order to exclude the activity released from lysosomes and/or other intracellular sites. To this purpose, each enzymatic assay was performed on aliquots of medium previously maintained in contact with cells for different times.

   (b) To exclude some intracellular enzymatic degradation of the fluorigenic substrates, the absence of fluorescence associated with the cells was evaluated at the end of each enzymatic assay by fluorescent microscopy and by fluorimetric analysis of the lysed cells in 0.25 M glycine (pH 10.7).

5. Notes for cell-live based assays using radioactive gangliosides: detection of *trans* activity of PM-associated β-galactosidases and β-glucocerebrosidases

(a) Radioactive standards for the second run need to be loaded on the HPTLC after the first run. Non-radioactive lipid standards could be also used. After the run, the HPTLC is incubated in a chamber in the presence of iodine. After 10 min, the iodine marked the lipid standards with an orange/yellow staining. On the colored spot, 10 pCi of tritium labeled substance (the origin does not matter) is loaded. This way, the HPTLC is ready for digital autoradiography and the standards are marked.

(b) Control experiments on the ceramide shed from fibroblasts are performed by feeding cells with radioactive Lactosylceramide and then checking the lipids remained attached to the plastic plate.

6. Notes for cell-live based assay by cell feeding with radioactive gangliosides

(a) Non-radioactive lipid standards can also be used. After the run, the HPTLC is incubated in a chamber in the presence of iodine. After 10 min the iodine marked the lipid standards with an orange/yellow staining. On the colored spot 10 pCi of tritium labeled substance (the origin does not matter) is loaded. This way, the HPTLC is ready for digital autoradiography and the standards are marked.

References

1. Piccinini M, Scandroglio F, Prioni S, Buccinna B, Loberto N, Aureli M, Chigorno V, Lupino E, DeMarco G, Lomartire A, Rinaudo MT, Sonnino S, Prinetti A (2010) Deregulated sphingolipid metabolism and membrane organization in neurodegenerative disorders. Mol Neurobiol 41(2–3):314–340. https://doi.org/10.1007/s12035-009-8096-6
2. Aureli M, Loberto N, Lanteri P, Chigorno V, Prinetti A, Sonnino S (2011) Cell surface sphingolipid glycohydrolases in neuronal differentiation and aging in culture. J Neurochem 116(5):891–899. https://doi.org/10.1111/j.1471-4159.2010.07019.x
3. Merrill AH Jr (2011) Sphingolipid and glycosphingolipid metabolic pathways in the era of sphingolipidomics. Chem Rev 111(10):6387–6422. https://doi.org/10.1021/cr2002917
4. Kolter T, Sandhoff K (1999) Sphingolipids-their metabolic pathways and the pathobiochemistry of nerodegenrative diseases. Angew Chem Int Ed 38:1532–1568
5. Aureli M, Bassi R, Loberto N, Regis S, Prinetti A, Chigorno V, Aerts JM, Boot RG, Filocamo M, Sonnino S (2012) Cell surface associated glycohydrolases in normal and Gaucher disease fibroblasts. J Inherit Metab Dis 35(6):1081–1091. https://doi.org/10.1007/s10545-012-9478-x
6. Aureli M, Samarani M, Loberto N, Bassi R, Murdica V, Prioni S, Prinetti A, Sonnino S (2013) The glycosphingolipid hydrolases in the central nervous system. Mol Neurobiol 50:1–12. https://doi.org/10.1007/s12035-013-8592-6
7. Valaperta R, Chigorno V, Basso L, Prinetti A, Bresciani R, Preti A, Miyagi T, Sonnino S (2006) Plasma membrane production of ceramide from ganglioside GM3 in human fibroblasts. FASEB J 20(8):1227–1229
8. Aureli M, Bassi R, Prinetti A, Chiricozzi E, Pappalardi B, Chigorno V, Di Muzio N, Loberto N, Sonnino S (2012) Ionizing radiations increase the activity of the cell surface glycohydrolases and the plasma membrane ceramide content. Glycoconj J 29:585. https://doi.org/10.1007/s10719-012-9385-2
9. Aureli M, Masilamani AP, Illuzzi G, Loberto N, Scandroglio F, Prinetti A, Chigorno V, Sonnino S (2009) Activity of plasma membrane beta-galactosidase and beta-glucosidase. FEBS Lett 583(15):2469–2473

10. Aureli M, Gritti A, Bassi R, Loberto N, Ricca A, Chigorno V, Prinetti A, Sonnino S (2011) Plasma membrane-associated glycohydrolases along differentiation of murine neural stem cells. Neurochem Res 37(6):1344–1354. https://doi.org/10.1007/s11064-012-0719-z
11. Aureli M, Samarani M, Murdica V, Mauri L, Loberto N, Bassi R, Prinetti A, Sonnino S (2014) Gangliosides and cell surface ganglioside glycohydrolases in the nervous system. Advances in neurobiology 9:223–244. https://doi.org/10.1007/978-1-4939-1154-7_10
12. Chigorno V, Cardace G, Pitto M, Sonnino S, Ghidoni R, Tettamanti G (1986) A radiometric assay for ganglioside sialidase applied to the determination of the enzyme subcellular location in cultured human fibroblasts. Anal Biochem 153(2):283–294
13. Magini A, Mencarelli S, Tancini B, Ciccarone V, Urbanelli L, Hasilik A, Emiliani C (2008) Identification and characterization of mature beta-hexosaminidases associated with human placenta lysosomal membrane. Biosci Rep 28(4):229–237. https://doi.org/10.1042/BSR20080075
14. Valaperta R, Valsecchi M, Rocchetta F, Aureli M, Prioni S, Prinetti A, Chigorno V, Sonnino S (2007) Induction of axonal differentiation by silencing plasma membrane-associated sialidase Neu3 in neuroblastoma cells. J Neurochem 100(3):708–719
15. Wendeler M, Sandhoff K (2009) Hexosaminidase assays. Glycoconj J 26(8):945–952. https://doi.org/10.1007/s10719-008-9137-5
16. Chiricozzi E, Niemir N, Aureli M, Magini A, Loberto N, Prinetti A, Bassi R, Polchi A, Emiliani C, Caillaud C, Sonnino S Chaperone therapy for GM2 Gangliosidosis: effects of Pyrimethamine on beta-Hexosaminidase activity in Sandhoff fibroblasts. Mol Neurobiol 50:159–167. https://doi.org/10.1007/s12035-013-8605-5

Chapter 19

Pharmacological Modulation of Glycosphingolipid Metabolism

Jin-ichi Inokuchi, Takashi Ode, and Miki Hara-Yokoyama

Abstract

The experimental approach to deplete cellular glycosphingolipids (GSLs) with the specific inhibitors of glycosphingolipid biosynthesis has the potential to identify functions of endogenous GSLs. Most GSLs are derived from glucosylceramide (GlcCer). D-*threo*-1-phenyl-2-decanoylamino-3-morpholino-1-propanol (D-PDMP) inhibits GlcCer synthase and has been used extensively to study the biological functions of living cells. D-PDMP inhibits mTORC1 activity, which is independent of its inhibitory activity on GlcCer synthase. We also developed an analog of D-PDMP, D-*threo*-1-phenyl-2-benzyloxycarbonylamino-3-pyrrolidino-1-propanol (D-PBPP) lacking the effect on mTORC1. Here, we summarize the effects of D-PDMP and D-PBPP on the metabolism of GSLs and cell growth.

Key words Glucosylceramide (GlcCer) synthase, Glycolsphingolipids, Gangliosides, D-PDMP, D-PBPP, mTORC1

1 Introduction

Efforts to deplete cellular glycosphingolipids (GSLs) by means of the specific inhibitors of GSL biosynthesis are useful for determining the functions of endogenous GSLs. Nearly all GSLs are derived from glucosylceramide (GlcCer), which is synthesized from ceramide and UDP-glucose. The sequential addition of additional monosaccharide and sialic acid to GlcCer results in a complex family of structures, such as the ganglio, globo, isoglobo, neolacto, and lacto series.

It has been demonstrated by Inokuchi and Radin [1] that an analog of ceramide, D-*threo*-1-phenyl-2-decanoylamino-3-morpholino-1-propanol (D-PDMP), could inhibit UDP-glucose N-acylsphingosine glucosyltransferase [EC 2.4.1.80] (GlcCer synthase). Thus, D-PDMP has been used extensively to study the biological function of GSLs in living cells, including in cell growth, cell adhesion [2–4]. Recent studies also reveals that D-PDMP exhibits a possible therapeutic intervention for atherosclerosis [5] and cardiac hypertrophy [6]. Apart from the inhibitory effect on

Sandro Sonnino and Alessandro Prinetti (eds.), *Gangliosides: Methods and Protocols*, Methods in Molecular Biology, vol. 1804, https://doi.org/10.1007/978-1-4939-8552-4_19, © Springer Science+Business Media, LLC, part of Springer Nature 2018

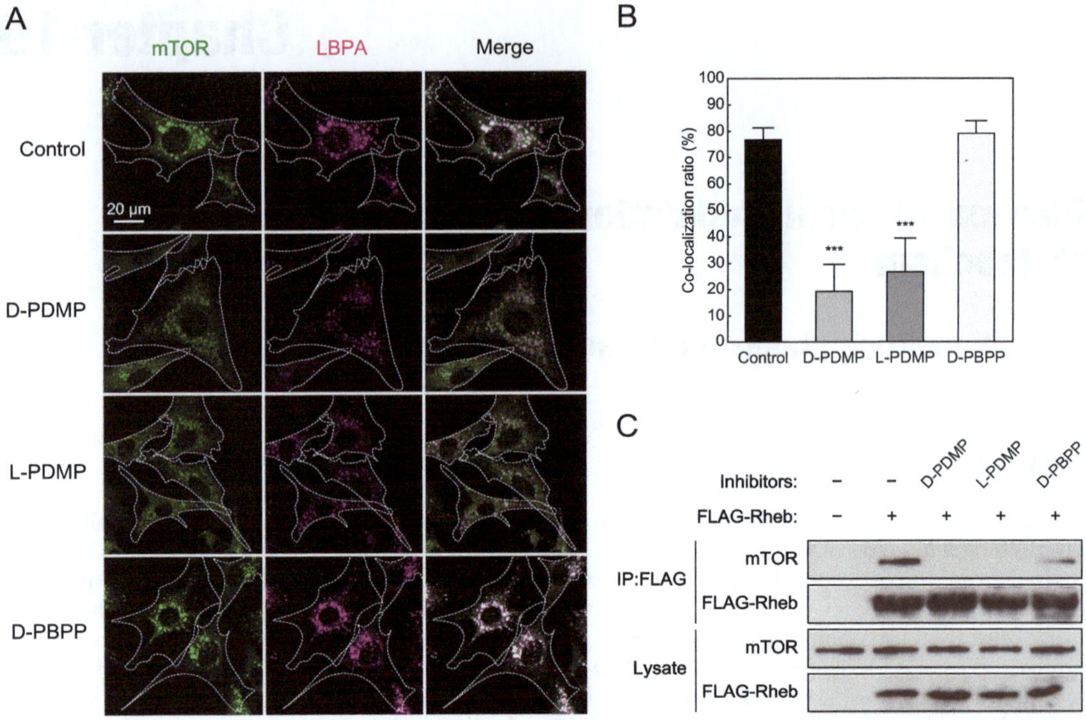

Fig. 1 mTOR is dissociated from late endosome/lysosome by D-PDMP and L-PDMP, but not by D-PBPP, treatments [11] (**A**) Intracellular localization of mTOR in the absence or the presence of 40 μM D-PDMP, L-PDMP, or D-PBPP in MC3T3-E1 cells. The cells were costained with anti-mTOR antibody (green) and anti-LBPA antibody (magenta). Cells are outlined with white dashed lines. (**B**) Quantification of the co-localization between mTOR and LBPA in MC3T3-E1 cells. Data are expressed as means ± SD. ***$P < 0.001$ compared with control group. (**C**) FLAG-tagged Rheb (FLAG-Rheb) was expressed in MC3T3-E1 cells in the absence or the presence of D-PDMP, L-PDMP, or D-PBPP. Proteins in the lysates and the immunoprecipitates with anti-FLAG antibody were analyzed by immunoblotting using anti-FLAG-antibody or anti-mTOR antibody

GlcCer synthase, D-PDMP has also been reported to alter the organization of late endosome/lysosome (LE/Ly) [7, 8] and the endoplasmic reticulum (ER) [9]. Furthermore, D-PDMP induces dissociation of mTOR from LE/Ly (Fig. 1) and inhibits the activity of mTOR (Fig. 2) [10]. This effect of D-PDMP is independent of its action as the inhibitor of GlcCer synthase, since two stereoisomers of PDMP, with and without the inhibitor activity, showed essentially the same effect [10]. Thus, D-PDMP could be useful to evaluate the biological potential of ceramide analogues on intracellular organelles.

Previously, we developed an analog of D-PDMP, D-*threo*-1-phenyl-2-benzyloxycarbonylamino-3-pyrrolidino-1-propanol (D-PBPP) [11] (Table 1). D-PBPP proved to be a potent inhibitor of GlcCer synthase without mTORC1 inhibition [10] (Fig. 2). Here, the effects of D-PDMP and D-PBPP on GlcCer synthase, the metabolism of GSLs and mTORC1 are summarized.

Table 1
Comparison of the effects of glucosylceramide synthase inhibitors, D-PDMP and D-PBPP, on cellular sphingolipid levels

Content (μg/mg protein)			
Control	**20 μM D-PDMP**	**20 μM D-PBPP**	**Sphingolipids**
Ceramide	0.58	0.71	0.52
Sphingomyelin	3.84	5.03	3.92
GlcCer	0.76	0.26	0.19
GM3	2.48	1.91	0.77

C6 glioma cells were treated with or without 20 μM D-PDMP or D-PBPP PDMP for 4 days. The lipid extraction and analyses were performed as described previously [11]. The data are the means of duplicate determinations and the differences in the values in each group were less than 10%

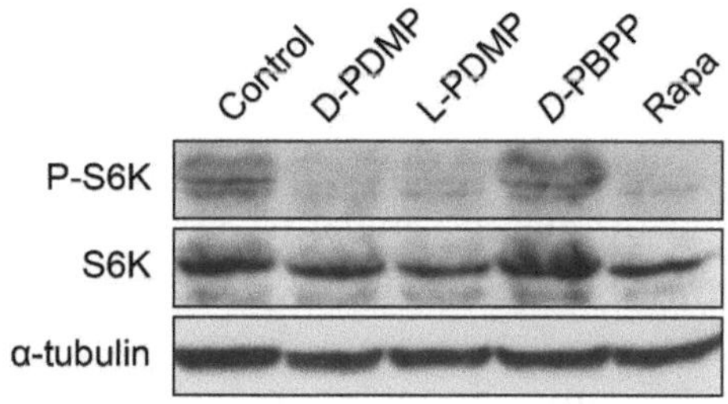

Fig. 2 The mTORC1 activity is inhibited by D-PDMP and L-PDMP, but not by D-PBPP [11]. MC3T3-E1 cells were cultured in the absence or the presence of 40 μM D-PDMP, L-PDMP, PBPP, or 100 nM rapamycin. Proteins in the lysates were analyzed by immunoblotting using anti-S6K antibody, anti-phosphorylated-S6K antibody, or anti-α-tubulin antibody

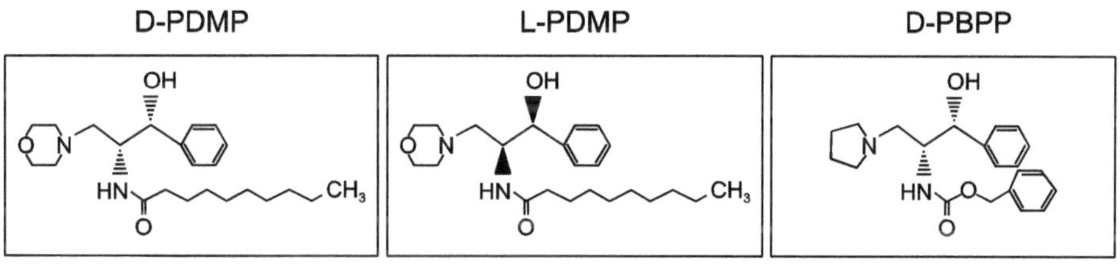

Fig. 3 Chemical structures of D-PDMP, L-PDMP, and D-PBPP

2 Materials

2.1 Reagents

1. D-PDMP: D-threo(1R,2R)-1-phenyl-2-decanoylamino-3-morpholino-1-propanol (Tokyo Chemical Industry Co., Ltd.) (Fig. 3)

   Specification: D-PDMP.HCl.H_2O.

   Molecular Weight: 445.

Molecular Formula: $C_{23}H_{41}ClN_2O_4$.

4 mM stock solution in water. Store aliquots at −30 °C. Filter through 0.22 μm membrane filter before and store at 4 °C as a stock solution.

2. L-PDMP: D-threo(1S,2S)-1-phenyl-2-decanoylamino-3-morpholino-1-propanol (Tokyo Chemical Industry Co., Ltd.) (Fig. 3)

   Specification: L-PDMP.HCl.H_2O.

   Molecular Weight: 445.

   Molecular Formula: $C_{23}H_{41}ClN_2O_4$.

4 mM stock solution in water. Store aliquots at −30 °C. Filter through 0.22 μm membrane filter before use and store at 4 °C as a stock solution.

3. D-PBPP:D-threo(1R,2R)-1-phenyl-2-benzyloxycarbonylamino-3-pyrrolidino-1-propanol (Tokyo Chemical Industry Co., Ltd.) (Fig. 3)

   Specification: D-PBMP.HCl.H_2O.

   Molecular Weight: 407.

   Molecular Formula: $C_{21}H_{27}ClN_2O_4$.

4 mM stock solution in water. Store aliquots at −30 °C.

2.2 Immunocytochemistry

1. 15-mm round glass coverslips.
2. 12-well plate.
3. Growth medium: modified MEM, 10% fetal bovine serum, 4 mM L-glutamine, 50 U/ml L-penicillin, 50 μg/ml streptomycin.
4. Osteogenic medium: growth medium supplemented with 10 mM α-glycerophosphate, 50 μg/ml ascorbic acid, and 10 nM dexamethasone.
5. Paraformaldehyde (PFA): 3% PFA in PBS. Dissolve 20 g PFA with ~40 ml 1 N NaOH by heating and then add 10 ml concentrated ×10 PBS. Check pH using pH test paper and adjust pH to 7.1–7.3. Fill up to 100 ml with water to make 20% PFA stock solution. Store at 4 °C.
6. PBS containing 100 mM glycine (PBSG).
7. Digitonin solution: Digitonin is dissolved in PBS (5 mg/ml). Store the aliquots at −30 °C.
8. Blocking solution, diluent solution: 1% BSA in PBS.
9. Plastic container.
10. Paper towel.

11. Anti-mTOR Ab (1:400; 2983 Cell Signaling Technology, MA, USA, rabbit monoclonal).
12. anti-LBPA Ab (1:20 dilution, prepared as described [8], mouse monoclonal).
13. Alexa 555-conjugated anti-mouse IgG (1:200 dilution).
14. Alexa 488-conjugated anti-rabbit IgG (1:200 dilution).
15. ProLong Gold.

2.3 Immunoprecipitation

1. Lysis buffer (2×): 40 mM Tris–HCl, pH 7.5, 300 mM NaCl, 20 mM iodoacetamide, 10 mM NaF, 10 mM sodium pyrophosphate, 20 mM EDTA, 20 mM α-glycerophosphate, 2% Triton X-100.
2. Sample buffer (3×): 0.175 M Tris–HCl (pH 6.8), 5% SDS, 0.3 M dithiothreitol, 0.02% bromophenol blue, 15% glycerol.
3. Anti-FLAG M2 affinity gel.

2.4 Immunoblotting

1. Polyvinylidene difluoride (PVDF) membrane.
2. Western blot transfer buffer: 0.025 M Tris, 0.192 M glycine, 0.1% SDS, 20% methanol.
3. Tris-buffered saline (TBS; ×10): 1.5 M NaCl, 0.1 M Tris–HCl, pH 7.4.
4. TBS containing 0.1% Tween 20 (TBST).
5. Blocking solution: 5% BSA in TBST.
6. Biotinylated anti-FLAG Ab (1:1000 dilution; GTX77459, GeneTex, CA, USA, rabbit polyclonal).
7. Anti-FLAG M2 Ab (1:2000 dilution, F1804 Sigma-Aldrich, mouse monoclonal).
8. Anti-mTOR Ab (1:1000; 2983 Cell Signaling Technology, MA, USA, rabbit monoclonal).
9. Anti-S6K Ab (1:500 dilution, rabbit polyclonal).
10. Anti-P-S6K Ab (1:500 dilution, rabbit polyclonal).
11. Anti-α-tubulin Ab (1:10,000 dilution, rabbit polyclonal).
12. Horseradish peroxidase (HRP)-conjugated anti-biotin antibody.
13. HRP-conjugated anti-mouse IgG.
14. HRP-conjugated anti-rabbit IgG.
15. ECL Prime Western Blotting Detection Reagent.

2.5 Thin Layer Chromatography

1. Solvent 1: chloroform–methanol–water (65:25:4).
2. Solvent 2: chloroform–methanol–0.5% $CaCl_2$ in water (55:45:10).

3. Solvent 3: chloroform–methanol–acetic acid–formic acid–water (42:18:7.2:2.4:1.2).
4. Solvent 4: chloroform–methanol–acetic acid (91:2:7).
5. Orcinol reagent: 0.2% orcinol in 2 M sulfuric acid.
6. Cupric acetate reagent: 3% cupric acetate–8% H_3PO_4.
7. Silica Gel 60 HPTLC plates.

2.6 Lipid Extraction and Purification

1. DEAE-Sephadex A-25 column (0.8 × 4.5 cm).
2. Chloroform–methanol–water (4:8:3, v/v/v).
3. Chloroform–methanol–water (30:60:8, v/v/v).
4. Chloroform–methanol–1 M aqueous Na-acetate (30:60:8, v/v/v).

3 Methods

3.1 Effects of Glucosylceramide Synthase Inhibitors, D-PDMP and D-PBPP, on Cellular Sphingolipids

1. Plate cells at 5 × 10^6 cells per culture dish of 15 cm diameter (145 cm^2)
2. On the next day, replace the culture medium in the presence or absence of an inhibitor.
3. Culture for 1–6 days.
4. Wash cells with PBS and scrape from the dishes.
5. Extract total lipids with chloroform/methanol/water (4:8:3, v/v/v) for two consecutive 24 h periods, followed by 4 h extraction at 40 °C.
6. The total lipid extract was dissolved in chloroform–methanol–water (30:60:8, v/v/v), passed through a DEAE-Sephadex A-25 column, and eluted successively with 5 volumes each of the same solvent (neutral lipid fraction), and chloroform–methanol–1 M aqueous Na-acetate (30:60:8) (acidic lipid fraction).
7. Decompose glycerolipids by saponification and remove salts (*see* **Note 1**)
8. Develop TLC on Silica Gel 60 HPTLC plates with solvent 1 for neutral GSLs, with solvent 2 for gangliosides, with solvent 3 for sphingomyelin, with solvent 4 for ceramide.
9. Visualize neutral GSLs with orcinol reagent, gangliosides with resorcinol-HCl reagent, sphingomyelin and ceramide with cupric acetate reagent.
10. Perform densitometry with a dual wavelength flying spot scanner

Examination by means of TLC of GSLs of C6 glioma cells after incubation disclosed that D-PBPP was more effective than

D-PDMP in lowering the GlcCer and GM3 levels, as expected from its greater effectiveness in vitro as a GlcCer synthase inhibitor [11] (Table 1). The accumulation of cellular ceramide and sphingomyelin occurred with D-PDMP; however, the D-PBPP treatment did not affect these sphingolipid levels (Table 1).

3.2 Dissociation of mTOR from LE/Ly Upon PDMP Treatment in MC3T3-E1 Cells (*See* Note 2)

1. Place 15-mm round glass coverslips in 12-well plate (*see* **Note 3**).
2. Seed MC3T3-E1 cells in growth medium on the coverslips (4×10^4 cells/well).
3. After 24 h, exchange the medium from growth medium to osteogenic medium containing 40 μM D-PDMP, L-PDMP or PBPP.
4. After the culture for 6–24 h, fix the cells with 3% PFA for 20 min.
5. Wash the coverslips with PBSG three times.
6. Permeabilize the cells with digitonin/PBSG for 5 min (*see* **Note 4**).
7. Wash the coverslips with PBS three times.
8. Block the coverslips with blocking solution for 1 h.
9. Treat the cells with anti-mTOR Ab together with anti-LBPA Ab for 1 h at room temperature (*see* **Note 5**).
10. Wash the coverslips with PBS containing 0.1% BSA three times
11. Treat the cells with Alexa 555-conjugated anti-mouse IgG and Alexa 488-conjugated anti-rabbit IgG for 1 h at room temperature as in **step 9**.
12. Wash the coverslips with PBS three times.
13. Rinse the coverslips with deionized water to remove PBS.
14. Leave the coverslips to air-dry.
15. Put 4 μl of ProLong Gold in the center of a slide glass and cover the solution with the coverslip.
16. Acquire images using confocal microscopy.

3.3 mTOR Dissociation from Its Activator Rheb Upon PDMP Treatment

1. Seed MC3T3-E1 cells in growth medium on 10-cm dishes (3×10^5 cells/dish).
2. After 24 h, add 40 μM D-PDMP, 40 μM L-PDMP, or 40 μM PBPP.
3. After 24 h, transfect the cells with the plasmid encoding FLAG-tagged Rheb using FuGENE6 (*see* **Note 6**).
4. After 48 h, wash the cells with PBS.
5. Add 10 ml of PBS and detach the cells with a scraper.
6. Transfer all the materials into a new tube.
7. Centrifuge for 5 min at 240 × *g*.

8. Withdraw the medium (*see* **Note 7**).
9. Add 150 μl of lysis buffer containing 0.5% Triton X-100 (*see* **Note 8**) and lyse the cells for 1 h on ice.
10. Transfer the lysate to microtube and centrifuge the lysate for 15 min at 15,000 × *g* at 4 °C.
11. Transfer the supernatant to a new microtube.
12. Incubate the supernatants with 20 μl of anti-FLAG M2 affinity gel for 2 h at 4 °C.
13. Recover the immune-complex by centrifugation for 2 min at 6230 × *g* at 4 °C.
14. Wash the immune-complex with PBS three times.
15. Add 30 μl of sample buffer to each sample.
16. Boil the immune-complex for 5 min.
17. Centrifuge the lysate for 2 min at 6230 × *g* at room temperature.
18. Transfer the supernatant to a new microtube.
19. Separate proteins in the supernatants with SDS-PAGE.
20. Immediately following SDS-PAGE, lay the gel on one PVDF membrane placed on a filter paper.
21. Cover another filter paper on the gel.
22. Set this assembly within the transfer system for electrophoretic transfer at 30 V for 17 h (*see* **Note 9**).
23. Block the membrane with blocking solution for 1 h.
24. Incubate the membrane with either biotinylated anti-FLAG Ab, anti-FLAG M2 Ab, or anti-mTOR Ab.
25. Wash the membrane with TBST.
26. Incubate the membrane with horseradish peroxidase (HRP)-conjugated anti-biotin antibody, HRP-conjugated anti-mouse IgG or HRP-conjugated anti-rabbit IgG for 1 h at room temperature.
27. Wash the membrane with TBST.
28. Visualized the signals using the ECL Prime Western Blotting Detection Reagent.

3.4 Inhibition of mTOR Activity Upon PDMP Treatment

1. Seed MC3T3-E1 cells in 6-well plate using growth medium (1.8×10^5 cells/well).
2. After 24 h, exchange the medium to osteogenic medium containing either 40 μM D-PDMP, 40 μM L-PDMP, 40 μM PBPP or 100 nM rapamycin.
3. After 24 h, withdraw the culture medium.

4. Add 100 μl of lysis buffer containing 1% Triton X-100 (*see* **Note 10**) per well and lyse the cells for 15 min on ice.
5. Transfer the lysate to a new tube and centrifuge the lysate for 10 min at 21,880 × *g* at 4 °C.
6. Transfer the supernatant to a new tube (80 μl).
7. Add 40 μl of sample buffer (3×).
8. Boil the sample for 5 min.
9. Separate proteins with SDS-PAGE and transfer to PVDF membrane (as in **steps 19–22** in Subheading 3.3).
10. Block the membrane with blocking solution for 1 h.
11. Incubate the membrane with anti-S6K Ab, anti-P-S6K Ab or anti-α-tubulin Ab.
12. Wash the membrane with TBST.
13. Incubate the membrane with HRP-conjugated anti-rabbit IgG for 1 h at room temperature.
14. Wash the membrane with TBST.
15. Visualized the signals using the ECL Prime Western Blotting Detection Reagent.

4 Notes

1. Expose the acidic and neutral fraction to mild alkaline hydrolysis by 100 mM NaOH in methanol at 37 °C for 2 h. Neutralize with 1 N HCl and then add equal volume of 50 mM NaCl. Apply the solutions to Sep-Pak C_{18} cartridge (Waters Co.), wash the cartridge with 40 ml distilled water, elute lipids with 10 ml methanol and 10 ml chloroform–methanol (1:1, v/v). Evaporate the eluates and dissolve the dried samples in chloroform–methanol (1:1, v/v) for TLC.
2. MC3T3-E1 cells derived from calvaria are the mouse preosteoblastic cell line [12]. MC3T3-E1 cells can differentiate into osteoblasts and mineralize in vitro.
3. Rinse coverslips with ethanol for sterilization and leave them to air-dry before use.
4. LBPA will be removed from the cells by Triton X-100 [8].
5. Place 50 μl of each antibody solution on Parafilm, which is at the bottom of a closed plastic container with a moist paper towel. Then cover the antibody solution with the coverslip. Care should be taken to contact the cells with the antibody.
6. In order to get comparable levels of FLAG-Rheb protein, the amount of the plasmid in the absence or the presence of

D-PDMP, L-PDMP or D-PBPP is 3.75, 7.5, 7.5, or 3.75 μg, respectively [10].

7. It is possible to store the cells at −80 °C.
8. Dilute 2× lysis buffer 4 times with water.
9. Constant voltage at 30 V when the distance between cathode and anode is 7 cm.
10. Dilute 2× lysis buffer 2 times with water.

References

1. Inokuchi J, Radin NS (1987) Preparation of the active isomer of 1-phenyl-2-decanoyl-amino-3-morpholino-1-propanol, inhibitor of murine glucocerebroside synthetase. J Lipid Res 28(5):565–571
2. Inokuchi J, Momosaki K, Shimeno H, Nagamatsu A, Radin NS (1989) Effects of D-threo-PDMP, an inhibitor of glucosylceramide synthetase, on expression of cell surface glycolipid antigen and binding to adhesive proteins by B16 melanoma cells. J Cell Physiol 141(3):573–583. https://doi.org/10.1002/jcp.1041410316
3. Inokuchi J, Jimbo M, Momosaki K, Shimeno H, Nagamatsu A, Radin NS (1990) Inhibition of experimental metastasis of murine Lewis lung carcinoma by an inhibitor of glucosylceramide synthase and its possible mechanism of action. Cancer Res 50(20):6731–6737
4. Radin NS, Inokuchi J (1991) Roles of glycosphingolipids revealed by D-PDMP, an inhibitor of glucosylceramide synthase. Trends Glycosci Glycotechnol 9:S37–S45
5. Chatterjee S, Bedja D, Mishra S, Amuzie C, Avolio A, Kass DA, Berkowitz D, Renehan M (2014) Inhibition of glycosphingolipid synthesis ameliorates atherosclerosis and arterial stiffness in apolipoprotein E−/− mice and rabbits fed a high-fat and -cholesterol diet. Circulation 129(23):2403–2413. https://doi.org/10.1161/CIRCULATIONAHA.113.007559
6. Mishra S, Bedja D, Amuzie C, Foss CA, Pomper MG, Bhattacharya R, Yarema KJ, Chatterjee S (2015) Improved intervention of atherosclerosis and cardiac hypertrophy through biodegradable polymer-encapsulated delivery of glycosphingolipid inhibitor. Biomaterials 64:125–135. https://doi.org/10.1016/j.biomaterials.2015.06.001
7. Rosenwald AG, Pagano RE (1994) Effects of the glucosphingolipid synthesis inhibitor, PDMP, on lysosomes in cultured cells. J Lipid Res 35(7):1232–1240
8. Makino A, Ishii K, Murate M, Hayakawa T, Suzuki Y, Suzuki M, Ito K, Fujisawa T, Matsuo H, Ishitsuka R, Kobayashi T (2006) D-threo-1-phenyl-2-decanoylamino-3-morpholino-1-propanol alters cellular cholesterol homeostasis by modulating the endosome lipid domains. Biochemistry 45(14):4530–4541. https://doi.org/10.1021/bi052104y
9. Sprocati T, Ronchi P, Raimondi A, Francolini M, Borgese N (2006) Dynamic and reversible restructuring of the ER induced by PDMP in cultured cells. J Cell Sci 119(Pt 15):3249–3260. https://doi.org/10.1242/jcs.03058
10. Ode T, Podyma-Inoue KA, Terasawa K, Inokuchi JI, Kobayashi T, Watabe T, Izumi Y, Hara-Yokoyama M (2017) PDMP, a ceramide analogue, acts as an inhibitor of mTORC1 by inducing its translocation from lysosome to endoplasmic reticulum. Exp Cell Res 350(1):103–114. https://doi.org/10.1016/j.yexcr.2016.11.011
11. Jimbo M, Yamagishi K, Yamaki T, Nunomura K, Kabayama K, Igarashi Y, Inokuchi JI (2000) Development of a new inhibitor of glucosylceramide synthase. J Biochem 127(3):485–491
12. Sudo H, Kodama HA, Amagai Y, Yamamoto S, Kasai S (1983) In vitro differentiation and calcification in a new clonal osteogenic cell line derived from newborn mouse calvaria. J Cell Biol 96(1):191–198

Chapter 20

Molecular Dynamics of Gangliosides

Takumi Yamaguchi and Koichi Kato

Abstract

Computational methodologies have immense potential to delineate the dynamic conformations of glycoconjugates including gangliosides, thereby characterizing the conformational adaptability of their glycans upon interacting with various target proteins. Replica-exchange molecular dynamics simulations have been employed to effectively explore the vast conformational spaces of large, branched carbohydrate moieties. When experimentally validated using NMR, molecular simulations can provide dynamical views of molecular recognition events involving the ganglioside glycans.

Key words Molecular dynamics simulation, GLYCAM, Conformational analysis, NMR

1 Introduction

Conformational flexibility is an important property of glycoconjugates including gangliosides. Their carbohydrate moieties possess significant degrees of motional freedom, which endows them with conformational adaptability in their interaction with various proteins [1, 2]. Computational methodologies such as Monte Carlo and molecular dynamics (MD) simulations have provided powerful tools for characterizing the conformational dynamics of glycans at the atomic level. In particular, recent developments with improved force fields have enabled accurate MD simulations on the time scale of microseconds and longer to delineate the molecular behavior of oligosaccharides surrounded by explicit water molecules [3]. Moreover, generalized ensemble algorithms such as replica-exchange MD (REMD) simulations have been employed to overcome the problems of insufficient sampling due to higher energy barriers between multiple minima in the conformational energy landscapes of large, branched oligosaccharides [4–7]. In this method, parallel replicas are run with different temperatures and the coordinates of replicas are exchanged during the calculation to efficiently explore the vast conformational spaces of biomacromolecules. In this chapter, a typical computational procedure for

Sandro Sonnino and Alessandro Prinetti (eds.), *Gangliosides: Methods and Protocols*, Methods in Molecular Biology, vol. 1804, https://doi.org/10.1007/978-1-4939-8552-4_20, © Springer Science+Business Media, LLC, part of Springer Nature 2018

REMD simulations of oligosaccharides in solution is outlined, giving the result of a conformational analysis of the GM1 pentasaccharide as an example.

2 Materials

1. The initial structure and topology file of the oligosaccharide are created using the tleap module of the AmberTools program package.
2. The GLYCAM06 force field is used for the oligosaccharide while the TIP3P model is employed to represent water [8].
3. REMD simulations are carried out by using the MPI version of the pmemd module of the AMBER program package [9].
4. The bonds involving hydrogen atoms are constrained using the SHAKE algorithm.
5. The electrostatic interactions are treated with the Particle Mesh Ewald algorithm. The typical cutoff distance is 12 Å for the nonbonded interactions.
6. The ptraj module of the AmberTools program package is used for analyzing the simulation results.

3 Methods

1. Create the initial structure of the glycan and place it in a box containing water molecules (*see* Subheading 4).
2. Add Na^+ ions to neutralize the system and save the topology file.
3. Carry out energy minimization by 500 steps of steepest descent.
4. Perform an equilibrium simulation of 2.0 fs for the time step in the *NPT* ensemble with $P = 1$ atm and $T = 300$ K.
5. Perform a production REMD simulation of 2.0 fs for the time step in the *NVT* ensemble using replicas with an exponential temperature distribution between 300 K and 500 K.
6. Extract trajectories at temperatures of interest.

4 Notes

1. Table 1 shows an example of the content of an input file for creating the initial structure and topology file of the GM1 pentasaccharide using the tleap module of the AmberTools 13 program. TIP3P models were added to the solvent layer to

Table 1
The content of an input file for creating the initial structure and topology file of GM1 pentasaccharide

```
source leaprc.GLYCAM_06h
glycan = sequence { ROH 4GB WLB }
set glycan tail glycan.3.O4
glycan = sequence { glycan 3VB 0LB }
set glycan tail glycan.3.O3
glycan = sequence { glycan 0SA }
solvateoct glycan TIP3PBOX 12.0
addions glycan Na+ 1
saveamberparm glycan GM1.top GM1.crd
quit
```

Table 2
The content of an input file for the REMD simulation of GM1 pentasaccharide

```
&cntrl
irest=0, ntx=1,
nstlim=1000, dt=0.002,
ntt=3, gamma_ln=1.0,
temp0=XXXXX, ig=RANDOM_NUMBER,
ntc=2, ntf=2, nscm=1000,
ntp=0, ntb=1,
cut=12.0,
ntpr=1000, ntwx=1000, ntwr=1000,
numexchg=1000,
iwrap=1,
/
```

ensure a distance of at least 12 Å from any atom. The three-letter codes for describing the carbohydrate residues relevant to the GLYCAM06 force field are summarized in GLYCAM-Web (http://glycam.org) developed by Woods and coworkers. The initial structures of glycans can also be created through this site. Table 2 displays an example of an input file for the REMD simulation. The exchange procedure for each replica is

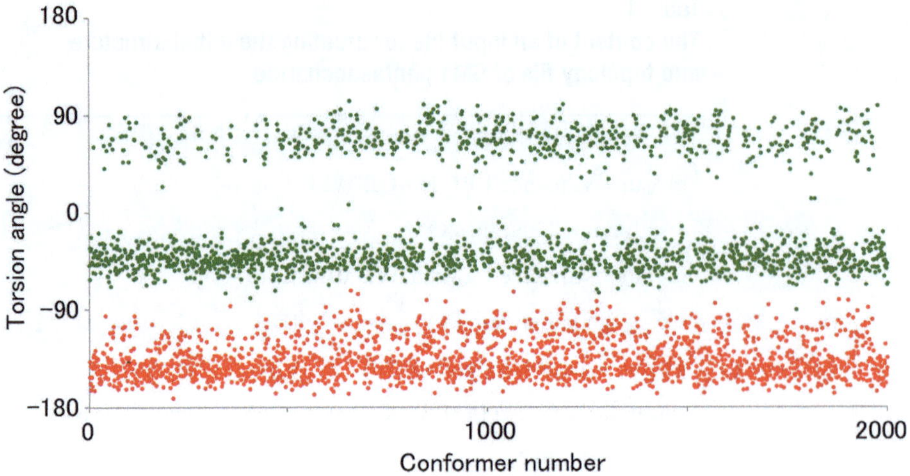

Fig. 1 The REMD trajectories of the torsion angles of the NeuAc–Gal glycosidic linkage in the GM1 pentasaccharide. 2000 conformers were extracted at equal intervals from the REMD results. Green and red dots represent the torsion angle ϕ and ψ, respectively

applied every 1000 MD steps. An informative tutorial for enhanced sampling, produced by the Amber developers, is available on the Amber home page (http://ambermd.org). Figure 1 shows the computed trajectories of the torsion angles of the NeuAc–Gal glycosidic linkage in 2000 conformers of the GM1 glycan extracted at equal intervals from the REMD simulations.

Although molecular simulations are potentially powerful, the results often depend on the simulation conditions, including the selected initial structure and/or calculation time period. Therefore, the validation of the simulation results based on experimental observations is desirable (**Note 1**). NMR spectroscopy offers useful validation tools giving conformational information on oligosaccharides through scalar couplings and the nuclear Overhauser effect (NOE), even though the NMR information is averaged over the dynamic conformational ensemble [10]. The recently emerging paramagnetism-assisted NMR approach provides extremely useful information on oligosaccharides for the validation of conformational ensembles derived from MD simulations [11].

In this method, a lanthanide chelator is chemically attached to the reducing terminus of the oligosaccharide for observing pseudocontact shifts (PCSs) as sources of atomic long-distance information [12, 13]. PCSs are observed as differences in $^{1}H/^{13}C$ chemical shifts in ^{1}H–^{13}C heteronuclear single-quantum coherence spectra between complexes with a paramagnetic ion and a diamagnetic reference typified by Tm^{3+} and La^{3+} ions, respectively. These experimentally observed PCS values are compared with those calculated from the conformational ensemble models of the oligosaccharide derived from MD simulations. The PCS values calculated for the conformational ensemble of GM1 pentasaccharide through conventional MD simulations with a total simulation time of 240 ns exhibit a significant discordance with the experimental data, probably because of insufficient conformation sampling. By contrast, an ensemble model composed of 2000 conformers extracted from the REMD simulations with a total simulation time of 1.9 μs provided PCS values that reproduce the experimentally observed PCS data well (Fig. 2). The conformational spaces computed for the NeuAc–Gal glycosidic linkage of this pentasaccharide by REMD simulations are significantly different from those in the conventional MD simulations [5].

The conformation of the GM1 molecule in membranes was also studied by MD simulations in conjunction with paramagnetism-assisted NMR experiments [14]. In this study, water-soluble and membrane-anchored paramagnetic probes were employed for the conformational analysis of GM1 embedded in membrane-mimicking phospholipid bicelles.

Figure 3 shows the experimentally validated conformational spaces occupied by GM1 pentasaccharide in addition to the conformations observed in crystal structures of its complexes with specific proteins such as cholera toxin [15]. The protein-bound conformations are all found in the conformational space of the uncomplexed pentasaccharide, suggesting that the specific conformations are selected from the dynamic ensemble upon interaction with the proteins (Fig. 3). The liaison of molecular simulations and experimental approaches thus provides dynamic views of molecular recognition events involving the ganglioside glycans.

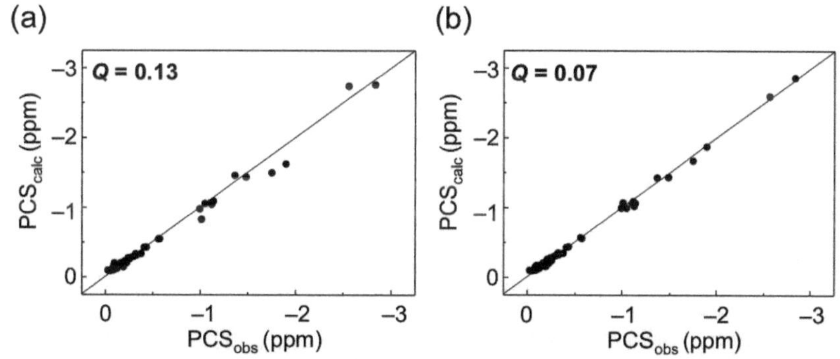

Fig. 2 Correlations between experimentally observed PCS values with Tm^{3+} and computationally calculated PCS data for the GM1 glycan. The PCS values were back-calculated from ensemble models derived from (**a**) conventional MD and (**b**) REMD simulations. The REMD simulations were performed using the AMBER12 program package. The simulations were carried out for 60 ns with 32 replicas with an exponential temperature distribution between 300 and 500 K. Q values, defined by the eq. $Q = \mathrm{rms}(PCS_{calc} - PCS_{obs})/\mathrm{rms}(PCS_{obs})$, are criteria of agreement between the experimental and calculated values. This figure was originally reproduced from reference 5

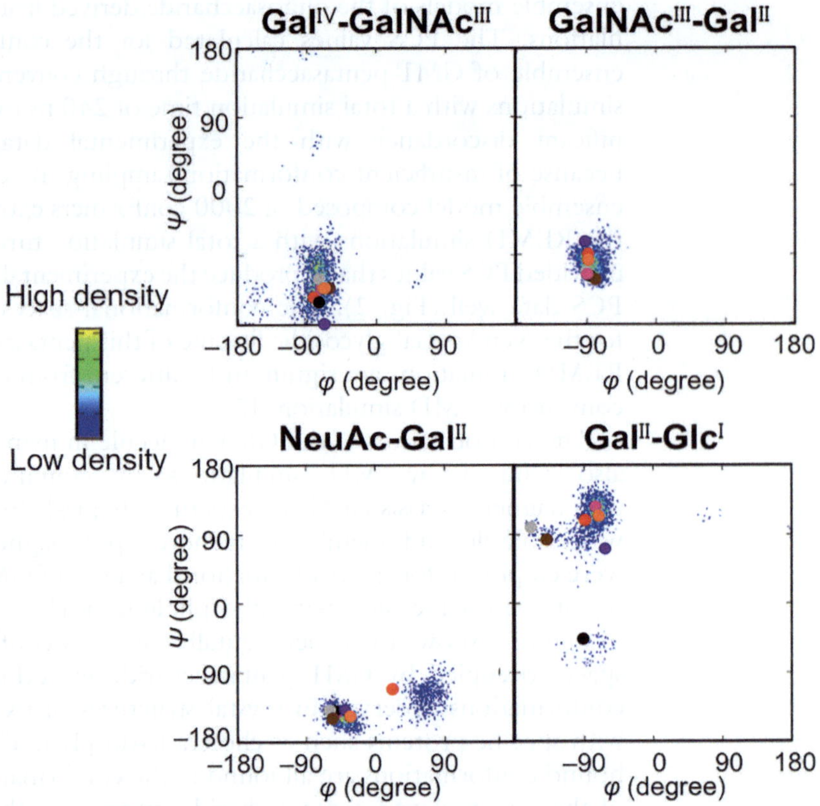

Fig. 3 Torsion angle density maps of the glycosidic linkages of the GM1 pentasaccharide obtained by the experimentally validated REMD simulations. The glycosidic torsions of the GM1 pentasaccharide in dimethyl sulfoxide calculated by NOE-based molecular mechanics (●), in aqueous solution calculated by NOE-based molecular modeling (●), and GM1-acetyl in aqueous solution calculated by conventional MD simulations (●) with those of the GM1 pentasaccharide in complex with cholera toxin [PDB code: 2CHB (●) and 3CHB (●)]; polyoma virus [3BWR (●)]; AB_5 toxin [4L6T (●)]. This figure was originally reproduced from reference 15

References

1. Yagi-Utsumi M, Yamaguchi T, Uekusa Y, Kato K (2017) CHAPTER 6 NMR characterization of the conformations, dynamics, and interactions of glycosphingolipids. In: NMR in Glycoscience and Glycotechnology. The Royal Society of Chemistry, pp 161–178
2. Kato K, Yagi H, Yamaguchi T (2017) NMR characterization of the dynamic conformations of oligosaccharides. In: Webb GA (ed) Modern magnetic resonance. Springer International Publishing, pp 1–18
3. Fadda E, Woods RJ (2010) Molecular simulations of carbohydrates and protein–carbohydrate interactions: motivation, issues and prospects. Drug Discov Today 15:596–609
4. Re S, Nishima W, Miyashita N, Sugita Y (2012) Conformational flexibility of N-glycans in solution studied by REMD simulations. Biophys Rev 4:179–187
5. Zhang Y, Yamaguchi T, Satoh T, Yagi-Utsumi M, Kamiya Y, Sakae Y, Okamoto Y, Kato K (2015) Conformational dynamics of oligosaccharides characterized by paramagnetism-assisted NMR spectroscopy in conjunction with molecular dynamics simulation. Adv Exp Med Biol 842:217–230
6. Yamaguchi T, Sakae Y, Zhang Y, Yamamoto S, Okamoto Y, Kato K (2014) Exploration of conformational spaces of high-mannose-type oligosaccharides by an NMR-validated simulation. Angew Chem Int Ed 53:10941–10944
7. Suzuki T, Kajino M, Yanaka S, Zhu T, Yagi H, Satoh T, Yamaguchi T, Kato K (2017) Conformational analysis of a high-mannose-type

oligosaccharide displaying glucosyl determinant recognised by molecular chaperones using NMR-validated molecular dynamics simulation. ChemBioChem 18:396–401

8. Kirschner KN, Yongye AB, Tschampel SM, González-Outeiriño J, Daniels CR, Foley BL, Woods RJ (2008) GLYCAM06: a generalizable biomolecular force field. Carbohydrates. J Comput Chem 29:622–655
9. Case DA, Cheatham TE, Darden T, Gohlke H, Luo R, Merz KM, Onufriev A, Simmerling C, Wang B, Woods RJ (2005) The Amber biomolecular simulation programs. J Comput Chem 26:1668–1688
10. Woods RJ, Tessier MB (2010) Computational glycoscience: characterizing the spatial and temporal properties of glycans and glycan-protein complexes. Curr Opin Struct Biol 20:575–583
11. Yamaguchi T, Kato K (2014) Paramagnetism-assisted nuclear magnetic resonance analysis of dynamic conformations and interactions of oligosaccharides. In: Endo T, Seeberger PH, Hart GW, Wong C-H, Taniguchi N (eds) Glycoscience: biology and medicine. Springer, Japan, Tokyo, pp 137–145
12. Kato K, Yamaguchi T (2015) Paramagnetic NMR probes for characterization of the dynamic conformations and interactions of oligosaccharides. Glycoconj J 32:505–513
13. Canales A, Jiménez-Barbero J (2017) CHAPTER 5 lanthanide-chelating carbohydrate conjugates to detect carbohydrate-protein interactions. In: NMR in Glycoscience and Glycotechnology. The Royal Society of Chemistry, pp 150–160
14. DeMarco ML (2012) Three-dimensional structure of glycolipids in biological membranes. Biochemistry 51:5725–5732
15. Yagi-Utsumi M, Kato K (2015) Structural and dynamic views of GM1 ganglioside. Glycoconj J 32:105–112

Chapter 21

Determination of Glycolipid Ligands of Carbohydrate-Binding Proteins Directly on Thin Layer Chromatograms

John L. Magnani

Abstract

Determination of a glycolipid ligand can quickly reveal the carbohydrate epitope recognized by carbohydrate-binding proteins (CBP) which may include antibodies, lectins, toxins, and virulence factors. Thin layer chromatography (TLC) is a simple, well-known, and heavily used method to separate glycolipids based on their migration through a thin layer of chromatography media, usually silica gel, coated on the surface of a rigid plate. Reacting carbohydrate-binding proteins directly on a TLC plate of separated glycolipids is a powerful technique to quickly determine functional carbohydrate epitopes present in an entire glycolipid extract of a target cell type or tissue.

Key words Immunostaining, Anti-carbohydrate antibodies, Glycolipid antigens, Thin layer chromatography (TLC), Lectins, Carbohydrate-binding proteins (CBP), Toxins

1 Introduction

Glycolipids contain carbohydrates that are both unique and others that are shared with glycoproteins. In contrast to glycoproteins, glycolipids in general contain one carbohydrate ligand for each molecule. Determination of a glycolipid ligand can quickly reveal the carbohydrate epitope recognized by protein receptors such as antibodies, lectins, toxins, and other carbohydrate-binding proteins (CBP). Thin layer chromatography is a simple, robust, and commonly used method for separating glycolipids extracted from cells or tissues by chromatography using a mobile phase of various solvents over a flat surface containing a media (stationary phase) such as silica gel. Using this method the entire glycome of such target cells or tissues as defined by their glycolipids can be separated and exposed on the surface of these coated plates, which can then be probed with the CBP to identify the carbohydrate ligand from that targeted glycome. This technique has the advantage over traditional screens of immobilized carbohydrates by ELISA's in microtiter plates or on glycoarrays, as those screens are only as

Sandro Sonnino and Alessandro Prinetti (eds.), *Gangliosides: Methods and Protocols*, Methods in Molecular Biology, vol. 1804, https://doi.org/10.1007/978-1-4939-8552-4_21, © Springer Science+Business Media, LLC, part of Springer Nature 2018

good as the carbohydrates contained within the screen. Given the large number and complexity of carbohydrates, the true native carbohydrate epitope may be missing from the library of any screen. In contrast, using whole cell extracts, the entire glycome from the targeted cell is screened by this TLC plate binding method, thereby identifying the natural carbohydrate epitope expressed by the glycolipids from the biological target tissue.

The first example of this method was developed using I^{125}-labeled cholera toxin as a probe for TLC plates of chromatographed tissue extracts from chick neural retina as well as purified gangliosides. In this original method, any bound I^{125}-labeled cholera toxin on the TLC plate was detected by autoradiography on X-ray film [1, 2]. Once developed, the method was applied to monoclonal antibodies directed against tumor markers. In contrast with current genomic methods for detecting tumor markers, using monoclonal antibodies allows detection of tumor markers through the eyes of the immune system and will detect post translational modifications of proteins such as glycosylation, which is missed by current genomic screens. The first example of this application was the analysis of a monoclonal antibody (NS-116-19cl9) that could distinguish gastrointestinal tumor cells from normal cells and was abbreviated 19-9 [3]. Using this TLC plate binding method, the antigen was identified as a monosialoganglioside [4] and the carbohydrate structure was determined from the purified glycolipid antigen as Sialylated Lacto-N-fucopentaose II [5], more commonly known as sialyl Le^a (Fig. 1). Based on the nomenclature of the antibody, this antigen became known as CA19-9 and has become an FDA-approved diagnostic marker for pancreatic cancer and is one of the most commonly used serum diagnostics for cancer [6]. While this TLC plate binding method was used to determine the carbohydrate structure of the epitope on glycolipid antigens, the diagnostic biomarker detected in patient's sera is this same carbohydrate antigen (sialyl Le^a) expressed on a mucin [7], thereby demonstrating that this method can determine carbohydrate ligands of CBP's that are shared by glycolipids and glycoproteins. This pancreatic cancer diagnostic marker (CA19-9), whose structure was determined by identifying the glycolipid antigen using this TLC plate binding method, has been in routine use for over 20 years and has undoubtedly played a role in the extension and quality of life for many pancreatic cancer patients. In addition, this structure (sialyl Le^a) was later determined to bind E-selectin, which explains its function as a tumor marker allowing metastasis of tumor cells from the bloodstream. The determination of the structure of this functional carbohydrate formed the basis to rationally design a new class of therapeutic drugs (glycomimetics) now in clinical trials for cancer and inflammatory diseases [8, 9].

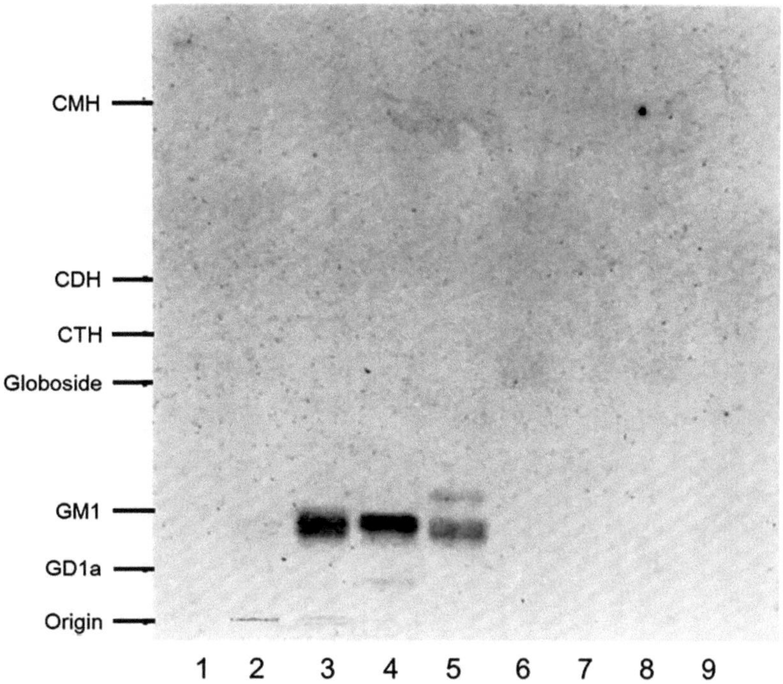

Fig. 1 Detection of the CA19-9 antigen (sialyl Le^a) among human tissue glycolipid extracts using antibody 19-9. Lane 1, intestinal mucosae; lane 2, primary colon adenocarcinoma; lane 3, pancreatic cancer liver met.; lane 4, colorectal cancer cell line SW1116; lane 5, meconium; lane 6, melanoma cell line WM9; lane 7, brain gangliosides; lane 8, normal spleen; lane 9, erythrocytes. The positions of cochromatographed standard glycolipids are shown on the left

The original method of probing TLC plates of glycolipids for direct binding has been modified over the years to increase sensitivity and ease of use. This chapter describes the current use of the method in which radioactivity is replaced with an enzymatic colorimetric detection of direct binding. The example given in this chapter describes its use for carbohydrate –binding monoclonal antibodies [10], (Fig. 2). The same method is compatible for analysis of other carbohydrate-binding proteins (CBP) such as lectins, toxins, and virulence factors as long as the CBP can be directly labeled with an enzyme (i.e., horseradish peroxidase, HRP) or detected by an antibody labeled with an enzyme (HRP). It should also be noted that other modifications of this assay have evolved over the years to include the direct binding of pathogens [11] as well as mammalian cells [12] directly to TLC plates of separated glycolipids. In the latter applications both the receptors and ligands are unknown which presents an interesting and highly exploratory method for discovery of new carbohydrate-binding proteins as well as their carbohydrate ligands

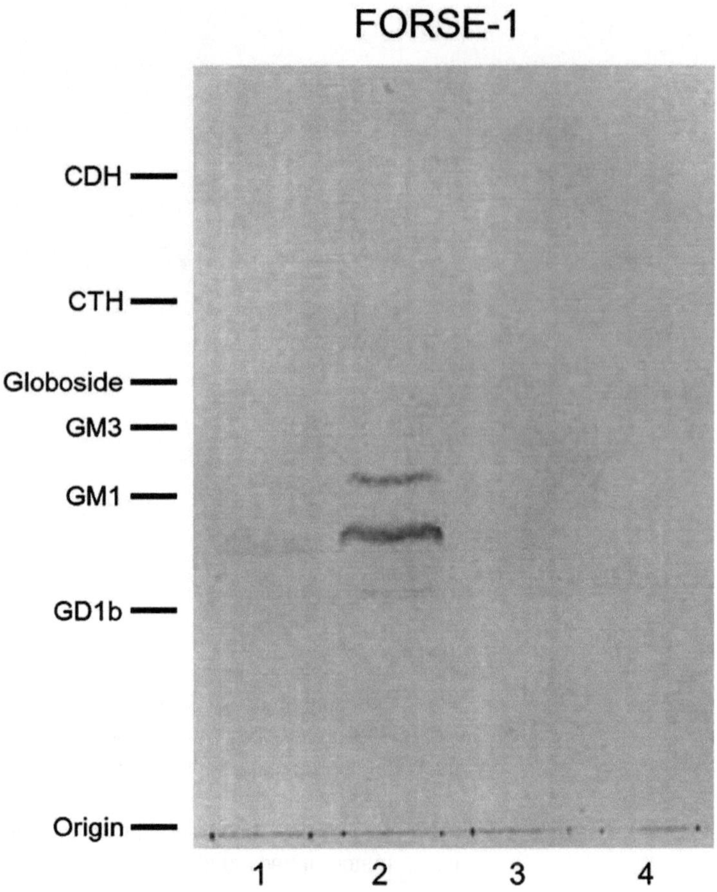

Fig. 2 Thin layer chromatogram immunostaining with antibody FORSE-1 that recognizes embryonic progenitor cells in the central nervous system (CNS). Results were obtained using the enzymatic procedure presented in this chapter. Lipid extracts were derived from embryonic rat brain and each lane contains the amount of extract obtained from 60 μl of packed cells. FORSE-1 recognizes the Le^x epitope and is found in the neutral glycolipid fraction (lane 2); lane 1, Folch lower phase extract; lane 3, sulfatide fraction; lane 4, monosialoganglioside fraction

2 Materials

1. Glass Chromatography Tank.
2. Aluminum-backed HPTLC silica gel plates with fluorescence indicator.
3. Glass microsyringe.
4. Glass petri dishes (150 mm).
5. Plastic petri dishes (150 mm).
6. Whatman #1 filter paper.

7. Heated blow dryer.
8. Spraying apparatus for orcinol spray.
9. Plastic squeeze bottle.
10. 0.1% solution of polyisobutylmethacrylate beads dissolved in acetone.
11. 0.25% potassium chloride in water.
12. Phosphate buffered saline (PBS) (0.01 M sodium phosphate, 0.15 M NaCl, pH 7.6).
13. Bovine serum albumin (BSA), 1% in PBS.
14. 0.1% orcinol monohydrate dissolved in 5% H_2SO_4.
15. Diaminobenzidine tetrahydrochloride (DAB) .
16. Chloroform–methanol–0.25%KCl (5:4:1).

3 Methods

3.1 Preparation of the Chromatography Tank

1. Add 150 ml of chloroform–methanol–0.25% KCl (5:4:1) to a glass chromatography tank (25 cm × 27 cm × 10 cm) lined with filter paper and covered tightly.
2. Allow solvent to absorb into the filter paper and the vapors in the tank to equilibrate for at least 2 h. For best results, prepare a new tank daily.

3.2 Preparation of the Silica Gel Plate

1. Cut the aluminum-backed silica gel plate (1) to the desired size using scissors. The optimum height should be 10 cm.
2. Draw a very light pencil line 1.5 cm parallel to the bottom of the plate being careful not to scratch the silica gel.
3. Using a glass microsyringe, spot the glycolipid sample along a 5 mm path on the pencil line. Separate these paths (lanes) by 2.5 mm.
4. Samples to be detected for carbohydrate content by orcinol or resorcinol spray should be separately located on the plate from those lanes to be reacted with carbohydrate-binding proteins.

3.3 Chromatography of the Silica Gel Plate

1. Using a pair of long metal forceps, grasp the top of the spotted silica gel plate and gently place in the tank oriented with the spotted sample just above the level of the solvent. Allow the top edge to lean against the rear of the developing chamber.
2. Cover the tank tightly and allow the plate to remain undisturbed until the ascending solvent line reaches the top of the plate (approx. 30–45 min).
3. As soon as the solvent reaches the top of the TLC plate, remove the plate by grasping the top edge with forceps and allow the plate to thoroughly air-dry under a chemical fume hood.

3.4 Staining the Standards for Carbohydrates

1. Cut off the portion of the TLC plate containing the purified glycolipid standards.
2. In a chemical fume hood, spray the plate containing the glycolipid standards briefly with 0.1% orcinol dissolved in 5% H_2SO_4.
3. Immediately place in a prewarmed oven at 120 °C for approximately 5–10 min.
4. If staining is weak, **steps 2** and **3** may be repeated.

3.5 Preparation of the Chromatographed TLC Plate for Reaction with the Carbohydrate-Binding Protein

1. Spot 2 μl of the carbohydrate-binding protein (0.5 mg/ml) at the bottom of the plate below the origin as a positive control. Do the same with bovine serum albumin (BSA) as a negative control. Allow to completely air dry.
2. Soak the TLC plate for 90 s in a glass petri dish (150 mm) filled with a 0.1% solution of polyisobutylmethacrylate beads predissolved in acetone (2). For best results the plate should enter the solution quickly, long side first, at a 45 degree angle (3).
3. Remove the plate and stand upright under a chemical fume hood to completely air dry. Measure the surface area of the TLC plate.
4. Spray the TLC plate vigorously with PBS containing 1% BSA and then submerge immediately in the same solution in a petri dish for 1 h.

3.6 Reaction of the Carbohydrate-Binding Protein Directly on the TLC Plate

1. Dilute the carbohydrate-binding protein (i.e., antibody) to 10 μg/ml in PBS containing 1% BSA. The amount needed is calculated based on the surface area of the TLC plate. The volume to be overlaid on the horizontal TLC plate surface is calculated as 65 $\mu l/cm^2$ of the TLC plate area.
2. Prepare a pedestal slightly smaller than the TLC plate with glass microscope slides by laying them flat on the bottom of a plastic petri dish (150 mm).
3. Remove the TLC plate from the 1% BSA in PBS blocking buffer with small forceps and gently drain buffer by touching the corner of the TLC plate to a paper towel. Immediately place the TLC horizontally by laying it flat on the glass slide pedestal with the silica gel side up. Do not allow the plate to dry during this and any of the following steps until the end of the protocol.
4. Using a pipette, gently cover the TLC plate evenly with the carbohydrate-binding protein solution (65 $\mu l/cm^2$) making certain that the liquid maintains its level on the top of the plate by surface tension and does not drain underneath the plate. The glass microscope slide pedestal under the TLC plate helps the process (4).

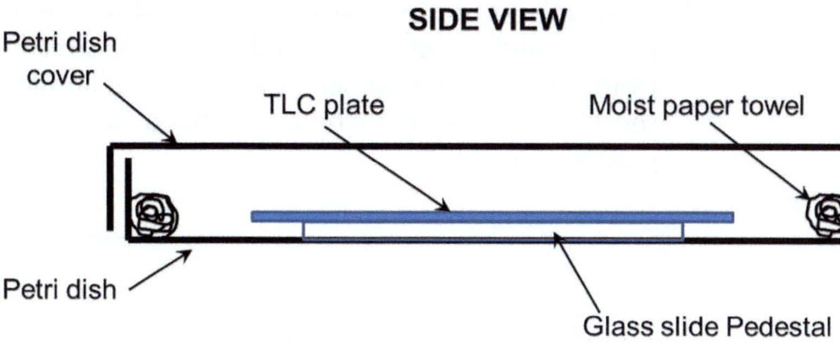

Fig. 3 Incubation chamber for binding to glycolipids on TLC plates

5. Pieces of paper towel saturated with water can be placed around the TLC plate, being careful not to disturb the plate or the glass pedestal to maintain humidity during the incubation step (optional depending on the length of incubation). Cover the plastic petri dish and allow the carbohydrate-binding protein solution to incubate with the surface of the TLC plate (typically 2 h for anti-carbohydrate antibodies). (Fig. 3)

3.7 Detection of Bound Protein Using a Secondary Peroxidase-Labeled Anti-Protein Antibody

1. Dilute the secondary peroxidase-labeled anti-carbohydrate binding protein (i.e., peroxidase-labeled goat anti-mouse IgG for an IgG primary antibody) to 1 μg/ml in PBS containing 1% BSA.
2. Cut the nozzle of a plastic squeeze bottle to a large opening to allow slow flow. Using this squeeze bottle filled with PBS to gently and thoroughly wash the TLC plate over an open sink under slow flow. This type of squeeze bottle should be used for all TLC plate washings (5).
3. Lay the TLC plate horizontally on top of a fresh glass slide pedestal as above and using a pipette, gently cover the plate evenly with the peroxidase-labeled anti-carbohydrate-binding protein solution prepared in **step 1** above (65 $\mu l/cm^2$ of the TLC plate surface area).
4. Cover the petri dish and incubate at room temperature for 1 h.

3.8 Addition of Colorimetric Enzyme Substrate to Detect Glycolipids Reacting with the Carbohydrate-Binding Protein

1. Prepare fresh solution of diaminobenzidine tetrahydrochloride (DAB) as a substrate for the horseradish peroxidase-labeled second antibody.
2. Thoroughly and gently wash the TLC plate with the squeeze bottle containing PBS described above over an open sink.
3. Lay the TLC plate on a fresh glass slide pedestal and gently pipette the DAB solution on the surface of the TLC plate. Allow reaction to proceed until the positive controls appear as colored spots and bands and before the background staining becomes too intense (about 5–10 min).

4. Wash gently but extensively with distilled water to stop the reaction and dry the plate under a gentle stream of warm air from a heated blow dryer. Any glycolipid antigen (or ligand) should be clearly visible by a color band and comparable to the positive control.

4 Notes

1. Aluminum-backed HPTLC plates are used as they can be easily cut with scissors for custom sizes. It is also useful to bend one of the top corners of the plate upward at a 90^0 angle. This allows the plate to be grabbed and manipulated by small forceps throughout the procedure.
2. The coating of the TLC plate with polyisobutylmethacrylate allows for greater sensitivity and substantially lowers background staining of the silica gel surface.
3. When submerging the plate into the polyisobutylmethacrylate in acetone solution, use one smooth continuous motion of submerging the long edge of the plate into the acetone solution at a 45° angle. This avoids potential irregularities in the coating that may occur.
4. Make sure that the TLC plate lay flat and parallel to the glass slide pedestal during the incubations. Incubation buffers must be slowly overlaid on the flat plate at a volume of 65 $\mu l/cm^2$ of the surface area of the plate. This is the proper volume to cover the plate with surface tension and avoid overloading the plate and spilling incubation buffer off of the plate.
5. An easy method to wash the TLC plate between incubations is to use a squeeze bottle with the nozzle cut to a wide opening, thereby allowing a gentle stream of buffer or water (last step) to flow over the surface of the silica gel. The shear forces from an unmodified squeeze bottle may be too strong and damage the silica gel surface.

References

1. Magnani JL, Smith DF, Ginsburg V (1980) Detection of gangliosides that bind cholera toxin: direct binding of 125I-labeled toxin to thin layer chromatograms. Anal Biochem 109:399–402
2. Magnani JL, Brockhaus M, Smith DF, Ginsburg V (1982) Detection of glycolipid ligands by direct binding of carbohydrate-binding proteins to thin-layer chromatograms. In: Methods in enzymology-complex carbohydrates, part D, vol 83. Academic Press, New York, NY, pp 235–241
3. Koprowski H, Herlyn M, Steplewski Z, Sears HF (1981) Specific antigen in serum of patients with colon carcinoma. Science 212(4490):53–55
4. Magnani JL, Brockhaus M, Smith DF, Ginsburg V, Blaszczyk M, Mitchell KF, Steplewski Z, Koprowski H (1981) A monosialoganglioside is a monoclonal antibody-defined antigen of colon carcinoma. Science 212(4490):55–56

5. Magnani JL, Nilsson B, Brockhaus M, Zopf D, Steplewski Z, Koprowski H, Ginsburg V (1982) A monoclonal antibody-defined antigen associated with gastrointestinal cancer is a ganglioside containing sialylated lacto-N-fucopentaose II. J Biol Chem 257(23):14365–14369
6. Bast RC, Gold P, Ming Chu T, Magnani JL (2013) Reflections on the discovery of carcinoembryonic antigen, prostate-specific antigen, and cancer antigens CA125 and CA19-9. Clin Chem 59(1):22–31
7. Magnani JL, Steplewski Z, Koprowski H, Ginsburg V (1983) Identification of the gastrointestinal and pancreatic cancer-associated antigen detected by monoclonal antibody 19-9 in the sera of patients as a mucin. Cancer Res 43:5489–5492
8. Telen MJ, Wun T, McCavit TL, De Castro LM, Krishnamurti L, Lanzkron S, Hsu LL, Smith WR, Rhee S, Magnani JL, Thackray H (2015) Randomized phase 2 study of GMI-1070 in SCD: reduction in time to resolution of vaso-occlusive events and decreased opioid use. Blood 125(17):2656–2664
9. DeAngelo DJ, Liesveld JL, Jonas BA, O'Dwyer ME, Bixby DL, Magnani JL, Thackray HM, Becker PS (2016) A phase I/II study of GMI-1271, a novel E-selectin antagonist, in combination with induction chemotherapy in relapsed/refractory and elderly previously untreated acute myeloid leukemia: results to date. Blood 128:4049
10. Allendorfer KL, Magnani JL, Patterson PH (1995) FORSE-1, an antibody that labels regionally restricted subpopulations of progenitor cells in the embryonic central nervous system, recognizes the Lex carbohydrate on a proteoglycan and two glycolipid antigens. Mol Cell Neurosci 6:381–395
11. Deal CD, Krivan HC (1994) Solid-phase binding of microorganisms to glycolipids and phospholipids. Methods Enzymol 236:346–353
12. Swank-Hill P, Needham LK, Schnaar RL (1987) Carbohydrate-specific cell adhesion directly to glycosphingolipids separated on thin-layer chromatography plates. Anal Biochem 163:27–35

Chapter 22

Identification of KDN-Gangliosides

Ken Kitajima

Abstract

KDN is an abbreviated name of 2-keto-3-deoxy-D-*glycero*-D-*galacto*-nononic acid, and belongs to sialic acid members like *N*-acetylneuraminic acid (Neu5Ac), and *N*-glycolylneuraminic acid (Neu5Gc). The aminoacyl group at C5 position of Neu5Ac is replaced by a hydroxyl group in KDN. Like Neu5Ac, KDN exists in various glycoconjugates including glycosphingolipids in vertebrates and gram-negative bacteria. Because of its unique properties, some methods are specifically applicable to KDN residue, although most detection methods for Neu5Ac are also applicable. In this chapter, methods for identification of KDN residues in glycosphingolipids are described, focusing on two methods that are often used, i.e., the fluorescent HPLC analysis and the TLC immunostaining with the antibodies specific to α2,3- and α2,8-KDN residues.

Key words Deaminoneuraminic acid, KDN, Sialic acid, KDN-ganglioside, Fluorometric HPLC analysis, Thiobarbituric acid method, Resorcinol method, Orcinol sulfuric acid method, Anti-KDN antibody, TLC immunostaining

1 Introduction

Sialic acids (Sias) are a group of 9-carbon urosonic acids and widely found in vertebrate glycoproteins and glycolipids and in bacterial polysaccharides. Sias had been defined as derivatives of *N*-acylneuraminic acids consisting of *N*-acetylneuraminic acid (Neu5Ac), *N*-glycolylneuraminic acid (Neu5Gc), and their modified ones with acetyl, methyl, and sulfate groups. KDN was discovered in 1986 as a new type of Sia that was distinct from the modified *N*-acylneuraminic acids, and named as KDN [1]. KDN contains completely the same structure as Neu5Ac, including the stereochemistry, except that the *N*-acetyl group is replaced by a hydroxyl group at C5 position (Fig. 1). Thus, KDN is also called deaminoneuraminic acid. This new sugar is correctively denoted as 3-deoxy- D-*glycero*-D-*galacto*-nonulosonic acid or 2-**k**eto-3-**d**eoxy-D-*glycero*-D-*galacto*-**n**ononic acid, after which the abbreviated name was defined as KDN. KDN was first found in polysialoglycoprotein (PSGP) from rainbow trout eggs [1], and now found as

Sandro Sonnino and Alessandro Prinetti (eds.), *Gangliosides: Methods and Protocols*, Methods in Molecular Biology, vol. 1804, https://doi.org/10.1007/978-1-4939-8552-4_22,

© Springer Science+Business Media, LLC, part of Springer Nature 2018

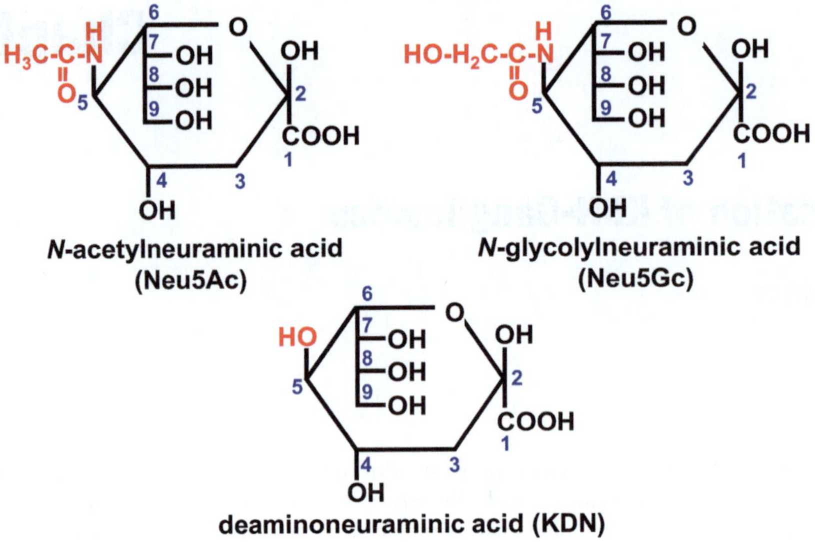

Fig. 1 Structure of KDN in free form

a constituent of proteins, lipids, and polysaccharides in various organisms [2]. Glycosphingolipids containing KDN are denoted as KDN-gangliosides. In 1991, (KDN)GM3 (Fig. 2) was first found in rainbow trout sperm [3]. Subsequently, (KDN)GD1a, (KDN,Neu5Ac)GD1a, (KDN)GD1α, and (KDN,Neu5Ac)GD1α (Fig. 2) were found in other tissues of rainbow trout [4, 5]. KDN residues in glycoconjugates show unique properties that are different from Neu5Ac residues. KDN-containing glycans have a tendency to make a complex by themselves, and show higher adsorptive property to anion-exchange and gel filtration resins of polysaccharide-base [3, 6]. Actually, (KDN)GM3 has a property to bind to Gg3 through the glycan-glycan interaction, and this interaction may be involved in sperm-egg binding of rainbow trout fertilization [7]. In view of the metabolism in vertebrates, KDN is biosynthesized from mannose 6-phosphate; in contrast, Neu5Ac is from *N*-acetylmannosamine 6-phosphate using the common enzymes basically all through the biosynthetic pathway to KDN residues in glycoconjugates [8–11]. Interestingly, in mammals, CMP-Sia synthetases have lost the activity to KDN [12, 13]. CMP-Neu5Ac is abundantly provided, while no CMP-KDN is available in mammals. Therefore, KDN-gangliosides rarely exist in mammals, but more in other vertebrates.

2 Materials

KDN is available from Sigma-Aldrich, Toronto Research Chemicals, and so on. KDN is also described as ketodeoxynonulosonic acid as a product name. KDN-gangliosides are not commercially available.

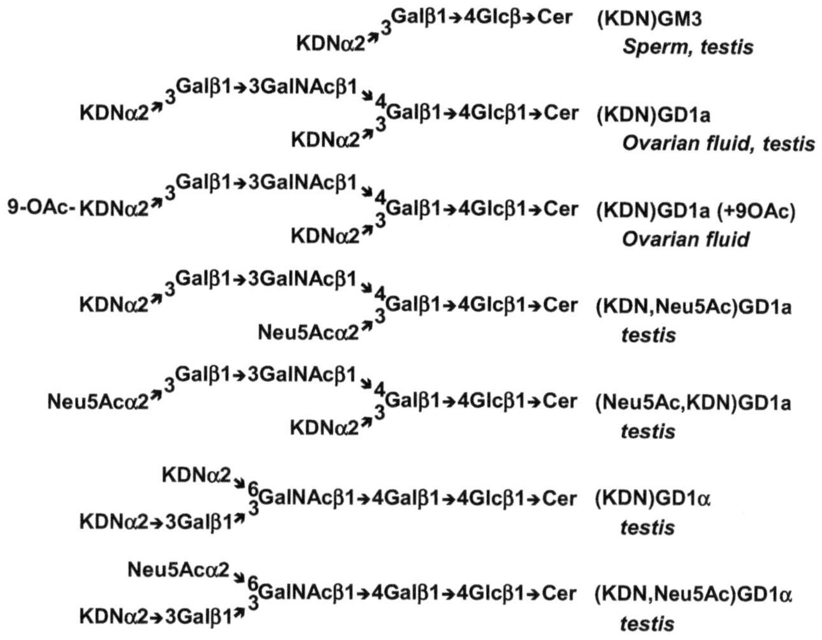

Fig. 2 KDN-gangliosides found in rainbow trout. (KDN)GM3, (KDN)GD1a, (9-*O*-acetyl)(KDN)GD1a, (KDN,Neu5Ac)GD1a, (Neu5Ac,KDN)GD1a, (KDN)GD1α, and (KDN,Neu5Ac)GD1α

(KDN)GM3 is abundantly present in rainbow trout sperm and testis, and can be prepared as described previously [3, 4]. Crude lipid fractions are prepared from various cells and tissues using conventional lipid extraction methods.

2.1 Fluorometric High-Performance Liquid Chromatography (Fluorometric HPLC) Analysis

1. 1,2-diamino-4,5-methylenedioxybenzene (DMB) (Dojindo, Kumamoto, Japan).
2. Trifluoroacetic acid–DMB reagent: 7 mM DMB, 5.0 mM trifluoroacetic acid, 1 M 2-mercaptoethanol, 18 mM sodium bisulfite, stable at 4 °C for 2 weeks [14].
3. Elution solvent: methanol–acetonitrile–water (7:9:84, v/v/v) [15].
4. TSK-gel ODS-120T HPLC column (250 mm × 4.6 mm i.d.)
5. A Jasco LC-900 HPLC system equipped with a Jasco FP-920 fluorescence detector.

2.2 Thin-Layer Chromatography (TLC) Immunostaining

1. Monoclonal antibody (mAb.)kdn3G recognizing (KDN) GM3 [3] (*see* **Note 1**).
2. mAb.kdn8kdn recognizing α2,8-linked oligoKDN [16] (*see* **Note 2**).
3. TLC plate (Polygram SilG, Macherey-Nagel, Germany) (*see* **Note 3**).
4. Blocking buffer: 1% polyvinylpyrrolidone (PVP), 1% bovine serum albumin (BSA), and 0.02% NaN_3 in 10 mM sodium phosphate buffer-0.15 M NaCl (PBS).

5. Washing buffer: 0.05% Tween 20 in PBS.
6. Substrate solution: 0.1 M Tris–HCl buffer (pH 7.5) containing 0.009% H_2O_2, 0.8 mg/ml 3,3′-diaminobenzidine (DAB), and 0.4 mg/ml LiCl.

3 Methods

3.1 Fluorometric HPLC

1. A lipidic sample containing KDN is hydrolyzed at 80 °C for 60 min with 0.1 M trifluoroacetic acid in a screw-capped glass tube to release KDN. The resultant insoluble materials, if present, are removed by centrifugation (10,000 × *g*, 15 min), and the supernatant is lyophilized or evaporated.
2. To the dried sample, 20 μl of 0.02 M trifluoroacetic acid and 20 μl of DMB reagent are added, and incubated at 50 °C for 2 h.
3. An aliquot of the reaction mixture (2–20 μl) is injected into a HPLC column (preequilibrated with elution solution) and eluted with the elution solution, operating isocratically at 1 ml/min at a column temperature of 26 °C, monitored by fluorometric detector (excitation, 373 nm; fluorescent, 448 nm) (*see* **Note 4**).
4. When the amount of KDN in a lipid sample to be tested is of large amount (~nmol), colorimetric analyses such as thiobarbituric acid (TBA) method, is applicable (*see* **Note 5**).

3.2 TLC Immunostaining

1. A lipid sample containing KDN (50 ng as total lipid amount) is applied to a silica gel TLC plate, and after developed with chloroform–methanol–0.2% aqueous $CaCl_2$ solution (55:45:10, v/v/v), the plate is subjected to TLC immunostaining [17, 18] (*see* **Note 3**). In parallel, the sample (10 μg as total lipid amount) is run under the same conditions to monitor a chromatographic profile by visualization with colorimetric reagents (*see* **Note 6**). For control, a lipid sample is pretreated with and without KDNase SM, which can specifically cleave glycosidic linkages of KDN, but not those of Neu5Ac. It is confirmed that components that disappear on the KDNase treatment contain KDN residues (*see* **Note 7**).
2. After air-drying, the plate is incubated at 37 °C for 1 h with the blocking buffer and then with mAb.kdn3G (7.5 μg/ml as IgG), at 37 °C, for 2 h. The plate is washed three times with the washing buffer, incubated at room temperature for 30 min with a biotin-conjugated IgG horse anti-mouse IgG antibody (Vector Laboratories, Inc.) under gentle shaking, and washed three times with the washing buffer. Then the plate is incubated the avidin–biotinylated horseradish peroxi-

dase complex solution (Vector Elite ABC Kit, Vector Laboratories, Inc.) at room temperature for 30 min under gentle shaking. The enzyme activity remaining on the plate is visualized by incubation with the substrate solution at room temperature (*see* **Note 8**).

3. The anti-KDN antibody-reactive component is indicated to be a KDN-ganglioside. Migration rate of a KDN-ganglioside is often slightly smaller than that of the corresponding ganglioside that contains Neu5Ac instead of KDN.

4 Notes

1. A monoclonal antibody kdn3G is a mouse IgG. This antibody was prepared by immunizing a mouse with (KDN)GM3, followed by establishing that hybridoma cell secreting anti-(KDN)GM3 antibody which were selected from fused cells between the immunized mouse splenocytes and myeloma cells [18]. This antibody also recognizes (KDN)GD1a and 9-*O*-acetylated (KDN)GD1a (Fig. 2) [4]. The antibody recognizes KDNα2,3Galβ1-, but does not discriminate 9-*O*-acetyl KDNα2,3Galβ1- from the non-*O*-acetylated form.
2. mAb.kdn8kdn recognizes α2,8-linked oligo/polyKDN structure with the degree of polymerization 2 and greater [19]. This antibody is of IgM type.
3. TLC plate types, coating procedures of the plate, and other modified procedures for TLC immunostaining are also applicable [20].
4. Relative retention time for each sialic acid derivative was expressed as the retention time relative to that of a pyruvic acid derivative set equal to 1.0 [15]. In this method, a small amount of KDN, ~ pmol, is detectable.
5. Different from Neu5Ac, KDN is not detected by the resorcinol reagent [1], which has been used for *N*-acylneuraminic acids (Neu5Acyl). The TBA method [21, 22] has been used for detecting free, but not bound, Neu5Acyl, and therefore, prior release of free Neu5Acyl from the corresponding residues are necessary for quantitation of Neu5Acyl by the TBA method. In contrast, KDN can be detected by the TBA method without prior hydrolysis, because even a bound form of KDN is oxidized by periodate treatment in the TBA method to give the pink chromophores. KDN and Neu5Acyl are discriminated by the TBA method without hydrolysis [23]. It should be noted that crude lipid fractions often contain some components to give pinky chromophores by the TBA method, which has the maximum absorption wavelength is 532 nm.

Attention has to be paid to this point when KDN is searched for by the TBA method. In the TBA reaction, bound KDN gives chromophores that have maximum absorption wavelength in 550 nm with molar absorption coefficient identical to that of Neu5Acyl.

6. For visualization with colorimetric reagents, an air-dried TLC plate is sprayed with a reagent and heated until the color is sufficiently developed: 80 °C for 30 min for the resorcinol reagent [24]; 120 °C for 30 min for 10% sulfuric acid–methanol [3]; 80 °C for 30 min for orcinol–sulfuric acid [20, 25]. KDN is visualized with 10% sulfuric acid–methanol and the orcinol–sulfuric acid, but not with the resorcinol reagent, while Neu5Ac is more sensitive to the resorcinol reagent.
7. Various known sialidases are known to fail to cleave the KDN glycosides. However, KDNase SM purified from *Sphingobacterium multivorum* is shown to act on various KDN-containing glycosides (KDNα2,3Gal/GalNAc, KDNα2, 6Gal/GalNAc, KDNα2,8Sia) in glycoconjugates to hydrolyze the KDN residues. Prominently, no cleavage occurs on the Neu5Ac or Neu5Gc residues [26, 27].
8. There are other methods for visualizing the immunostained components. Peroxidases and alkaline phosphatases are used as enzymes conjugated with the secondary antibody, and their products obtained from their substrates are detected by color development and by sensitive chemiluminescence.

References

1. Nadano D, Iwasaki M, Endo S, Kitajima K, Inoue S, Inoue Y (1986) A naturally occurring deaminated neuraminic acid, 3-deoxy-D-glycero-D-galacto-nonulosonic acid (KDN). Its unique occurrence at the nonreducing ends of oligosialyl chains in polysialoglycoprotein of rainbow trout eggs. J Biol Chem 261:11550–11557
2. Inoue S, Kitajima K (2006) KDN (deaminated neuraminic acid): dreamful past and exciting future of the newest member of the sialic acid family. Glycoconj J 23:277–290
3. Yu S, Kitajima K, Inoue S, Inoue Y (1991) Isolation and structural elucidation of a novel type of ganglioside, deaminated neuraminic acid (KDN)-containing glycosphingolipid, from rainbow trout sperm. The first example of the natural occurrence of KDN-ganglioside, (KDN)GM3. J Biol Chem 266:21929–21935
4. Yu S, Kitajima K, Inoue S, Muto Y, Kasama T, Handa S, Inoue Y (1993a) Structure of novel gangliosides, deaminated neuraminic acid (KDN)-containing glycosphingolipids, isolated from rainbow trout ovarian fluid. Biochemistry 32:9221–9229
5. Yu S, Kitajima K, Inour S, Khoo KH, Morris HR, Dell A, Inoue Y (1995) Expression of new KDN-gangliosides in rainbow trout testis during spermatogenesis and their structural identification. Glycobiology 5:207–218
6. Kanamori A, Inoue S, Iwasaki M, Kitajima K, Kawai G, Yokoyama S, Inoue Y (1990) Deaminated neuraminic acid-rich glycoprotein of rainbow trout egg vitelline envelope. Occurrence of a novel α-2,8-linked oligo(deaminated neuraminic acid) structure in *O*-linked glycan chains. J Biol Chem 265:21811–21819
7. Yu S, Kojima N, Hakomori SI, Kudo S, Inoue S, Inoue Y (2002) Binding of rainbow trout sperm to egg is mediated by strong carbohydrate-to-carbohydrate interaction between (KDN)GM3 (deaminated neuraminyl ganglioside) and Gg3-like epitope. Proc Natl Acad Sci U S A 99:2854–2859

8. Angata T, Matsuda T, Kitajima K (1998) Synthesis of neoglycoconjugates containing deaminated neuraminic acid (KDN) using rat liver α2,6-sialyltransferase. Glycobiology 8:277–284
9. Angata T, Nakata D, Matsuda T, Kitajima K, Troy FA 2nd. (1999a) Biosynthesis of KDN (2-keto-3-deoxy-D-*glycero*-D-*galacto*-nononic acid). Identification and characterization of a KDN-9-phosphate synthetase activity from trout testis. J Biol Chem 274:22949–22956
10. Angata T, Nakata D, Matsuda T, Kitajima K (1999) Elevated expression of free deaminoneuraminic acid in mammalian cells cultured in mannose-rich media. Biochem Biophys Res Commun 261:326–331
11. Go S, Sato C, Furuhata K, Kitajima K (2006) Oral ingestion of mannose alters the expression level of deaminoneuraminic acid (KDN) in mouse organs. Glycoconj J 23:411–421
12. Nakata D, Münster AK, Gerardy-Schahn R, Aoki N, Matsuda T, Kitajima K (2001) Molecular cloning of a unique CMP-sialic acid synthetase that effectively utilizes both deaminoneuraminic acid (KDN) and *N*-acetylneuraminic acid (Neu5Ac) as substrates. Glycobiology 11:685–692
13. Fujita A, Sato C, Münster-Kühnel AK, Gerardy-Schahn R, Kitajima K (2005) Development of a simple and efficient method for assaying cytidine monophosphate sialic acid synthetase activity using an enzymatic reduced nicotinamide adenine dinucleotide/oxidized nicotinamide adenine dinucleotide converting system. Anal Biochem 337:12–21
14. Hara S, Takemori Y, Yamaguchi M, Nakamura M, Ohkura Y (1987) Fluorometric high-performance liquid chromatography of *N*-acetyl- and *N*-glycolylneuraminic acids and its application to their microdetermination in human and animal sera, glycoproteins, and glycolipids. Anal Biochem 164:138–145
15. Sato C, Inoue S, Matsuda T, Kitajima K (1998) Development of a highly sensitive method for detecting α2,8-linked oligo/polysialic acid residues in glycoproteins blotted on the membrane. Anal Biochem 261:191–197
16. Kanamori A, Inoue S, Xulei Z, Zuber C, Roth J, Kitajima K, Ye J, Troy FA, Inoue Y (1994) Monoclonal antibody specific for α2,8-linked oligo deaminated neuraminic acid (KDN) sequences in glycoproteins. Preparation and characterization of a monoclonal antibody and its application in immunohistochemistry. Histochemistry 101:333–340
17. Higashi H, Fukui Y, Ueda S, Kato S, Hirabayashi Y, Matsumoto M, Naiki M (1984) Sensitive enzyme-immunostaining and densitometric determination on thin-layer chromatography of *N*-glycolylneuraminic acid-containing glycosphingolipids, Hanganutziu-Deicher antigens. J Biochem (Tokyo) 95:1517–1520
18. Yu S, Kitajima K, Inoue Y (1993b) Monoclonal antibody specific to alpha-2-3-linked deaminated neuraminyl beta-galactosyl sequence. Glycobiology 3:31–36
19. Sato C, Kitajima K, Inoue S, Seki T, Troy FA 2nd, Inoue Y (1995) Characterization of the antigenic specificity of four different anti-(α2→8-linked polysialic acid) antibodies using lipid-conjugated oligo/polysialic acids. J Biol Chem 270:18923–18928
20. Go S, Go S, Veillon L, Ciampa MG, Mauri L, Sato C, Kitajima K, Prinetti A, Sonnino S, Inokuchi JI (2017) Altered expression of ganglioside GM3 molecular species and a potential regulatory role during myoblast differentiation. J Biol Chem 292:7040–7051
21. Aminoff D (1961) Methods for the quantitative estimation of *N*-acetylneuraminic acid and their application to hydrolysates of sialomucoids. Biochem J 81:384–392
22. Uchida Y, Tsukada Y, Sugimori T (1977) Distribution of neuraminidase in Arthrobacter and its purification by affinity chromatography. J Biochem (Tokyo) 82:1425–1433
23. Kitajima K, Inoue S, Kitazume S, Inoue Y (1992) Analytical methods for identifying and quantitating deamidated sialic acid (2-keto-3-deoxy-D-glycero-D-galactonononic acid) and α2,8-linked poly(oligo)nonulosonate residues in glycoconjugates. Anal Biochem 205:244–250
24. Svennerholm L (1957) Quantitative estimation of sialic acids. II. A colorimetric resorcinol-hydrochloric acid method. Biochim Biophys Acta 24:604–611
25. Chisada S, Yoshimura Y, Sakaguchi K, Uemura S, Go S, Ikeda K, Uchima H, Matsunaga N, Ogura K, Tai T, Okino N, Taguchi R, Inokuchi J, Ito M (2009) Zebrafish and mouse α2,3-sialyltransferases responsible for synthesizing GM4 ganglioside. J Biol Chem 284:30534–30546
26. Kitajima K, Kuroyanagi H, Inoue S, Ye J, Troy FA 2nd, Inoue Y (1994) Discovery of a new type of sialidase, "KDNase," which specifically hydrolyzes deaminoneuraminyl (3-deoxy-D-glycero-D-galacto-2-nonulosonic acid) but not N-acylneuraminyl linkages. J Biol Chem 269:21415–21419
27. Nishino S, Kuroyanagi H, Terada T, Inoue S, Inoue Y, Troy FA, Kitajima K (1996) Induction, localization, and purification of a novel sialidase, deaminoneuraminidase (KDNase), from *Sphingobacterium multivorum*. J Biol Chem 271:2909–2913

Chapter 23

Glycoreplica Peptides

Teruhiko Matsubara, Toshinori Sato, and Takao Taki

Abstract

"Glycoreplica peptides" are prepared using a phage display peptide library and monoclonal antibodies that recognize the carbohydrate epitopes of glycoconjugate antigens. The peptides obtained not only mimic the shapes of original glycoconjugate antigens but also have some of their functions. We herein describe how to identify the amino acid alignments of glycoreplica peptides using phage display selection against carbohydrate-binding proteins. Target-specific peptides and proteins may be selected from the large repertory of a peptide/protein library using phage display technology. Glycoreplica peptides have the potential to become alternatives to carbohydrate ligands such as mimotopes for vaccinations and carbohydrate-derived drugs for carbohydrate-related diseases.

Key words Phage display technology, Random peptide library, Lectin, Anti-carbohydrate antibody, Glycosidase inhibitor

1 Introduction

A limited amount of (sialyl)glycoconjugates are isolated from natural products, and the structures of sialyloligosaccharides are too complex to synthesize organically and/or enzymatically. Instead of glycoconjugates, a series of short peptides that bind to the receptor-binding sites of carbohydrate-binding proteins are potential ligands for elucidating the biological functions of glycoconjugates and inhibiting carbohydrate–protein interactions. These peptides have been named as "glycoreplica peptides" [1, 2] and "carbohydrate-mimetic peptides (CMPs)" [3–5], and their amino acid alignments may be identified by phage display selection against carbohydrate-binding proteins (Fig. 1). In this chapter, we describe the procedure for selecting peptides from a randomized library using phage display technology (Fig. 2).

Concanavalin A (ConA) is a well-known lectin that is commercially available for the biological investigation of glycoconjugates and binds to α-mannose (α-Man) and α-glucose (α-Glc) [6, 7]. Phage display selection was simultaneously performed against

Sandro Sonnino and Alessandro Prinetti (eds.), *Gangliosides: Methods and Protocols*, Methods in Molecular Biology, vol. 1804, https://doi.org/10.1007/978-1-4939-8552-4_23, © Springer Science+Business Media, LLC, part of Springer Nature 2018

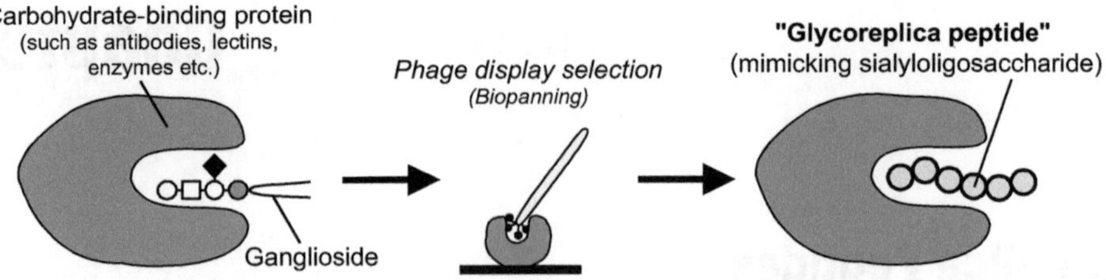

Fig. 1 Identification of glycoreplica (carbohydrate-mimetic) peptides by phage display selection (biopanning) from a random peptide library. Glycoreplica peptides are identified by biopanning against carbohydrate-binding proteins. Glycoreplica peptides bind to the receptor-binding sites of carbohydrate-binding proteins by mimicking oligosaccharide structures

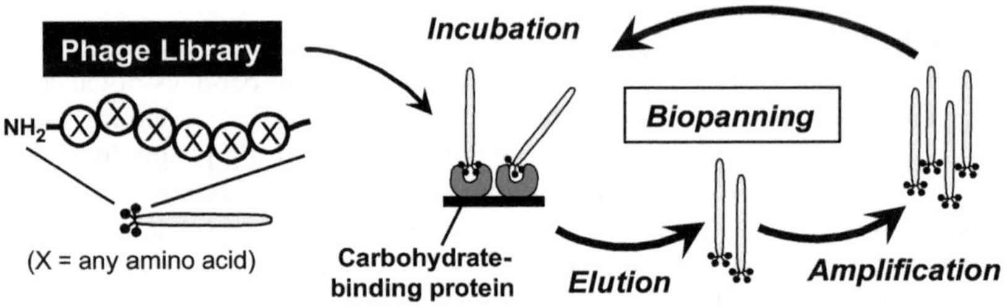

Fig. 2 Schematic representation of the biopanning procedure used to obtain glycoreplica peptides. A filamentous M13 phage carrying a peptide library (X_n) that is displayed on the ***N*** terminus of coat protein III. The phage library is incubated with the target carbohydrate-binding protein, and unbound phages are removed by washing. After the elution of phages bound to the receptor-binding site of the protein in the presence of the carbohydrate ligand, the eluted phages are amplified by infection with the host ***E. coli*** strain. This cycle is repeated several times, and target-specific phages are then collected

ConA by two groups, and short peptides were identified from octapeptide [8] and hexapeptide libraries [9]. The peptides identified contained a consensus sequence of Tyr-Pro-Tyr (YPY) and inhibited the sugar–ConA interaction. Based on this methodology, ganglioside GD1α-mimetic peptides were identified by biopanning against the anti-GD1α antibody KA17 (Fig. 3) [2]. This peptide, WHWRHRIPLQLAAGR, was considered to mimic the sialyl oligosaccharide moiety of GD1α, and tripeptide WHW-modified liposomes inhibited adhesion between RAW117-H10 cells and HSE cells [10]. In addition, glycoreplica peptides identified by biopanning against anti-carbohydrate antibodies are available as carbohydrate antigens in vaccinations [11].

Using commercially available phage vectors, it is also possible to make original libraries and sublibraries (*see* Subheading 2.1, **step 2**). Influenza virus hemagglutinin (HA) recognizes sialylgalactose-containing glycoconjugates on the cell surface in the initial stage of the infection process [12]. Sialic acid-mimic peptides were identified from a pentadecapeptide library by selection with HA, and a mutational sublibrary of the sialic acid-mimic peptide has been prepared [13]. Secondary selection was performed to improve

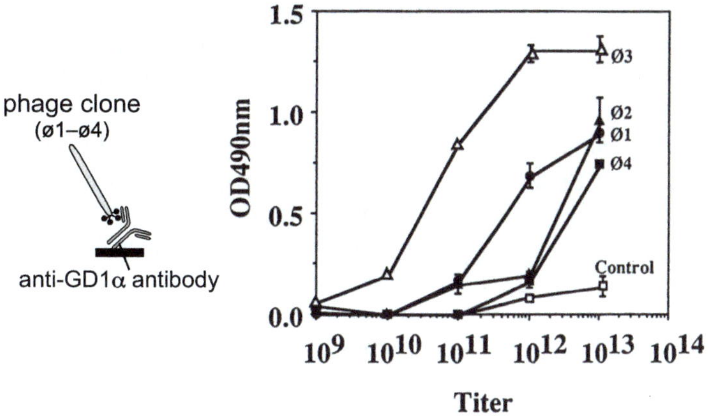

Fig. 3 Recognition of glycoreplica peptides by an anti-carbohydrate antibody. Binding of the selected phage clones against the anti-GD1α-antibody KA17 assessed by phage ELISA (reproduced ref. 2 with permission from Elsevier)

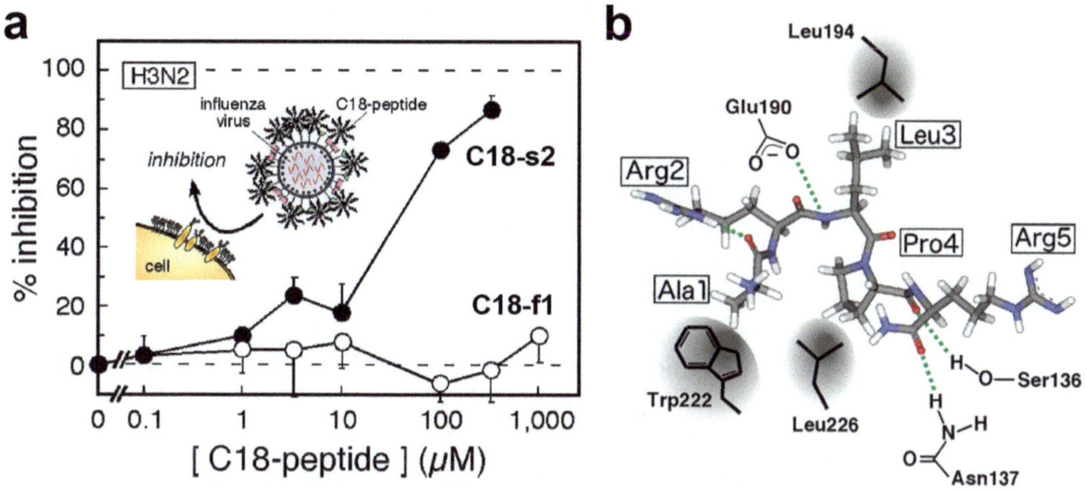

Fig. 4 Inhibition of viral infection by glycoreplica peptides. (**a**) Inhibition of the influenza H3N2 (A/Aichi/2/68) virus infection of MDCK cells by *N*-stearoyl s2 peptide (C18-s2). s2, ARLPRTMVHPKPAQP; f1, ARLSPTMVHPNGAQP. (**b**) A proposed complex of the ARLPR peptide with H3HA predicted by a docking simulation. The HA structure was obtained from PDB (entry 1HGG) (reproduced ref. 13 with permission from Elsevier)

binding affinities for HAs, and the corresponding *N*-stearoyl peptide strongly inhibited infections by influenza virus (Fig. 4). Glycoreplica peptides are useful as carbohydrate-derived drugs for carbohydrate-related diseases [14, 15].

2 Materials

A typical filamentous M13 or fd phage carrying a peptide library displays foreign peptides (X_n, n = 6–15) on the N terminus of coat protein III (Fig. 2) [16]. General recipes and protocols for the purification, amplification, and titering of filamentous phages are basically performed according to the manufacturer's instructions (New England Biolabs, Inc.) and cited references [17, 18].

2.1 Phage Library and Host *E. coli* Strain

1. Ph.D.™ Phage Display Peptide Library Kits are purchased from New England Biolabs, Inc. There are heptapeptide (7-mer), cyclic heptapeptide (cyclic 7-mer), and dodecapeptide (12-mer) sequences. These random peptide libraries typically contain ~2×10^9 independent clones supplied in Tris-buffered saline (TBS) with 50% glycerol (2×10^{13} plaque-forming units (pfu)/mL). Store at −20 °C.
2. The original phage library with a customized sequence may be prepared using a display vector with the Ph.D.™ Peptide Display Cloning System (New England Biolabs, Inc.). This kit contains an M13KE gIII cloning vector (7223 bp) and extension primer. Store at −20 °C.
3. The host *E. coli* strain ER2738 (New England Biolabs, Inc.) is supplied in LB with 50% glycerol. Store at −20 °C.

2.2 Media and Solutions for General M13 Phage Methods

1. LB medium: For 1 L, 10 g Bacto typtone, 5 g yeast extract, and 5 g NaCl. Autoclave and store at room temperature.
2. LB-Tet plates: Autoclave 100 mL of LB medium +1.5 g agar. Cool to <70 °C and add 0.1 mL (1/1000 volume) of a tetracycline stock (20 mg/mL in ethanol or 50% glycerol). Pour into plates. Store at 4 °C.
3. IPTG–Xgal solution: Dissolve 0.25 g isopropyl-β-D-thiogalactoside (IPTG) and 0.2 g 5-bromo-4-chloro-3-indoyl-β-D-thiogalactoside (Xgal) in 5 mL of dimethyl sulfoxide (DMSO). Store at −20 °C.
4. LB–IPTG–Xgal plates: Autoclave 1 L of LB medium +15 g agar. Cool to <70 °C and add 1 mL of IPTG–Xgal solution. Pour into plates. Store at 4 °C.
5. Top agar: For 200 mL, 2 g Bacto typtone, 1 g yeast extract, 1 g NaCl, 0.2 g $MgCl_2$-$6H_2O$, and 1.4 g agar. Autoclave and store at room temperature.
6. PEG–NaCl: For 600 mL, autoclave 100 g polyethylene glycol #6000, 117 g NaCl, and 475 mL of water in a 1-L bottle. Cool to <70 °C and mix completely. Store at room temperature.
7. TBS: 50 mM Tris–HCl and 150 mM NaCl, pH 7.5. Autoclave and store at room temperature.
8. TBS–gelatin: Autoclave 100 mL of TBS + 0.1 g gelatin. Store at room temperature.
9. 0.02% NaN_3–TBS: Dilute 5% NaN_3–TBS with TBS and filter-sterilize using a 0.22-μm filter.

2.3 ER2738 Cell Culture

1. Streak out ER2738 from the glycerol culture on an LB-Tet plate to obtain single colonies. Incubate at 37 °C overnight. Store at 4 °C. Use within 1 month.
2. Overnight culture: Inoculate 5–10 mL of LB-Tet with a single colony of ER2738 (**step 1**) and incubate at 37 °C overnight with shaking. Store at 4 °C until used.
3. Early-log phase culture: Inoculate 20 mL of LB with a single colony of ER2738 and incubate at 37 °C for 4–5 h. Store at 4 °C until used. Alternatively, dilute 0.2 mL (1/100 volume) of an overnight culture with 20 mL of LB.
4. Mid-log phase culture: Inoculate 5–10 mL of LB with a single colony of ER2738 and incubate at 37 °C with shaking until the mid-log phase (OD_{600} ~ 0.5). Alternatively, dilute 0.1 mL (1/100 volume) of an overnight culture with 10 mL of LB and incubate at 37 °C for 2–3 h with shaking. Store at 4 °C until used.

2.4 Phage Titering

1. Prepare a mid-log phase ER2738 culture (Subheading 2.3, **step 4**).
2. Prewarm three LB–IPTG–Xgal plates at 37 °C until ready for use.
3. Prepare three diluted solutions with tenfold serial dilution of the amplified phage (typically, 10^{-7}, 10^{-8}, and 10^{-9}); Dilute 1 μL of the original phage solution with 99 μL of TBS–gelatin to give a 10^{-2} (1/100) dilution. Continue this dilution to obtain 10^{-6} and 10^{-8} dilutions. Dilute 10 μL of the 10^{-6} and 10^{-8} dilutions with 90 μL of TBS–gelatin to give 10^{-7} and 10^{-9} dilutions. The typical dilution range for the eluted phage (unamplified) is 10^2–10^4 (10^{-2}–10^{-4} dilutions).
4. Dispense 200 μL/tube of the mid-log phase ER2738 culture into three 1.5-mL tubes.
5. Melt Top Agar in a microwave. Dispense approximately 3 mL/tube into three culture tubes.
6. For phage infection, mix 10 μL of each diluted solution (**step 3**) with the ER2738 culture (**step 4**) and incubate at room temperature for >1 min.
7. Transfer the infected cells (210 μL/tube) to a culture tube containing Top Agar (45–70 °C) (**step 5**), and immediately pour onto a prewarmed LB–IPTG–Xgal plate (**step 2**).
8. Cool the plate for >5 min and incubate at 37 °C overnight.
9. Count blue plaques on plates with 20–200 plaques and calculate the number of phages (pfu) in the diluted solution (*see* **Note 1**). Estimate the phage concentration of the original solution (pfu/mL).
10. Alternatively, the virion concentration (virion/mL) is estimated from absorbance at 260 nm (*see* **Note 2**).

2.5 Solutions for the Biopanning Procedure

1. 5% bovine serum albumin (BSA)–TBS: Dissolve 5 g BSA in 95 mL of TBS. Store at −20 °C.
2. TBST: TBS containing 0.1% Tween 20.
3. Elution buffer (glycine–HCl and 1 mg/mL BSA, pH 2.2): Dissolve 10 mg BSA and 1 mg phenol red (optional) in 0.1 M glycine–HCl buffer (pH 2.2) and filter using a 0.22-μm pore size membrane. Store at 4 °C.
4. 1 M Tris–HCl buffer (pH 9.1): Dissolve 12.1 g Tris–HCl in 80 mL of water and adjust pH with HCl. Make up to 100 mL with water. Autoclave and store at room temperature.

3 Methods

3.1 Immobilization of Carbohydrate-Binding Proteins

The target carbohydrate-binding protein is adsorbed directly on a 96-well microtiter plate (**step 1**) or covalently immobilized on a carboxylic acid-carrying polystyrene microplate (**step 2**) [13, 19]. Alternatively, a biotinylated protein is immobilized on a streptavidin-coated microplate (**step 4**) [1, 2, 8, 9].

1. Direct physical adsorption: Fill the wells of the microplate for the immunoassay (MaxiSorp™ surface, Thermo Scientific™ Nunc™) with 100 μL of the protein solution (50–100 μg/mL) and incubate at room temperature for 2 h (or at 4 °C overnight). Wash the wells with TBS and follow by blocking (**step 3**).
2. Covalent immobilization: Prior to the addition of the protein solution, activate the carboxy group on the bottom of the wells with 60 μL of the 1-ethyl-3-(3-dimethylaminopropyl) carbodiimide hydrochloride (EDC)/*N*-hydroxysuccinimide (NHS) solution (0.1 g/mL EDC and 0.1 g/mL NHS in water) for 10 min (code MS-8796F, Sumitomo Bakelite Co., Ltd., Japan). Fill the wells with 100 μL of the protein solution (typically 50–100 μg/mL) and incubate at room temperature for 2 h. After EDC/NHS activation, wash the wells with water and fill with 100 μL of 1 mM ethanolamine to deactivate the excess reactive succinimidyl groups remaining in the wells. Wash the wells with TBS and follow by a blocking (**step 3**).
3. Blocking wells with BSA: Fill the target protein-immobilized wells (**step 1** or **2**) with 300 μL of 1–5% BSA–TBS and incubate at room temperature for 2 h (or at 4 °C overnight). Wash the wells with TBS and use immediately for biopanning (protein-coated wells).
4. Streptavidin-coated wells: Fill the wells with 100 μL of streptavidin (100 μg/mL) in 100 mM $NaHCO_3$ at pH 8.6 and incubate at 4 °C overnight. Treat with blocking solution (100 mM

$NaHCO_3$, 5 mg/mL BSA, 0.1 μg/mL streptavidin, and 0.02% NaN_3) at room temperature for 1 h. Wash the wells with TBS and use immediately for biopanning.

3.2 Biopanning Against Carbohydrate-Binding Proteins

1. Dilute 10^8–10^{10} pfu of the phage (0.01–1 μL of the original library) with 100 μL of TBS. Fill the protein-coated wells (Subheading 3.1) with the phage solution and incubate for 1–2 h. Wash 3–5 times with TBS (or TBST).
2. Optional: For pre-elution, fill the wells with eluent solution at a low concentration (1 μM–1 mM), incubate for 1 h, and wash off. Phages that bind loosely to the target protein are removed by this step.
3. Elute phages that bind to the carbohydrate receptor-binding site of the target protein with 100 μL of the carbohydrate ligand solution at a high concentration (1–10 mM) for 1–2 h. In order to ensure that the glycoreplica peptide is obtained, carbohydrate ligand solution is recommended in this step (*see* **Note 3**).
4. Titer a small aliquot of the eluted phages (1–10 μL) and estimate the number of phages collected (output phage) (*see* Subheading 2.4).
5. Amplify the remaining of the eluted phages by an incubation with 20 mL of an early-log ER2738 culture in a 250-mL Erlenmeyer flask at 37 °C for 4–5 h with vigorous shaking (>180 rpm).
6. Transfer the culture to a 50-mL conical tube and centrifuge the tube at 5000 × *g* at 4 °C for 15–20 min.
7. Transfer the supernatant to another tube that contains 3 mL (1/6 volume) of PEG/NaCl. Mix gently by inverting the tube 100 times and allow the phage to precipitate at 4 °C for at least 4 h (overnight preferable).
8. Centrifuge the tube at 5000 × *g* at 4 °C for 15–20 min to pellet the phage. Remove the supernatant.
9. Suspend the phage pellet in 1 mL of TBS, transfer the suspension to a 1.5-mL tube, and centrifuge the tube at 20,000 × *g* for 10 min to remove residual cells.
10. Transfer the supernatant to a fresh 1.5-mL tube and reprecipitate with 0.15 mL (1/6 volume) of PEG/NaCl as described in **step** 7. Allow the phage to precipitate at 4 °C for at least 1 h.
11. Centrifuge the tube at 20,000 × *g* at 4 °C for 10 min to pellet the phage. Remove the supernatant.
12. Suspend the phage pellet in 200 μL of 0.02% NaN_3–TBS and centrifuge the tube at 20,000 × *g* for 10 min.

13. Transfer the supernatant to a fresh 0.5-mL tube (amplified phage). Store at 4 °C.
14. Titer a small aliquot of the amplified phage (1 μL) and estimate the phage concentration of this solution (*see* Subheading 2.4).
15. Prepare protein-coated wells and perform the second round using 10^8–10^{10} pfu of the amplified phage from the first round as the input phage (**steps 1–14**).
16. Typically repeat **step 15** 3–5 times until the ratio of the output phage to the input phage is saturated (*see* **Note 4**).

3.3 Plaque Amplification for Phage Cloning

When target protein-binding phages are enriched (Subheading 3.2), individual phage clones are isolated and the amino acid alignment of glycoreplica peptides is identified. The phage genomes (single-stranded DNA) of the isolated phage clones are used as a template for sequencing.

1. For 16 clones, dilute 0.16 mL (1/100 volume) of an ER2738 overnight culture with 16 mL of LB and dispense 1 mL of the diluted culture into 2-mL tubes.
2. Stab a blue plaque with a pipette tip (or toothpick) and inoculate the plaque into the diluted culture. Incubate tubes at 37 °C for 4–5 h (no longer) with shaking.
3. Centrifuge the tube at 20,000 × *g* for 10 min and transfer the supernatant to a fresh tube. This amplified phage stock may be stored at 4 °C for several weeks.
4. Amplify the amplified phage clones by an incubation with 20 mL of a mid-log ER2738 culture in a 50-mL conical tube at 37 °C for 4–5 h with shaking.
5. Perform **steps 6–13** in Subheading 3.2 to obtain the cloned phage solution (200 μL each).

3.4 DNA Sequencing

1. Purify the single-strand DNA of the phage clone from 50–100 μL of the cloned phage solution (Subheading 3.3) using a QIAprep Spin M13 kit (Qiagen) according to the manufacturer's instructions.
2. Obtain the DNA sequence with the ABI PRISM 3100 Genetic Analyzer using a −96 gIII sequencing primer (5′-^{HO}CCCTCATAGTTAGCGTAACG-3′, New England Biolabs, Inc.) with the ABI PRISM BigDye Terminator Cycle Sequencing Kit (Applied Biosystems).

3.5 Phage ELISA

In order to confirm the binding of isolated phage clones to the target protein, phage ELISA is used. A wild-type M13 phage clone (or primary library) is often employed as a control. Prepare wells at each concentration in triplicate.

1. Prepare protein-coated wells blocked with 1% BSA–TBS (Subheading 3.1).
2. Prepare 10-fold (or 3.3-fold) serial dilutions of the cloned phage (Subheading 3.3) in 100 μL of TBS.
3. Fill protein-coated wells with the phage solution and incubate for 1–2 h. Wash three times with TBS (or TBST) and then block with 200 μL of 2–5% BSA–TBS.
4. Fill the wells with 100 μL of a 1:1000 dilution of an anti-fd bacteriophage antibody (produced in rabbits), incubate for 1 h, and wash off. Alternatively, use horseradish peroxidase conjugated to an anti-M13 monoclonal antibody (GE Healthcare) and skip **step 5**.
5. Fill the wells with 100 μL of a 1:2000–1:10,000 dilution of a peroxidase-conjugated anti-rabbit IgG antibody, incubate for 1 h, and wash off.
6. Dissolve a 5-mg *o*-phenylenediamine (OPD) tablet in 12.5 mL of citric acid-phosphate buffer at pH 5 (0.4 mg OPD/mL).
7. Add 8 μL of 30% H_2O_2 to the OPD solution and immediately fill the well with 100 μL of the solution to develop color.
8. After 10–15 min, add 100 μL of 3 M H_2SO_4 to quench the reaction.
9. Measure absorbance at 492 nm using a microplate reader to evaluate the amount of the phage clone bound to the target protein.

4 Notes

1. The phage titer of the amplified phage is typically approximately 10^{12} pfu/mL from a 20-mL culture (e.g., 1.2×10^{12} pfu/mL is calculated from 120 plaques in a 10^{-8} plate).
2. Virion (physical particle) concentrations may be estimated from a broad peak at 260–280 nm [18]. M13KE phage concentration (virion/mL) = $A_{269} \times 6 \times 10^{16}/(7200$ bases of single-stranded DNA). When A_{269} is 1.0, the concentration is 8.3×10^{12} virions/mL. Typically, 1 pfu = 10–20 virions (unpublished data).
3. Alternatively, a general elution buffer (glycine–HCl, 1 mg/mL BSA, pH 2.2) may be used to eluate phages. After a 15-min incubation, neutralize immediately with 15 μL of 1 M Tris–HCl buffer (pH 9.1). However, the elution buffer elutes all phages that bind to the target protein.

4. If biopanning is successful, the yield of the phage at the final cycle may be up to 0.1–1%, where yield (%) = [(output phage)/(input phage)] × 100 [13, 19].

Acknowledgment

This study was supported by the Japan Society for the Promotion of Science Kakenhi grants (no. 14380411 and 17300159 to T.S.; no. 15 K01806 to T.M.).

References

1. Taki T, Ishikawa D, Hamasaki H, Handa S (1997) Preparation of peptides which mimic glycosphingolipids by using phage peptide library and their modulation on beta-galactosidase activity. FEBS Lett 418(1–2):219–223
2. Ishikawa D, Kikkawa H, Ogino K, Hirabayashi Y, Oku N, Taki T (1998) GD1alpha-replica peptides functionally mimic GD1alpha, an adhesion molecule of metastatic tumor cells, and suppress the tumor metastasis. FEBS Lett 441(1):20–24
3. Monzavi-Karbassi B, Hennings LJ, Artaud C, Liu T, Jousheghany F, Pashov A, Murali R, Hutchins LF, Kieber-Emmons T (2007) Preclinical studies of carbohydrate mimetic peptide vaccines for breast cancer and melanoma. Vaccine 25(16):3022–3031
4. Hatakeyama S, Sugihara K, Shibata TK, Nakayama J, Akama TO, Tamura N, Wong SM, Bobkov AA, Takano Y, Ohyama C, Fukuda M, Fukuda MN (2011) Targeted drug delivery to tumor vasculature by a carbohydrate mimetic peptide. Proc Natl Acad Sci U S A 108(49):19587–19592
5. Matsubara T (2012) Potential of peptides as inhibitors and mimotopes: selection of carbohydrate-mimetic peptides from phage display libraries. J Nucleic Acids 2012:740982
6. Schwarz FP, Puri KD, Bhat RG, Surolia A (1993) Thermodynamics of monosaccharide binding to concanavalin a, pea (Pisum sativum) lectin, and lentil (Lens culinaris) lectin. J Biol Chem 268(11):7668–7677
7. Derewenda Z, Yariv J, Helliwell JR, Kalb AJ, Dodson EJ, Papiz MZ, Wan T, Campbell J (1989) The structure of the saccharide-binding site of concanavalin a. EMBO J 8(8):2189–2193
8. Oldenburg KR, Loganathan D, Goldstein IJ, Schultz PG, Gallop MA (1992) Peptide ligands for a sugar-binding protein isolated from a random peptide library. Proc Natl Acad Sci U S A 89(12):5393–5397
9. Scott JK, Loganathan D, Easley RB, Gong X, Goldstein IJ (1992) A family of concanavalin A-binding peptides from a hexapeptide epitope library. Proc Natl Acad Sci U S A 89(12):5398–5402
10. Takikawa M, Kikkawa H, Asai T, Yamaguchi N, Ishikawa D, Tanaka M, Ogino K, Taki T, Oku N (2000) Suppression of GD1alpha ganglioside-mediated tumor metastasis by liposomalized WHW-peptide. FEBS Lett 466(2–3):381–384
11. Kieber-Emmons T (1998) Peptide mimotopes of carbohydrate antigens. Immunol Res 17(1–2):95–108
12. Wiley DC, Skehel JJ (1987) The structure and function of the hemagglutinin membrane glycoprotein of influenza virus. Annu Rev Biochem 56:365–394
13. Matsubara T, Onishi A, Saito T, Shimada A, Inoue H, Taki T, Nagata K, Okahata Y, Sato T (2010) Sialic acid-mimic peptides as hemagglutinin inhibitors for anti-influenza therapy. J Med Chem 53(11):4441–4449. https://doi.org/10.1021/jm1002183
14. Taki T, Ishikawa D, Ogino K, Tanaka M, Oku N, Asai T, Popa I, Portoukalian J (2008) A new approach for drug discovery from glycobiology and phage-displayed peptide library technology. Biochim Biophys Acta 1780(3):497–503
15. Ladner RC, Sato AK, Gorzelany J, de Souza M (2004) Phage display-derived peptides as therapeutic alternatives to antibodies. Drug Discov Today 9(12):525–529
16. Smith GP, Petrenko VA (1997) Phage Display. Chem Rev 97(2):391–410
17. Noren KA, Noren CJ (2001) Construction of high-complexity combinatorial phage display peptide libraries. Methods 23(2):169–178

18. Smith GP, Scott JK (1993) Libraries of peptides and proteins displayed on filamentous phage. Methods Enzymol 217:228–257
19. Matsubara T, Onishi A, Yamaguchi D, Sato T (2016) Heptapeptide ligands against receptor-binding sites of influenza hemagglutinin toward anti-influenza therapy. Bioorg Med Chem 24(5):1106–1114. https://doi.org/10.1016/j.bmc.2016.01.039

INDEX

Sandro Sonnino and Alessandro Prinetti (eds.), *Gangliosides: Methods and Protocols*, Methods in Molecular Biology, vol. 1804, https://doi.org/10.1007/978-1-4939-8552-4, © Springer Science+Business Media, LLC, part of Springer Nature 2018

N

O

P

Q

R

S

MIX
Papier aus verantwortungsvollen Quellen
Paper from responsible sources
FSC® C105338

If you have any concerns about our products,
you can contact us on
ProductSafety@springernature.com

In case Publisher is established outside the EU,
the EU authorized representative is:
Springer Nature Customer Service Center GmbH
Europaplatz 3, 69115 Heidelberg, Germany

Printed by Libri Plureos GmbH
in Hamburg, Germany